Surgery in and around the Orbit

W0050596

Nederlandse Vereniging voor
Mondziekten, Kaak- en Aangezichtschirurgie

Peter J. J. Gooris • Maarten P. Mourits
J. Eelco Bergsma
Editors

Surgery in and around the Orbit

CrossRoads

Springer

Editors

Peter J. J. Gooris
Affiliate Professor
Department of Oral
and Maxillofacial Surgery
Amphia Hospital Breda
Breda, The Netherlands

Maarten P. Mourits
Professor Emeritus
Department of Ophthalmology
Amsterdam University Medical
Centers, Location AMC
Amsterdam, The Netherlands

J. Eelco Bergsma
Professor
Department of Oral
and Maxillofacial Surgery
Amphia Hospital Breda
Breda, The Netherlands

ISBN 978-3-031-40699-7   ISBN 978-3-031-40697-3   (eBook)
https://doi.org/10.1007/978-3-031-40697-3

BOOA - Foundation

© The Editor(s) (if applicable) and The Author(s) 2023, corrected publication 2024
This book is an open access publication.

Open Access   This book is licensed under the terms of the Creative Commons Attribution 4.0
International License (http://creativecommons.org/licenses/by/4.0/), which permits use, sharing,
adaptation, distribution and reproduction in any medium or format, as long as you give
appropriate credit to the original author(s) and the source, provide a link to the Creative Commons
license and indicate if changes were made.
The images or other third party material in this book are included in the book's Creative
Commons license, unless indicated otherwise in a credit line to the material. If material is not
included in the book's Creative Commons license and your intended use is not permitted by
statutory regulation or exceeds the permitted use, you will need to obtain permission directly
from the copyright holder.
The use of general descriptive names, registered names, trademarks, service marks, etc. in this
publication does not imply, even in the absence of a specific statement, that such names are
exempt from the relevant protective laws and regulations and therefore free for general use.
The publisher, the authors, and the editors are safe to assume that the advice and information in
this book are believed to be true and accurate at the date of publication. Neither the publisher nor
the authors or the editors give a warranty, expressed or implied, with respect to the material
contained herein or for any errors or omissions that may have been made. The publisher remains
neutral with regard to jurisdictional claims in published maps and institutional affiliations.

This Springer imprint is published by the registered company Springer Nature Switzerland AG
The registered company address is: Gewerbestrasse 11, 6330 Cham, Switzerland

Paper in this product is recyclable.

"CrossRoads" is dedicated to all patients who have shown so much confidence in our multidisciplinary team, and to the doctors who consider following in our footsteps.

Acknowledgments

We would like to express our gratitude to all the authors who have contributed to this book with so much enthusiasm.

We are much indebted to our colleague Prof. Dr. Eelco Bergsma who has been the coauthor of this book, spending much time with us and exhibiting so much patience while completing the digital workup of the book, thereby making this work possible.

Yvette Braaksma-Besselink and Dr. Hinke Marijke Jellema, orthoptists: we would like to thank you both for your contribution and insightful information on the intriguing aspects of diplopia.

Dr. Carl-Peter Cornelius: thank you so much for the enthusiastic elaboration on so many aspects of the anatomy of the orbit and the exceptional accompanying illustrations. It has been a great pleasure working with you.

This book would not be complete without your expertise and extensive input.

Prof. Dr. Thomas Dodson: your participation has been an honor. Though from a far distance, we are grateful for your Seattle University of Washington contribution. Thomas, your membership in the Peer Review Group is much respected.

Prof. Dr. Eddy Becking and Prof. Dr. Jonathan Roos: also thanks to you for being on board of the Peer Review Group.

Dr. Gertjan Mensink and Dr. Rob Noorlag: it has been a pleasure to work with you on the topic of orbital cellulitis.

Dr. Bram van der Pol and Dr. Geert-Jan Rutten: thank you for the neurosurgical input on the highly specialized area of orbital roof fractures.

Prof. Dr. Wilmar Wiersinga: as an internationally well-respected endocrinologist with a special interest in patients with Graves' disease, your contribution to the etiology and management of this disease is invaluable.

Dr. Maartje de Win: your contribution on imaging aspects is much appreciated; it is an indispensable and essential part of diagnostics, planning, and control in orbital trauma and pathology. We would also like to thank Roel Kloos, ophthalmologist at the department of Ophthalmology of the Amsterdam University Medical Centers for sharing many figures in several chapters.

To my fellow colleagues of the Dutch "Orbit Team" of the Amsterdam University Medical Centers, location AMC. Prof. Dr. Leander Dubois, Prof. Dr. Eddy Becking, Dr. Ruud Schreurs, Prof. Dr. Thomas Maal, Dr. Jesper

Jansen, and Juliana Sabelis, we are very grateful for all the work you have done with the emphasis on "teamwork." You have supplied this book with important issues on the current workup of the patient with an orbital wall fracture as well as with detailed information on treatment aspects including the preoperative digital planning concept.

We would like to thank Karthik Periyasamy of the Springer Editorial team for the continuous and structural support during the creation of the book.

We appreciate the financial support to realize this book by the scientific organization of the Dutch Association of Oral and Maxillofacial Surgery: Bevordering   Bijzondere   Wetenschappelijke   Onderwijs   en Onderzoekactiviteiten BOOA. We also want to thank the MKA-Groep, The Netherlands for their financial support.

A special word of thanks to all the patients referred to in this book.

Without your input, there would be no such book.

Peter J. J. Gooris

Maarten P. Mourits

The original version of this book has been revised. A correction to this book can be found at https://doi.org/10.1007/978-3-031-40697-3_23

Contents

Contributors

Alfred G. Becking, MD, DDS, PhD., FEBOMFS Department of Oral and Maxillofacial Surgery, Amsterdam University Medical Centers, Location AMC, Amsterdam, The Netherlands

J. Eelco Bergsma, MD, DDS, PhD Department of Oral and Maxillofacial Surgery, Amphia Hospital Breda, Breda, The Netherlands

Amsterdam University Medical Centers, Location AMC, Amsterdam, The Netherlands

Acta dental school, Amsterdam, The Netherlands

Yvette Braaksma-Besselink Department of Ophthalmology, Amsterdam University Medical Centers, Location AMC, Amsterdam, The Netherlands

Carl-Peter Cornelius, MD, DDS Department of Oral and Maxillofacial Surgery, Facial Plastic Surgery, Ludwig Maximilians University, Munich, Bavaria, Germany

Jasjit K. Dillon, MD, DDS, BDS, FDSRCS Department of Oral and Maxillofacial Surgery, University of Washington, Seattle, WA, USA

Oral and Maxillofacial Surgery Harborview Medical Center, University of Washington, Seattle, WA, USA

Thomas Dodson, DMD, MPH, FACS Department of Oral and Maxillofacial Surgery, University of Washington, Seattle, WA, USA

Leander Dubois, MD, DDS, PhD Department of Oral and Maxillofacial Surgery, Amsterdam University Medical Centers, Amsterdam, The Netherlands

Peter J. J. Gooris, MD, DDS, PhD, FEBOMFS Department of Oral and Maxillofacial Surgery, Amphia Hospital Breda, Breda, The Netherlands

Amsterdam University Medical Centers, Amsterdam, The Netherlands

Department of Oral and Maxillofacial Surgery, University of Washington, Seattle, WA, USA

Jesper Jansen, MD, DDS, PhD Department of Oral and Maxillofacial Surgery, Amsterdam University Medical Centers, Amsterdam, The Netherlands

Hinke Marijke Jellema, PhD Department of Ophthalmology, Amsterdam University Medical Centers, Location AMC, Amsterdam, The Netherlands

Thomas J. J. Maal, PhD, MSc 3D Lab Department of Oral and Maxillofacial Surgery, Radboud University Medical Center, Nijmegen, The Netherlands

Gertjan Mensink, MD, DDS, PhD Department of Oral and Maxillofacial Surgery, Amphia Hospital Breda, Breda, The Netherlands

Maarten P. Mourits, MD, PhD Amsterdam University Medical Centers, Location AMC, Amsterdam, The Netherlands

Rob Noorlag, MD, DDS, PhD Department of Oral and Maxillofacial Surgery, University Medical Center Utrecht, Utrecht, The Netherlands

Bram van der Pol, MD Department of Neurosurgery, St. Elisabeth-Tweesteden Hospital Tilburg, Tilburg, The Netherlands

J. Roos, MD, PhD Department of Ophthalmology, University of Milan and Mario Negri Institute, Milan, Italy

Geert-Jan Rutten, MD, PhD Department of Neurosurgery, St. Elisabeth-Tweesteden Hospital Tilburg, Tilburg, The Netherlands

Juliana F. Sabelis 3D Lab Department of Oral and Maxillofacial Surgery, Amsterdam University Medical Centers, Amsterdam, The Netherlands

Ruud Schreurs, MSc, PhD 3D Lab Department of Oral and Maxillofacial Surgery, Amsterdam University Medical Centers, Amsterdam, The Netherlands

Wilmar M. Wiersinga, MD, PhD Department of Internal Medicine-Endocrinology, Amsterdam University Medical Centers, Location AMC, Amsterdam, The Netherlands

Maartje M. L. de Win, MD, PhD Department of Radiology and Nuclear Medicine, Amsterdam University Medical Centers, Amsterdam, The Netherlands

Part I

Introduction

The Orbit: Introduction to "CrossRoads". Multidisciplinary Versus Solo Approach in Complex Cases

Maarten P. Mourits, Peter J. J. Gooris, and J. Eelco Bergsma

Already in 1920, in a report to the UK Minister of Health, the concept of team-based care was conceived [1]. Now, PubMed shows no less than 61,496 hits when one searches for "multidisciplinary." This short chapter is certainly not a plea to approach each and every medical problem by a team of specialists, but in cross-disciplinary caseload, today interdisciplinary care is mandatory. As we will see, when, for example, treating patients with Graves' disease, a multidisciplinary approach is highly recommended [2, 3].

Pathology in and around the orbit is typically an area where many medical disciplines intersect: ophthalmology, oral and maxillofacial surgery, ENT, plastic surgery, and endocrinology. Despite the expertise of every single specialist, tunnel vision is a threat and can lead to misdiagnosis.

With the introduction of medical specialists from different fields working closely together, this tendency for tunnel vision is coming to an end. In 1994, Stoll et al. [4] argued that "orbital complications of various pathogenesis" are best treated by interdisciplinary teamwork. The advantage of working with a multidisciplinary team will equate to a superior outcome for the patient: The end result of that input is much more than the sum of its parts. Also, the multidisciplinary consultation hour can serve as a real goldmine for the participating doctors during which much can be achieved. In order for this system to work effectively, each doctor needs to maintain their knowledge and associated skills by treating a minimum number of patients annually.

In the Netherlands, every doctor is allowed to perform any medical treatment, provided he/she does it according to the rules of the current "medical art"

In the Amphia Hospital in Breda, Netherlands, we have set up a collaboration between oral and maxillofacial surgeons, endocrinologists, and ophthalmologists. Over the past 15 years, this has proved to be very beneficial in treating ocular-related pathology. Our main focus has been Graves' orbitopathy and orbital fractures, but

M. P. Mourits (✉)
Amsterdam University Medical Centers, Location AMC, Amsterdam, The Netherlands

P. J. J. Gooris
Department of Oral and Maxillofacial Surgery, Amphia Hospital Breda, Breda, The Netherlands

Amsterdam University Medical Centers, Amsterdam, The Netherlands

Department of Oral and Maxillofacial Surgery, University of Washington, Seattle, WA, USA

J. E. Bergsma
Department of Oral and Maxillofacial Surgery, Amphia Hospital Breda, Breda, The Netherlands

Amsterdam University Medical Centers, Amsterdam, The Netherlands

Acta dental school, Amsterdam, The Netherlands

© The Author(s) 2023
P. J. J. Gooris et al. (eds.), *Surgery in and around the Orbit*,
https://doi.org/10.1007/978-3-031-40697-3_1

related problems often revealed themselves during collaboration.

For those doctors who pursue the same goal as we have or those who are interested in both orbital surgery and ophthalmic orbital pathology, we hope our book "CrossRoads" is a valuable resource. It is not our aim to give a full description of all diseases, disorders, and treatments that exist in these fields; however, we have tried to provide an overview of some of the problems we have encountered and, hence, raise the interest for further reading.

References

1. Watkins J, Straughton K, King N. There is no 'I' in team but there may be a PA. Future Healthc J. 2019;6:177–80.
2. Bogusiak K, Puch A, Arkuszewski P. Goldenhar syndrome: current perspectives. World J Pediatr. 2017;13:405–15.
3. Wiersinga WM. Management of Graves' ophthalmopathy. Nat Clin Pract Endocrinol Metab. 2007;3:396–404.
4. Stoll W, Busse H, Wessels N. Detailed results of orbital and optic nerve decompression. NHO. 1994;42:685–90.

Open Access This chapter is licensed under the terms of the Creative Commons Attribution 4.0 International License (http://creativecommons.org/licenses/by/4.0/), which permits use, sharing, adaptation, distribution and reproduction in any medium or format, as long as you give appropriate credit to the original author(s) and the source, provide a link to the Creative Commons license and indicate if changes were made.

The images or other third party material in this chapter are included in the chapter's Creative Commons license, unless indicated otherwise in a credit line to the material. If material is not included in the chapter's Creative Commons license and your intended use is not permitted by statutory regulation or exceeds the permitted use, you will need to obtain permission directly from the copyright holder.

Part II

Anatomical Aspects of the Orbit

Carl-Peter Cornelius and Peter J. J. Gooris

Learning Objectives

To acquire a/an

- Comprehensive knowledge of the bony composition within and around the precincts of the orbits.
- Understanding of the topographical relationships of the orbit to the middle cranial fossa, cavernous sinus, infratemporal and pterygopalatine fossa, the paranasal air sinuses, and the lateral nasal wall.
- Impression of the individual variability of passageways in and out the orbit, which becomes specifically apparent in large-scale data pools of morphometric imaging (CT, MRI, CBCT) studies.
- Habitude for a thorough diagnostic workup to anticipate anatomic normal-variants for increased finesse in surgical planning and postoperative analysis.

C.-P. Cornelius (✉)
Department of Oral and Maxillofacial Surgery, Facial Plastic Surgery, Ludwig Maximilians University, Munich, Bavaria, Germany
e-mail: peter.cornelius@med.uni-muenchen.de

P. J. J. Gooris
Department of Oral and Maxillofacial Surgery, Amphia Hospital Breda, Breda, The Netherlands

Amsterdam University Medical Centers, Amsterdam, The Netherlands

Department of Oral and Maxillofacial Surgery, University of Washington, Seattle, WA, USA

Introduction

The orbits are paired mirror symmetric cavities of bone on either side of an intermediary compartment that is made up by the external and inner nose as well as the ethmoid and sphenoid sinuses. The orbits contain the eyeballs, i.e., the optical and light/image processing units, that represent the entry of the visual organ. Besides the eyeballs or globes with their refractive apparatus and retinal receptors, the visual organ consists of auxiliary adnexa, such as the lacrimal system, the extraocular muscles, and the adipose body including a multitude of neural connections and vascular supply. The posterior end of each orbital cavity (orbital apex) centers over the bony surfaces in the close vicinity of the superior orbital fissure and the posterior sinkhole of the inferior orbital fissure. These apex areas of the orbit are identical with the regions in juxtaposition of the lateral surfaces of the cubic body of the sphenoid bone. This coincidence fostered the idea of this fundamental anatomy chapter to describe the osseous construction of the orbits systematically building up from scratch at the orbital apex in the manner of a technical blueprint and exploded view drawings of the individual components. This process will result in a depiction of the overall assembly of both orbits and an outline of their spatial and topographical relationships at the transition between the maxillae, central and lateral midface components, and the cranial base

© The Author(s) 2023

P. J. J. Gooris et al. (eds.), *Surgery in and around the Orbit*,
https://doi.org/10.1007/978-3-031-40697-3_2

and vault. Finally, the position, shape, dimensions, and important variations of the gates and passageways within this framework of bones and paranasal sinuses are addressed.

Method

Classic textbook and atlas descriptions were used to model the basic bony anatomy of the orbit and serve as introduction to outline the details of the passageways as well as topographical aspects. The recent relevant literature on the bony openings was identified and a small selection assigned to review the details of the bony openings. Photographs of post mortem human subject dissections, illustrations and digital drawings were synthesized to present a comprehensive survey of topics spanning the breadth of current knowledge on the bony constituents inside and around the orbit.

I. Osteology—Bony Building Blocks

Overall Skeletal and Topographical Configuration

The orbits are formed by facial and cranial bones. Each orbit is assembled of seven bones: sphenoid, frontal ethmoid, lacrimal, palatine, maxilla, zygoma [1]. These skeletal building blocks outline a cone- or pear-shaped cavity with a thick marginal rim framing the aperture at its base in contrast to the thin-walled cone construction. In terms of a geometric concept, the orbit can be translated into a pyramid with a quadrangular base, giving rise to four concentric walls which carry a three-sided spire or apex on the top end (Fig. 2.1a–f). The open base or aditus orbitae is projecting fronto-laterally and the apex posteromedially toward the optic foramen. The aditus is framed by thick and prominent marginal rims. These are well defined except for the medial side, which is discontinuous due to the interposition of the fossa for the lacrimal sac between its lower and upper part.[1] The medial rim or nasal orbital

margin consists of the frontonasal process of the maxilla, the lacrimal bone, and the maxillary process of the frontal bone. The frontonasal process extends upward into the anterior lacrimal crest and forms the medial orbital margin in the lower part (MOMLP). The supraorbital rim continues into the posterior lacrimal crest in a more backward plane and creates a second parallel bone ridge, the upper part of the medial orbital margin (MOMUP) with the lacrimal fossa in between at a fluted bevel. The superior and inferior orbital margins curve distinctly posterior, so that the lateral rim is least projecting in the whole orbital circumference. The four walls of the human orbit or the internal orbit, respectively, are formed by the seven bones named above. The roof of the orbit is composed largely of the orbital plate of the frontal bone anteriorly and of the lesser wing of the sphenoid (LWS) with a minor part in the posterior part. The triangular shape of the roof narrows toward the orbital apex. The anterior portion of the frontal bone is containing the frontal sinuses which can extend far up into the squamous part of the frontal bone and far back over the orbital roof when extremely pneumatized. The floor of the anterior cranial fossa forms the endocranial side of both orbital roofs. The fossa for the lacrimal gland is a shallow depression in the anterolateral aspect of the roof next to the zygomaticofrontal suture (ZFS). A small depression in the anteromedial portion of the roof, the trochlear fovea, is the site of attachment for the fibrocartilaginous ring (pulley) girdling the tendon of the superior oblique muscle.

The junctions of the walls in the superomedial, superolateral, inferolateral, and inferomedial quadrants or pyramidal corners are rounded in reality. The overall configuration of the internal orbit (i.e., the inside) comes closer to the pyramid model than the external orbit due to its integration into frames, pillars, support, and pneumatized paranasal structures. With regard to their anatomical subunits, the entire orbital margin is occasionally divided into three sections, a supraorbital rim, an inferomedial or maxillary rim, and an inferolateral or zygomatic rim.

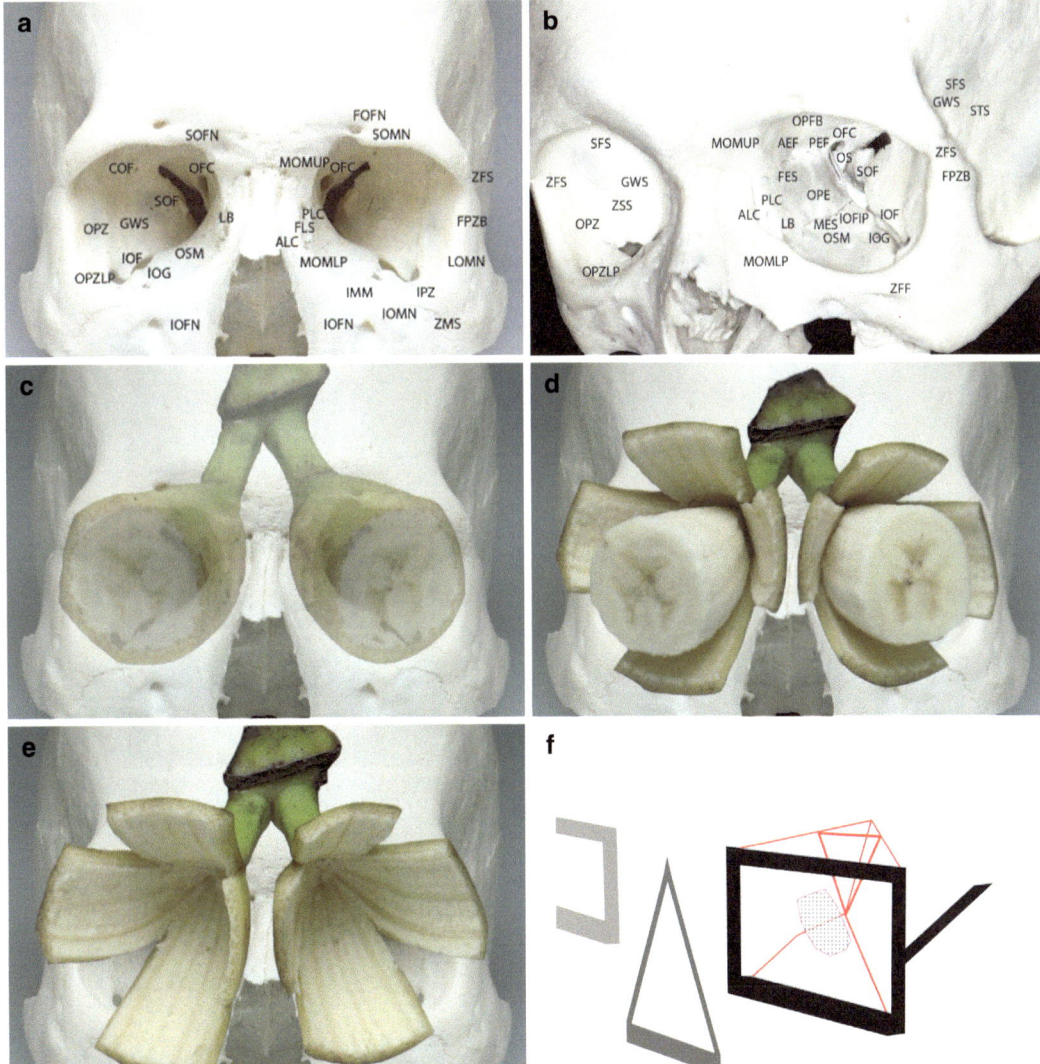

Fig. 2.1 (**a**) Orbits—anterior view—rim circumference. The broad blending of the anterior and posterior lacrimal crest results in discontinuity and ambiguity at the medial orbital rim. The anterior view limits the visibility over the internal orbital surfaces. *ALC* anterior lacrimal crest, *COF* cranioorbital foramen, *FLS* fossa for lacrimal sac, *FOFN* frontal orbital foramen/notch, *FPZB* frontal process of zygomatic bone, *GWS* greater wing of sphenoid, *IMM* infraorbital margin of maxilla, *IOMN* infraorbital margin, *IOF* inferior orbital fissure, *IOFN* infraorbital foramen, *IOG* infraorbital groove, *IPZ* infraorbital process of zygoma, *LB* lacrimal bone, *LOMN* lateral orbital margin, *MOMLP* medial orbital margin—lower part, *MOMUP* medial orbital margin—upper part, *OFC* optic foramen/canal, *OPLZP* orbital plate of zygoma lower part, *OPZ* orbital plate of zygoma, *OSM* orbital surface of maxilla, *PLC* posterior lacrimal crest, *SOF* superior orbital fissure, *SOFN* supraorbital foramen/notch, *SOMN* supraorbital margin, *ZFS* zygomatico–frontal suture, *ZMS* zygomatico—maxillary suture.

(**b**) Orbits—left antero-oblique view to get insight into the inferomedial and anterolateral orbital walls concurrently. *AEF* anterior ethmoidal foramen, *FES* frontoethmoidal suture, *IOFIP* IOF isthmus promontory, *MES* maxilloethmoidal suture line, *OPE* orbital plate of ethmoid, *OPFB* orbital plate of frontal bone, *OS* optic strut, *PEF* posterior ethmoidal foramen, *PLC* posterior lacrimal crest, *SFS* sphenofrontal suture, *STS* spheno—temporal suture, *ZFF* zygomaticofacial foramen, *ZSS* zygomaticosphenoid suture. (**c–e**) Banana visualization of the geometric concept—both orbits. Likewise pealing bananas, the complex 3D orbital walls are translated into a 2D structure (Idea: C. P. Cornelius, Realization: Klaus Völcker, Regensburg). (**f**) Geometric scheme of the orbit—a quadrangular pyramid converts into a tetrahedron posteriorly. Accordingly, the frontal cross section in the apex is triangular. Dot matrix = posteromedial bulge (With permission from Orbital Fractures ISBN 0-88937-139-3 and ISBN 3-8017-(now Hogrefe Publishing), www.hogrefe.com)

Sphenoid Bone—Constituent of the Upper, Medial, and Lateral Orbital Wall

The sphenoid bone with its cuboidal body at the center and three pairs of outstretched wings or leg-like processes, respectively, can be appreciated as the principal constructive element of the skull base. It has often been considered as the most complex polymorphous bone of the human skeleton (Figs. 2.2a–d and 2.3).

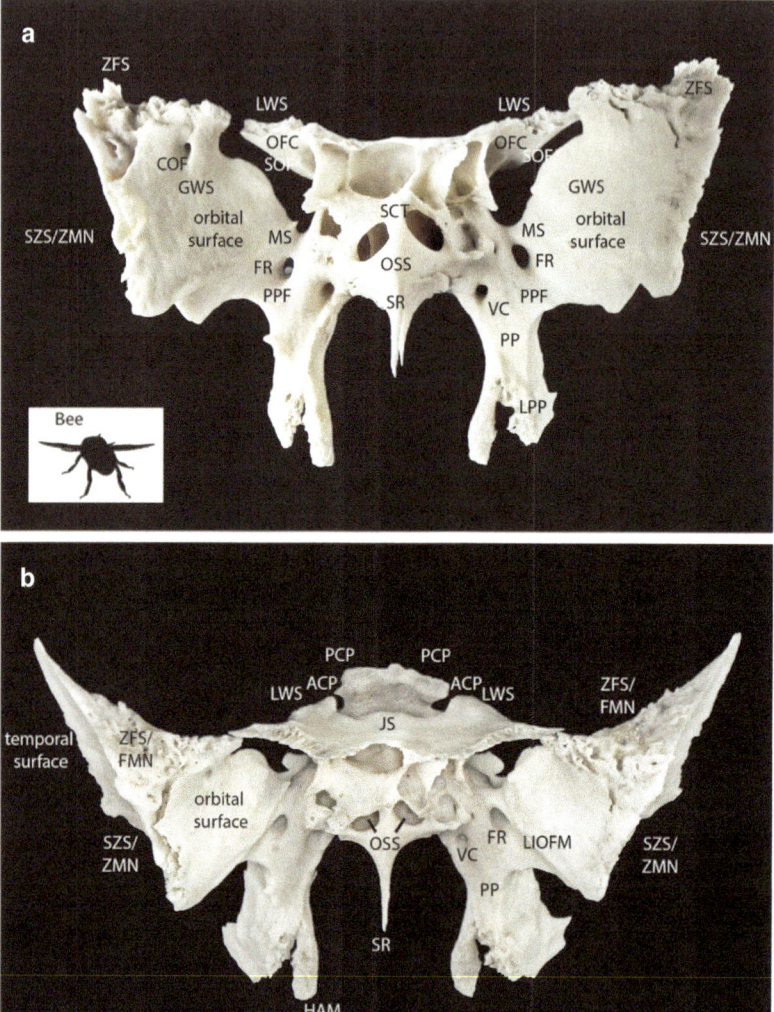

Fig. 2.2 Main features of the sphenoid from different aspects (**a**–**d**). The apices of the orbit are identical with the bony core regions around the SOF on both sides of the sphenoid body. (**a**) Sphenoid—anterior view—inset lower left corner: flying insect or bee like appearance. Note: 'Posterior ethmoid air cells' invading the upper anterior surface of the sphenoid body – correlating to 'Onodi' cells. *FR* foramen rotundum, *LPP* lateral pterygoid plate, *LWS* lesser wing of sphenoid, *MS* maxillary strut, *OSS* opening of sphenoid sinus, *PP* pterygoid process, *PPF* pterygopalatine fossa, *SR* sphenoid rostrum, *SZS/ZMN* sphenozygomatic suture/zygomatic margin, *VC* vidian canal/pterygoid canal. (Bee inset with permission from https://surgeryreference.aofounda-tion.org). (**b**) Sphenoid—Superoanterior view. *ACP* anterior clinoid process, *PCP* posterior clinoid process, *HAM* hamulus, *JS* jugum sphenoidale/sphenoid yoke, *LIOFM* lateral inferior orbital fissure margin, *SR* sphenoid rostrum, *ZFS/ZMN* zygomaticofrontal suture/zygomatic margin. (**c**) Sphenoid—Superior view—Inset lower left corner: flying bat like appearance. *CAG* carotid artery groove, *DS* dorsum sellae, *FO* foramen ovale, *FSM* foramen spinosum, *HF* hypophyseal fossa, *SQM* squamous margin. (Bat inset with permission from https://surgeryreference.aofoundation.org). (**d**) Sphenoid—Posterior view. *DS* dorsum sellae, *GLO* groove in the lateral orbit, *LPP* lateral pterygoid plate, *PCS* prechiasmatic sulcus, *PF* pterygoid fossa

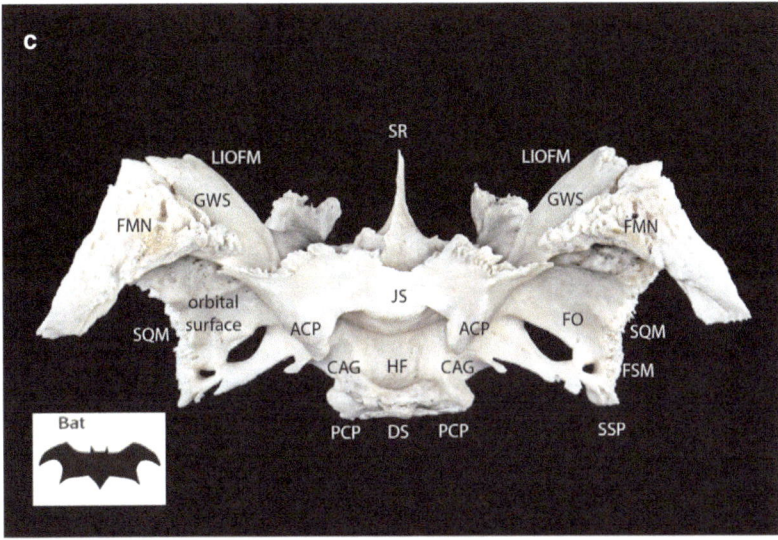

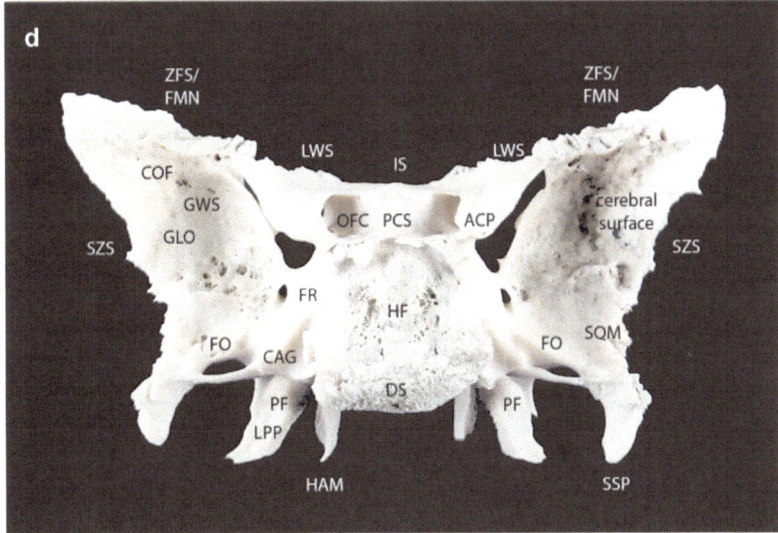

Fig. 2.2 (continued)

Its front view silhouette has the appearance of a wasp (Fig. 2.2a). Because of this appearance, a common term for it was "sphecoid" bone, according to Ancient Greek σφήξ (sphḗx ≈ "wasp" or "hornet") which was used every now and then from the beginnings of medical science in Greco-Roman antiquity until the nineteenth century. The tale (historically unconfirmed) is still bandied that the c in sphenoid was erroneously confused with an *n* resulting in sphenoid which has the meaning of σφην (sphēnoeid ≈ "wedge") in the publishing process of an anatomy textbook and never set right again since then [3, 4]. The triangular lesser wings (alae minores) of the sphenoid (LWS) extend from the superolateral aspect of the body to form the

upper border of the superior orbital fissure (SOF), the most posterior part of the orbital roof, and the posterior ridge of the floor of the anterior cranial fossa on either side (Fig. 2.2a). The greater wing (ala major) of the sphenoid (GWS) separates the orbit from the middle cranial fossa and is part of the vertical pterygomaxillary buttress. Each GWS is attached on the lateral aspect of the body by a radix (root). The GWS can be conceived as assembled of three divisions (anterolateral/posterolateral/inferior) which are arranged along and below a vertical compact pillar with a triangular horizontal cross section, nowadays called the sphenoid door jamb (SDJ see later). It ensues a complicated 3-dimensional configuration of two adjacent corners exhibiting several external and cerebral (internal/endocranial) surfaces. The external medial surface of the anterolateral division corresponds to the plane posterior portion of the lateral orbital wall. The external lateral surface of the same division is docked close to its anterior border at a T-junction by the posterolateral division. The external surfaces of both the anterolateral and posterolateral divisions blend to form parts of the concave medial boundary of the temporal fossa and the external lateral orbital wall. The inferior GWS division spreads underneath the other two divisions starting posteriorly from their vertical intersection (Fig. 2.2c, d). The upper surface of the inferior division conforms to the floor of the medial cranial fossa which is bordered anteriorly and laterally by the cerebral surfaces of the vertical GWS divisions moreover. The bottom surface of the inferior GWS division circumscribes the infratemporal fossa. Axial cross sections unveil the posterior GWS as a central trigone with a spongious bony space between the orbital, temporal, and cranial cortical surfaces (Fig. 2.15b). This potential space for surgical decompression is termed the sphenoid door jamb (SDJ). The superior orbital fissure (SOF) is a gap between the LWS and

the GWS. The lower end of the SOF is bounded by the maxillary strut; this is a bony bridge across the foramen rotundum and integral part of the upper GWS radix. The paired pterygoid processes project bilaterally from the inferior aspect of the sphenoid body and its connection to the GWS downwards. It is the base of the lateral pterygoid plate (lamina) that fuses with the root of the GWS in a longitudinal direction. The medial and lateral pterygoid plates band together along the upper portion of their anterior delineations to form the posterior concavity of the pterygopalatine fossa (PPF). The anterior bottom portions of the pterygoid plates remain set asunder forming the pterygoid notch. The notch gap is interposed by the solid inferoposterior end (pyramidal process) of the perpendicular plate of the palatine bone. The suture line arrangements differ and are collectively addressed as the pterygopalatine fissure.

The pterygoid fossa is the vertically oriented open recess between the pterygoid process plates viewed from the posterior aspect. The sphenoid connects to all of the other cranial base bones (Fig. 2.3) and to all bones of the facial skeleton except for the lacrimal and nasal bones.

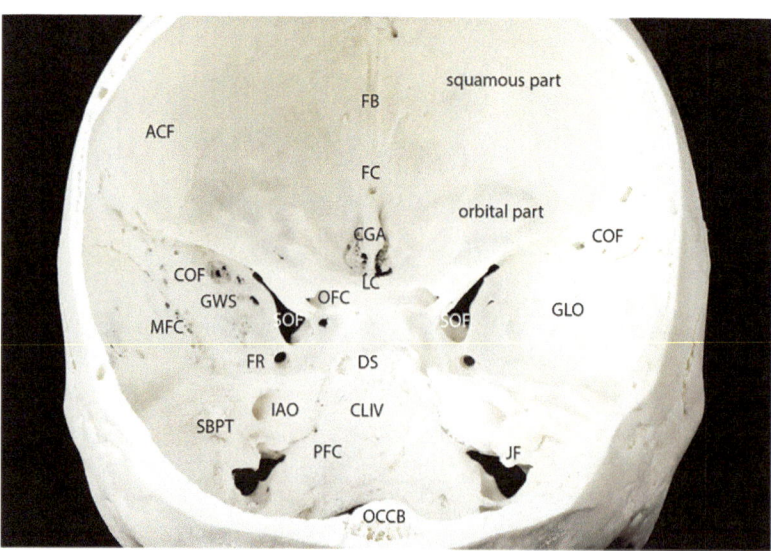

Fig. 2.3 Sphenoid within the cranial fossae—Endocranial surface view. Skull pitched downward posteriorly to bring the OFC, SOF, and Foramen rotundum into view. COF left > right. Note: foramen caecum (FC), crista galli (CGA), and cribriform plate (LC); bony spicule on the floor surface of the anterior cranial fossa. Identify Pterion region inside skull and groove in the lateral orbit (GLO— [2]). The sphenoid entity made up of the body, orbital api- ces, GWS, LWS, and PPs assumes the role of a backbone that defines the spatial relationships at the transition from the face to the cranial base. *ACF* anterior cranial fossa, *CLIV* clivus, *FB* frontal bone, *IAO* internal acoustic opening, *JF* jugular foramen, *MCF* middle cranial fossa, *OCCB* occipital bone—internal protuberance, *PFC* posterior cranial fossa, *SBPT* superior border of petrous part of temporal bone

Bony Orbital Apex—A Three-Walled/Tetrahedron Spire

The orbital apex within the sphenoid bone needs an elaborated description.

So far there is no unanimously consented definition of the sagittal extent of the orbital apex [5]. For the purposes here, it is regarded as the most posterior projection of the orbital cavity that starts in front of the level of the maxillary strut or foramen rotundum with a narrow triangular base in the frontal cross section (Fig. 2.4a, b) and ends with a pointed tip within the openings of the optic foramen or SOF,

respectively. The roof side of the orbital apex is formed by the LWS, the medial side bordered by the lateral surface of the sphenoid body, and the lateral side adjoined by the GWS. The orbital floor with the orbital surface of the palatine bone at its posterior portion comes to a halt at the posterior (rear) IOF sinkhole and does not reach into the orbital apex (Fig. 2.4a, b). In general, two openings can be distinguished within the orbital apex, the optic foramen (OF) representing the intraorbital end of the optic canal and the SOF. The optic strut (OS) constitutes the inferolateral wall of the OC and separates the latter from the SOF (Fig. 2.5a, b).

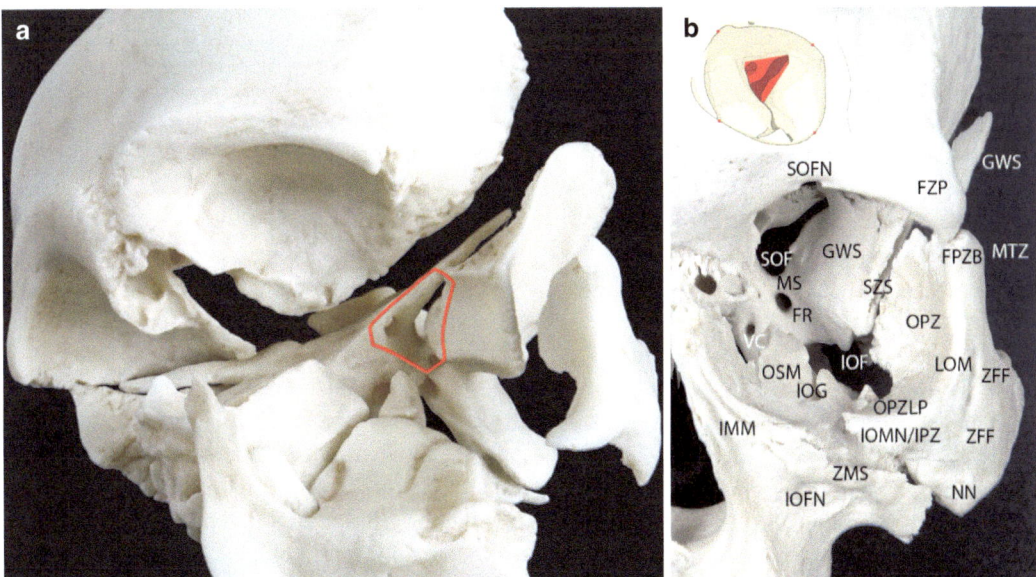

Fig. 2.4 (a) Exploded model of the left orbit. Red color marking of the orbital apex which is three-walled spire in contrast to the four-walled pyramid conforming the mid-orbit and anterior orbit. (b) Left orbit—Anterior view. Sutures disrupted, palatine, and ethmoid bones removed to spotlight the triangular orbital apex in the back and to expose the posterior (rear) IOF sinkhole. Inset (upper left corner): Geometric outline of the orbital apex. *LOM* lateral orbital margin, *FZP* fronto—zygomatic process, *IOMN/IPZ* infraorbital margin/infraorbital process of zygoma, *NN* nomen nominandum, *ZFF* zygomaticofacial foramen, *ZMS* zygomatico—maxillary suture. (Inset with permission from https://surgeryreference.aofoundation.org. Copyright by AO Foundation, Switzerland)

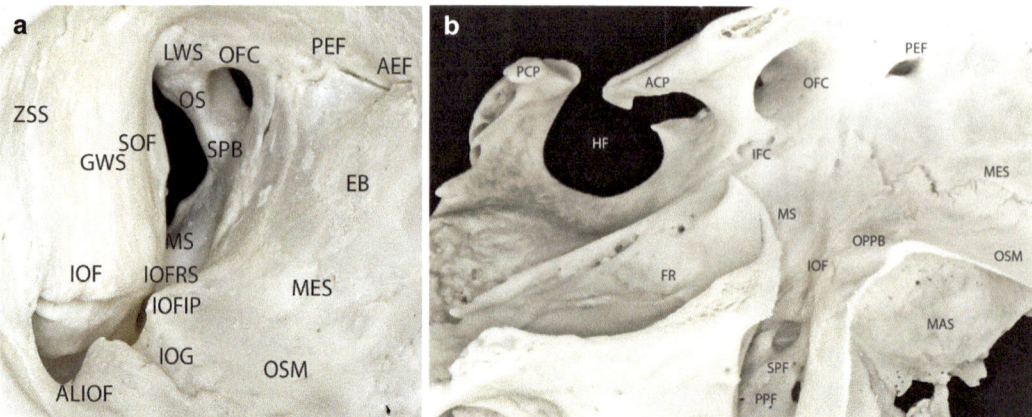

Fig.2.5 (**a**) Architecture of the entire orbital fissure system (right—capital "L"—mirrored image. (—for unmirrored L on the left side—see Fig. 2.4b) extending into the orbital apex and optic canal—anterior close-up view of the major bony pathways in and out of the orbit. *ALIOF* anterior loop of inferior orbital fissure, *EB* ethmoid bone, *IOFRS* inferior orbital fissure rear sinkhole, *IOFIP* inferior orbital fissure isthmus promontory, *MES* (~ *IOS*) maxilloethmoidal suture line (~ inferomedial orbital strut), *SPB* sphenoid body. (**b**) Junction of orbital apex and lateral sellar region/middle cranial fossa (right)—lateral view. The infraoptic groove (sulcus), tubercle, or canal is the origin of the common annular tendon (Zinn's ring). *IFC* infraoptic canal, *MAS* maxillary antrum/sinus, *OPPB* orbital plate of orbital process of palatine bone, *SPF* sphenopalatine foramen

Frontal Bone—Orbital Roof Constituents

The unpaired frontal bone features external and cerebral surfaces on the sides of a large squamous vertical forehead part (squama frontalis), a nasal part, and two horizontally oriented orbital parts. Each orbital part passes backward behind the supraorbital rim and configures the superior wall (roof) of the orbit in unison with the floor of the anterior cranial fossa to a major extent (Fig. 2.6). The posterior margin of the orbital part articulates with the LWS, which forms a minor most posterior roof portion congruous with the roof of the apex. The posterolateral orbital part margin connects with the upper edge of the GWS along the sphenofrontal suture line (Fig. 2.7). The anterolateral orbital part brings up the zygomatic process and joins with the orbital plate and the temporal process of the zygoma along the frontozygomatic suture line. The orbital roof has the outline of an isoceles triangle that bends up into a concavity. The lacrimal fossa is

a shallow depression in the anterolateral roof for the lacrimal gland. The trochlear fovea conforms to the anteromedial adherence zone of trochlear fiber condensations.

The ethmoidal notch is a cove-like slot in between the two orbital plates with a quadrilateral outline, from an endocranial view it is located in the central region of the anterior cranial fossa. Anteriorly the frontal notch turns into the nasal process of the frontal bone (frontal beak) with a serrated margin on either side of the superior nasal spine, a sharp downward process in the midline. The nasal margins articulate with the nasal bones and the frontonasal maxillary processes. The anterior and superior borders of the lacrimal bone connect to the frontonasal maxillary process and to a small strip of the notch margin just behind its anterior corner. In the intact, articulated cranium the interorbital slot is filled by the cribriform plate and the crista galli of the ethmoid bone. The margins of the notch contain several partially opened sinus cells, which have their counterparts in the upper surface of the ethmoid. The sandglass- or two-funnel-shaped fron-

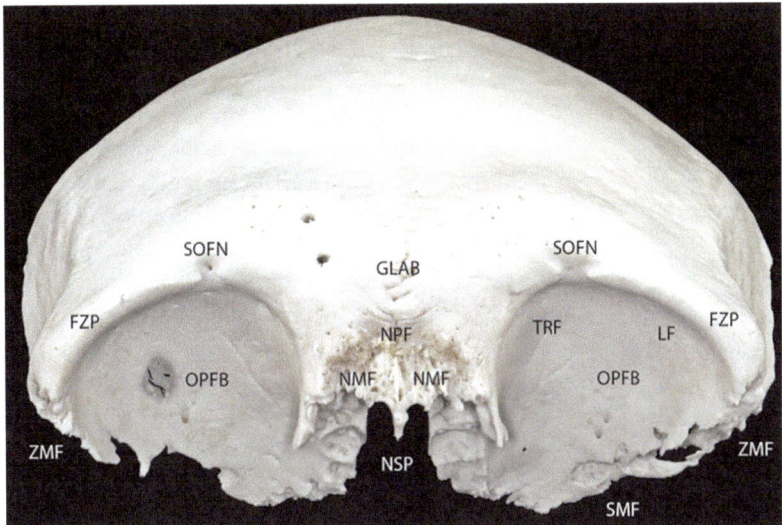

Fig. 2.6 Frontal bone—Anteroinferior view. Orbital surfaces and orbital rims are passing backward. Nasal margin roughly serrated with nasal spine situated medially. Supraorbital openings shaped as notches; Ethmoidal notch bordered with roofs of ethmoidal air cells. *GLAB* glabella, *LF* lacrimal fossa, *NMF* nasal margin of frontal bone, *NPF* nasal part of frontal bone, *NSP* nasal spine, *SMF* sphenoidal margin of frontal bone, *TRF* trochlear fovea, *ZMF* zygomatic margin of frontal bone

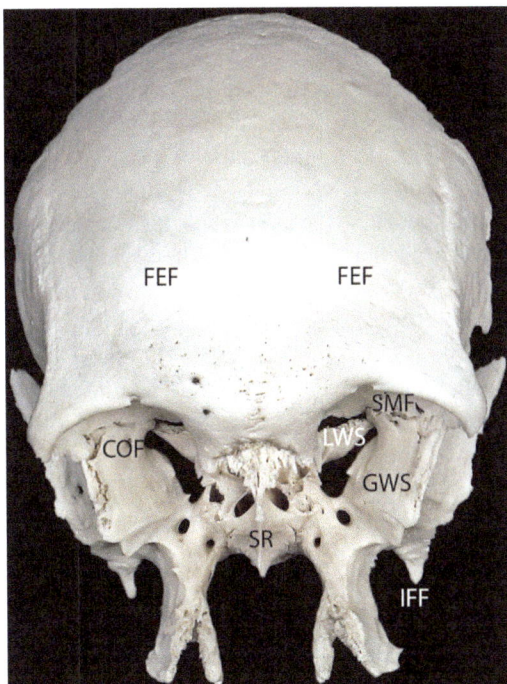

Fig. 2.7 Articulation of frontal bone and sphenoid—Anterior view. The orbital roof is built up including the LWS. The GWS provides the posterior part of the lateral orbital wall. *FEF* frontal eminence of frontal bone, *IFF* infratemporal fossa

tonasal communication and drainage tract presents with an isthmus (frontal ostium) caudal to the bottom of the medial floor of each sinus. The cranial funnel of the tract is accommodated within the confines of the anterolateral notch margin and proceeds further downwards in the niches between the nasal margins and the superior nasal spine to reach into the frontal recess or the ethmoidal infundibulum [6–8]. The extensions of the frontal sinus over the orbital roofs and behind the squamous part of the frontal bone can be intensively pneumatized. The two tables of bone can be extremely thin and may contain dehiscences, so that the periorbita in the roof is in direct contact with the dura mater. The aeration patterns of the frontal sinus show great gender and interindividual variations in number, dimensions, outline, symmetry, septation, and laterality.

The superior orbital wall (roof) to its greatest extent consists of the orbital part of the frontal bone (Fig. 2.8). The most posterior minor portion at the apex is formed by the lesser wing of the sphenoid. The orbital roof takes a triangle shape bent up into a dome.

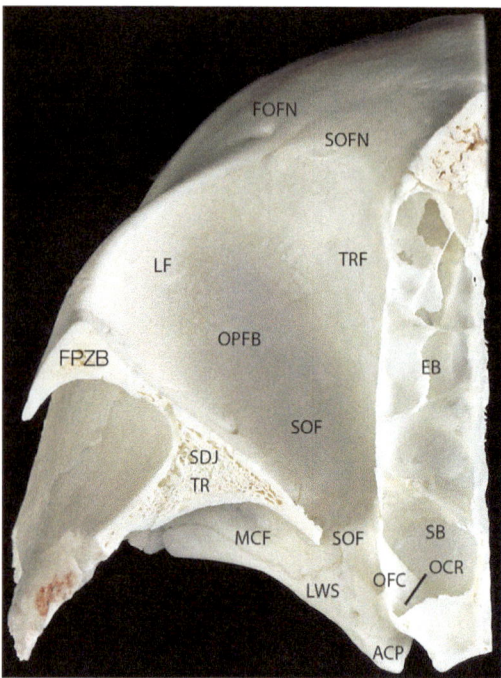

Fig. 2.8 Orbital roof (right) – Inferior view. *OCR* opticocarotid recess, *SB* sphenoid bone, *SDJ* sphenoid door jamp, *TR* trigone (GWS)

Ethmoid, Lacrimal Bone and Frontonasal Maxillary Process—Constituents of the Medial Orbital Wall

The medial wall of the orbit is part of the centrofacial or naso-orbito-ethmoido-nasal unit. It is built up by the sphenoid, ethmoid, lacrimal bone, and the maxilla in a posterior to anterior order (Fig. 2.9).

The lateral wall of the sphenoid body is lined up in front with the lamina papyracea, the paper-thin orbital plate of the ethmoid, and the lateral surfaces of the lacrimal bone and the naso-frontal maxillary process (NFP). The unpaired ethmoid (ethmoidal bone) is located between the two orbits. It contains the ethmoid sinus cells and is forming parts of the nasal and the orbital cavity. The ethmoid bone is built up of the perpendicular

plate along the sagittal plane with the crista galli at the upper end (Fig. 2.10a–e). The perpendicular plate corresponds to the bony nasal septum. The cribriform plate extends on each side of the upper perpendicular plate margin in kind of the bar of a T-beam profile. The ethmoidal labyrinths refer to honeycomb arrays of small air cells hanging downwards from the lateral bounds of the cribriform plates. The ethmoidal air cells vary in number and volume on each side and extend in the orbital (i.e., edges of the ethmoidal notch) and squamous part of the frontal bone (Fig. 2.11), occasionally into the sphenoid and into frontonasal process of the maxilla, too. The paramedian portions of the cribriform plate form the roof of the nasal cavity. The rudimentary supreme, the superior, and the middle nasal turbinates (conchae) go down from the medial side of each labyrinth. The boundaries of an ethmoidal labyrinth are often reviewed for didactical illustration in analogy to the sides of a matchbox cover standing on its small longitudinal side (Fig. 2.10a). The box cover is closed posteriorly with the frontal surface of the sphenoid body. In addition to the already open anterior entrance the bottom side of the box is absent in accordance to the open sides of the middle nasal meatus. The lateral wall of the box represents the orbital plate, which is identical with the large rectangular center piece of the medial orbital wall. The orbital plate is paper-thin (lamina papyracea) and transparent, so that the contours of ethmoidal air cells become visible through it (Fig. 2.12). The lower lateral wall of the box or in other terms the inferior ethmoidal cells are adjacent to the medial roof side of the maxillary sinus. The medial wall of the labyrinthine box corresponds to the upper and middle turbinates in alignment at the lateral wall of the inner nose (turbinal wall). The curved basal (ground) lamellae of the superior and middle turbinates fuse with each other and intermingle with the air cell septa network buttressing the paper plate medial orbital wall. The basal lamina of the middle turbinate also reinforces the maxilloethmoidal suture resulting in a firm bony thickening,

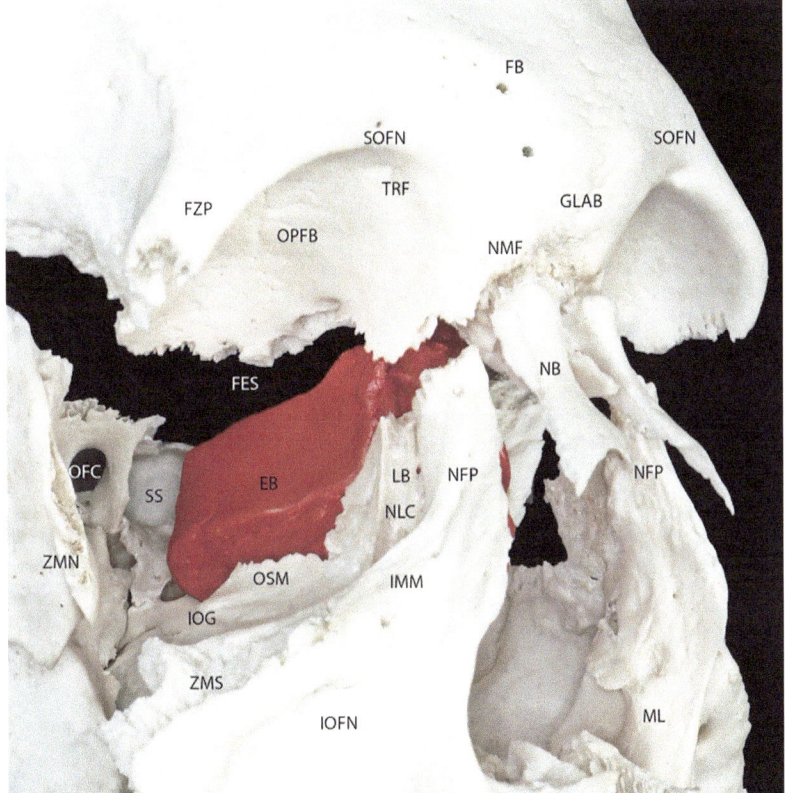

Fig. 2.9 Bony constituents of the right medial orbital wall—Right inferolateral view. The ethmoid (red colored) with the lamina papyracea makes up the large center part of the ensemble of the sphenoid, ethmoid, lacrimal bone, and frontonasal process of the maxilla. *ML* maxillary line, *NB* nasal bone, *NFP* nasofrontal process of maxilla, *NLC* nasolacrimal canal, *SS* sphenoid sinus, *ZMN* zygomatic margin

which is referred to as the inferomedial orbital strut (IOS) (Fig. 2.24a, b).

Clinical Implications

Significance of morphologic properties of the medial orbital wall:

Few and large-volume ethmoidal air cells extending over larger areas of the lamina papyracea are supposed to predispose more fracture susceptibility of the medial orbital wall, in contrast to a high-density reinforcing honeycomb structure [9].

The small trapezoidal lacrimal bones have a nasal and an orbital surface. They are located right in front of the orbital plate of the ethmoid and close to the medial orbital wall across the nasofrontal maxillary process (NFP). Concurrently, they complete the lateral wall of the anterior ethmoidal cells and are part of the lateral wall of the nose where this is overlying the tip of the middle turbinate. A vertical ridge, the posterior lacrimal crest, protrudes along the orbital surface and divides it into a smooth plane posteriorly and into a carved out longitudinal groove (lacrimal sulcus). The anterior margin of this sulcus joins to the likewise grooved posterior margin of the NFP and completes the entrance into the nasolacrimal canal. The orifice at the initial canal section is widened and creating a fossa for the lacrimal sac. The upper lateral slope of the NFP carries the anterior lacrimal crest in parallel to the anterior circumference of the canal inlet zone.

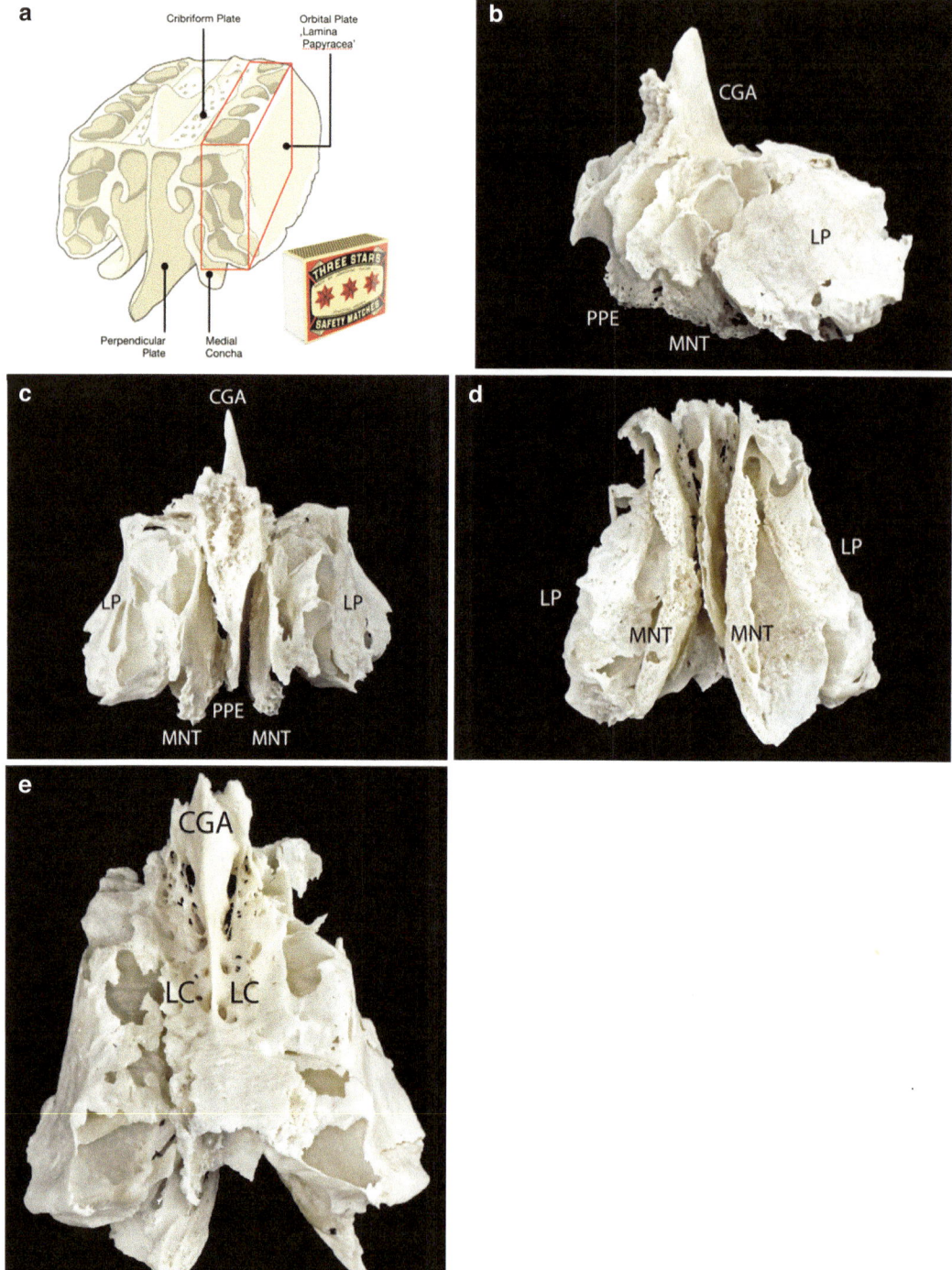

Fig. 2.10 (**a**) Graphic layout of the ethmoid—Left anterolateral view. Illustration of main components. Red contour silhouette indicates geometry of an ethmoidal labyrinth. Inset: Matchbox comparison to review the surfaces of a labyrinth (see text). (Ethmoid sketch with permission from https://surgeryreference.aofoundation.org). (**b**) (Upper right): Ethmoid—Left anterolateral view. Bony analogue to Fig. 2.9. Note: Anterior and middle ethmoidal air cells with open roof. *LP* lamina papyracea, *MNT* middle nasal turbinate (Concha), *PPE* perpendicular plate of ethmoid. (**c**) (Middle left): Ethmoid—Posterior view. Note: Posterior ethmoidal air cells open. (**d**) (Middle right): Ethmoid—Inferior view. (**e**) (Low): Ethmoid—Posterosuperior view. Labyrinthine air cells with open roofs. *LC* lamina cribrosa/cribriform plate

Fig. 2.11 Frontal and ethmoid bones—Frontal and ethmoid bones from superior view showing the upper cerebral surfaces. Articulation assembly. The ethmoid air cells are capped by apposition with the roofs of the ethmoid notch edges

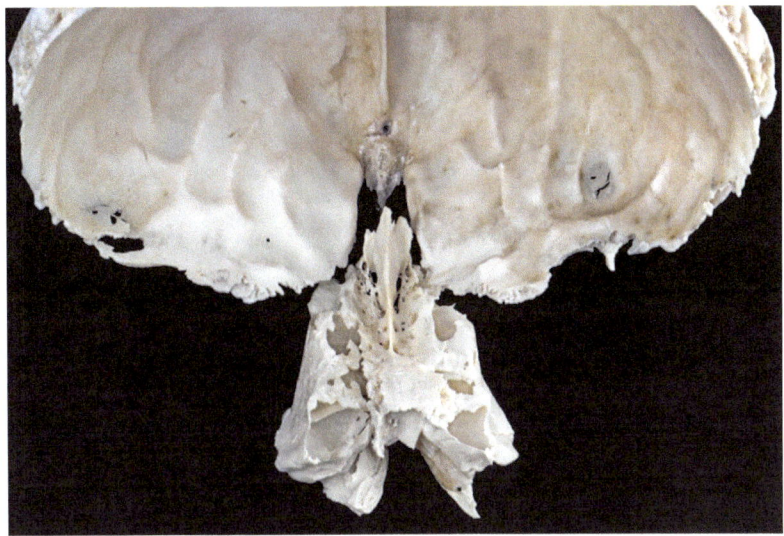

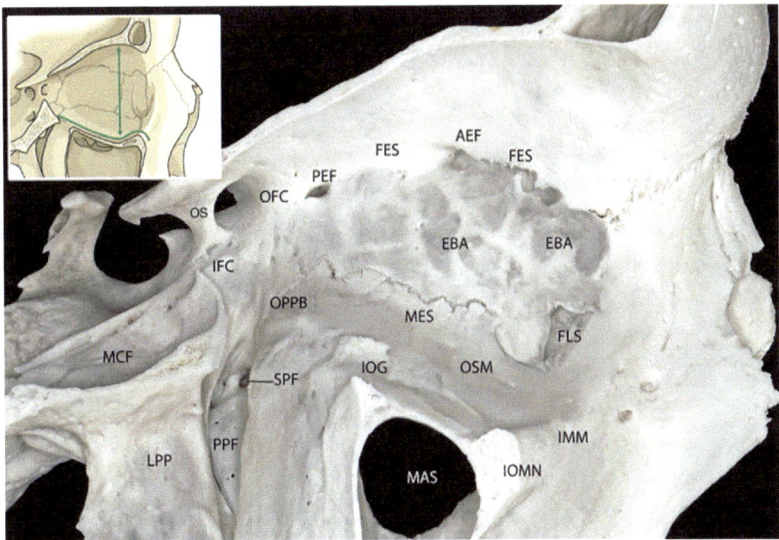

Fig. 2.12 Medial orbital wall—lateral to medial view offering a look into the OFC and the pterygopalatine fossa along the medial IOF margin and retrotuber maxillary region. Inset (upper left corner)—Vertical extent of the postentry zone. Behind the infraorbital and supraorbital rim (vertical green arrow). Lazy-S shape of ascending orbital floor - undulation allusive only (green line). *EBA* ethmoid bone/air cells, *LPP* lateral pterygoid plate. (Inset with permission from https://surgeryreference.aofoundation.org)

The posterior lacrimal crest may be elongated together with a small portion of the lower smooth plane into a hooked process. This lacrimal hamulus interdigitates with a corresponding notch of the NFP at the caudal periphery of the canal entrance.

Palatine Bone and Maxilla—Major Constituents of the Orbital Floor

The palatine bone is interposed between the back of the maxilla and the base as well as the medial and lateral plates of the pterygoid process (Fig. 2.13a). It

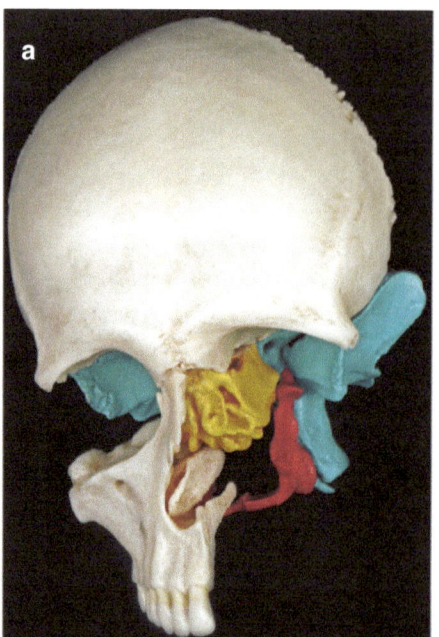

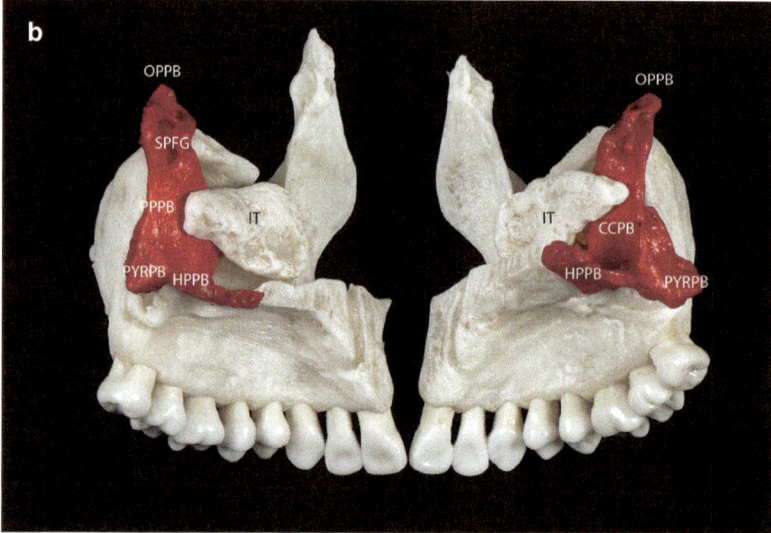

Fig. 2.13 (**a**) Osseous model set up demonstrating the origin of the posterior ledge—the orbital process of the palatine bone (red) is the rearmost portion of the orbital floor with its orbital plate ('facies orbitalis'/OPPB) (= posterior ledge), it is contiguous to the upper end of the perpendicular plate. Ethmoid bone (yellow), sphenoid (turquoise) with GWS, LWS, and pterygoid process— outer lamina and hamulus. (**b**) Posterior view of the maxillae (swung out) to appreciate the orbital process of palatine bones (red) as robust constituent of the posterior orbital floor. *CCPB* conchal crest of palatine bone, *HPPB* horizontal process of palatine bone, *IT* inferior turbinate, *PPPB* perpendicular process of palatine bone, *SPFG* sphenopalatine foramen/groove. (**c**) Right palatine bone— view from the front. *OPPB* orbital process of palatine

bone, *PYRPB* pyramidal process of palatine bone, *SPNPB* sphenopalatine notch (palatine bone), *SPPB* sphenoidal process of palatine bone. (**d**) Right palatine bone—view from above in anteroposterior orientation. *PNS* posterior nasal spine. (**e**) Pair of isolated palatine bones oriented in an oblique antero-posterior direction resembles antlers. The smooth surface of the orbital plate at the top of the right orbital process (OPPB) breaks off at a distinct edge posteriorly and converges into the front wall of the rear sink of the inferior orbital fissure (IOFRS). MPPB, Maxillary Process of Palatine Bone; GPG, Greater Palatine Groove. OPPB Detail–Inset (upper right corner): edged demarcation line between the posterior end of orbital floor and IOF rear sink (green line grid) surfaces

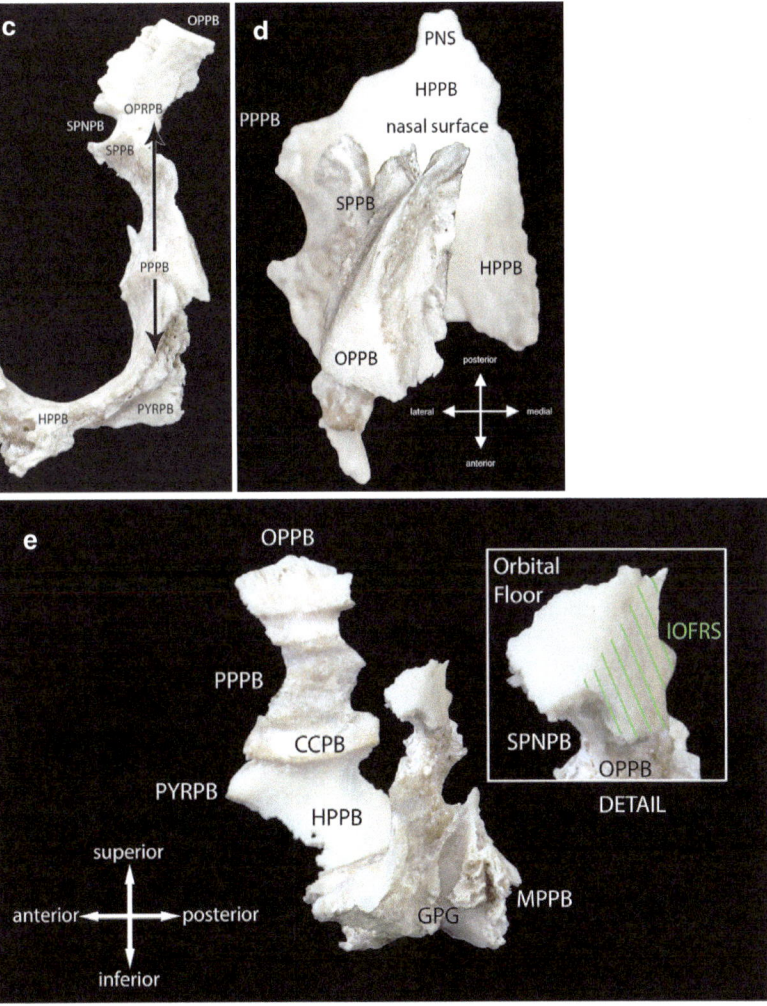

Fig. 2.13 (continued)

contributes to the posterior lateral wall of the nasal cavity, the hard palate, the PPF, and the orbital floor in an intriguing configuration. Basically, the palatine bone consists of a vertical part (perpendicular plate) with a medial oblique orientation and a horizontal part or plate providing the posterior hemi-hard palate portion. The upper end of the perpendicular plate is overtopped by the orbital and sphenoidal processes. The orbital process of the palatine bone (OPPB) is situated at a higher level than the sphenoid process and projecting anterosuperiorly. It is based on a neck narrowed by the half-oval sphenopalatine notch (SPNPB) at the posterior aspect which also sets it apart from the sphenoidal process that is attached at the inferior margin of the notch and diverges medially and upwards. The

orbital process presents with five surfaces in total. The anterior surface connects with upper medial end of the posterior surface of the maxilla. It is common to refer to the superior surface of the orbital process of the palatine bone, as to the orbital plate of the palatine bone (OPPB) more briefly. The OPPB is triangular in shape and sloping slightly upwards in combination with a downward tilting laterally (Fig. 2.13b–d). Since the OPPB conforms the rear end of the orbital floor, that is often preserved in defect fractures and may serve as support in surgical reconstruction therefore, it is simply addressed as the "posterior ledge" in surgical jargon. The remaining three surfaces are directed either posteriorly to abut the anterior sphenoid body, or medially to join with the ethmoid and laterally toward the pterygopalatine fossa (PPF). The orbital process may enclose an air cell, which is open either to the sphenoid sinus, the posterior ethmoidal cells or with both at once [10–12]. The sphenoidal process attaches to the medial base of the pterygoid process. The anterior border of the sphenoidal process bounds the posterior SPNPB margin, while the anterior notch margin is molded by the posterior neck of the orbital process (OPPB).

The sphenopalatine foramen (SPF) has a posteromedial angulation to the sagittal plane. The SPF results from the union of the inferior surface of the sphenoid body with the superomedial marginal SPNPB circumference. The communication between the PPF and the posterior nasal cavity is via the SPF. In topographical relations, the bony SPF surroundings form the medial wall of the posterior (rear) IOF sinkhole and the vertically oriented surface descending from the posterior OPPB border.

> ### Clinical Implications
> The position of the posterior ledge in terms of linear distances between the infraorbital margin above the infra-orbital foramen and landmark points at the OPPB, the anterior SOF and ventrolateral entrance of the OFC has been assessed in a CT-based cohort study as a guideline for the safe surgical approach in the repair of inferomedial orbital wall fractures [13].
>
> The potential of an OPPB resection for expansion of the surgical corridor in the endoscopic approach to the orbital apex has been checked in a morphometric analysis of the area dimensions in macerated skulls [14].

The orbital surface of the maxilla (OSM) makes up the largest part of the orbital floor [15]. It is the upper of a total of four surfaces of the body of the maxilla (corpus maxillae) which contains the maxillary sinus or the antrum (of Highmore), respectively. Hence, the OSM is also the ceiling of the antrum. The outline markings of the orbital surface fit a smooth triangular plate, which has an inclination angle of 45° to the horizontal plane. The posterior OSM border converts into the rounded anteromedial IOF margin and its consecutive vertical wall, which is interchangeable with the infratemporal surface of the maxillary body. At its front, the anterior OSM blends into the infraorbital margin (IMM). The IMM extends into the lower part of the medial orbital margin (MOMLP) and further on into the nasofrontal maxillary process (NFP). Laterally, the IMM terminates with the zygomatic process. The lower part of the medial orbital plate of the zygoma (OPZLP) provides a small anterolateral flange to the orbital floor just behind the lower lateral corner of the rim. A small bony depression behind the lower medial corner of the rim and immediately lateral alongside the nasolacrimal canal is the origin to the inferior oblique muscle.

The NFP has a plate-like shape and is forming part of the sidewall of the nose along the upper lateral border of the piriform aperture. At the cranio-medial end, the NFP articulates with the frontal bone and the nasal bone (Fig. 2.14).

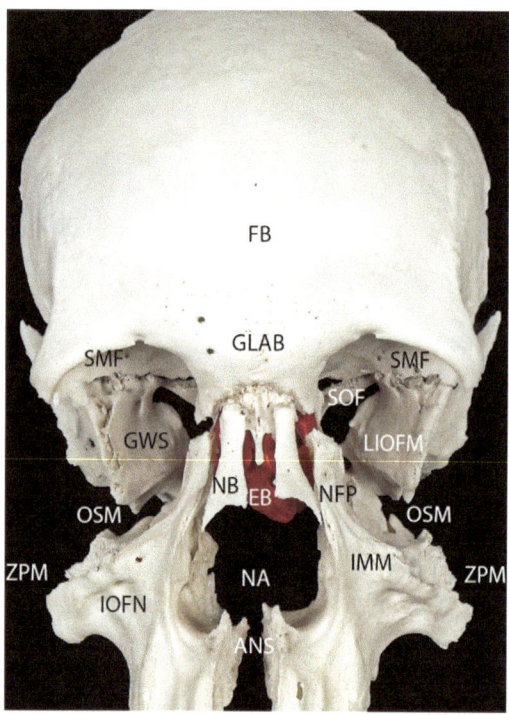

Fig. 2.14 Maxillary building blocks in central midface— Anterior view showing the maxillary processes and relationships to frontal bone, nasal bones, and ethmoid (red colored). *ANS* anterior nasal spine, *NA* nasal aperture, *ZPM* zygomatic process of maxilla

The posterior margin raises the anterior lacrimal crest (ALC) before it transforms into the lacrimal fossa and the entrance into the nasolacrimal groove and canal (NLC), which emanate in combination with the contours and borders of the lacrimal bone.

The zygomatic or malar process is passing laterally for articulation with the zygomatic bone. This process rests on the zygomaticomaxillary crest, a porch jutting out with a slightly arched border from the junction of the anterior and infratemporal surfaces of the maxillary body that extends upward at the level of the first upper molar. The upper surface of this process is a rough and serrated triangular platform that has a downward angulation to the lateral side. The borders of this platform bring up the zygomaticomaxillary suture lines together with the edges of the medial undersurface of the zygoma (Figs. 2.15a, b, 2.16b, and 2.17a, b). The orbital floor is shorter in anteroposterior extent than the three other orbitals walls and thus is missing in the orbital apex (Fig. 2.15a, b).

The posterior or infratemporal surface of the maxillary body covers the curving vertical area between the posterior aspect of the nasal surface to the lateral margin of the zygomaticomaxillary crest (Fig. 2.15c). The maxillary tuberosity is bowing out backward from the lower lateral portion of this surface. A small crescentic rough site at the medial posteroinferior aspect at the junction of the posterior and medial surface is attached to the maxillary process at the lateral surface of the perpendicular plate of the palatine bone. The greater palatine canal is formed by the apposition of two obliquely oriented grooves descending on each side of the bony interface.

The melding seamline along the upper border of posterior surface is the link to the orbital plate and corresponds to the anteromedial IOF margin. This rounded hillslope at the back of the seamline into the IOF rear sinkhole (IOFRS) is engraved by the infraorbital groove (sulcus) [16]. The infraorbital groove (IOG) often begins with a deep U-shaped hollowing in the middle of the posterior maxillary floor. It passes forward over varying distances before it transforms into a canal structure [17]. The infraorbital canal (IOC) moves downwards and gets suspended on a bony beam resembling a mesentery protruding from underneath the orbital floor [18]. This conduit ends in the anterior antral wall usually with a single infraorbital foramen (IOFN). In adults, the foramen is located 7–10 mm below the infraorbital margin and above the canine fossa purportedly on a plumb line passing through the mid-pupil.

Clinical Implications

The position and relative length of the infraorbital groove and canal are supposed to be predisposing factors to the occurrence and pattern of fractures within the orbital floor [19].

The medial or nasal surface is dominated by the maxillary hiatus, a large irregular aperture to the maxillary sinus, which is covered and downsized by the perpendicular plate of the palatine bone posteriorly, the inferior turbinate inferiorly, and the ethmoidal labyrinth—in particular by the uncinate process—together with the lacrimal bone superoanteriorly. The medial OSM border is assembled alongside the maxilla-ethmoidal suture line as part of the inferomedial orbital strut (IOS) and comes up against the lower borders of the lamina papyracea and the lacrimal bone [20]. The two paired maxillae are the largest bones in the midface. Uniting below the nasal aperture at the level of the anterior nasal spine and along the midline of the hard palate, they aggregate to the upper jaw (Fig. 2.14).

Apart from the body including the frontonasal and zygomatic process, there are two other components contributing to the overall architecture of a maxilla, the horizontal palatine process, and the alveolar process curvature. Altogether, these components are involved in the structural organization of the orbital, nasal/paranasal and oral cavities.

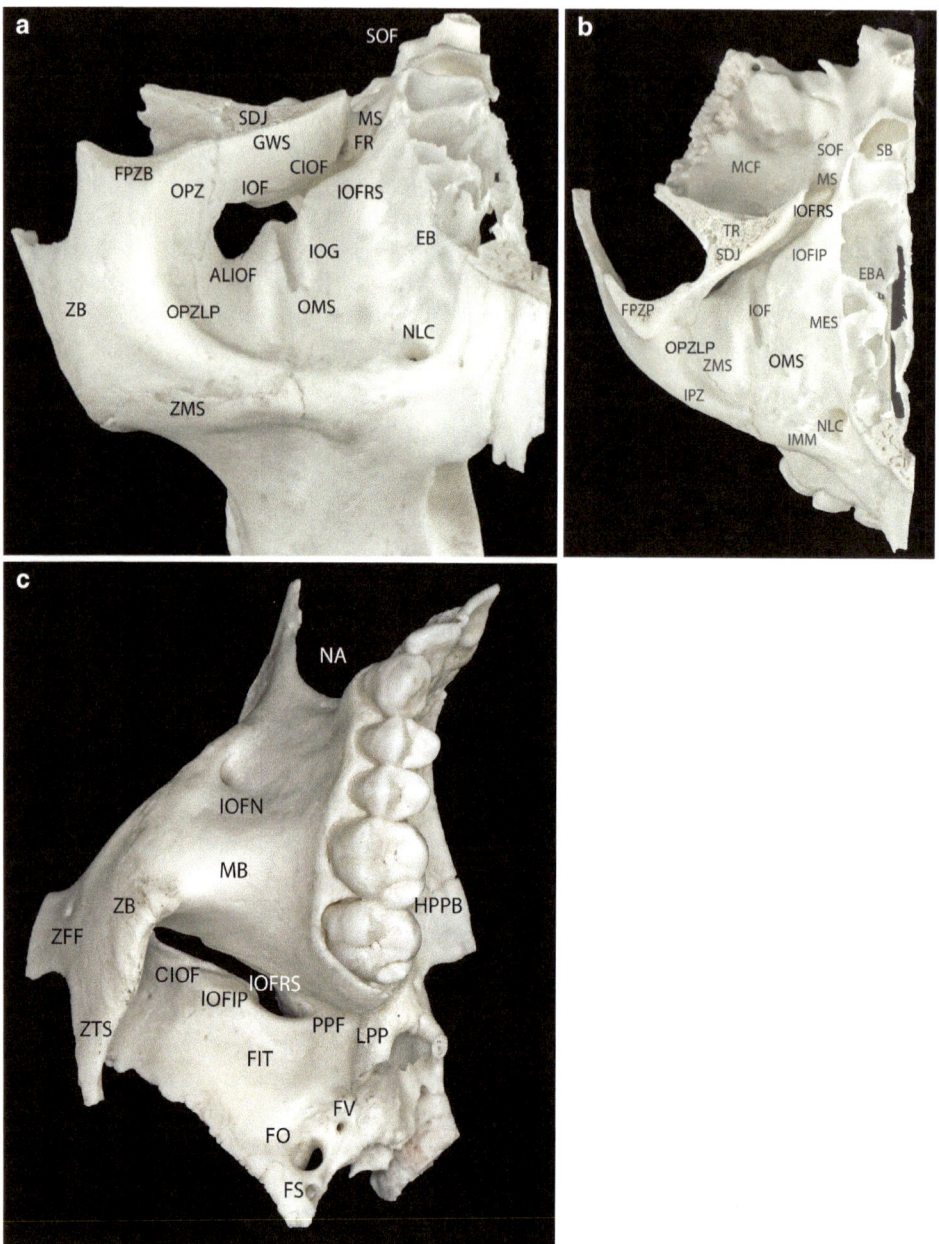

Fig. 2.15 (**a**) Orbital floor (right)—anterocranial view (lower margin tilted forward). The IOF communicates with the infratemporal and pterygopalatine fossa. Rather than a simple punched perforation in a shelf the IOF is configured as a ravine with steep sides, sinkholes, and affluents such as the foramen rotundum, pterygoid canal, and inferior orbital groove. *CIOF* confluence IOF isthmus, *OPZLP* orbital plate of zygoma lower part, *IOFRS* inferior orbital fissure rear sinkhole. (**b**) Orbital floor (right) from above. The posterior end of the IOF ends in a sinkhole in front of the maxillary strut. Thus, the orbital floor does not contribute to the apex (= posterior orbit). nasal/paranasal and oral cavities. (**c**) Assembly of maxilla, zygoma, and sphenoid bone (right)—inferolateral aspect showing retrotubar and infratemporal region. The ravine-like character of the IOF with robust posterior bony borders (TR and medial IOF margin) is confirmed. *FIT* fossa infratemporalis, *FV* foramen vesalii/sphenoidal emissary foramen, IOFIP inferior orbital fissure promontory, LPP lateral pterygoid plate, *MB* maxillary bone, *ZB* zygomatic bone, *ZTS* zygomatica–temporal suture

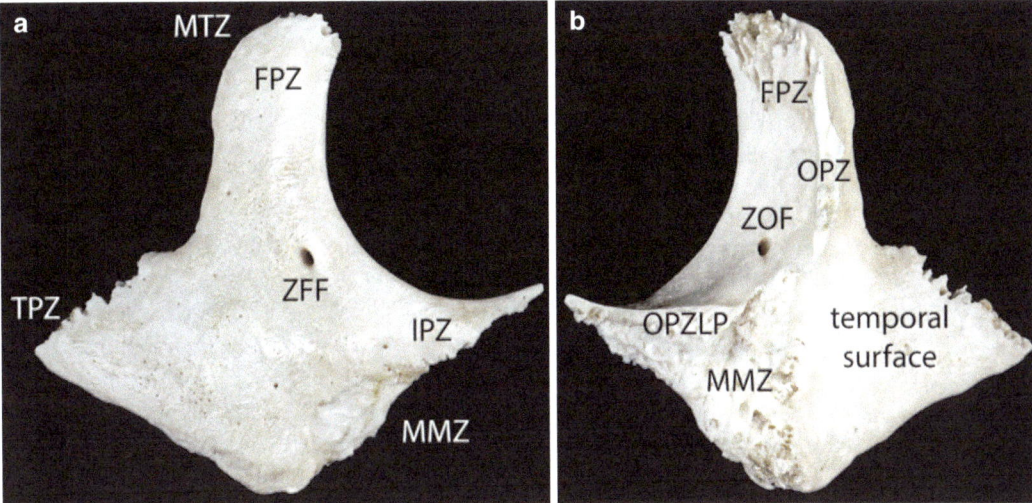

Fig. 2.16 (**a**) Right Zygoma—Lateral view. Convex outer zygomatic surface. Note zygomatic-facial foramen (ZFF) next to the reversal point along the curved margin in the lower lateral orbital quadrant. A slight roll-shaped elevation below the ZFF gives origin to the zygomaticus major muscle. *MMZ* maxillary margin of zygoma, *TPZ* temporal process of zygoma. (**b**) Right Zygoma–Medial view. The two portions of the overall orbital plate, the OPZ and the OPZLP (i.e., the anterolateral orbital floor) divide the orbital surface from the temporal surface along the sphenozygomatic suture line (SZS - see Fig. 2.4b). The rough keel-shaped area, the maxillary margin (MMZ) is part of the broad bony interface with the maxilla (ZPM—see Fig. 2.14), termed the zygomatic-maxillary suture (ZMS) line. *ZOF* zygomatico-orbital foramen

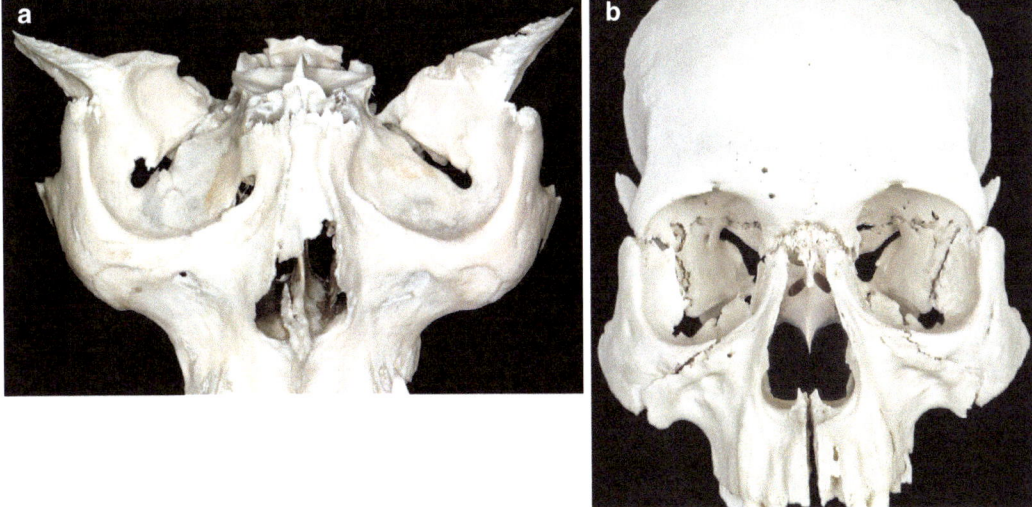

Fig. 2.17 (**a**) Zygomatic bones in closed lock position between sphenoid and maxillae—anterior view of the midface, LWS, and frontal bone removed. Follow the connections along sphenozygomatic suture (SZS) lines and zygomatic-maxillary suture (ZMS) lines/junctions. The anterior IOF loops are interposed in between the OPZ and OPZLP. (**b**) Assembly of midface building blocks completed—Anterior view. Note: Ethmoid, vomer, and palatine bones removed

Zygoma—Constituent of the Lateral Orbital Wall

The zygoma, malar, or cheek bone is the prominent cornerstone of the upper lateral midface (Figs. 2.4b and 2.16a, b). The orbital plate (facies orbitalis) of the zygoma (OPZ) is directed posteromedially at an angle of 45° toward the sagittal plane. In conjunction with the vertical frontal process (FPZ), the OPZ completes the lateral orbital wall and makes up the lateral orbital margin (LOM) including the rim around the lower lateral quadrant. Thereto the OPZ joins up with the anterior GWS border along the sphenozygomatic suture line (SZS) (Fig. 2.17a), while the zygomatic-frontal-suture line (ZFS) is the anterosuperior contact zone between the two according frontal processes from above and below (Fig. 2.17b). Anterior to the massive central trigone of the GWS, the lateral orbital wall turns into a monocortical layer with the SZS line located in the thinnest portion. The indented bay between the margins of the anterior loop of the IOF partitions a small horizontally oriented lower part (OPZLP) from the overall orbital plate. The OPZLP conforms the anterolateral orbital floor (Figs. 2.15a, b and 2.17a, b).

The orbital tubercle or Whitnall's tubercle [21] is a small roundly protuberance of 2 mm or 3 mm diameter on the inner OPZ immediately (2–4 mm) in the marginal territory behind the orbital rim and about 10–11 mm beneath the ZFS. The tubercle gives attachment to the lateral retinacular suspension complex. The marginal tubercle (MTZ) corresponds to a spine at the posterior upper edge of a FPZ widening that is occasionally present somewhat below the ZFS.

The solid zygomatic body has a rhombus shape and exhibits three further extensions in a medial, caudal, and posterior direction. The infraorbital process (IPZ) courses medially, while the adjacent maxillary margin is beveled diagonally downwards and laterally to merge with the tapering inferior tip of the zygoma (malar tuberosity), at last the temporal process reaches out backward to convey into the anterior zygomatic arch. The outer surface of the zygoma is commonly addressed as the malar or oftentimes inappropriately as lateral surface. The inner surface has been summed up as the temporal or infratemporal surface without drawing any clear distinction between the rear sides neither of the orbital plate or frontal process, the zygomatic body, or the temporal process.

A set of foramina with a network of interconnecting zygomatic channels perforates the OPZ as well as the malar and infratemporal surfaces of the zygoma. These foramina show a lot of variations from complete absence to multiplicity [22, 23]. The single or plural orifice(s) of the zygomatic-orbital foramen (ZOF) open(s) at the inner surface of the anterior inferolateral orbital quadrant next to the IOF loop (Fig. 2.15b). The zygomatic-facial foramen (ZFF) is located within the central range of the outer zygoma surface (Fig. 2.16a). The zygomatic-temporal foramen (ZTF) occupies an ascended position behind the frontal process or the zygomatic body.

Clinical Implications

The sphenozygomatic suture line (SZS) is an essential reference and reliable guide in the reduction of the fractured zygoma and consecutively for reestablishing of the outer facial frame. For this purpose, the lateral orbital wall is under intraoperative control from inside the orbit for accurate restoration into a flat continuous plane [24, 25].

Anterior Orbit—Midorbit—Posterior Orbit (Apex)

It is still often used practice to divide the length of the orbit into three-thirds according to the anteroposterior extension. This division is intuitive and not based on acknowledged anatomical references or appropriate metric distances between consented measuring points or planes. The orbital floor which is shorter than the other three walls is often the only zone taken into con-

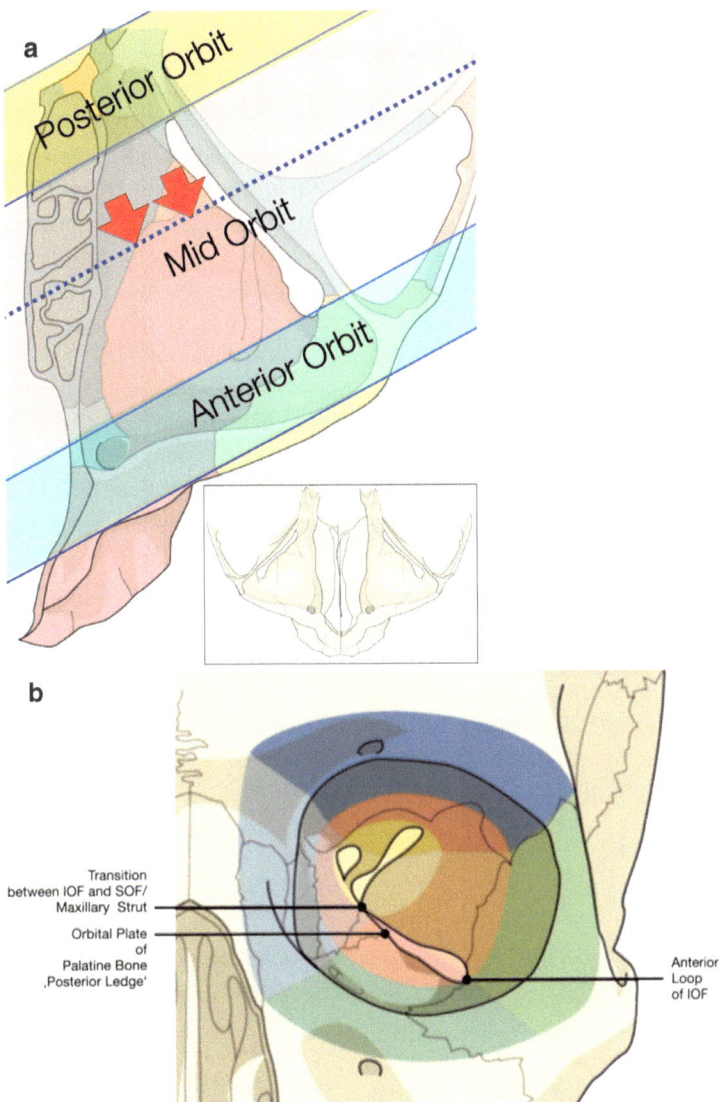

Fig. 2.18 (**a**) Left orbital floor from above—Borderlines between the anterior orbit (blue green)—midorbit (pink) and posterior orbit (yellow) in relation to the IOF. The dotted line indicates altered proportions, if the anterior border of the orbital plate of the palatine process serves as a point of reference. Inset (right lower corner)—orienting view over both orbits. (**b**) Left orbit frontal view—points of reference along the IOF and ensuing concentric. (**a** and **b** with permission from https://surgeryreference.aofoundation.org)

sideration and divided into three parts. Within the perspectives of the entire orbit, this must lead to confusion, since the apex with its triangular cross section is usually regarded as the posterior part, whereas the orbital floor extends over the remaining two thirds of the orbital depth. A more pragmatic approach is the partitioning into an anterior orbit, a midorbit, and a posterior orbit (apex orbitae) based on reproducible anatomic landmarks.

The IOF appears most suitable for the provision of reproducible topographic indicators (Fig. 2.18a, b). A frontal plane at the level of a tangent to the tip of the anterior IOF loop provides the boundary between anterior orbit and the beginning of the midorbit. A second frontal plane placed at the transition of the IOF into the SOF anterior to the maxillary strut separates the midorbit from the posterior orbit or the apex orbitae.

In other words, the anterior orbit lies in front and the posterior orbit in the rear of the IOF while the midorbit corresponds to the anteroposterior extent of the IOF. Obviously, this way of division does not result in an even metric tripartition but puts up the midorbit as the major component. The IOF landmarks need extrapolation around the entire orbital circumference, so that a series of concentric circles for a topographical allocation inside the orbit ensues (Fig. 2.18b).

II. Craniomaxillofacial Transition and Passageways

Fissures, Canals, Grooves, Foramina, Notches, Fossae—Bony Openings within the Precincts of the Orbit

Bony openings (canals, grooves, fissures, foramina, notches) lend pathways for neural and vascular linkages from the internal orbit to the cranial cavity, infratemporal fossa, paranasal sinuses, inner nose, and the face (Details see Chap. 3).

Optic Canal (OC)

The optic foramen/canal (OC) opens into the superomedial corner of the orbital apex, where the posterior medial wall extension meets with the roof. The canal has an elliptical cross section (approx. Diameters 4–9.5 mm) over a length between 5.5 mm and 11.5 mm. The optic foramen may lie posterior to the anterior face of sphenoid sinus, though this position can differ on the sides [26].

The posterior canal end is located medial to the anterior clinoid process in the middle cranial fossa. The conventional OC route (frequency 90%) courses along the lateral wall of the sphenoid and through the medial wall of the orbital apex. The mean angle of entry is about 60° to the coronal plane then. As a variant (frequency 10%), the OC enters the orbit through the roof of the apex at a more acute angle of 32° to the coronal plane [27]. Asymmetry of the optic canals is not uncommon, however [28]. The canal is formed by the sphenoid body inferomedially, by the anterior root of the lesser wing of the sphenoid (LWS) superiorly and the optic strut inferolaterally (Figs. 2.5a, b). The optic strut (OS) is a bony abutment linking the sphenoid body and the anterior clinoid process, which on account of its original designation as "sphenoid strut" is considered as a posterior [29] or lateral root [30] of the LWS. The anteroposterior OS location in terms of attachment to the sphenoid body relative to the prechiasmatic sulcus varies and was classified as presulcal, sulcal, and postsulcal [31, 32]. The OS angulation varies over a wide range between 30 degrees to almost 60 degrees. The OS dimensions measure about up to 8 mm in length, up to 5 mm in width and between 3 mm to 4 mm in thickness. Optic canal variations include a duplication and a keyhole anomaly at the inferolateral OS wall [33]. The occurrence of a sphenoethmoidal cell [34, 35] can further affect the OC disposition. Onodi cells are particular posterior ethmoid air cells that pneumatize the sphenoid body superior and lateral to the sphenoid sinus (Fig. 2.2a, b) directly related to the OC and optic nerve (CN II), respectively. The medial OC wall can protrude the lateral portion of Onodi cell or the OC can even pass through the middle of the cell as a bony tube. The prevalence of Onodi cells is reported as high as 60% in clinico-anatomical studies.

Excessive pneumatization can also involve the OS and the anterior clinoid process by lateral extension from the opticocarotid recess [36]. As the name implies, the opticocarotid recess [37] refers to a small space between the prominences of the OC and the carotid canal in the clinoid segment (C5) of the internal carotid artery [38].

The OS separates the OC from the superior orbital fissure (SOF). The inferior OS border corresponds to the superomedial SOF border. The minute infraoptic tubercle corresponds to a thickening located just beneath the anterior base of the OS [39]. It is the origin of the common annular tendon (Zinn's ring).

Superior Orbital Fissure (SOF)

The SOF extends inferolaterally to the OC and OS. It is frequently described as a club- or L-shaped gap interposed between the LWS and the GWS or the orbital roof and the lateral wall, respectively (Figs. 2.5a, b, 2.7, and 2.8). It slopes alongside the lateral apex wall inferomedially, where it levels along the sphenoid body to reach the top of the maxillary strut (MS). The MS is a transverse bony bridge above the foramen rotundum. At the lower end, the SOF margins blend into the posterior outlines of the inferior orbital fissure (IOF) (Fig. 2.19). The posterior (rear) sinkhole of the IOF (IOFRS) separates the lower SOF end from the orbital process of the palatine bone anteriorly, however.

The inferomedial portion of the SOF has a widened configuration next to the sphenoid body that becomes narrower toward the superolateral end between the LWS and GWS [29, 40–42]. The lateral SOF margin is projecting somewhat more anteriorly than the medial margin, so that the SOF outlines do not lie in a strictly coronal plane.

Midway between the narrow and broad SOF partition the lateral (GWS) margin of the SOF may exhibit the lateral rectus spine (spina rectus lateralis—SRL). If present (frequency 60%), the form of this bony spur can vary from pointed as a tubercle over a tongue-shaped, rectangular, or irregular projection to two spines [43]. The spur serves as attachment site of the lateral part of the common annular tendon (Zinn's ring), more specifically the tendon of the lateral rectus muscle (LRM) and/or an additional tendinous LRM slip. An irregular projection or a double spine may be due to a bipartite tendon insertion.

The medial SOF margin is formed by the OS in its upper part, the lateral surface of the sphenoid body borders the lower part. The cavernous segment (C4) of the internal carotid artery (ICA) is indicated by the carotid groove (sulcus) running an S-shaped course (i.e., "carotid siphon") along the lateral surface of the sphenoid body. This part of the ICA is implemented in the cavernous sinus and covered by its lining membrane.

Coming from a bend at the posterior clinoid process, the C4 ICA segment passes forward to curve upward posterior to the medial SOF edge and the OS on to the medial side of the anterior clinoid process.

The crescentiform SOF shape has been classified in up to nine different basic types [44, 45] with different prevalence and metrics. The morphological types appear to have a bearing on the topography of soft tissue structures within the

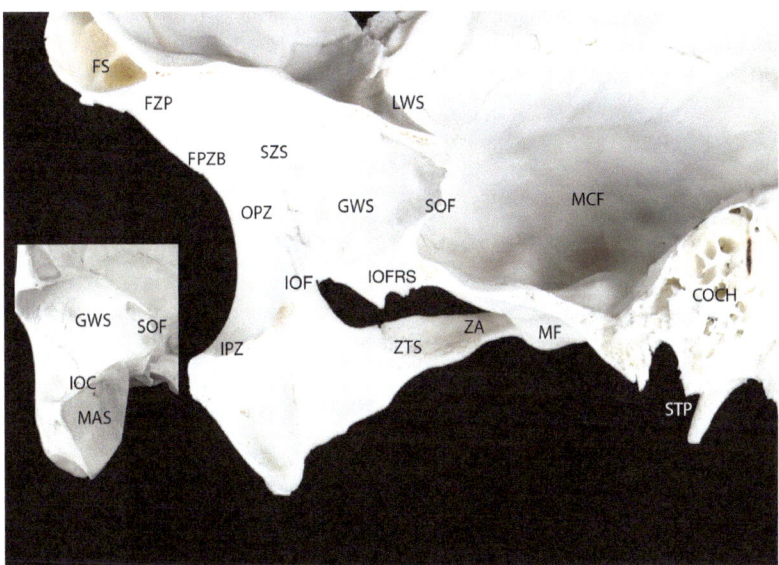

Fig. 2.19 Lateral orbital wall—Medial View. Inset—Oblique medial view. The lateral orbital wall consists of the GWS posteriorly and the orbital surface (‚facies') of the zygoma anteriorly. *COCH* cochlea, *FS* frontal sinus, *MF* mandibular fossa, *STP* styloid process, *ZA* zygomatic arch

SOF opening [46]. The SOF is the major communication between the internal orbit and the medial cranial fossa. The dura mater lining of the middle cranial fossa and the cavernous sinus are continuous with the periorbita via the SOF. The periorbita within the orbital apex and the fibrous components enfolding the densely packed neurovascular structures passing through the OC and SOF fuse to Zinn's ring or more precisely to the common circular connective tissue funnel, which is the origin of the four extraocular rectus muscles. Zinn's ring subdivides the SOF into three individual hubs according to a superolateral, central, and inferior section (Chap. 3).

Cranio-Orbital Foramen (COF)

The COF is laying either in the GWS or the orbital surface of the frontal bone close to the superolateral extremity of the SOF near or within the sphenofrontal suture line. The COF may also merge with the tip of the SOF ending in kind of an incompletely separate foramen. Synonyms in use for the COF are orbito-meningeal foramen, lacrimal foramen, sphenofrontal foramen, and anastomotic foramen [47]. The foramina may be single or multiple, whereby multiple refers to 1 or 2 accessory foramina. Numerous possible permutations result from unilateral or bilateral occurrence and from differences of the number between the sides [48–50]. COF are not consistently present; reported incidences range from 30% to 85% with an average of 50% [51]. The shape of a COF is usually circular, rarely oval. The COF diameters usually vary between 0.3 and 2.5 mm. If accessory foramina are present, the diameter of the main COF is comparatively larger.

Larger COF communicate with the cranial fossa through a short conduit and join the orbit either to the anterior (A-subtype) or middle (M-subtype) cranial fossa [52, 53]. Smaller foramina are mostly blind-ending. The orbital branch of the middle meningeal artery (MMA), in general with a comitant vein, is transmitted through the COF (Figs. 2.2a, d, 2.3, and 2.7) or as an alternative through the superolateral SOF. This vessel is inconstant (frequency 40–50%) and

anastomoses with the lacrimal artery (LA) branch of the ophthalmic artery (OA) as a common branching pattern (see Chap. 3).

The vascular pattern of arteries entering or leaving the orbit through the COF or the SOF demonstrates great diversity [52, 54]. Variants apart from the above-mentioned meningo-lacrimal connection (MMA–LA) are a meningo-ophthalmic anastomosis (MMA–OA) for the collateral vascular supply of the orbit or there are recurrent vessel courses back from the orbit to the middle cranial fossa originating from the LA or OA to the MMA or through a network of small branches for the vascularization of the meninges and tentorium. Meningo-lacrimal and meningo-ophthalmic arteries may run in parallel with discrepancies in diameter.

An anomaly paramount to note is the origin of the OA from the MMA as the sole vascular source of the orbit (estimated frequency 1%), when the usual OA exit from the ICA is either absent or obliterated [54–58]. An unusually large-sized COF (diameter 3–4 mm) should be considered as a warning sign that a vessel of the same size might be the major or the sole blood supply to the orbit [59]. Without any doubt, the exact COF location is of relevance to prevent complications (hemorrhage/erroneous ligation of a terminal vessel connection/amaurosis) during subperiorbital dissection of the deep lateral orbital wall.

A plenty of morphometric studies have addressed the COF distances to the lateral or medial SOF end, the FZS, the superior orbital notch, and Whitnall's orbital tubercle [53, 60]. The FZS appears as the most suitable landmark for surgical purposes and its distances to COF have been recorded in almost every of the existing studies showing values between 22 and 35 mm depending on the ethnicity of the sample.

Clinical Implications
Caveat:

If the provenance of a larger vessel passing across the COF or the narrow superolateral SOF has not been unequivocally clarified, one should refrain from performing a ligation.

The presence of a groove on the lateral wall of the orbit (Fig. 2.3) running between the SOF or COF and the IOF (frequency 30–70%) has been associated with the course of an anastomosis between the MMA and infraorbital blood vessels [2, 61]. This could not be confirmed later and the groove has been considered as an abrupt thinning of the bone [47].

Inferior Orbital Fissure (IOF)

The IOF prototype outline corresponds to a silhouette reminiscent of a cat-tongue chocolate or a double-ended spoon with a short intermediate handle inside the orbit. Though overall, it is a complex 3D opening of varying shapes [62] (Fig. 2.15a–c). The IOF separates the floor portion of the midorbit from the lateral orbital wall and provides passageways and portals for vessels, nerves, or fat pads to the pterygopalatine, infratemporal, and temporal fossa [63, 64]. The long IOF axis runs a posteromedial to anterolateral route starting at the maxillary strut and extending to the tip of a loop in between the upper lateral and lower anteromedial orbital surfaces of the zygoma. Occasionally, this part of the zygomatic-orbital flange is absent, so that the loop ends after the IOF margins formed by the maxilla and GWS have joined at an anterior junction. Posteromedially, the body of the sphenoid and the palatine bone contribute to the rear IOF sinkhole (IOFRS). The foramen rotundum opens into this sinkhole, which is continuous with the lower SOF along a floored level above the maxillary strut.

The narrowing in the center of the IOF originates from a crescent-shaped overhang ("isthmus promontory") (IOFIP) of the orbital floor next to the surface of the palatine bone.

Morphometric data of the IOF are rather scarce. An often-cited study performed on dry skulls went for average IOF dimensions [62]. The IOF covered a mean area of 60 ± 40 mm^2 with a range of 2mm^2 to 232 mm^2. The longest distance from the posterior sinkhole to the anterior loop (= diagonal length) had a mean of 18.2 ± 4.9 mm (range 2.8–32.4 mm). The span between the isthmus promontory to the GWS margin (= narrowest borders) amounted to a mean of 1.9 mm ± 1.3 mm (range 0.2–6.6 mm). The width of the anterior IOF loop (= widest distance) was 5.7 ± 2.6 mm (range 1.5–17.4 mm). The range of values allowed a distinction into eight IOF types, which had different frequencies.

However, these findings need to be treated with caution because the measurements derived from photographs taken from an anterior view at an angle of $78°$ degrees to the horizontal. This angulation obstructs the view in particular into the rear IOF sinkhole due to the isthmus promontory sticking out in the foreground.

The obvious consequence is that the values do not reflect the true size and shape variants. A more realistic outcome might have been yielded from a perpendicular view onto the IOF from a basal sight onto to the IOF outline in the temporal fossa (Fig. 2.15c) or a look from inside the orbit either with an 90 °angulated endoscope or an axial cut of 3D reformatted CT scans.

It is of note that the term posterior or rear IOF sinkhole in this text replaces what has been previously labelled posterior basin of the IOF in related papers of our group [65, 66]. Originally, the designation basin of the inferior orbital fissure had been referred to a conceptual area of thick bone in the lateral orbit amenable to removal in decompression surgery [67]. Pertaining to this original meaning the IOF basin consists of a bony area corresponding to the lower part of the orbital flange of the zygoma including the posterior portion of lateral orbital rim at the same vertical level and reaching medially all the way to the zygomaticomaxillary suture or even over it into the lateral portion of the maxillary sinus.

Foramen Rotundum (FR)— Maxillary Strut (MS)

Unlike suggested by its name, the foramen rotundum is not a foramen but a short canal structure that penetrates the common base of the lateral pterygoid plate and the GWS (Figs. 2.2a, d, 2.3, 2.4b, 2.5b, and 2.7) in the upper portion of the PPF and near to the transitional region between the orbit and the cavernous sinus via the SOF. The medial canal border is formed by the lateral wall of the sphenoid sinus. The FR is the path of communication for the maxillary nerve (CN V2) between the middle cranial and the pterygopalatine fossa (PPF). The FR/canal is not easily accessible in intact dry skulls, so that quantitative measurements are all CT based on axial and coronal sections on patient series. The length of the canal has an average of 6.3 mm (range 2.1 mm–10.8 mm) with a mean diameter of 2 mm (0.8–4.4 mm) and an intercanal distance to the pterygoid canal (PC) of 2.6 mm (range 0–8.8 mm). These indicative measurements from a retrospective study on 50 patients [68] show disparities to similar studies [69–71], which are explained by difficulties to determine the anterior and posterior limits of the canal accurately due to angulated planes of the in- and outlets. The canal usually runs a course from behind in an anterolateral downward direction and can assume various types. These FR/canal types differ with regard to their placement in the sphenoid bone, at or within the sinus wall or as a conduit inside the sinus [72]. The vertical and transverse position of the FR/canal can be located above the base of the lateral pterygoid plate coinciding with the base of the lateral pterygoid plate (frequency 50%) or with a medial (47%) or lateral offset (3%) from the midline [72]. A rare but not negligible variant is a branching of a canal from the lateral wall of the foramen rotundum that opens into the orbit (length 5 mm, diameter 0.2–0.5 -1.0 mm). This canal is straight and directed slightly superolaterally and likely transmits the zygomatic nerve and/or a portion of the infraorbital nerve [73].

The maxillary strut (MS) relates to the bony bridge across the foramen rotundum (FR). An accessory small foramen going through the MS is rarely present and not to be confused with the FR [74].

Pterygopalatine Fossa (PPF)— Sphenopalatine Foramen (SPF)— Pterygoid Canal (Vidian) (VC)

By convention, the PPF is the larger space anteriorly bounded by the curving backside of the maxilla, posteriorly by the fused front of the pterygoid process plates together with the base of the sphenoid bone and anteromedially by the perpendicular plate of the palatine bone [12, 75].

The PPF (Fig. 2.5, 2.12, and 2.15) represents a central transit hub connecting within the facial skeleton [76] via the:

- Pterygomaxillary fissure—to the infratemporal fossa (maxillary artery),
- Foramen rotundum (FR)—to the MCF (CN V2),
- Inferior orbital fissure (IOF)—to the posterior midorbit (CN V2),
- Sphenopalatine foramen—to the posterior nasal cavity (sphenopalatine artery/posterior superior nasal nerve branches of pterygopalatine ganglion—PPG),
- Pterygoid (Vidian) canal—to the MCF (greater and deep petrosal nerves),
- Greater ("pterygopalatine") (GPC) and lesser palatine canal—to the oral cavity (major and minor palatine neurovascular bundles).

From the lateral view, the PPF looks like a narrow space tapering inferiorly to form the greater palatine canal.

The lateral opening of the PPF, named the pterygomaxillary fissure (PMF), has a sharp-edged posterior border. This is formed by the lateral margin of the anterior surface of the base of the pterygoid process and inferiorly by the fused pterygoid plates. The anterior PMF border is more evasive owing to the convex contour of the posterior wall of the maxillary sinus [12]. A variety of PMF types and sizes has been reported [77].

The sphenopalatine foramen (SPF) connects the superior part of the PPF with the posterior nasal cavity. The SPF and its thin-walled constituents, the sphenopalatine notch (SPN), and the inferior margin of the sphenoid body are angulated in an anterolateral plane, so that the SPF is located anterior to the opening of the Vidian canal into the PPF [12]. The Vidian Canal (VC) traverses the base of the medial pterygoid plate from the anterior border of the foramen lacerum to the exit in the posterior PPF (Figs. 2.2a, b, 2.4b, 2.5b, 2.7, and 2.17b).

Analogous to the FR, three VC types can be distinguished as completely embedded in the sphenoid bone under the floor, through the floor or protruding into the sphenoid sinus and within the sinus [72, 78]. These conditions are undoubtedly correlated to the type and degree of SS pneumatization [79]. A myriad of studies is dealing with the morphometry of the VC, FR, and their surroundings (e.g., [71, 72, 80–87]) due to their utmost importance as a landmark in endoscopic skull base surgery. A metanalysis of this immense data pool is beyond the scope of this article.

Ethmoidal Foramina (EF)

The EF are the funnel-shaped openings of the ethmoid canals (EC). The EF are laid out in an anterior–posterior row alongside the frontoethmoidal suture (FES) line (Figs. 2.5a, 2.11, and 2.12). As a common finding, there are two foramina [88]. They may be supernummary up to a maximum of six that can appear unilaterally, bilaterally, or in side asymmetry [89–91]. The most anterior and the most posterior EF are addressed as the anterior EF (AEF) and the posterior EF (PEF). The topography of the holes not only varies in the sagittal direction but vertically, too, with an intra-sutural (frequency 85%) or extrasutural position, below or above the FES [92, 93].

Clinical Implications
Awareness of the proximity of the posterior most foramen (PEF) to the optic canal (OFC) is regarded as critically important in the prevention of accidental optic nerve lesions and amaurosis during dissection of the periorbital lining along the medial orbital wall.

Injuries of the ethmoid arteries may result in massive hemorrhage, retroorbital hematomas, and an orbital compartment syndrome.

Predictably deroofing ethmoidal canals (EC) at and in particular above the FES level carries increased risk for bleeding and accidental entry into the anterior cranial fossa and cribriform plate.

A guideline to prevent interference with the OFC and CN II during subperiorbital dissection of the medial orbital wall been summarized in "the rule of halves," which indicates the relationships of the ALC to the AEF, the AEF to the PEF, and the PEF to the OFC in the brief formula: 24 mm–12 mm–6 mm [94] (see Chap. 3 - Periorbital Dissection, Fig. 3.43a, b). A vast number of morphometric studies have scrutinized the reliability of this mnemonic rule, not all of them fully consolidating it (e.g., [60, 95–97]). Differences in the morphometric parameters arose from gender and age [98]. Ethnicity was deemed as a major variable for the presence of supernumerary EF [89]. Orbits of Asian or African descent had greater EF numbers and a shorter length of the orbit with the consequence of an increased density of the EF in comparison with Caucasians [99].

Insofar it has been repeatedly emphasized that the ratio between the EF distances from anterior landmarks and the length of the medial orbital wall must be accounted for in a meaningful evaluation [88, 90]. The preference should be given to case-based individual measurements, because exceptions from the rule are to be expected [48, 49]. The ethmoidal canals (EC) or sulci (incomplete canals) provide connections between the orbit, the ethmoid nasal roof and the anterior cranial fossa.

They commence with funnel-shaped openings to pass the ethmoidal labyrinth and enter the olfactory groove on the top surface of the cribriform plate. The frontoethmoidal suture line (FES) marks the level of the ethmoidal roof. The cribriform plate, however, may lie up to 10-mm caudal to the FES. The EC run a diagonal course in anterior direction containing the homonymous neurovascular structures [100]. Their 3D topography and dimensions (diameter, length) leave ample room for variations. A mean of 8.2 mm (range 4–12 mm) was reported for the average length of the anterior ethmoid canal (AEC) and a mean of 7.6 mm (range 2 mm −13 mm) (mean 7.6 mm) for the posterior ethmoid canal (PEC) [101].

Infraorbital Foramen (IOFN)

The infraorbital canal opens with a same named foramen at the anterior facial wall (Figs. 2.1a, b, 2.14, 2.15c, 2.17a, b, and 2.20). The foramen is typically single and located 7–10 mm below the infraorbital margin. The canal follows coursing underneath the anterior orbital floor and turns into the infraorbital groove (sulcus) further backward, which runs below the floor surface to its end at the IOF rear sink. This transformation from a true groove into a canal occurs halfway along the course within the orbital floor. Otherwise, the IOC can be closed over the whole length, so that an IOG is absent or the groove is covered with a very thin, transparent osseous layer pretending a "pseudocanal" [102, 103]. An abundance of normative data on the morphological characteristics and dimensions of the IOG/IOC as well as the occurrence, number and location of accessory infraorbital foramina, and side channels has been produced over the last decade [104–114].

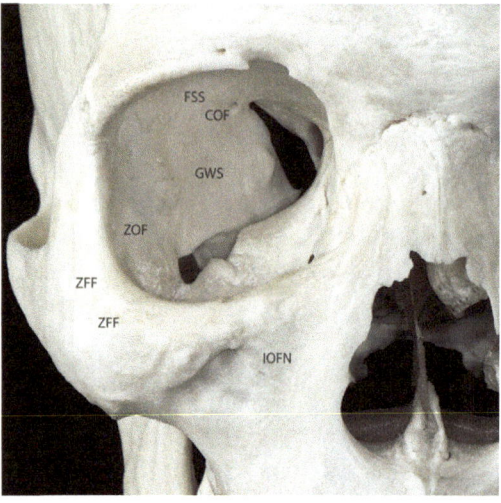

Fig. 2.20 Foramina in inferolateral orbital quadrant and surface of zygomatic body—Anterior view. Two ZFF, one ZOF, ZTF not visible

Supraorbital and Frontal Foramina/Notches (SOFN/FOFN)

The supraorbital margin can embody supraorbital and frontal foramina and/or passages formed as incisurae (notches) (Figs. 2.1a, 2.4b, 2.6, 2.7, 2.8, 2.14, and 2.20) or it may have plane and even contours without breaks or perforations [115]. Foramina or notches either occur bilaterally, but also alternate between the two sides of an individual [116, 117]; they appear single or in clusters of multiple openings with overlapping assignment as SOFN or FOFN. SOFN have an oval shape, the horizontal diameter being larger than the vertical. Supraorbital notches are usually wider than supraorbital foramina.

The mean SOFN distances from to the facial midline, frontozygomatic suture, or temporal crest of frontal bone display a wide variation with regard to gender, age, and ethnicities [116, 118–120]. The frontal SOFN plane can be angulated at all three spatial levels. Supraorbital notches are usually completed by fascial bands encircling their floor. The fascial bands exhibit variation patterns—osteofibrotic partially containing bone and/or either divided by additional horizontal or vertical septations [121]. The vertical height position of SOFN differs—notches are of course contiguous with the margin; foramina are located superior to the rim, sometimes in high positions up to 19 mm above [115].

Clinical Implications

The course and depth of the bony canal subsequent to such high positioned supraorbital foramina is of interest in surgical approaches to the superior orbital rim and orbital roof requiring osteotomies to release and mobilize the supraorbital nerve. The course of the canal in terms of a horizontal or steep inclination toward the end in the orbital roof in conjunction with the thickness of the rim determines the anteroposterior length (depth). The depth was reported accordingly to range between 2 mm and 12 mm [122].

While SOFN are a constant finding FOFN are facultative at varying low frequencies [123]. The average distance between the medial FOFN edge from the lateral SOFN edge ranges between 2 and 15 mm [117]. Either foramina or notches may co-occur bilaterally or differ at the two sides. The FOFN number, size, and shapes of apertures are most diverse.

Zygomaticoorbital (ZOF), Zygomaticofacial (ZFF), and Zygomaticotemporal (ZTF) Foramina

The openings and subsequent intrabony courses of the ZOF, ZFF, and ZTF (Fig. 2.15a–c) follow various patterns [22]. As the name indicates, ZOF refers to single or multiple ($\leq$ 6 in skulls of African American descent, [124]) openings in the inner surface of the anterior inferolateral orbital quadrant of the orbit (Figs. 2.16b, and 2.20). Each ZOF represents the entrance to a separate or an interconnected canal which exits at the facial (ZFF) and/or temporal (ZTF) zygomatic surface. Hence, there can be an array of independent canals (ZOF– ZFF and ZOF–ZTF) only as well as a principal interconnected canal system with Y-kind divisions/subdivisions standing either alone or with additional independent connections [22]. The ZFF was mapped in an investigation of 429 (i.e., 858 zygomas) adult skulls [125] from nine ethnic groups from geographies around the world and both genders. An across-line laser module was used to generate two reproducible reference lines superimposed onto each zygoma and measure the distances to the ZFF. The ZFF occurrence per zygoma varied from no identifiable foramen (frequency 16.3%) to one foramen (49.8%) over two (29%) and three (3.4%) to a maximum of four (1.4%) foramina. The incidence and ZFF sites differed considerably among the ethnicities but were consistent between female and male subjects. The distance of the ZFF from the infraorbital margin, however, was larger in male (mean 7.1 mm) than in female (6.2 mm) subjects.

The ZFF sites were related to the intersection of a horizontal line going through the deepest point of the infraorbital rim in parallel with the Frankfurt plane and a perpendicular vertical line going through the posterior end of the zygomaticofrontal suture (ZFS). The majority of ZFF were concentrated to a field in the upper and lower quadrants of the intersection with outliers in the posterior (upper > lower) quadrants (total approximate size 30 mm x 25 mm). If the ZFF sites were reevaluated within the limits of a circle of 25 mm in diameter and its center 5 mm anterior to the intersection of the reference lines, a percentage of 93% ZFF on the right zygoma and 94% ZFF on the left were enclosed in this risk spot area irrespective of their ethnicity. A surgical safe zone was then defined beyond the boundaries of this circle and delineated clinically using surface landmarks.

The position and size of the intraorbital ("entry") opening (= ZFFin – equal to the ZFO) and the malar exit of the ZFF (= ZFFout) were the focus of a morphometric study in $n = 10$ fresh-frozen PMHS Caucasian heads [126]. The ZFFin number (total $n = 23$) per side side ranged from 0 to 3: none on 2 sides (frequency 10%), 1 on 14 sides (70%), 2 on 3 sides (15%), and 3 on 1 side (5%). The mean ZFFin diameter was 1.1 $\pm$ 0.5 mm. The ZFFin were located 5.1 $\pm$ 2.0 mm superior to the inferior margin of the orbit and 4.3 $\pm$ 1.6 mm medial to the lateral margin of the orbit. The ZFFout number (total $n = 22$) per side ranged from 0 to 2: none on 2 sides (10%), 1 on 12 sides (70%), and 2 on 4 sides (20%). The ZFFout were located 1.2 $\pm$ 2.9 mm inferior to the inferior margin of the orbit and 1.1 $\pm$ 3.0 mm lateral to the lateral margin of the orbit. With 1.4 $\pm$ 0.6 mm diameter, the ZFFout was larger than ZFFin. Surprisingly, the length as well as the continuity or reticulum of the canal(s) in between the foramina was not dealt with in detail as opposed to what the study title suggests.

A clear understanding of the zygomatic foramina, their prevalence, and variations as well as their nerve crossings is essential for a number of surgical procedures [127]:

- Localization and guidance to the inferior orbital fissure (IOF)—[128].
- Dissection and manipulation along the lateral orbital wall.
- Osteosynthesis/plate and screw placement in orbital/zygoma/midface trauma.
- Orbital decompression.
- Design and bone cuts of orbito-zygomatic osteotomies [129, 130].
- Advancement or on-lay augmentation of the inferolateral orbital rim.
- Transmaxillary approaches to the orbit.
- Subperiosteal face lift.
- Retrobulbar anesthesia [131].

Nasolacrimal Fossa (NLF)/ Groove (NLG)/ Canal (NLC)

The proper nasolacrimal fossa (NLF) is strictly circumscribed and demarcated by the anterior (ALC) and the posterior lacrimal crest (PLC) (Fig. 2.21a). The lateral lower edge of the fossa pointing outward into the anteromedial orbital floor is termed the lacrimal notch (LN). The lacrimo-maxillary suture (LMS) runs in parallel in between the ALC and the PLC and corresponds to the union between the nasofrontal process of the maxilla (NFP) and the lacrimal bone [132]. The LMS corresponds to the so-called "maxillary line" (ML—[133]), a curvilinear (mucosal) eminence along the inside of the lateral nasal wall that runs from the anterior attachment of the middle turbinate to the anterior end (lacrimal process) of the inferior turbinate (Fig. 2.21b). One of the more detailed NLF descriptions based on endoscopic findings and CT scans [134] is too elaborate to be reproduced here in all detail. A clear distinction between a nasolacrimal fossa and a nasolacrimal groove (NLG) has never been

thoroughly defined and the designations sometimes appear to be used synonymously. One description in terms of a vertically oriented groove addresses the anterior edge of the lacrimal bone depression as the lacrimal margin (LM). This LM is suggested to extend inferiorly to an end point at the conchal crest—a horizontal ridge which is projecting from the lateral sidewall to give an attachment for the inferior turbinate. The bony walls below the notch down to the level of this crista are then referred to as the lacrimal groove [135], what obviously does not fall short of the whole length of the nasolacrimal canal (NLC).

So, the NLC apparently encompasses the entity of NLF, NLG, and the bony tract underneath the inferior turbinate down to the terminal aperture ("lacrimal ostium") in the lateral wall of the inferior nasal meatus. In fact, there is no consented definition on the NLC vertical extent concerning its exact upper and lower ends. The morphometric measurements as well as the morphologic features vary at all NLC levels.

The NLF from ALC to PLC in a Caucasian population [132] consisting of $n = 47$ orbits from $n = 24$ PMHS (postmortem human subjects) had a mean width of 8.8 mm (range 6–12 mm). The average distance of the LMS was 4.3 mm from the ALC and 4.5 mm from the PLC in 25% of the orbits under investigation, what means they were almost equidistant from the mid-vertical line. In 42.5% of the orbits, the LMS deviated anteriorly to the ALC and in 32.5% the LMC was located posteriorly toward the PLC. These deviations exceeded 1 standard deviation in 34%. An LMS located closer to the PLC implies that the proportions within the NLF are shifted to the thicker frontonasal maxillary bone, which is an unfavorable condition for dacryocystorhinostomies. In populations with East Asian descent, the NLF is formed predominantly by a thick maxillary bone layer [136]. As an extremely rare finding case, the entire NLF can be formed by the maxillary bone [137]. On the other hand, the anterior ethmoid air cells may extend beyond the PLC producing unusual thinness.

The anteroposterior diameters of the nasolacrimal canal entrance (i.e., NLF) ranked between

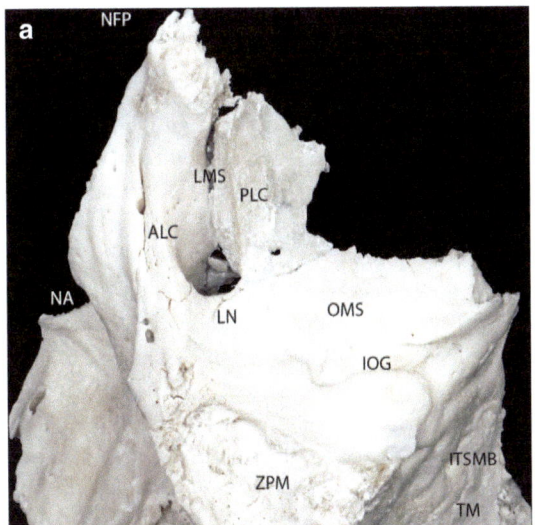
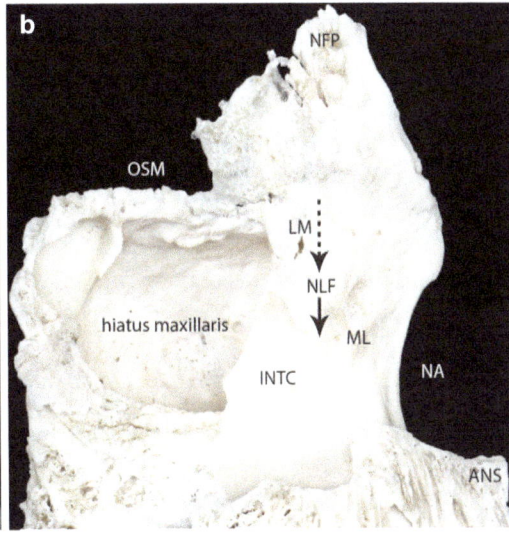

Fig. 2.21 (**a**) Nasofrontal maxillary process (NFP) and nasolacrimal entrance (NLC, NLF, NLG, LN)/lacrimal bone—Superolateral view. Note: Lacrimo-maxillary suture (LMS) line, ALC and PLC. (**b**) Nasofrontal maxil-lary process (NFP) and lacrimal margin (LM)—medial/endonasal view. Note: vertical extent of the nasolacrimal groove (NLG) and course of the maxillary line (ML). *INTC* inferior nasal turbinate crest

5.7 and 7.2 mm and the corresponding transverse diameters between 4.7 and 6.1 mm [138, 139]. The NLG in the configuration outlined above [135] had an average vertical length of 9.6 ± 2.1 mm. It displayed various morphological patterns (S-, boat-, hourglass-, cylinder-, barrel-, or funnel-shapes) ensuing in a different width according to the vertical height level with an average of 5.88 ± 1.53 mm at the upper one-third, 8.04 ± 2.05 mm at the middle one-third, and 5.94 ± 1.28 mm at lower one-third.

The morphometric parameters of the overall NLC have recently been reevaluated in a CBCT investigation (patients n = 100, age range 18–83 years with a mean of 42 years) including a summary of several preexisting studies [140].

The NLC diameters were assessed at the level of the infraorbital margin. Not surprisingly both the anteroposterior diameter with a mean of 6.6 ± 1.53 mm as well as the transverse diameter with a mean of 4.3 ± 1.0 mm had a similar magnitude as the former NLF measurements. The corresponding NLC sectional area averaged to 7.39 ± 3.29 mm^2 and the angle between lines through the length of the NLC and in parallel to

the nasal floor was 73.5 ± 6.8°. This angulation was influenced by gender (female > male) and age. All the other parameters did not show these dependencies. The length of the NLC remained unconsidered.

The narrowest part along the length of the NLC was identified at the entrance to the canal in a small series of PMSH dissections [139].

A multiplanar CT study is particularly note-worthy in so far as it provides accurate 3 D information on the NLC at six height levels along the vertical axis [141]. The major and minor diameters of the elliptical canals, the cross-sectional areas, and volume were assessed and correlated with gender, a younger (mean age 25 years) versus an elderly (mean age 60 years) age group and Black/White American cohorts (total n = 72 individuals). The NLC length was longer in men (12.3 mm) when compared with women (10.8 mm), just as the NLC volume was greater in men (327 mm^3) than in women (244 mm^3). Neither minimum canal diameters (mean 3.5 ± 0.8 mm) nor mini-mum cross-sectional areas (mean 11.7 ± 2.5 mm^2) demonstrated significant differences depending on gender, ethnicity, or age group. In the elderly age group, a trend was noticeable for greater canal cross-sectional diameters, possibly indicating an expansion of the canal aperture with aging. The diameters measured at two specific positions of the six height locations of the canal showed differences in elderly in comparison with younger patients. In the elders, the major axis diameters were increased at the superior NLC (apex) end with no alteration of the minor axis. Next to the inferior NLC (base) end, the minor axis diameters were larger, while the major axis kept constant. From a geometric aspect, the cross section of the upper canal became progressively more elongated and elliptical with aging, whereas the lower canal was getting circular. Moreover, the cross-sectional area at the canal base was greater in the people of color cohort than in Caucasians. No gender differences in the diameter or of the cross-sectional area along the canal length were noted.

Another recent CT study on the NLC using semi-automated segmentation techniques [142] refined the metric results above and added a classification into 5 morphological variants which is reminiscent to the NLG morphology patterns: A—cylindrical type with consistent diameter from top to bottom; B—lower-thicker type, wider at bottom than top; C—upper-thicker type, wider at top than bottom; D—spindle type, with extended middle portion; and E—hourglass type, with a particularly reduced middle portion.

The NLC lumen, basically modeled as a cylindrical bony pipe, is lined by the mucosal tissue layers of the nasolacrimal duct (NLD) which additionally modify the internal constrictions and openings of the lacrimal drainage system.

Clinical Implications

Average (normal) morphometric data of the NLF, NLG, and NLC are a prerequisite to identify mechanical risk factors in the etiology of primary acquired nasolacrimal duct obstruction (PANDO).

Paranasal Sinuses

This brief account has the intent to give a cursory summary about the juxtaposition of paranasal sinuses and the osseous structures of the orbit. The paranasal air sinuses (PAS) are paired mucosa-lined, aerated bony cavities, adjoining the orbits at four sides—the roof, the medial wall, the floor, and the apex.

The frontal sinus is located above the orbital roof, the ethmoid and sphenoid sinuses are adjacent to the medial wall and the apex, while the roof of the maxillary sinus is identical with the floor of the orbit. The paranasal sinuses communicate with the nasal cavity through small apertures or orifices. The anatomy of the paranasal sinuses is most complex with endless numbers of variations in size, shape, and symmetry as individual as fingerprints, so that exceptions are more common than rules [143]. The frontal sinus (FPS) may overlie large areas over the orbital roof and extend vertically far up into the squamous part frontal bone. The frontal sinus opens through a superior funnel medially into an inferior funnel and the nasofrontal duct that passes the anterior ethmoid and drains into the frontal recess of the middle nasal meatus.

The thin lateral wall of the ethmoid sinus (EPS) or the lamina papyracea (LP) is the major constituent of the medial orbital wall. The orbital roof connects with the ethmoid roof along the fronto-ethmoid suture line (FES) at the upper LP (Fig. 2.12). Alongside the maxillary-ethmoidal suture line, the lower LP border including the inferior labyrinthine cells joins with the transverse bone plate built by the orbital floor in coincidence with the maxillary sinus roof. A Haller cell, an additional basal ethmoid cell, may be interposed between the maxillary sinus and the lamina papyracea at their transition. The Haller cell may extend into the superomedial maxillary recess in medial direction allowing for a transmaxillary access to the ethmoid without the risk of entry into the orbit. Sometimes, the anterior ethmoid sinus cell group continues into the lacrimal bone or even underneath the frontonasal process of the maxilla. The anterior ethmoid cell group drains into the middle nasal meatus by way of the infundibulum. The posterior cell group drains into the superior meatus. The sphe-

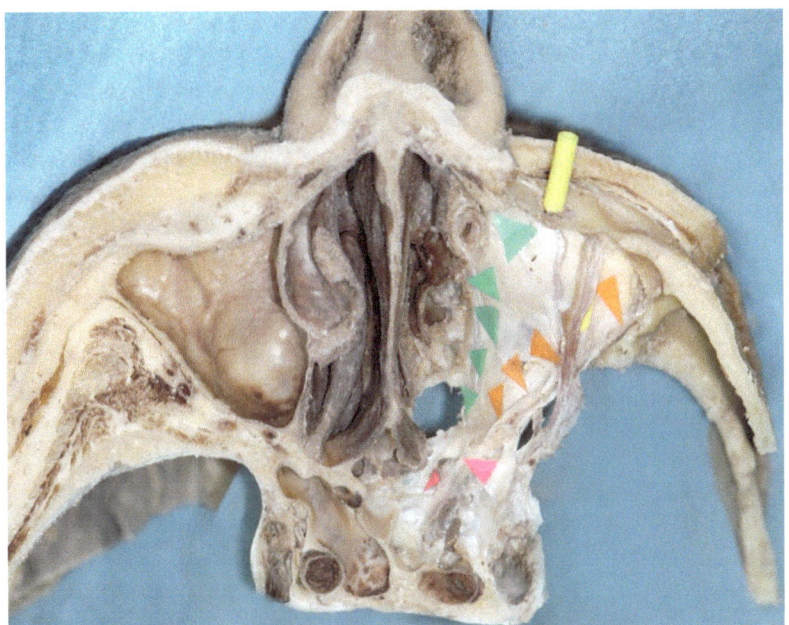

Fig. 2.22 Axial section through the maxillary sinuses at midlevel—Inferior view into the antral roofs, nasal cavity, ethmoid air cells, sphenoid sinus, and surrounding bone structures. Left side—Intact mucosal lining in antral roof and around nasal turbinate; Note: IOF contents. Right side—Periorbit exposed after bone removal in the lower (medio- inferolateral) orbital circumference. Infraorbital nerve and accompanying vessels (yellow tubes) run below the periorbita but there are perforators connecting to intraperiorbital vessels below the inferior rectus muscle (IRM). NLC/NLD and opened ethmoidal air cells along medial orbital wall (green arrows); Horner's orbital muscle [145]along IOF (orange arrows), Zinn's ring (pink arrows)

noid sinus (SPS) commonly aerates the sphenoid body [144]. The two-sided sinuses are usually separated and do not communicate with each other. The pneumatization can go far beyond the confines of the body into OS, ACP, GWS, and clivus. Large-size SPS in particular are characterized by bony imprinting and embossments. The lateral SPS wall, which is the partition to the orbital apex, is bulged inwards by the bony prominences of the ICA and CN II with the opticocarotid recess in between. Posteriorly, the lateral SPS wall relates to the cavernous sinus. The SPS roof is part of the anterior cranial fossa and provides the outlines of the sella turcica with a protruding pituitary gland fossa. The SPS floor may be indented by the passage belonging to the foramen rotundum (FR) and to the maxillary nerve (CN V2) as well as by the pterygoid/Vidian

canal (VC). Each SPS drains through an ostium of its own on its anterior wall into the sphenoethmoidal recess. The maxillary sinus (MPS) is the largest paranasal sinus (Fig. 2.22). It occupies the body of the maxilla. It is uncommon that the zygoma is pneumatized by the MPS though the MPS floor often reaches down to the level of the alveolar process base and may invade into toothless rim portions. Canine, premolar, and molar tooth roots may protrude into the MPS floor. The MPS roof or orbital floor is engraved by the IOG and/or tubuled by the IOC that forms a crease and is suspended in a mesentery-like bony plication on its way to the infraorbital foramen. The MPS medial wall forms the lateral wall of the nose. The MPS is draining over the medial wall into the semilunar hiatus of the middle nasal meatus.

Clinical Implications

Antrum Implosion or Silent sinus syndrome is a spontaneous, progressive disorder due to asymptomatic chronic maxillary sinusitis resulting in atelectasis of the maxillary sinus and in bony demineralization with long-term displacement of the orbital floor (see *Chap. 8.)*

Typical signs are facial asymmetry, enophthalmos, hypoglobus, and recession of the upper orbito-palpebral eyelid junction with unilateral manifestation in middle aged-patients between 30 and 40 years. The diagnosis is suspected clinically. CT imaging is pathognomonic and will confirm the underlying obstruction of the osteomeatal outlet and opacification/chronic mucosal inflammation of the maxillary sinus, leading to hypoventilation and negative sinus pressure. Antrostomy and reconstruction of the orbital floor using titanium implants are established treatment options [146].

The signs of a silent sinus syndrome may occur months after non-displaced orbital floor fractures [147], or even after repair of such fractures [148]. Posttraumatic is indistinguishable from spontaneous silent sinus and is presumably attributable to the same pathophysiology—subatmospheric pressure and antrum implosion. Secondary enophthalmos after repair of orbital floor fractures with resorbable implants should consider maxillary atelectasis and its etiology as a potential confounding factor besides the potential inadequacy of the implant material.

Lateral Wall of the Nose

The lateral nasal wall can be broken down into an arrangement of three or four turbinates in association with three meatus and the PAS outlets (Fig. 2.23a–c). The turbinates and meatus all run anteroposteriorly and direct the air flow. A turbinate or concha corresponds to a scroll of bone projecting from the lateral nasal wall (Fig. 2.23a,

b). The turbinates are named in ascending sequence as inferior, middle, superior, and supreme turbinate. The meatus are the passages beneath and lateral to the turbinates and are named as the turbinates passing along their superior aspect.

The inferior meatus is the space above the nasal floor and below the inferior turbinate. The inferior nasal turbinate (INT) is an independent spongy bone converging to a pointed posterior end. Anteriorly, the INT joins with the conchal crest, an oblique ridge inside the frontonasal process of the maxilla. The medial convex INT surface is numerously perforated and finely grooved for the harboring of a pronounced vascular network.

The concave lateral INT surface features three processes along its upper border. The lacrimal process projects anterosuperiorly, articulates with the lacrimal bone and assists in the formation of the lower portion of the nasal lacrimal canal (NLC). The NLC exit is found in the anterior portion of the lateral wall of the inferior meatus. The ethmoidal process ascends from the midportion of the superior INT border to join the uncinate process of the ethmoid. The thin lamina of the maxillary process turns inferolaterally and forms part of the medial wall of the maxillary sinus.

The middle, superior, and supreme nasal turbinates are components of the ethmoid.

The middle nasal turbinate (MNT) attaches to the lateral edge of the cribriform plate as superior part of its supporting overall basal (ground) lamella. The sphenopalatine foramen is located within the reach of the posterior end of the MNT.

The superior nasal turbinate (SNT) is approximately half as long as the MNT; it also extends into the ethmoid roof and attaches to the posterior part of the cribriform plate.

The supreme nasal turbinate is an uppermost rudimentary concha which occurs unilaterally or bilaterally in about 60% of individuals. The PAS openings into the meatus have been described in the previous paragraph. If a supreme turbinate and meatus are present, the posterior ethmoid cell group may open there. The secondary features of the lateral nasal wall are revealed at best if the INT and MNT are removed. These include the agger nasi, the uncinate process, the ethmoid

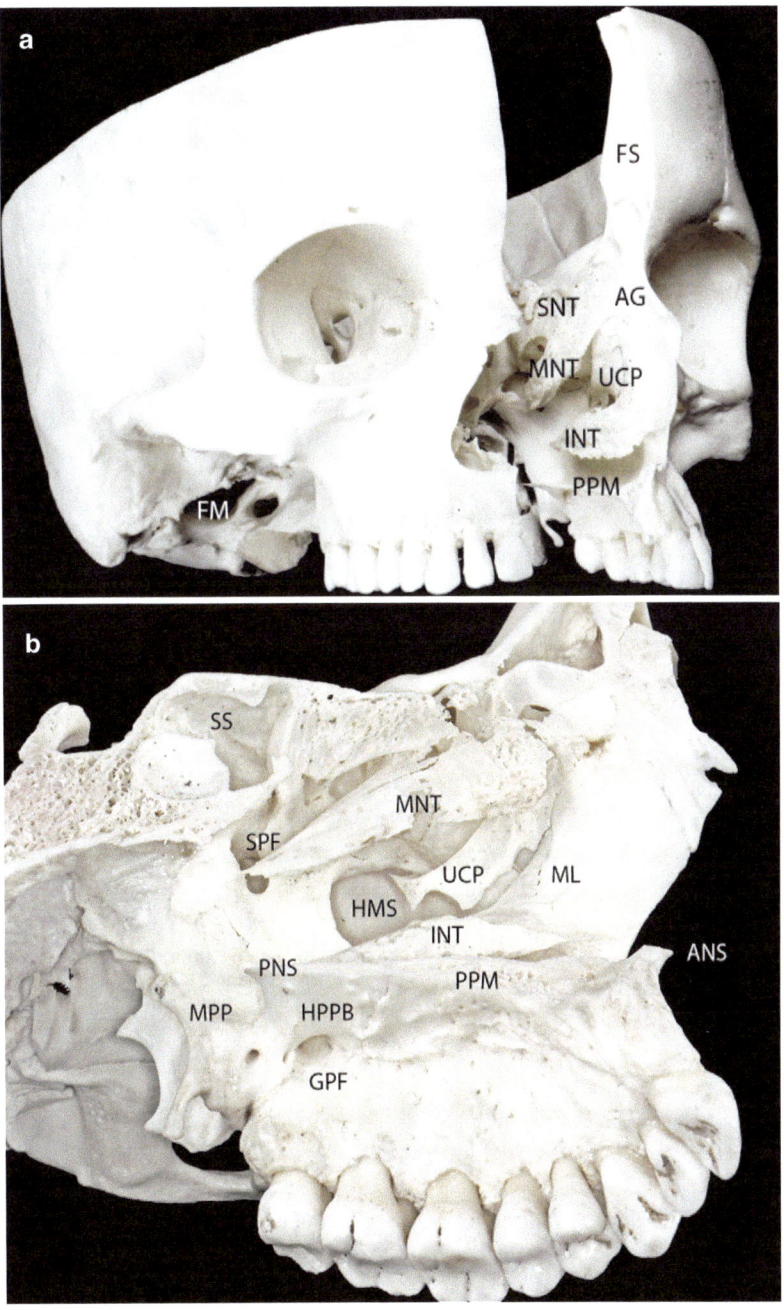

Fig. 2.23 (**a**) Skull split along midsagittal plane—antero-lateral view to gain insight to the nasal turbinates and meatus. *AG* agger nasi, *INT* inferior nasal turbinate (concha), *JF* jugular fossa/foramen, *MNT* middle nasal turbinate (concha), *PPM* palatine process of maxilla, *SNT* superior nasal turbinate (concha), *UCP* uncinate process. (**b**) Lateral nasal wall—medial view at anteroinferior angulation. Note: Uncinate process (UCP) projecting across the ostium (hiatus) of the maxillary sinus. *GPF* greater palatine foramen, *MPP* medial pterygoid plate. (**c**) Uncinate process (UCP) and bulla ethmoidalis (BE)—Transantral inferolateral view after removal of the anterior and lateral maxillary sinus walls

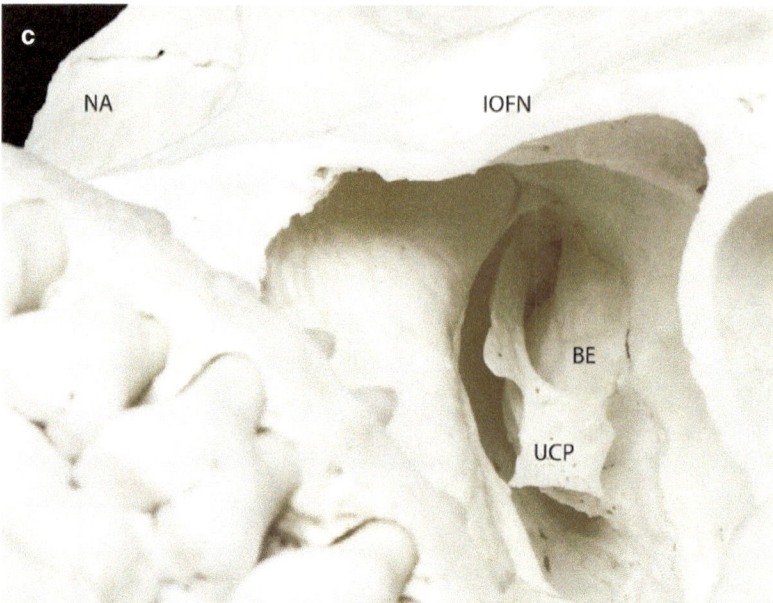

Fig. 2.23 (continued)

bulla, and the ethmoid infundibulum including the hiatus semilunaris. The agger nasi (AN) is a mound like prominence directly in front of the conchal crest and the MNT. The AN represents a vestige of the middle naso-turbinal. It is frequently aerated with an opening into the anterior middle meatus and into the ethmoid infundibulum. On the posterolateral side, the agger nasi may be fusing with the lacrimal bone and/or the medial orbital wall.

The hooked uncinate process (UCP) arises from the anterior ethmoid lateral to the anterior MNT attachment (Fig. 2.23b, c). The UCP is directed posteroinferiorly and projects at variable length across the wide medial ostium of the maxillary sinus. The MNT usually conceals the UCP. The ethmoid bulla refers to a single or more large ethmoidal cells that protrude from the anterolateral wall of the MNT. The space configured between the anterior inferior convex border of the bulla and the superior free edge of the uncinate process is the hiatus semilunaris. This hiatus is a two-dimensional opening leading into the ethmoidal infundibulum. The ethmoid infundibulum is a three-dimensional space extending downward and posteriorly between the lateral nasal wall and

the uncinate process medially. The posterosuperior boundary is the ethmoid bulla. The lateral side is consisting of the lamina papyracea and the frontonasal process, rarely of the lacrimal bone, too. Depending on the superior attachment of the uncinate process either laterally to the lamina papyracea or medially to the ethmoid roof, the infundibulum is closed as a terminal recess or continuous with the frontal recess.

Internal Orbital Buttresses

A set of three buttresses running in parallel stabilize the orbital floor in sagittal direction (Fig. 2.24a, b), the inferomedial orbital strut (IOS) along the maxilloethmoidal suture line. [20, 149, 150], the intermediary bony underpinning of the infraorbital groove/canal, and the reinforcement along the medial IOF margin laterally on par with the lateral floor strut (LFS). The involvement and fragmentation of these buttresses are an indicator for the severity of the trauma. The sagittal buttresses integrate deliberately into the overall framework of facial buttresses (Fig. 2.24c).

Fig. 2.24 (**a**) Sagittal internal orbital buttresses in parallel after removal of thin interjacent bone plates: IOS along the articulation of orbital floor and medial orbital wall with a strong anterior portion around the nasolacrimal duct, weaker midportion because of the lamina papyracea as well as aeration by the ethmoidal bulla and solid wide posterior palatine bone portion; midway floor strut surrounding the infraorbital groove and canal; lateral floor strut along anterior/lateral IOF edge. Detail (upper right corner)—Overview midface skeleton. *IOC* infraorbital canal, *MFS* midway floor strut, *STS* sphenotemporal suture. (**b**) Sagittal internal orbital buttresses and soft tissue relations (left zygoma, anterior, and middle skull base removed to expose the LFS). (**c**) Orbital buttresses—a vertical (red), sagittal (green), and transverse (blue) framework. (With permission from https://surgeryreference.aofoundation.org)

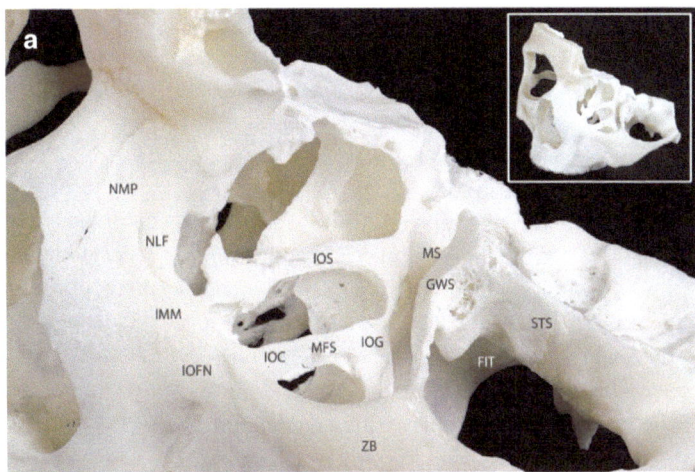

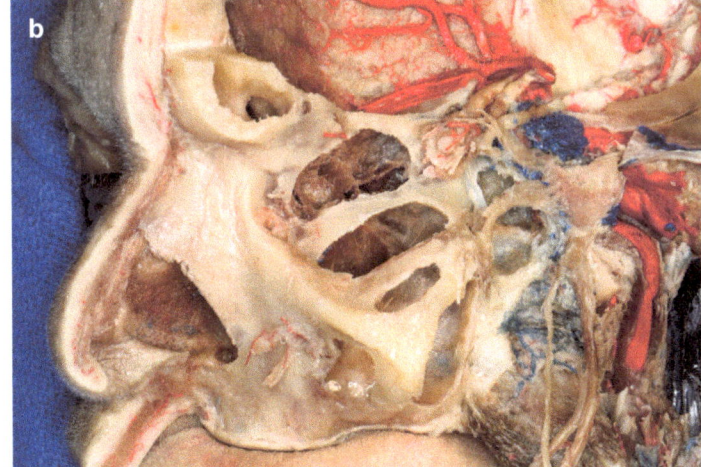

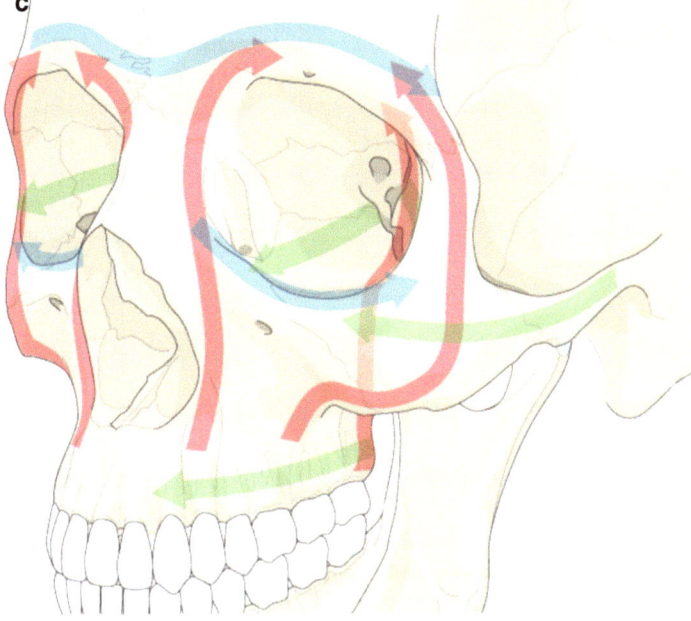

Orbital Dimensions, Volume, Surface Contours, 3D Globe Position—Interrelationship

Numerous factors determine the 3D position of the ocular globe inside the orbit [151, 152], its anterior projection, and radial relationships to the orbital rims; these are the overall dimension and volume as well as the geometry (width and angulation of the medial and lateral walls) and surface contours of the internal orbit, such as slopes, angles, planes, concavities, and convexities—in particular the so-called posterior medial bulge within the inferomedial wall transition [153–155].

The overall dimensions and volumes of the orbit show great variations owing to age, gender, and ethnicity [156–160]. The adult orbital aditus may show some typical average values approximating to 4 cm horizontal × 3.5 cm vertical in size with wide ranging deviations (Fig. 2.25). The anteroposterior depth is 4.5–5 cm approximately and the overall volume of the orbital cavity measures up to 30 cm^3 including the ocular globe, a sphere of 22–27 mm in diameter, up to 7.5 cm^3 in volume [161], and 69–85 mm in equatorial circumference.

The sagittal projection of the globe (corneal apex plane and/or equator) is referenced relative to the sagittal projection of the orbital rims, traditionally to the retruded lateral orbital rim only (exophthalmometry), more recently using 360° polar plots (CT morphometry) to measure distances to preassigned points all over the outer rim periphery [162]. Another CT method is the 3D characterization of the globe position in the orbit in relation to the center of the globe [163].

The globe projection in the coronal plane or the vertical/horizontal globe position is correlated to the delineation, shape, and curvatures of the orbital openings and the interorbital distance. The shape variations of the orbital aditus (Fig. 2.25a) have been classified into manifold categories, e.g., oval, rectangular, rhomboid, trapezoid, and subdivisions with a greater percentage of symmetry types in females (29.1%) than in males (23.81%) in a dry skull (n = 184) study [164].

For all aditus types, consistently the inclination of the medial part of the infraorbital rim (MOMLP) is parallel to the angle of the orbital floor while the lateral part (IPZ) and the adjacent floor both run in a horizontal plane.

The objective of linear morphometric measurements studies of the orbital skeleton is to provide guidelines for safe distances in periorbital dissection along each orbital wall [165] before the danger zone, viz. the orbital apex, is entered. Often quoted but just as often discredited mean maximum (standard) distances for adults (n = 24 dry skulls from India) on the orbital floor were 25 mm from the infraorbital foramen, in the orbital roof 30 mm from the supraorbital notch, in the lateral wall 25 mm from the frontozygomatic suture and in the medial wall (to the PEF) 30 mm from the anterior lacrimal crest [94]. Over time by the investigation of small series of dry skulls [166, 167] or PMHS specimens [168] and ultimately with the application of modern imaging techniques (CT—[169, 170], CBCT–[165]; MRI—[171, 172]) an extensive set of the reference points/distance lines/orbital cavity length/floor length and width, etc., was accumulated to delineate the spatial arrangement of the internal orbit and systematically assess its metrics.

Since orbital wall fractures typically involve the inferior (floor) and medial orbital wall their natural surface profile was thoroughly scrutinized and on the basis of CT scans translated into a data matrix to recognize the subtle slopes, depressions, and curvatures [173] enabling a true to original bony reconstruction [174]. In descriptive topographical terms, the lowest lateral point of the orbital floor is located next to the anterior IOF loop. The lowest portion of the orbital floor coincides with a short-pathed, circumferentially running concavity just behind the orbital rims. Originating from the periorbital dissection sequence this widening is named as "postentry zone" (Fig. 2.12 Detail). The orbital floor steadily ascends from the bottom of the postentry zone until it reaches the convex top site of the orbital plate of the palatine bone (OPPB—posterior ledge, Fig. 2.13a–e).

This curved process drops down abruptly backward along the infratemporal surface of the

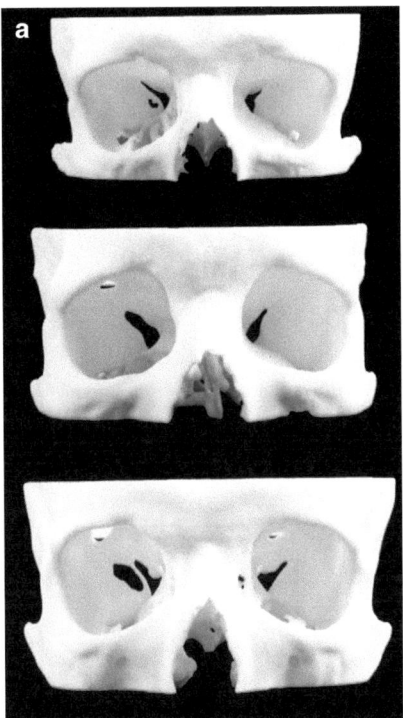

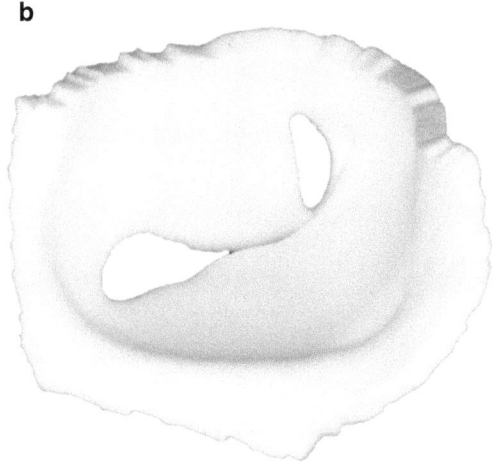
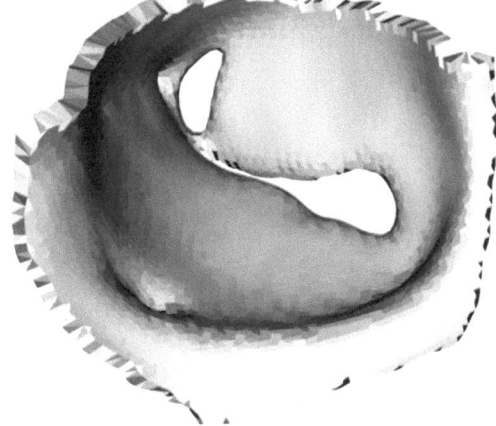

Fig. 2.25 (**a**) Orbital aditus—Stereolithography models at the same scale—Variations of the width and height ratio determine the shape and inclinations of the aperture. (**b**) Internal orbit—virtual reconstruction based on atlas segmentation—Focus on surface profile of the inferomedial maxilla into the posterior IOF sinkhole (Fig .2.12 and 2.26a). In a paramedian sagittal plane, the orbital floor from the deepest point to the OPPB assumes an undulating shape, which has been addressed as the "Lazy-S configuration".

orbital walls with posteromedial bulge and IOF isthmus promontory. Left orbit—screenshot. Right Orbit 3D printed epoxy resin Model [Courtesy of S. Schlager and M.C. Metzger, Albert-Ludwigs-Universität Freiburg, Freiburg i. Brsg.; Germany]

maxilla into the posterior IOF sinkhole (Fig .2.12 and 2.26a). In a paramedian sagittal plane, the orbital floor from the deepest point to the OPPB assumes an undulating shape, which has been addressed as the "Lazy-S configuration".

In the posterior transition of the orbital floor into the medial orbital wall, the bony surface profile is protruded by a focal convexity, i.e. the posterior medial bulge (PMB) (Figs. 2.25b and 2.26a–d). The PMB has reached the status of an undeletable engram (Fig. 2.1f). The PMB is considered the most essential support area to maintain the vertical globe height and anterior globe projection. It is referred to as "Beat

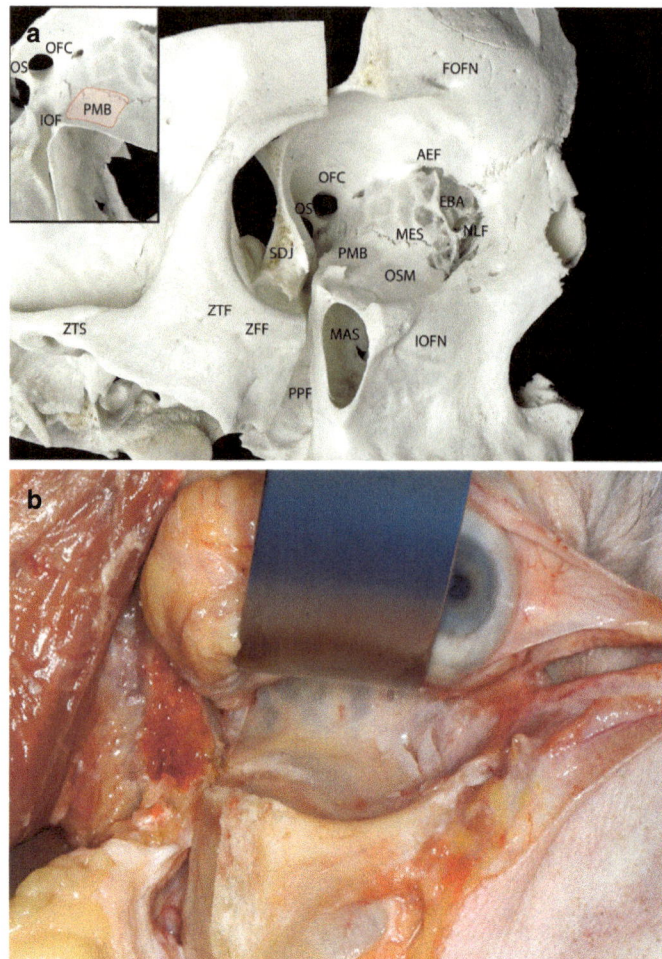

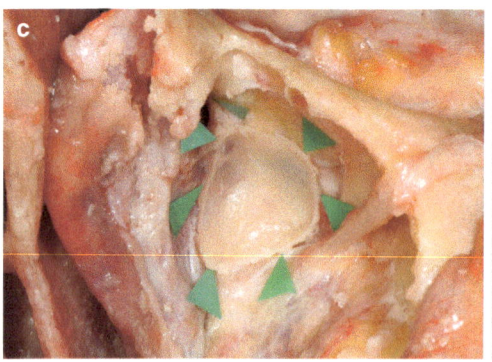

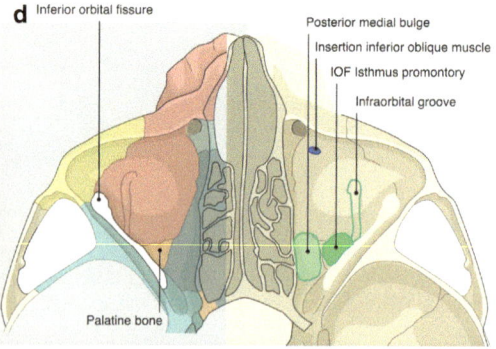

Fig. 2.26 (**a**) Orbit (right)—staggered partitions after midline sagittal plane section. "Lazy-S" becomes obvious along the medial orbital floor. The lateral orbital wall reveals as a flat plane. Inset (upper left corner of Fig. 2.26a.)—Convexity of the posterior inferomedial wall with, key area" outlined provisionally (red rectangle). *PMB* Posterior Medial Bulge ('Key Area'). (**b**) Exposure of "Lazy- S" and "key area"—by superior retraction of the orbital soft tissue contents and swinging of the eyelids anteriorly after removal of zygoma and GWS. (**c**) Left orbital floor—Transantral view.

Delineation of the 'posterior medial bulge' (PMB—green arrows). Mucosal lining and circumferential bony floor portions removed including medial maxillary antral wall/ uncinate process, infraorbital nerve released from canal/ foramen and transposed to the left over the zygomatic body. (**d**) Both orbital floors and bony precincts after axial section at intermediate height level of the orbits— View from above. Left- bony constituents. Right— Approximate topography of relevant landmarks during periorbital dissection. (With permission from https://surgeryreference.aofoundation.org

Hammer's key area," because this author [152] has emphasized the tremendous importance for the appropriate surgical reconstruction of the internal orbit time and again. The PMB has become some sort of trademark characteristic which is perpetuated in the design of commercially available anatomically preformed orbital titanium meshes/plates [155] and 3D models or molds [175]. The design information for 3D modeling of the inferomedial orbital circumference and the subsequent design of preformed orbital floor/medial orbital wall implants is gathered from segmentation or meanwhile from statistical shape analysis of CT data in unaffected orbits [155, 173, [176–184].

The graphic visualization of the variance models in all their details and beauty of their "heat maps" is fascinating but it should be borne in mind that the shape is reductionistic in pursuit of the target to come up with a few plate types which fit to a maximum of individual surface profiles. To that end, the shape requires technical modifications; for instance, undercuts are filled, openings, and depressions sealed and outliers eliminated. Therefore, it seems nothing more than a logical consequence for estimating the versatility and in particular the range of application of a preformed orbital mesh/plate that the manipulation specifics of the underlying statistical shape model are known to the end user in addition to the size and stratification of the population and the datasets according to laterality, gender, age, and ethnic group. At best, the statistical shape model as well as the STL files of the preformed meshes should be handy during preoperative virtual planning of fracture repair. Gross incongruency between the virtual (statistical shape) model and the patient's intact orbital wall anatomy would obviate the implant positioning from the outset.

The PMB, the sagittal "Lazy-S," the width of the orbital cavity between the lateral and medial wall and the inferior orbital rim are consistently reflected in the CT analysis studies though the difficulties to procure accurate and reproducible anatomic boundaries of the PMB are noteworthy. The PBM is still regarded a "fuzzy" characteristic of the orbital surface profile, since the distinct contour outlines are difficult to capture in the automated shape analysis [155].

The volume of the orbital cavity is a superordinate parameter having direct impact on the orbital contents and on changes in ocular globe position. Thus, quantitative orbital volume measurements (i.e., volume orbitometry) are crucial in the diagnostic workup and date back more than 3 decades [185]. The methods for orbital volume measurements are still striving to reach perfection, nowadays enhanced by computer-assisted programs and automatization [27, 157, 171, 186–191]. The orbital volume and morphology is correlated to anthropometric measurements (facial height and width, interorbital distance, etc.) [192]. On the other hand, the size of the adult orbital volume is associated with eyeball volume and with the grey matter volume of the visual cortex in a linear relationship [193].

The orbital volume has been shown to range between 16 cm^3 and 30 cm^3.

In a study characterizing the orbital volume in a normal population [194] from Chile (n = 398 orbits), the mean total orbital volume was 24.5 ± 3.1 cm^3 (range 16.9–35.0 cm^3). There were no differences in laterality but gender differences (male > female) and a steady orbital volume increase beyond the age of 30 years. Interestingly enough, if the orbits were divided into three zones in nearly the same way as in our proposition (Fig. 2.18a, b), the anterior zone had a mean volume of 17.30 ± 2.6 cm^3 (range 9.4–25.6 cm^3), which represented 70.7% of the total orbital volume.

The volume of the central zone (comparable to the midorbit) was 5.4 ± 1.8 cm^3 (range 2.2–12.3 cm^3) accounting for 22.1% of the total orbital volume. The posterior zone volume was 5.4 ± 1.8 cm^3 (range 2.2–12.3 cm^3) corresponding to a percentage of 8.2% of the total orbital volume. In line with previous correlations between fracture location, viz. volume increase in different regions of the orbital cavity [195], it was hypothesized that a volume increase in the anterior orbital zone has a negligible effect on the globe position. However, a volume increase in the central zone, which includes Beat Hammer's key area, was held responsible for the occurrence of hypoglobus. A volume increase in the posterior area was implicated to generate enophthalmos, eventually.

Intraindividual volume differences in the orbit to a percentage of 7–8% were considered to represent a normal range [196] for a long period.

A recent CT study (*n* = 93 subjects) has shown that the intraindividual difference of orbital volumes amounts to approximately 2% [154]. A dry skull study evaluating 242 orbits using alginate impressions and their water displacement demonstrated an average orbital volume of 26.75 ml for the right side and 26.65 ml for the left side [197]. The mean volume difference was indicated with 0.8 ml (range 0.02 to 3.64 ml) corresponding to around 3%. Despite numerous investigations into average values [160, 173, 177, 198], age-related changes of the surface profile of the internal orbit (orbital floor and the medial orbital wall) have eluded quantitative assessment so far. From a qualitative perspective, the inferior orbital rim contours are resorbed enlarging the height of the aditus orbitae [199–201], while the attaching orbital floor flattens toward its posterior end, so that its lowermost portion tends to move backward [153] (Fig. 2.27).

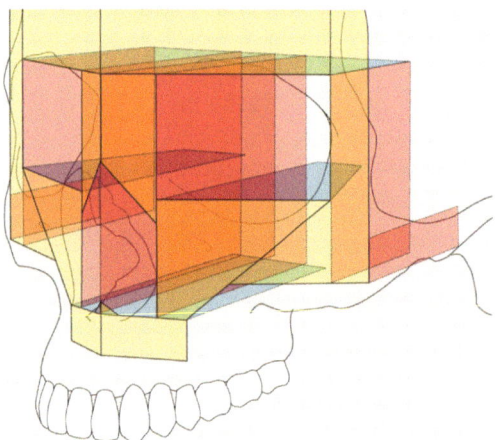

Fig. 2.27 Multiplanar blueprint design of midface and orbits. All cavities are outlined by a 3-dimensional framework of planes. In contrast to anatomic dissection or studies on intact dry skulls modern imaging techniques allow to visualize and investigate every structural element within the craniofacial skeleton relative to accurately reproducible references (Illustration based on [202]. With permission of https://surgeryreference.aofoundation.org)

Summary/Conclusion

To acquire macroscopic anatomical knowledge by traditional PMHS dissections is often purported as no longer appropriate in today's medical student education, because of competition with unprecedented modern learning opportunities and information overflow (conventional textbooks and atlases, videos, interactive audiovisual programs, digital apps, e-mail newsletters, journals, Wikipedia, online or presence academic lectures, plastinated specimens, etc.). Trainees as well as professionals from all surgical specialties return into anatomical dissection rooms, however, with the aim to enhance their practice and to directly experience the visual and haptic properties of soft embalmed or even better frozen-thawed tissues [203]. CMF specialists involved in orbital surgery make no exception to optimize their expertise and skills or to perform preclinical tests of novel techniques for access, navigation, repair, reconstruction, and even allotransplantation (e.g., [53, 64, 102, 204–208]).

This article is intended to go ahead and support any kind of learning ambitions with a reappraisal of common textbook knowledge of the bony orbital precincts, systematic, and topographical connotations [see Basic Literature] and an introduction into the enormous morphologic and morphometric variability of the orbital openings and passageways as described in recent publications.

Textbooks and atlases as well as digital apps often draw a picture representing the average or abstraction of limited samples.

Enormously large collections of dry skulls have been evaluated to get detailed insights into the entire spectrum of normal variations of the bony architecture depending on ethnicity, age, gender, body height and somatotype. Such demographic criteria may remain inhomogeneous or incomplete leading to controversial results opposite to the needs of a health care with regard to a rapidly increasing globalization. The advent of modern imaging techniques with a limitless morphologic data production (CT, CBCT, MRI, ultrasound) as well as internal views of the body by endoscopes necessitated to acquire a more refined interpretation and understanding of the anatomy with paralleling computer and/or artificial intelligence assisted post-processing (segmentation- [209]) into multiplanar views (Fig. 2.27), internal surface, 3D formats (Fig. 2.25b), and volume calculations). Indeed, the contemporary literature is replete with publications on morphological and morphometric findings of the bony orbit, going into previously inconceivable subdivisions (e.g., [173, 210]) and sub-millimeter differentiation. As outlined above, the quality of any such studies differs substantially and it is not clear what will remain actively referenced in the future. For sure, there will be no standstill until a "big data/evidence-based anatomy" [211, 212] with consecutive translation into clinical settings allowing for interactive man/machine reading of findings and images will be established.

Acknowledgments Great thanks go to Gerhard Poetzel and Rudolf Herzig, the photographers at the OMFS Departments, LMU Munich, for their excellent support.

Disclosure The authors declare no conflict of interest in relation to the content of this article. Dissections of anatomic specimen were conducted in compliance with state and local laws.

Copyright All figures, photographs and illustrations, the provenance of which is not specifically indicated in the captures are printed with permission and copyright retained by the two authors and their group as copyright holders.

References

1. Martins C, Costa ESIE, Campero A, et al. Microsurgical anatomy of the orbit: the rule of seven. Anat Res Int. 2011;2011:468727. https://doi.org/10.1155/2011/468727.
2. Royle G. A groove in the lateral wall of the orbit. J Anat. 1973;115(Pt 3):461–5.
3. Schünke M, Schulte E, Schumacher U, Stefan C, MacPherson B. Head, neck and neuroanatomy – Thieme atlas of anatomy. 3rd ed. Stuttgart Georg Thieme Publishers; 2020.
4. Costea C, Turliuc S, Cucu A, et al. The "polymorphous" history of a polymorphous skull bone: the sphenoid. Anat Sci Int. 2018;93(1):14–22. https://doi.org/10.1007/s12565-017-0399-5.
5. Engin Ö, Adriaensen G, Hoefnagels FWA, Saeed P. A systematic review of the surgical anatomy of the orbital apex. Surg Radiol Anat. 2021;43(2):169–78. https://doi.org/10.1007/s00276-020-02573-w.
6. Dassi CS, Demarco FR, Mangussi-Gomes J, Weber R, Balsalobre L, Stamm AC. The frontal sinus and frontal recess: anatomical, radiological and surgical concepts. Int Arch Otorhinolaryngol. 2020;24(3):e364–75. https://doi.org/10.1055/s-0040-1713923.
7. Messerklinger W. On the drainage of the normal frontal sinus of man. Acta Otolaryngol. 1967;63(2):176–81. https://doi.org/10.3109/00016486709128748.
8. Stammberger HR, Kennedy DW, Anatomic Terminology Group. Paranasal sinuses: anatomic terminology and nomenclature. Ann Otol Rhinol Laryngol Suppl. 1995;167:7–16.
9. Song WK, Lew H, Yoon JS, Oh MJ, Lee SY. Role of medial orbital wall morphologic properties in orbital blow-out fractures. Invest Ophthalmol Vis Sci. 2009;50(2):495–9. https://doi.org/10.1167/iovs.08-2204.
10. Gray H. Systematic. In: Lewis WH, editor. Anatomy of the human body. 20th ed. Philadelphia, PA/New York, NY: Lea and Febiger; 1918.
11. Whitnall SE. Anatomy of the human orbit and accessory organs of vision, Facsimile reprint of original edition 1921. New York: Robert E Krieger Publishing Company; 1979.
12. Daniels DL, Mark LP, Ulmer JL, et al. Osseous anatomy of the pterygopalatine fossa. AJNR Am J Neuroradiol. 1998;19(8):1423–32.
13. Gooris PJJ, Muller BS, Dubois L, et al. Finding the ledge: sagittal analysis of bony landmarks of the orbit. J Oral Maxillofac Surg. 2017;75:2613–27. https://doi.org/10.1016/j.joms.2017.07.156.
14. Mueller SK, Freitag SK, Bleier BS. Morphometric analysis of the orbital process of the palatine bone and its relationship to endoscopic orbital apex surgery. Ophthalmic Plast Reconstr Surg. 2018;34(3):254–7. https://doi.org/10.1097/IOP.0000000000000940.
15. Moon SJ, Lee WJ, Roh TS, Baek W. Sex-related and racial variations in orbital floor anatomy. Arch Craniofac Surg. 2020;21:219–24. https://doi.org/10.7181/acfs.2020.00143.
16. Elhadi AM, Zaidi HA, Yagmurlu K, et al. Infraorbital nerve: a surgically relevant landmark for the pterygopalatine fossa, cavernous sinus, and anterolateral skull base in endoscopic transmaxillary approaches. J Neurosurg. 2016;125(6):1460–8. https://doi.org/10.3171/2015.9.JNS151099.
17. Przygocka A, Szymański J, Jakubczyk E, Jędrzejewski K, Topol M, Polguj M. Variations in the topography of the infraorbital canal/groove complex: a proposal for classification and its potential usefulness in orbital floor surgery. Folia Morphol (Warsz). 2013;72(4):311–7. https://doi.org/10.5603/fm.2013.0052.
18. Carstocea L, Rusu MC, Pascale C, Sandulescu M. Three-dimensional anatomy of the transantral

intraseptal infraorbital canal with the use of cone-beam computed tomography. Folia Morphol (Warsz). 2020;79(3):649–53. https://doi.org/10.5603/FM.a2019.0109.

19. Takahashi Y, Nakano T, Miyazaki H, Kakizaki H. An anatomical study of the orbital floor in relation to the infraorbital groove: implications of predisposition to orbital floor fracture site. Graefes Arch Clin Exp Ophthalmol. 2016;254(10):2049–55. https://doi.org/10.1007/s00417-016-3455-2.

20. Kim JW, Goldberg RA, Shorr N. The inferomedial orbital strut: an anatomic and radiographic study. Ophthalmic Plast Reconstr Surg. 2002;18(5):355–64. https://doi.org/10.1097/00002341-200209000-00007.

21. Whitnall SE. On a tubercle on the malar bone, and on the lateral attachments of the tarsal plates. J Anat Physiol. 1911;45:426–32.

22. Kim HS, Oh JH, Choi DY, et al. Three-dimensional courses of zygomaticofacial and zygomaticotemporal canals using micro-computed tomography in Korean. J Craniofac Surg. 2013;24(5):1565–8. https://doi.org/10.1097/SCS.0b013e318299775d.

23. Coutinho DCO, Martins-Júnior PA, Campos I, Custódio ALN, Silva M. Zygomaticofacial, Zygomaticoorbital, and Zygomaticotemporal Foramina. J Craniofac Surg 2018;29(6):1583–7. https://doi.org/10.1097/scs.0000000000004530.

24. Gruss JS, Mackinnon SE. Complex maxillary fractures: role of buttress reconstruction and immediate bone grafts. Plast Reconstr Surg. 1986;78(1):9–22.

25. Rohner D, Tay A, Meng CS, Hutmacher DW, Hammer B. The sphenozygomatic suture as a key site for osteosynthesis of the orbitozygomatic complex in panfacial fractures: A biomechanical study in human cadavers based on clinical practice. Plast Reconstr Surg. 2002;110(6):1463–71. https://doi.org/10.1097/01.prs.0000029360.61857.ae.

26. Aujla JS, Curragh DS, Patel S, Selva D. Orbital apex anatomy: relationship between the optic foramen and anterior face of sphenoid sinus—a radiological study. Eye (Lond). 2021;35(9):2613–8. [Online Publication ahead of print]. https://doi.org/10.1038/s41433-020-01289-w.

27. Ugradar S, Goldberg R, Rootman D. Anatomic variation of the entrance of the optic canal into the orbit. Orbit. 2019;38(4):305–7. https://doi.org/10.1080/01676830.2018.1528619.

28. Zhang X, Lee Y, Olson D, Fleischman D. Evaluation of optic canal anatomy and symmetry using CT. BMJ Open Ophthalmol. 2019;4(1):e000302. https://doi.org/10.1136/bmjophth-2019-000302.

29. Natori Y, Rhoton AL Jr. Microsurgical anatomy of the superior orbital fissure. Neurosurgery. 1995;36(4):762–75. https://doi.org/10.1227/00006123-199504000-00018.

30. Curragh DS, Valentine R, Selva D. Optic strut terminology. Ophthalmic Plast Reconstr Surg. 2019;35(4):407–8. https://doi.org/10.1097/iop.0000000000001394.

31. Kerr RG, Tobler WD, Leach JL, et al. Anatomic variation of the optic strut: classification schema, radiologic evaluation, and surgical relevance. J Neurol Surg B Skull Base. 2012;73(6):424–9. https://doi.org/10.1055/s-0032-1329626.

32. Suprasanna K, Ravikiran SR, Kumar A, Chavadi C, Pulastya S. Optic strut and para-clinoid region—assessment by multi-detector computed tomography with multiplanar and 3 dimensional reconstructions. J Clin Diagn Res. 2015;9(10):TC06–9. https://doi.org/10.7860/JCDR/2015/15698.6615.

33. Bertelli E. Metoptic canal, duplication of the optic canal and Warwick's foramen in human orbits. Anat Sci Int. 2014;89(1):34–45. https://doi.org/10.1007/s12565-013-0197-7.

34. Onodi A, Strass HI. The optic nerve and the accessory cavities of the nose: contribution to the study of canalicular neuritis and atrophy of the optic nerves of nasal origin. Ann Otol Rhinol Laryngol. 1908;17(1):1–115. https://doi.org/10.1177/000348940801700101.

35. Onodi A. Des rapports entre le nerf optique et le sinus sphenoidal. La cellule ethmoidale postérieure en particulier. Revue Hebd Laryng d'Otol Rhinol. 1903;25:72–140.

36. Wang J, Bidari S, Inoue K, Yang H, Rhoton A Jr. Extensions of the sphenoid sinus: a new classification. Neurosurgery. 2010;66:797–816. https://doi.org/10.1227/01.NEU.0000367619.24800.B1.

37. Andrianakis A, Tomazic PV, Wolf A, et al. Optico-carotid recess and anterior clinoid process pneumatization—proposal for a novel classification and unified terminology: an anatomic and radiologic study. Rhinology. 2019;57(6):444–50. https://doi.org/10.4193/Rhin19.194.

38. Bouthillier A, van Loveren HR, Keller JT. Segments of the internal carotid artery: a new classification. Neurosurgery. 1996;38(3):425–32; discussion 432–423. https://doi.org/10.1097/00006123-199603000-00001.

39. Froelich S, Abdel Aziz KM, van Loveren HR, Keller JT. The transition between the cavernous sinus and orbit. In Dolenc VV, Rogers L (Eds.) Cavernous sinus—developments and future perspectives. p. 27–33, Wien Springer Verlag New York; 2009.

40. Dallan I, Castelnuovo P, de Notaris M, et al. Endoscopic endonasal anatomy of superior orbital fissure and orbital apex regions: critical considerations for clinical applications. Eur Arch Otorhinolaryngol. 2013;270(5):1643–9. https://doi.org/10.1007/s00405-012-2281-3.

41. Lang J. Skull base and related structures- atlas of clinical anatomy. Stuttgart, New York: Schattauer; 1995. p. 114–5.

42. Shi X, Han H, Zhao J, Zhou C. Microsurgical anatomy of the superior orbital fissure. Clin Anat. 2007;20(4):362–6. https://doi.org/10.1002/ca.20391.

43. Bisaria KK, Kumar N, Prakesh M, et al. The lateral rectus spine of the superior orbital fissure. J Anat. 1996;189(Pt 1):243–5.

44. Govsa F, Kayalioglu G, Erturk M, Ozgur T. The superior orbital fissure and its contents. Surg Radiol Anat. 1999;21:181–5. https://doi.org/10.1007/BF01630898.

45. Sharma PK, Malhotra VK, Tewari SP. Variations in the shape of the superior orbital fissure. Anat Anz. 1988;165(1):55–6.

46. Reymond J, Kwiatkowski J, Wysocki J. Clinical anatomy of the superior orbital fissure and the orbital apex. J Craniomaxillofac Surg. 2008;36(6):346–53. https://doi.org/10.1016/j.jcms.2008.02.004.

47. Erturk M, Kayalioglu G, Govsa F, Varol T, Ozgur T. The cranio-orbital foramen, the groove on the lateral wall of the human orbit, and the orbital branch of the middle meningeal artery. Clin Anat. 2005;18(1):10–4. https://doi.org/10.1002/ca.20020.

48. Abed SF, Shams P, Shen S, Adds PJ, Uddin JM. A cadaveric study of ethmoidal foramina variation and its surgical significance in Caucasians. Br J Ophthalmol. 2012a;96(1):118–21. https://doi.org/10.1136/bjo.2010.197319.

49. Abed SF, Shams P, Shen S, Adds PJ, Uddin JM, Manisali M. A cadaveric study of the cranio-orbital foramen and its significance in orbital surgery. Plast Reconstr Surg. 2012;129(2):307e–11e. https://doi.org/10.1097/PRS.0b013e31821b6382.

50. Simao-Parreira B, Cunha-Cabral D, Alves H, Silva SM, Andrade JP. Morphology and navigational landmarks of the cranio-orbital foramen in a Portuguese population. Ophthalmic Plast Reconstr Surg. 2019;35(2):141–7. https://doi.org/10.1097/IOP.0000000000001188.

51. Tomaszewska A, Zelaźniewicz A. Morphology and morphometry of the meningo-orbital foramen as a result of plastic responses to the ambient temperature and its clinical relevance. J Craniofac Surg. 2014;25(3):1033–7. https://doi.org/10.1097/scs.0000000000000552.

52. Macchi V, Regoli M, Bracco S, et al. Clinical anatomy of the orbitomeningeal foramina: variational anatomy of the canals connecting the orbit with the cranial cavity. Surg Radiol Anat. 2016;38(2):165–77. https://doi.org/10.1007/s00276-015-1530-8.

53. Siemionow M, Bozkurt M, Zor F, et al. A new composite eyeball-periorbital transplantation model in humans: an anatomical study in preparation for eyeball transplantation. Plast Reconstr Surg. 2018;141(4):1011–8. https://doi.org/10.1097/prs.0000000000004250.

54. Diamond MK. Homologies of the meningeal-orbital arteries of humans: a reappraisal. J Anat. 1991;178:223–41.

55. Liu Q, Rhoton AL Jr. Middle meningeal origin of the ophthalmic artery. Neurosurgery. 2001;49(2):401–6; discussion 406–407. https://doi.org/10.1097/00006123-200108000-00025.

56. Louw L. Different ophthalmic artery origins: embryology and clinical significance. Clin Anat. 2015;28(5):576–83. https://doi.org/10.1002/ca.22470.

57. Michalinos A, Zogana S, Kotsiomitis E, Mazarakis A, Troupis T. Anatomy of the ophthalmic artery: a review concerning its modern surgical and clinical applications. Anat Res Int. 2015;2015:591961. https://doi.org/10.1155/2015/591961.

58. Singh S, Dass R. The central artery of the retina. I. Origin and course. Br J Ophthalmol. 1960;44(4):193–212. https://doi.org/10.1136/bjo.44.4.193.

59. Perrini P, Cardia A, Fraser K, Lanzino G. A microsurgical study of the anatomy and course of the ophthalmic artery and its possibly dangerous anastomoses. J Neurosurg. 2007;106(1):142–50. https://doi.org/10.3171/jns.2007.106.1.142.

60. Yoon J, Pather N. The orbit: A re-appraisal of the surgical landmarks of the medial and lateral walls. Clin Anat. 2016;29(8):998–1010. https://doi.org/10.1002/ca.22787.

61. Santo Neto H, Penteado CV, de Carvalho VC. Presence of a groove in the lateral wall of the human orbit. J Anat. 1984;138(4):631–3.

62. Ozer MA, Celik S, Govsa F. A morphometric study of the inferior orbital fissure using three-dimensional anatomical landmarks: application to orbital surgery. Clin Anat. 2009;22(6):649–54. https://doi.org/10.1002/ca.20829.

63. De Battista JC, Zimmer LA, Theodosopoulos PV, Froelich SC, Keller JT. Anatomy of the inferior orbital fissure: implications for endoscopic cranial base surgery. J Neurol Surg B Skull Base. 2012;73(2):132–8. https://doi.org/10.1055/s-0032-1301398.

64. Lin BJ, Ju DT, Hsu TH, et al. Endoscopic transorbital approach to anterolateral skull base through inferior orbital fissure: a cadaveric study. Acta Neurochir. 2019;161(9):1919–29. https://doi.org/10.1007/s00701-019-03993-3.

65. Cornelius CP, Probst F, Metzger MC, Gooris PJJ. Anatomy of the orbits: skeletal features and some notes on the periorbital lining. Atlas Oral Maxillofac Surg Clin North Am. 2021;29(1):1–18. https://doi.org/10.1016/j.cxom.2020.10.001.

66. Cornelius CP, Mayer P, Ehrenfeld M, Metzger MC. The orbits-anatomical features in view of innovative surgical methods. Facial Plast Surg. 2014;30(5):487–508. https://doi.org/10.1055/s-0034-1394303.

67. Goldberg RA, Kim AJ, Kerivan KM. The lacrimal keyhole, orbital door jamb, and basin of the inferior orbital fissure. Three areas of deep bone in the lateral orbit. Arch Ophthalmol. 1998;116(1):1618–24. https://doi.org/10.1001/archopht.116.12.1618.

68. Martin-Duverneuil N, Sarrazin JL, Gayet-Delacroix M, Marsot-Dupuch K, Plantet MM. Le foramen rond. Anatomie et exploration radiologiques. Pathologie [The foramen rotundum Anatomy and radiological explorations Pathology]. J Neuroradiol. 2000;27(1):2–14.

69. Ginsberg LE, Pruett SW, Chen MY, Elster AD. Skull-base foramina of the middle cranial fossa: reassessment of normal variation with high-resolution CT. AJNR Am J Neuroradiol. 1994;15(2):283–91.

70. Kim HS, Kim DI, Chung IH. High-resolution CT of the pterygopalatine fossa and its communications. Neuroradiology. 1996;38(Suppl 1):S120–6. https://doi.org/10.1007/BF02278138.

71. Sepahdari AR, Mong S. Skull base CT: normative values for size and symmetry of the facial nerve canal, foramen ovale, pterygoid canal, and foramen rotundum. Surg Radiol Anat. 2013;35(1):19–24. https://doi.org/10.1007/s00276-012-1001-4.

72. Mohebbi A, Rajaeih S, Safdarian M, Omidian P. The sphenoid sinus, foramen rotundum and vidian canal: a radiological study of anatomical relationships. Braz J Otorhinolaryngol. 2017;83(4):381–7. https://doi.org/10.1016/j.bjorl.2016.04.013.

73. Bertelli E, Regoli M. Branching of the foramen rotundum. A rare variation of the sphenoid. Ital J Anat Embryol. 2014;119(2):148–53.

74. Bisaria KK, Kumar N, Jaiswal AK, Sharma PK, Mittal M, Bisaria SD. An accessory foramen deep in the infraorbital fissure. J Anat. 1996a;189(Pt 2):461–2.

75. Rusu MC, Didilescu AC, Jianu AM, Păduraru D. 3D CBCT anatomy of the pterygopalatine fossa. Surg Radiol Anat. 2013;35(2):143–59. https://doi.org/10.1007/s00276-012-1009-9.

76. Derinkuyu BE, Boyunaga O, Oztunali C, Alimli AG, Ucar M. Pterygopalatine fossa: not a mystery! Can Assoc Radiol J. 2017;68(2):122–30. https://doi.org/10.1016/j.carj.2016.08.001.

77. Puche-Torres M, Blasco-Serra A, Campos-Pelaez A, Valverde-Navarro AA. Radiological anatomy assessment of the fissura pterygomaxillaris for a surgical approach to ganglion pterygopalatinum. J Anat. 2017;231(6):961–9. https://doi.org/10.1111/joa.12690.

78. Omami G, Hewaidi G, Mathew R. The neglected anatomical and clinical aspects of pterygoid canal: CT scan study. Surg Radiol Anat. 2011;33(8):697–702. https://doi.org/10.1007/s00276-011-0808-8.

79. Gibelli D, Cellina M, Gibelli S, et al. Relation between volume of sphenoid sinuses and protrusion of Vidian nerve: possible applications to Vidian neurectomy. Surg Radiol Anat. 2020;42(5):583–7. https://doi.org/10.1007/s00276-019-02408-3.

80. Bryant L, Goodmurphy CW, Han JK. Endoscopic and three-dimensional radiographic imaging of the pterygopalatine and infratemporal fossae: improving surgical landmarks. Ann Otol Rhinol Laryngol. 2014;123(2):111–6. https://doi.org/10.1177/0003489414523707.

81. Inal M, Muluk NB, Arikan OK, Şahin S. Is There a relationship between optic canal, foramen rotundum, and Vidian canal? J Craniofac Surg. 2015;26(4):1382–8. https://doi.org/10.1097/scs.0000000000001597.

82. Karci B, Midilli R, Erdogan U, Turhal G, Gode S. Endoscopic endonasal approach to the vidian nerve and its relation to the surrounding structures: an anatomic cadaver study. Eur Arch Otorhinolaryngol. 2018;275(10):2473–9. https://doi.org/10.1007/s00405-018-5085-2.

83. Kasemsiri P, Solares CA, Carrau RL, et al. Endoscopic endonasal transpterygoid approaches: anatomical landmarks for planning the surgical corridor. Laryngoscope. 2013;123(4):811–5. https://doi.org/10.1002/lary.23697.

84. Kurt MH, Bozkurt P, Bilecenoglu B, Kolsuz ME, Orhan K. Morphometric analysis of vidian canal and its relations with surrounding anatomic structures by using cone-beam computed tomography. Folia Morphol (Warsz). 2020;79(2):366–73. https://doi.org/10.5603/FM.a2019.0094.

85. Papavasileiou G, Hajiioannou J, Kapsalaki E, Bizakis I, Fezoulidis I, Vassiou K. Vidian canal and sphenoid sinus: an MDCT and cadaveric study of useful landmarks in skull base surgery. Surg Radiol Anat. 2020;42(5):589–601. https://doi.org/10.1007/s00276-019-02414-5.

86. Vescan AD, Snyderman CH, Carrau RL, et al. Vidian canal: analysis and relationship to the internal carotid artery. Laryngoscope. 2007;117(8):1338–42. https://doi.org/10.1097/MLG.0b013e31806146cd.

87. Vuksanovic-Bozaric A, Vukcevic B, Abramovic M, Vukcevic N, Popovic N, Radunovic M. The pterygopalatine fossa: morphometric CT study with clinical implications. Surg Radiol Anat. 2019;41(2):161–8. https://doi.org/10.1007/s00276-018-2136-8.

88. Felding UA, Karnov K, Clemmensen A, et al. An applied anatomical study of the ethmoidal arteries: computed tomographic and direct measurements in human cadavers. J Craniofac Surg. 2018;29(1):212–6. https://doi.org/10.1097/scs.0000000000004157.

89. Piagkou M, Skotsimara G, Dalaka A, et al. Bony landmarks of the medial orbital wall: an anatomical study of ethmoidal foramina. Clin Anat. 2014;27(4):570–7. https://doi.org/10.1002/ca.22303.

90. Regoli M, Ogut E, Bertelli E. An osteologic study of human ethmoidal foramina with special reference to their classification and symmetry. Ital J Anat Embryol. 2016;121(1):66–76.

91. Takahashi Y, Kakizaki H, Nakano T. Accessory ethmoidal foramina: an anatomical study. Ophthalmic Plast Reconstr Surg. 2011a;27(2):125–7. https://doi.org/10.1097/IOP.0b013e318201c8fd.

92. Kazak Z, Celik S, Ozer MA, Govsa F. Three-dimensional evaluation of the danger zone of ethmoidal foramina on the frontoethmoidal suture line on the medial orbital wall. Surg Radiol Anat. 2015;37(6):935–40. https://doi.org/10.1007/s00276-015-1429-4.

93. Takahashi Y, Kakizaki H, Nakano T, Asamoto K, Ichinose A, Iwaki M. An anatomical study of the positional relationship between the ethmoidal foramina and the frontoethmoidal suture. Ophthal-

mic. Plast Reconstr Surg. 2011;27(6):457–9. https://doi.org/10.1097/IOP.0b013e318222eb82.

94. Rontal E, Rontal M, Guilford FT. Surgical anatomy of the orbit. Ann Otol Rhinol Laryngol. 1979;88(3Pt1):382–6. https://doi.org/10.1177/000348947908800315.

95. Abed SF, Shams PN, Shen S, Adds PJ, Uddin JM. Morphometric and geometric anatomy of the Caucasian orbital floor. Orbit. 2011;30(5):214–20. https://doi.org/10.3109/01676830.2010.539768.

96. Shin KJ, Gil YC, Lee SH, Song WC, Koh KS, Shin HJ. Positional relationship of ethmoidal foramens with reference to the nasion and its significance in orbital surgery. J Craniofac Surg. 2016;27(7):1854–7. https://doi.org/10.1097/scs.0000000000002911.

97. Vadgaonkar R, Rai R, Prabhu LV, Rai AR, Tonse M, Vani PC. Morphometric study of the medial orbital wall emphasizing the ethmoidal foramina. Surg Radiol Anat. 2015;37(7):809–13. https://doi.org/10.1007/s00276-014-1410-7.

98. Morales-Avalos R, Santos-Martínez AG, Ávalos-Fernández CG, et al. Clinical and surgical implications regarding morphometric variations of the medial wall of the orbit in relation to age and gender. Eur Arch Otorhinolaryngol 2016;273(9):2785-2793. https://doi.org/10.1007/s00405-015-3862-8.

99. Mueller SK, Bleier BS. Osteologic analysis of ethnic differences in supernumerary ethmoidal foramina: implications for endoscopic sinus and orbit surgery. Int Forum Allergy Rhinol. 2018;8(8):655–8. https://doi.org/10.1002/alr.22059.

100. Gufler H, Preiß M, Koesling S. Visibility of sutures of the orbit and periorbital region using multidetector computed tomography. Korean J Radiol. 2014;15(6):802–9. https://doi.org/10.3348/kjr.2014.15.6.802.

101. Cankal F, Apaydin N, Acar HI, et al. Evaluation of the anterior and posterior ethmoidal canal by computed tomography. Clin Radiol. 2004;59(11):1034–40. https://doi.org/10.1016/j.crad.2004.04.016.

102. Nam Y, Bahk S, Eo S. Anatomical study of the infraorbital nerve and surrounding structures for the surgery of orbital floor fractures. J Craniofac Surg. 2017;28(4):1099–104. https://doi.org/10.1097/SCS.0000000000003416.

103. Nguyen DC, Farber SJ, Um GT, Skolnick GB, Woo AS, Patel KB. Anatomical study of the intraosseous pathway of the infraorbital nerve. J Craniofac Surg. 2016;27(4):1094–7. https://doi.org/10.1097/SCS.0000000000002619.

104. Bahşi I, Orhan M, Kervancıoğlu P, Yalçın ED. Morphometric evaluation and surgical implications of the infraorbital groove, canal and foramen on cone-beam computed tomography and a review of literature. Folia Morphol (Warsz). 2019;78(2):331–43. https://doi.org/10.5603/FM.a2018.0084.

105. Chrcanovic BR, Abreu MH, Custódio AL. A morphometric analysis of supraorbital and infraorbital foramina relative to surgical landmarks. Surg Radiol

Anat. 2011;33(4):329–35. https://doi.org/10.1007/s00276-010-0698-1.

106. Fontolliet M, Bornstein MM, von Arx T. Characteristics and dimensions of the infraorbital canal: a radiographic analysis using cone beam computed tomography (CBCT). Surg Radiol Anat. 2019;41(2):169–79. https://doi.org/10.1007/s00276-018-2108-z.

107. Hwang K, Lee SJ, Kim SY, Hwang SW. Frequency of existence, numbers, and location of the accessory infraorbital foramen. J Craniofac Surg. 2015;26(1):274–6. https://doi.org/10.1097/SCS.0000000000001375.

108. Hwang SH, Kim SW, Park CS, Kim SW, Cho JH, Kang JM. Morphometric analysis of the infraorbital groove, canal, and foramen on three-dimensional reconstruction of computed tomography scans. Surg Radiol Anat. 2013;35(7):565–71. https://doi.org/10.1007/s00276-013-1077-5.

109. Martins-Júnior PA, Rodrigues CP, De Maria ML, Nogueira LM, Silva JH, Silva MR. Analysis of anatomical characteristics and morphometric aspects of infraorbital and accessory infraorbital foramina. J Craniofac Surg. 2017;28(2):528–33. https://doi.org/10.1097/SCS.0000000000003235.

110. Nanayakkara D, Peiris R, Mannapperuma N, Vadysinghe A. Morphometric analysis of the infraorbital foramen: the clinical relevance. Anat Res Int. 2016;2016:7917343. https://doi.org/10.1155/2016/7917343.

111. Rahman M, Richter EO, Osawa S, Rhoton AL Jr. Anatomic study of the infraorbital foramen for radiofrequency neurotomy of the infraorbital nerve. Neurosurgery. 2009;64(5 Suppl 2):423–7.; discussion 427–428. https://doi.org/10.1227/01.NEU.0000336327.10368.79.

112. Rusu MC, Săndulescu M, Cârstocea L. False and true accessory infraorbital foramina, and the infraorbital lamina cribriformis. Morphologie. 2020;104(344):51–8. https://doi.org/10.1016/j.morpho.2019.12.003.

113. Tezer M, Öztürk A, Gayretli Ö, Kale A, Balcioğlu H, Şahinoğlu K. Morphometric analysis of the infraorbital foramen and its localization relative to surgical landmarks. Minerva Stomatol. 2014;63(10):333–40.

114. Xu H, Guo Y, Lv D, et al. Morphological structure of the infraorbital canal using three-dimensional reconstruction. J Craniofac Surg. 2012;23(4):1166–8. https://doi.org/10.1097/SCS.0b013e31824dfcfd.

115. Beer GM, Putz R, Mager K, Schumacher M, Keil W. Variations of the frontal exit of the supraorbital nerve: an anatomic study. Plast Reconstr Surg. 1998;102(2):334–41. https://doi.org/10.1097/00006534-199808000-00006.

116. Barker L, Naveed H, Adds PJ, Uddin JM. Supraorbital notch and foramen: positional variation and relevance to direct brow lift. Ophthalmic Plast Reconstr Surg. 2013;29(1):67–70. https://doi.org/10.1097/IOP.0b013e318279fe41.

117. Webster RC, Gaunt JM, Hamdan US, Fuleihan NS, Giandello PR, Smith RC. Supraorbital and supratrochlear notches and foramina: anatomical variations and surgical relevance. Laryngoscope. 1986;96(3):311–5. https://doi.org/10.1288/00005537-198603000-00014.

118. Lim NK, Kim YH, Kang DH. Precision analysis of supraorbital transcranial exits using three dimensional multidetector computed tomography. J Craniofac Surg. 2019;30(6):1894–7. https://doi.org/10.1097/scs.0000000000005557.

119. Nanayakkara D, Manawaratne R, Sampath H, Vadysinghe A, Peiris R. Supraorbital nerve exits: positional variations and localization relative to surgical landmarks. Anat Cell Biol. 2018;51(1):19–24. https://doi.org/10.5115/acb.2018.51.1.19.

120. Tsutsumi S, Ono H, Ishii H, Yasumoto Y. Visualization of the supraorbital notch/foramen using magnetic resonance imaging. J Clin Neurosci. 2019;62:212–5. https://doi.org/10.1016/j.jocn.2019.01.005.

121. Fallucco M, Janis JE, Hagan RR. The anatomical morphology of the supraorbital notch: clinical relevance to the surgical treatment of migraine headaches. Plast Reconstr Surg. 2012;130(6):1227–33. https://doi.org/10.1097/PRS.0b013e31826d9c8d.

122. Shimizu S, Osawa S, Utsuki S, Oka H, Fujii K. Course of the bony canal associated with high-positioned supraorbital foramina: an anatomic study to facilitate safe mobilization of the supraorbital nerve. Minim Invasive Neurosurg. 2008;51(2):119–23. https://doi.org/10.1055/s-2008-1042434.

123. Saylam C, Ozer MA, Ozek C, Gurler T. Anatomical variations of the frontal and supraorbital transcranial passages. J Craniofac Surg. 2003;14(1):10–2. https://doi.org/10.1097/00001665-200301000-00003.

124. Zhao Y, Chundury RV, Blandford AD, Perry JD. Anatomical description of zygomatic foramina in African American skulls. Ophthalmic Plast Reconstr Surg. 2018;34(2):168–71. https://doi.org/10.1097/iop.0000000000000905.

125. Ferro A, Basyuni S, Brassett C, Santhanam V. Study of anatomical variations of the zygomaticofacial foramen and calculation of reliable reference points for operation. Br J Oral Maxillofac Surg. 2017;55(10):1035–41. https://doi.org/10.1016/j.bjoms.2017.10.016.

126. Iwanaga J, Badaloni F, Watanabe K, Yamaki KI, Oskouian RJ, Tubbs RS. Anatomical study of the zygomaticofacial foramen and its related canal. J Craniofac Surg. 2018;29(5):1363–5. https://doi.org/10.1097/scs.0000000000004457.

127. Wartmann CT, Loukas M, Tubbs RS. Zygomaticofacial, zygomaticoorbital, and zygomaticotemporal foramina. Clin Anat. 2009;22(5):637–8. https://doi.org/10.1002/ca.20822.

128. Gupta T, Gupta SK. The ZMF: Is it a reliable intraoperative guide for the IOF? Clin Anat. 2009;22(4):451–5. https://doi.org/10.1002/ca.20783.

129. Martins C, Li X, Rhoton AL, Jr. Role of the zygomaticofacial foramen in the orbitozygomatic craniotomy: anatomic report. Neurosurgery 2003;53(1):168–72; discussion 172–163. https://doi.org/10.1227/01.neu.0000068841.17293.bb.

130. Rhoton AL. The orbit. Neurosurg Rhoton's Anat. 2003;53:331–62; Rhoton cranial anatomy and surgical approaches, Part 2, Chapter 7

131. Patel P, Belinsky I, Howard D, Palu RN. Location of the zygomatico-orbital foramen on the inferolateral orbital wall: clinical implications. Orbit. 2013;32(5):275–7. https://doi.org/10.3109/01676830.2013.799703.

132. Shams PN, Abed SF, Shen S, Adds PJ, Uddin JM. A cadaveric study of the morphometric relationships and bony composition of the caucasian nasolacrimal fossa. Orbit. 2012;31(3):159–61. https://doi.org/10.3109/01676830.2011.648809.

133. Chastain JB, Cooper MH, Sindwani R. The maxillary line: anatomic characterization and clinical utility of an important surgical landmark. Laryngoscope. 2005;115(6):990–2. https://doi.org/10.1097/01.Mlg.0000163764.01776.10.

134. Fayet B, Racy E, Assouline M, Zerbib M. Surgical anatomy of the lacrimal fossa a prospective computed tomodensitometry scan analysis. Ophthalmology. 2005;112(6):1119–28. https://doi.org/10.1016/j.ophtha.2005.01.012.

135. Ipek E, Esin K, Amac K, Mustafa G, Candan A. Morphological and morphometric evaluation of lacrimal groove. Anat Sci Int. 2007;82(4):207–10. https://doi.org/10.1111/j.1447-073X.2007.00185.x.

136. Ye H, Feng Y, Yin S. Anatomy study of the lacrimal bone in dacryocystorhinostomy. Lin Chung Er Bi Yan Hou Tou Jing Wai Ke Za Zhi. 2007;21(17):774–6.

137. Bisaria KK, Saxena RC, Bisaria SD, Lakhtakia PK, Agarwal AK, Premsagar IC. The lacrimal fossa in Indians. J Anat. 1989;166:265–8.

138. Ali MJ, Schicht M, Paulsen F. Morphology and morphometry of lacrimal drainage system in relation to bony landmarks in Caucasian adults: a cadaveric study. Int Ophthalmol. 2018;38(6):2463–9. https://doi.org/10.1007/s10792-017-0753-6.

139. Takahashi Y, Nakamura Y, Nakano T, et al. The narrowest part of the bony nasolacrimal canal: an anatomical study. Ophthalmic Plast Reconstr Surg. 2013;29(4):318–22. https://doi.org/10.1097/IOP.0b013e31828de0b0.

140. Okumuş Ö. Investigation of the morphometric features of bony nasolacrimal canal: a cone-beam computed tomography study. Folia Morphol (Warsz). 2020;79(3):588–93. https://doi.org/10.5603/FM.a2019.0099.

141. Ramey NA, Hoang JK, Richard MJ. Multidetector CT of nasolacrimal canal morphology: normal variation by age, gender, and race. Ophthalmic Plast Reconstr Surg. 2013;29(6):475–80. https://doi.org/10.1097/IOP.0b013e3182a230b0.

142. Lee S, Lee UY, Yang SW, et al. 3D morphological classification of the nasolacrimal duct: Anatomical study for planning treatment of tear drainage obstruction. Clin Anat. 2021;34(4):624–33. https://doi.org/10.1002/ca.23678. [Publ online ahead of print]

143. Anon JB, Rontal M, Zinreich SJ. Anatomy of the paranasal sinuses. Stuttgart New York: Georg Thieme Verlag; 1996.

144. Anusha B, Baharudin A, Philip R, Harvinder S, Shaffie BM. Anatomical variations of the sphenoid sinus and its adjacent structures: a review of existing literature. Surg Radiol Anat. 2014;36(5):419–27. https://doi.org/10.1007/s00276-013-1214-1.

145. Wilden A, Feiser J, Wohler A, Isik Z, Bendella H, Angelov DN. Anatomy of the human orbital muscle (OM): Features of its detailed topography, syntopy and morphology. Ann Anat 2017;211:39–45. https://doi.org/10.1016/j.aanat.2017.01.008.

146. Clarós P, Sobolewska AZ, Cardesa A, Lopez-Fortuny M, Claros A. Silent sinus syndrome: combined sinus surgery and orbital reconstruction—report of 15 cases. Acta Otolaryngol. 2019;139(1):64–9. https://doi.org/10.1080/00016489.2018.1542161.

147. Canzi G, Morganti V, Novelli G, Bozzetti A, Sozzi D. Posttraumatic delayed enophthalmos: analogies with silent sinus syndrome? case report and literature review. Craniomaxillofac Trauma Reconstr. 2015;8(3):251–6. https://doi.org/10.1055/s-0034-1399799.

148. Brown SJ, Hardy TG, McNab AA. "Silent sinus syndrome" following orbital trauma: a case series and review of the literature. Ophthalmic Plast Reconstr Surg. 2017;33(3):209–12. https://doi.org/10.1097/iop.0000000000000713.

149. Goldberg RA, Shorr N, Cohen MS. The medial orbital strut in the prevention of postdecompression dystopia in dysthyroid ophthalmopathy. Ophthalmic Plast Reconstr Surg. 1992;8(1):32–4. https://doi.org/10.1097/00002341-199203000-00005.

150. Bleier BS, Lefebvre DR, Freitag SK. Endoscopic orbital floor decompression with preservation of the inferomedial strut. Int Forum Allergy Rhinol. 2014;4(1):82–4. https://doi.org/10.1002/alr.21231.

151. Goldberg RA, Relan A, Hoenig J. Relationship of the eye to the bony orbit, with clinical correlations. Aust N Z J Ophthalmol. 1999;27(6):398–403. https://doi.org/10.1046/j.1440-1606.1999.00243.x.

152. Hammer B. Orbital fractures. Diagnosis, operative treatment, secondary corrections. Seattle Toronto Bern: Hofgrefe & Huber Publishers; 1995.

153. Borumandi F, Hammer B, Noser H, Kamer L. Classification of orbital morphology for decompression surgery in Graves' orbitopathy: two-dimensional versus three-dimensional orbital parameters. Br J Ophthalmol. 2013;97(5):659–62. https://doi.org/10.1136/bjophthalmol-2012-302825.

154. Lieger O, Schaub M, Taghizadeh E, Buchler P. How symmetrical are bony orbits in humans? J Oral Maxillofac Surg. 2019;77(1):118–25. https://doi.org/10.1016/j.joms.2018.08.018.

155. Noser H, Hammer B, Kamer L. A method for assessing 3D shape variations of fuzzy regions and its application on human bony orbits. J Digit Imaging. 2010;23(4):422–9. https://doi.org/10.1007/s10278-009-9187-7.

156. Graillon N, Boulze C, Adalian P, Loundou A, Guyot L. Use of 3D orbital reconstruction in the assessment of orbital sexual dimorphism and its pathological consequences. J Stomatol Oral Maxillofac Surg. 2017;118(1):29–34. https://doi.org/10.1016/j.jormas.2016.10.002.

157. Kokemueller H, Zizelmann C, Tavassol F, Paling T, Gellrich NC. A comprehensive approach to objective quantification of orbital dimensions. J Oral Maxillofac Surg. 2008;66(2):401–7. https://doi.org/10.1016/j.joms.2006.05.062.

158. Ousterhout DK. Curve analysis of the aging orbital aperture. Plast Reconstr Surg. 2003;111(2):953. https://doi.org/10.1097/00006534-200302000-00093.

159. Ugradar S, Lambros V. Orbital volume increases with age: A computed tomography-based volumetric study. Ann Plast Surg. 2019;83(6):693–6. https://doi.org/10.1097/sap.0000000000001929.

160. Prevost A, Muller S, Lauwers F, Heuzé Y. Quantification of global orbital shape variation. Clin Anat 2023. https://doi.org/10.1002/ca.24007.

161. Acer N, Sahin B, Ucar T, Usanmaz M. Unbiased estimation of the eyeball volume using the Cavalieri principle on computed tomography images. J Craniofac Surg. 2009;20(1):233–7. https://doi.org/10.1097/SCS.0b013e3181843518.

162. Eckstein LA, Shadpour JM, Menghani R, Goldberg RA. The relationship of the globe to the orbital rim. Arch Facial Plast Surg. 2011;13(1):51–6. https://doi.org/10.1001/archfacial.2010.102.

163. Willaert R, Shaheen E, Deferm J, Vermeersch H, Jacobs R, Mombaerts I. Three-dimensional characterisation of the globe position in the orbit. Graefes Arch Clin Exp Ophthalmol. 2020;258(7):1527–32. https://doi.org/10.1007/s00417-020-04631-w.

164. Lepich T, Dabek J, Jura-Szoltys E, Witkowska M, Piechota M, Bajor G. Orbital opening shape and its alphanumerical classification. Adv Clin Exp Med. 2015;24(6):943–50. https://doi.org/10.17219/acem/31458.

165. Sarkar S, Baliga M, Prince J, Ongole R, Natarajan S. Morphometric and volumetric measurements of orbit with cone-beam computed tomography. J Oral Maxillofac Surg. 2021;79(3):652–64. https://doi.org/10.1016/j.joms.2020.10.026.

166. Karakas P, Bozkir MG, Oguz O. Morphometric measurements from various reference points in the orbit of male Caucasians. Surg Radiol Anat. 2003;24(6):358–62. https://doi.org/10.1007/s00276-002-0071-0.

167. McQueen CT, DiRuggiero DC, Campbell JP, Shockley WW. Orbital osteology: a study of the surgical landmarks. Laryngoscope. 1995;105(8 Pt 1):783–8. https://doi.org/10.1288/00005537-199508000-00003.

168. Danko I, Haug RH. An experimental investigation of the safe distance for internal orbital dissection. J Oral Maxillofac Surg. 1998;56(6):749–52. https://doi.org/10.1016/s0278-2391(98)90812-6.

169. Kang HS, Han JJ, Oh HK, Kook MS, Jung S, Park HJ. Anatomical studies of the orbital cavity using three-dimensional computed tomography. J Craniofac Surg. 2016;27(6):1583–8. https://doi.org/10.1097/scs.0000000000002811.

170. Singh J, Rahman RA, Rajion ZA, Abdullah J, Mohamad I. Orbital Morphometry: A Computed Tomography Analysis. J Craniofac Surg 2017;28(1):e64-e70). https://doi.org/10.1097/scs.0000000000003218.

171. Bontzos G, Mazonakis M, Papadaki E, et al. Orbital volume measurements from magnetic resonance images using the techniques of manual planimetry and stereology. Natl J Maxillofac Surg. 2020;11(1):20–7. https://doi.org/10.4103/njms.NJMS_9_20.

172. Detorakis ET, Drakonaki E, Papadaki E, Pallikaris IG, Tsilimbaris MK. Effective orbital volume and eyeball position: an MRI study. Orbit. 2010;29(5):244–9. https://doi.org/10.3109/01676831003664319.

173. Metzger MC, Schön R, Tetzlaf R, et al. Topographical CT- data analysis of the human orbital floor. Int J Oral Maxillofac Surg. 2007;36(1):45–53. https://doi.org/10.1016/j.ijom.2006.07.013.

174. Schön R, Metzger MC, Zizelmann C, Weyer N, Schmelzeisen R. Individually preformed titanium mesh implants for a true-to-original repair of orbital fractures. Int J Oral Maxillofac Surg. 2006;35(11):990–5. https://doi.org/10.1016/j.ijom.2006.06.018.

175. Weadock WJ, Heisel CJ, Kahana A, Kim J. Use of 3D printed models to create molds for shaping implants for surgical repair of orbital fractures. Acad Radiol. 2020;27(4):536–42. https://doi.org/10.1016/j.acra.2019.06.023.

176. Metzger MC, Bittermann G, Dannenberg L, et al. Design and development of a virtual anatomic atlas of the human skull for automatic segmentation in computer-assisted surgery, preoperative planning, and navigation. Int J Comput Assist Radiol Surg. 2013;8(5):691–702. https://doi.org/10.1007/s11548-013-0818-6.

177. Kamer L, Noser H, Schramm A, Hammer B. Orbital form analysis: problems with design and positioning of precontoured orbital implants: a serial study using post-processed clinical CT data in unaffected orbits. Int J Oral Maxillofac Surg. 2010;39(7):666–72. https://doi.org/10.1016/j.ijom.2010.03.005.

178. Bittermann G, Metzger MC, Schlager S, et al. Orbital reconstruction: prefabricated implants, data transfer, and revision surgery. Facial Plast Surg. 2014;30(5):554–60. https://doi.org/10.1055/s-0034-1395211.

179. Doerfler HM, Huempfner-Hierl H, Kruber D, Schulze P, Hierl T. Template-based orbital wall fracture treatment using statistical shape analysis. J Oral Maxillofac Surg. 2017;75(7):1475.e1471–8. https://doi.org/10.1016/j.joms.2017.03.048.

180. Gass M, Füssinger MA, Metzger MC, et al. Virtual reconstruction of orbital floor defects using a statistical shape model. J Anat. 2022;240(2):323–9. [June 2021 – e Publication ahead of print]

181. Huempfner-Hierl H, Doerfler HM, Kruber D, Hierl T. Morphologic comparison of preformed orbital meshes. J Oral Maxillofac Surg. 2015;73(6):1119–23. https://doi.org/10.1016/j.joms.2015.01.031.

182. Jansen J, Dubois L, Schreurs R, et al. Should virtual mirroring be used in the preoperative planning of an orbital reconstruction? J Oral Maxillofac Surg. 2018;76(2):380–7. https://doi.org/10.1016/j.joms.2017.09.018.

183. Kim MJ, Lee MJ, Jeong WS, Hong H, Choi JW. Three-dimensional computer modeling of standard orbital mean shape in Asians. J Plast Reconstr Aesthet Surg. 2020;73(3):548–55. https://doi.org/10.1016/j.bjps.2019.09.027.

184. Schreurs R, Klop C, Maal TJJ. Advanced diagnostics and three-dimensional virtual surgical planning in orbital reconstruction. Atlas Oral Maxillofac Surg Clin North Am. 2021;29(1):79–96. https://doi.org/10.1016/j.cxom.2020.11.003.

185. Bite U, Jackson IT, Forbes GS, Gehring DG. Orbital volume measurements in enophthalmos using three-dimensional CT imaging. Plast Reconstr Surg. 1985;75(4):502–8. https://doi.org/10.1097/00006534-198504000-00009.

186. Jansen J, Schreurs R, Dubois L, Maal TJ, Gooris PJ, Becking AG. Orbital volume analysis: validation of a semi-automatic software segmentation method. Int J Comput Assist Radiol Surg. 2016;11(1):11–8. https://doi.org/10.1007/s11548-015-1254-6.

187. Mottini M, Wolf CA, Seyed Jafari SM, Katsoulis K, Schaller B. Stereographic measurement of orbital volume, a digital reproducible evaluation method. Br J Ophthalmol. 2017;101(10):1431–5. https://doi.org/10.1136/bjophthalmol-2016-309998.

188. Osaki TH, de Castro DK, Yabumoto C, et al. Comparison of methodologies in volumetric orbitometry. Ophthalmic Plast Reconstr Surg. 2013;29(6):431–6. https://doi.org/10.1097/IOP.0b013e31829d028a.

189. Safi AF, Richter MT, Rothamel D, et al. Influence of the volume of soft tissue herniation on clinical symptoms of patients with orbital floor fractures. J Craniomaxillofac Surg. 2016;44(12):1929–34. https://doi.org/10.1016/j.jcms.2016.09.004.

190. Strong EB, Fuller SC, Chahal HS. Computer-aided analysis of orbital volume: a novel technique. Ophthalmic Plast Reconstr Surg. 2013;29(1):1–5. https://doi.org/10.1097/IOP.0b013e31826a24ea.

191. Wagner ME, Gellrich NC, Friese KI, et al. Model-based segmentation in orbital volume measurement with cone beam computed tomography and evaluation against current concepts. Int J Comput Assist Radiol Surg. 2016;11(1):1–9. https://doi.org/10.1007/s11548-015-1228-8.

192. Nitek S, Wysocki J, Reymond J, Piasecki K. Correlations between selected parameters of the human skull and orbit. Med Sci Monit. 2009;15(12):Br370–7.

193. Pearce E, Bridge H. Is orbital volume associated with eyeball and visual cortex volume in humans? Ann Hum Biol. 2013;40(6):531–40. https://doi.org/10.3109/03014460.2013.815272.

194. Andrades P, Cuevas P, Hernandez R, Danilla S, Villalobos R. Characterization of the orbital volume in normal population. J Craniomaxillofac Surg. 2018;46(4):594–9. https://doi.org/10.1016/j.jcms.2018.02.003.

195. Oh SA, Aum JH, Kang DH, Gu JH. Change of the orbital volume ratio in pure blow-out fractures depending on fracture location. J Craniofac Surg. 2013;24(4):1083–7. https://doi.org/10.1097/SCS.0b013e31828b6c2d.

196. Forbes G, Gehring DG, Gorman CA, Brennan MD, Jackson IT. Volume measurements of normal orbital structures by computed tomographic analysis. AJR Am J Roentgenol. 1985;145(1):149–54. https://doi.org/10.2214/ajr.145.1.149.

197. Tandon R, Aljadeff L, Ji S, Finn RA. Anatomic variability of the human orbit. J Oral Maxillofac Surg. 2020;78(5):782–96. https://doi.org/10.1016/j.joms.2019.11.032.

198. Nagasao T, Hikosaka M, Morotomi T, Nagasao M, Ogawa K, Nakajima T. Analysis of the orbital floor morphology. J Craniomaxillofac Surg 2007;35(2):112–9. https://doi.org/10.1016/j.jcms.2006.12.002.

199. Kloss FR, Gassner R. Bone and aging: effects on the maxillofacial skeleton. Exp Gerontol. 2006;41(2):123–9. https://doi.org/10.1016/j.exger.2005.11.005.

200. Mendelson BC, Hartley W, Scott M, McNab A, Granzow JW. Age-related changes of the orbit and midcheek and the implications for facial rejuvenation. Aesthet Plast Surg. 2007;31(5):419–23. https://doi.org/10.1007/s00266-006-0120-x.

201. Pessa JE, Chen Y. Curve analysis of the aging orbital aperture. Plast Reconstr Surg. 2002;109(2):751–5; discussion 756–760. https://doi.org/10.1097/00006534-200202000-00051.

202. Gentry LR, Manor WF, Turski PA, Strother CM. High-resolution CT analysis of facial struts in trauma: 1. Normal anatomy. AJR Am J Roentgenol. 1983;140(3):523–32. https://doi.org/10.2214/ajr.140.3.523.

203. Cotofana S, Gavril DL, Frank K, Schenck TL, Pawlina W, Lachman N. Revisit,reform, and redesign: a novel dissection approach for demonstrating anatomy of the orbit for continuing professional development education. Anat Sci Educ 2021;14(4):505–12. https://doi.org/10.1002/ase.2006.

204. Davidson EH, Wang EW, Yu JY, et al. Total human eye allotransplantation: developing surgical protocols for donor and recipient procedures. Plast Reconstr Surg. 2016;138(6):1297–308. https://doi.org/10.1097/PRS.0000000000002821.

205. Dubois L, Jansen J, Schreurs R, et al. Predictability in orbital reconstruction: a human cadaver study. Part I: Endoscopic-assisted orbital reconstruction. J Craniomaxillofac Surg. 2015;43(10):2034–41. https://doi.org/10.1016/j.jcms.2015.07.019.

206. Dubois L, Schreurs R, Jansen J, et al. Predictability in orbital reconstruction: a human cadaver study. Part II: Navigation-assisted orbital reconstruction. J Craniomaxillofac Surg. 2015;43(10):2042–9. https://doi.org/10.1016/j.jcms.2015.07.020.

207. Dubois L, Essig H, Schreurs R, et al. Predictability in orbital reconstruction. a human cadaver study, part III: Implant-oriented navigation for optimized reconstruction. J Craniomaxillofac Surg. 2015;43(10):2050–6. https://doi.org/10.1016/j.jcms.2015.08.014.

208. Ferrari M, Schreiber A, Mattavelli D, et al. The inferolateral transorbital endoscopic approach: a preclinical anatomic study. World Neurosurg. 2016;90:403–13. https://doi.org/10.1016/j.wneu.2016.03.017.

209. Hsung TC, Lo J, Chong MM, Goto TK, Cheung LK. Orbit segmentation by surface reconstruction with automatic sliced vertex screening. IEEE Trans Biomed Eng. 2018;65(4):828–38. https://doi.org/10.1109/tbme.2017.2720184.

210. Fitzhugh A, Naveed H, Davagnanam I, Messiha A. Proposed three-dimensional model of the orbit and relevance to orbital fracture repair. Surg Radiol Anat. 2016;38(5):557–61. https://doi.org/10.1007/s00276-015-1561-1.

211. Standring S. A brief history of topographical anatomy. J Anat. 2016;229(1):32–62. https://doi.org/10.1111/joa.12473.

212. Tomaszewski KA, Henry BM, Pekala PA, Standring S, Tubbs RS. The new frontier of studying human anatomy: Introducing evidence-based anatomy. Clin Anat. 2018;31(1):4–5. https://doi.org/10.1002/ca.2294.

Suggested Reading

Bron AJ, Tripathi RC, Tripathi BJ. Wolff's anatomy of the eye and orbit. 8th ed. London: Chapman and Hall Medical; 1997.

Dauber W. Pocket atlas of human anatomy—founded by Heinz Feneis. 5th ed. Stuttgart New York: Georg Thieme Verlag; 2007.

Doxanas M, Anderson RL. Clinical orbital anatomy. Baltimore: Williams and Wilkins; 1984.

Dutton JJ, Waldrop TG. Atlas of clinical and surgical orbital anatomy. 2nd ed. Elsevier Saunders; 2011.

Ellis E III, Zide MF. Surgical approaches to the facial skeleton. 2nd ed. Philadelphia, PA: Lippincott Williams & Wilkins; 2006.

Holmes S, Blythe J. Orbit and orbital trauma. In: Brennan PA, Standring SM, Wiseman SM, editors. Gray's surgical anatomy. 1st ed. Amsterdam: Elsevier; 2020.

Janfaza P, Nadol JB, Galla RJ, Fabian RL, Montgomery WW. Surgical anatomy of the head and neck. Philadelphia, PA: Lippincott Williams & Wilkins; 2001.

Lang J. Clinical anatomy of the head, neurocranium, orbit, and craniocervial regions. Heidelberg/New York: Springer Verlag; 1983.

Logan B, Reynolds P, Rice S, Hutchings R. McMinn's color atlas of head and neck anatomy. 5th ed. Amsterdam: Elsevier; 2017.

Peris-Celda M, Martinez-Soriana F, Rhoton AL Jr. Rhoton's Atlas of head, neck, and brain: 2d and 3d images. New York, NY: Thieme Medical Publishers, Inc.; 2018.

Pernkopf E. Atlas of topographical and applied human anatomy. Volume 1, Head and neck. Philadelphia, PA: Urban & Schwarzenberg; 1980.

Rhoton AL. The middle cranial base and cavernous sinus. In: Dolenc VV, Rogers L (Eds.) Cavernous sinus—developments and future perspectives. p. 3–26, Wien Springer Verlag New York; 2009.

Rhoton AL, Natori Y. The orbit and sellar region: microsurgical anatomy and operative approaches. New York (NY): Thieme Medical Publishers; 1996.

Rhoton AL Jr, Hardy DG, Chambers SM. Microsurgical anatomy and dissection of the sphenoid bone, cavernous sinus and sellar region. Surg Neurol. 1979;12:63–104.

Rootman J. Orbital surgery- a conceptual approach. 2nd ed. Philadelphia, PA: Lippincott Williams & Wilkins; 2014.

Turvey TA, Golden BA. Orbital anatomy for the surgeon. Oral Maxillofac Surg Clin N Am. 2012;24(4):525–36. https://doi.org/10.1016/j.coms.2012.08.003.

Standring S. Gray's anatomy. 42nd ed. Philadelphia, PA: Elsevier; 2021.

Standring S. Gray's anatomy. 39th ed. Edinburgh: Churchill Livingstone; 2005.

Von Arx T, Lozanoff S. Clinical oral anatomy—a comprehensive review for dental practitioners and researchers. Switzerland: Springer International Publishing; 2017.

Von Lanz T, Wachsmuth W. Praktische Anatomie, Band 1, Teil 1, Kopf, Teil B. In: Lang J, Wachsmuth W, editors. Gehirn und Augenschädel. Hrsgb. New York: Springer Berlin Heidelberg; 1979.

Zide BM, Jelks G. Surgical anatomy of the orbit. New York: Raven Press; 1985.

Zide BM. Surgical anatomy around the orbit. The system of zones. Philadelphia, PA: Lippincott Williams and Wilkins; 2006.

Open Access This chapter is licensed under the terms of the Creative Commons Attribution 4.0 International License (http://creativecommons.org/licenses/by/4.0/), which permits use, sharing, adaptation, distribution and reproduction in any medium or format, as long as you give appropriate credit to the original author(s) and the source, provide a link to the Creative Commons license and indicate if changes were made.

The images or other third party material in this chapter are included in the chapter's Creative Commons license, unless indicated otherwise in a credit line to the material. If material is not included in the chapter's Creative Commons license and your intended use is not permitted by statutory regulation or exceeds the permitted use, you will need to obtain permission directly from the copyright holder.

Anatomy of the Orbit: Overall Aspects of the Peri- and Intra Orbital Soft Tissues

3

Peter J. J. Gooris and Carl-Peter Cornelius

Learning Objectives
- Appreciate anatomical aspects of soft tissue around the orbit including the eyelids and lacrimal system
- Distinguish the different locations of adipose tissue in and around the orbit
- Recall major anatomic topographic and functional details of the globe itself
- Understand the extraocular musculature system
- Apprehend the essentials of the neurovascular pathways involved in ophthalmic function

The original version of the chapter has been revised. A correction to this chapter can be found at https://doi.org/10.1007/978-3-031-40697-3_23

P. J. J. Gooris (✉)
Department of Oral and Maxillofacial Surgery, Amphia Hospital Breda, Breda, The Netherlands

Amsterdam University Medical Centers, Amsterdam, The Netherlands

Department of Oral and Maxillofacial Surgery, University of Washington, Seattle, WA, USA
e-mail: p.gooris@gmail.com

C.-P. Cornelius
Department of Oral and Maxillofacial Surgery, Facial Plastic Surgery, Ludwig Maximilians University, Munich, Bavaria, Germany
e-mail: peter.cornelius@med.uni-muenchen.de

Introduction

The eye can be considered as a most important organ. It is the window of the human perception system: open your eyes and record the visible information; it offers personal detail: for your eyes only!

As it is such an essential and delicate organ, the globe needs adequate protection which is facilitated by the eyelid curtain system and ample humidification through the lacrimal system.

To provide a maximum of capability for the gathering of visual information, a highly effective intrinsic muscular system is attached to the globe which allows motility in all directions. Adipose tissue of the orbit facilitates physical motility of eye-related structures. The role of neurovascular supply is essential for vision and functional movement.

In view of the special task of this precious organ, this chapter will elaborate on anatomical aspects directly related to the globe. Anatomical soft tissue aspects of the apex of the orbit will be discussed.

The Protective Curtain System of the Ocular Globe

The Eyelids

The eyelids resemble to soft tissue plications, that form a mobile shutter to cover the anterior surface of the eyeball from above and below. The

© The Author(s) 2023, corrected publication 2024
P. J. J. Gooris et al. (eds.), *Surgery in and around the Orbit*,
https://doi.org/10.1007/978-3-031-40697-3_3

eyelids, are referred to as part of the facial soft tissue envelope, nonetheless they are indispensable for the orbit having functions such as mechanical protection and maintenance of the globe by the lacrimal system with lubrication, cleansing, and drainage of the region.

The aperture between upper and lower eyelid margins is called the palpebral fissure, which measures approximately 8–10 mm, widest at the midpoint when open. The upper eyelid just covers the upper aspect of the cornea for approximately 2 mm. The horizontal length is approximately 30 mm. Medially and laterally the eyelids meet at an angle of approximately 60°.

When we follow the palpebral fissure sideways, the lateral canthal angle is positioned a little ±2 mm higher than the medial canthal angle.

Each eyelid can be divided into an external, anterior and internal, posterior lamella.

The outer or anterior lamella is coated with delicate skin on top of the orbicularis oculi muscle (OOM): myocutaneous lamella. In this anterior lamella, eyelashes are located.

The inner or posterior lamella is built up of the tarsal plates and covered by the palpebral conjunctiva, thus corresponding to a tarsoconjunctival lamella. This surface is in touch with the globe (Fig. 3.1).

The eyelid margin itself measures about 2 mm in thickness.

A line on the lid margins visible at the transition zone between the anterior and posterior lamellae is formed by the marginal projection of the pars ciliaris of Riolan's muscle, called the Gray line (intermarginal sulcus) [1, 2]

The orifices of Meibomian glands open posterior to the Gray line and are positioned in the internal lamella.

Tarsal Plates

The superior and inferior tarsal cartilage plates give structural support to the eyelids. They are the central component within a fibrous framework of canthal tendons, fascial attachments, and suspension ligaments.

The tarsal plate itself consists of a dense fibrous connective tissue layer and supplies rigidity to the eyelid. The tarsi are ± 1–1.5 mm thick. Coursing medial and lateral, the tarsal plates' height decreases. At their endings, they finally pass into the medial and lateral canthal ligament, fibrous connective tissue bands, acting as suspensory structures for the canthi. Meibomian sebaceous glands which produce meibum, an oily substance

Fig. 3.1 Schematic sagittal view upper eyelid. (With permission from S. Steenen)

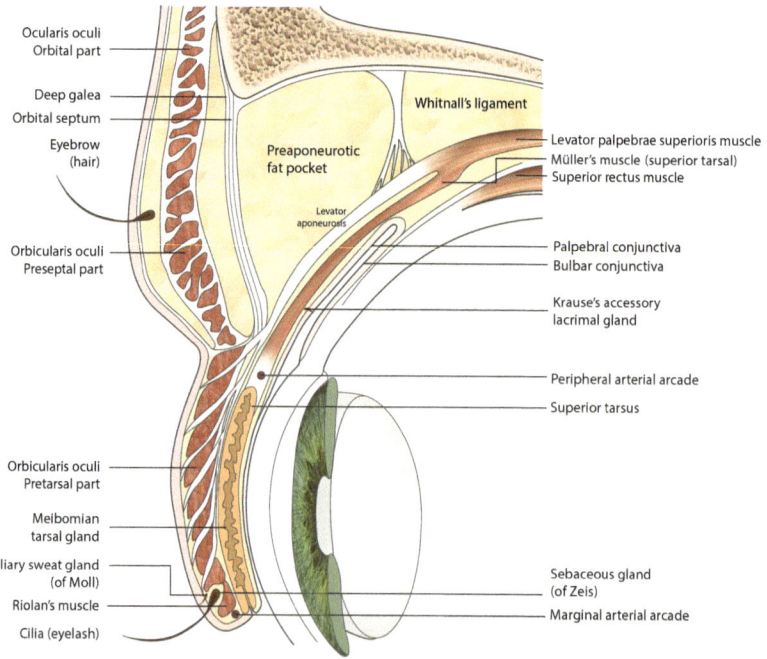

that prevents evaporation of the eye's tear film are present in the upper and lower tarsal plate.

The tarsi are supported medially by the medial rectus, capsulopalpebral fascia, and Horner's muscle.

Motor Innervation Eyelids

Motor innervation is dual: CN III, CN VII, and sympathetic nerve fibers.

CN III, a branch of its superior division is responsible for innervation of the main upper eyelid retractor, the levator palpebrae superior muscle. A branch of the inferior CN III division innervates the inferior rectus muscle which through the communication with the adjacent capsulopalpebral fascia is responsible for lower eyelid retraction during downward gaze.

CN VII supplies innervation to the orbicularis muscle (temporal and zygomatic divison CN VII—main eyelid protractor); the frontalis-, procerus- and corrugator supercilii musculature: brow depression and contribute—support to upper eyelid protraction.

Sympathetic fibers innervate the superior tarsal muscle, Müller's muscle, which contributes to upper eyelid retraction and the inferior tarsal muscle, supportive for lower eyelid retraction.

The sensory innervation of the upper eyelid is multiple; terminal branches of the ophthalmic division (CN V1) are involved (frontal nerve), the supraorbital nerve, the supratrochlear nerve, the infratrochlear nerve and the lacrimal nerve [3]. The innervation will be described in more detail in subsequent subchapters.

Clinical Implications
The Gray line formed by the projecting pars ciliaris of Riolan's muscle [2] separates the anterior from the posterior lamella; it indicates an avascular plane, which is a useful anatomic marker, for easy surgical access.

Surgical approaches to the orbit can have implications on the periorbital soft tissues.

Eyelid skin incisions can interfere with sensory innervation; approach to the orbital rim in case of orbital floor reconstruction, surgery through muscular planes should be carried out at different levels to minimize postoperative scar contractures. In case of a lateral canthotomy extension, too wide horizontal skin incisions can cause iatrogenic damage to the frontal branches of the CN VII.

Upper Eyelid

Musculature

Below the thin loosely attached dermis covering the eyelids, depending on the level we encounter the preseptal periocular striated OOM, the (aponeurosis) levator palpebrae superior and Müller's superior tarsal muscle.

At a level in approximation to the lid margin, we encounter the pretarsal OOM, the muscle of Riolan and the tarsal plate. The inner aspect is covered by palpebral conjunctiva (Fig. 3.1).

Jean Riolan [2] first described a set of muscle fibers encircling the meibomian glands collectively known as the muscle of Riolan. In 2002, Lipham et al. [1] subsequently showed through histologic analysis that the muscle of Riolan is a component of the orbicularis muscle, relating eyelid blink to meibum secretion. In fact, the muscle of Riolan represents the pretarsal orbicularis muscle.

The OOM is a widespread array of muscle fibers that lies beneath the skin, over the orbital septum and the tarsoligamentous sling [4–6]. It is an integral part of the superficial musculoaponeurotic system (SMAS). The OOM is separated from the overlying dermis by a modest fibroadipose tissue layer through which fibrous septa perforate merging with the underlying muscle layer (Fig. 3.2). It is a striated muscle that runs parallel to the eyelid margin and is innervated by

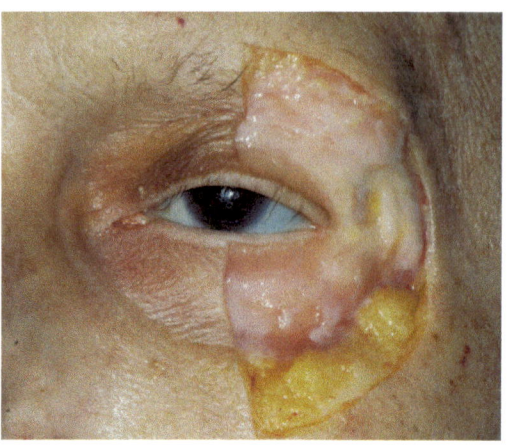

Fig. 3.2 Extent and division of orbicularis oculi muscle fibers over lateral half of the eyelids and orbit. Suborbicularis oculi fat (SOOF) in the subcutaneous plane positioned anterior of the orbicularis oculi muscle (inferolateral quadrant)—anatomic specimen

multiple (zygomatic and frontotemporal) branches of CN VII, a plexus running deep to the OOM, arriving from lateral.

The OOM is divided into three major concentric or ring-shaped partitions, from peripheral to central, the orbital, preseptal, and pretarsal portions.

The orbital OOM arises medially from origins along the superior and inferior orbital rims as well as the medial canthal tendon and forms a wide loop around the lateral circumference of the orbital aperture. Superomedially, the orbital portion covers the corrugator supercilii muscle and the anterior temporoparietal fascia. In the inferior temporal and cheek area, the orbital orbicularis extends to the origin of the masseter and the zygomaticus minor muscle at the surface of the zygomatic body. Along the infraorbital rim and the maxillary frontal process, the origins of the upper lip and nasal ala elevators are surrounded by the OOM.

The preseptal and the pretarsal OOM portions overlie the orbital septum and the tarsal plates. At the medial canthal tendon and around the lacrimal canaliculi and lacrimal sac, both the portions divide into two components, anterior or superficial and posterior or deep heads. Both superficial heads connect to the medial canthal tendon and beyond to the anterior lacrimal crest as well as to

the nasofrontal maxillary process. The deep head of the preseptal OOM portion, or Jones' muscle, attaches latero-posteriorly to the fascia of the lacrimal sac (Fig. 3.3a, b). The deep pretarsal orbicularis head, known as Horner's muscle, pars lacrimalis or tensor tarsi, passes posterior to the canaliculi and the lacrimal sac to attach to the posterior lacrimal crest (Fig. 3.3a–c) (see section "Canthal Ligaments").

Laterally, the preseptal orbicularis from the lower and upper eyelid inserts onto the horizontal lateral palpebral raphe which attaches to the zygoma.

The preseptal portion of the orbicularis muscle functions in voluntary and involuntary eyelid closure.

The pretarsal inferior and superior orbicularis portion join the lateral canthal ligament (LCL) at Whitnall's tubercle.

The levator palpebrae superioris (LPSM) and Müller's superior muscle retract or elevate the upper eyelid; they maintain the eyelid in position.

The LPSM is a striated muscle [7], which is separable from the underlying superior rectus muscle (Figs. 3.4 and 3.5) apart from the medial border, where both muscles are adherent to a common fascial sheath. The LPSM originates from the lesser wing of the sphenoid (LWS) above Zinn's ring and extends forward where it blends with the fibers of the superior rectus muscle. About 1 cm behind the orbital septum, it fans out as a thin membranous sheet into the upper eyelid.

At the level just behind the superior orbital rim and location of Whitnall's ligament, the levator divides into two layers: a superior layer which continues into the aponeurosis and an inferior layer which blends forward into Müller's muscle (Figs. 3.1 and 3.6). Interconnecting fibrous tissue between the covering sheaths of the levator and superior rectus muscle allows for corresponding movements during vertical gaze.

Below the level of Whitnall's ligament (WL) and before reaching the upper edge of the tarsal plate, the LPSM [8–10] is widening horizontally into a broad fibrous aponeurotic sheath, the levator aponeurosis, arching over the globe with two tendinous extremities, the lateral and medial horns.

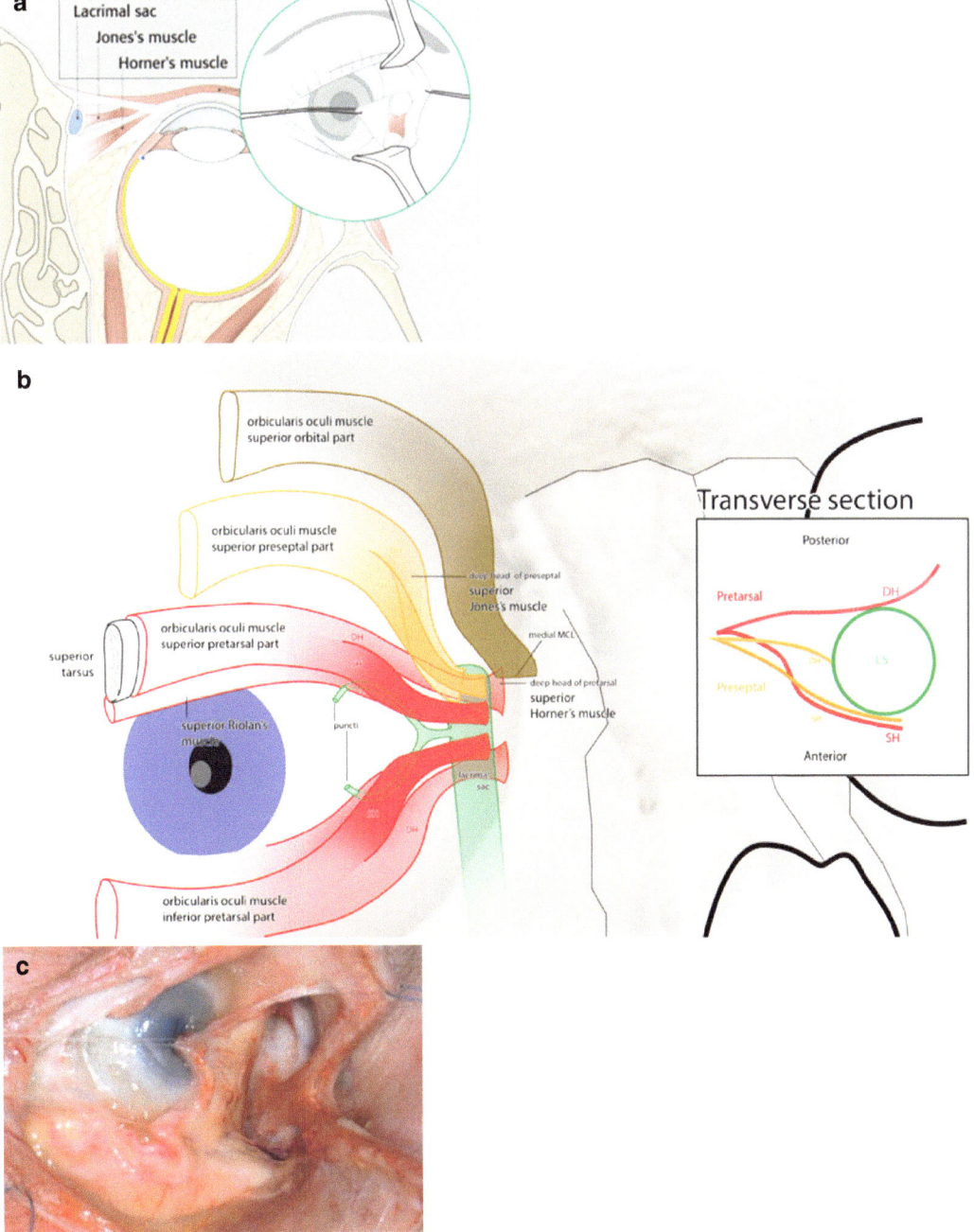

Fig. 3.3 (**a**) Axial view of the medial extension of the deep preseptal (Jones') and deep pretarsal (Horner's) portion of the orbicularis oculi muscle in relation to the lacrimal sac. (With permission from https://surgeryreference.aofoundation.org) (**b**) Schematic frontal view – medial extension of superior orbital, preseptal and pretarsal orbicularis muscles including the superficial head (SH) and deep head (DH) of the latter two; the (superior) muscle of Riolan is part of the deep head of the (superior) pretarsal (DHPT) orbicularis muscle and relates to the DHPT attachment to the posterior lacrimal crest. Inset: Axial view - Anterior / posterior relationships of Horner's and Jones' muscle. The deep head of the preseptal muscle (DHPS) (pale yellow pattern) is anteriorly located to the deep head of the pretarsal muscle (DHPT) (light red pattern). (With permission from S. Steenen). (**c**)Anatomic specimen Lateral view of the anterior medial wall of the orbit. Horner's muscle shown after combined retrocaruncular lower fornix swinging eyelid approach. Plica semilunaris flap retracted medially; Horner's muscle attached to the posterior lacrimal crest

Fig. 3.4 Anatomic specimen – Deep structures to upper eyelid. Eyelid cut into halves median sagittally and pulled sideways for exposure. Superior rectus muscle hooked up, LPSM cut and elevated with tweezers, Whitnall's ligament—whitish band going across-parallel to eyelid border—

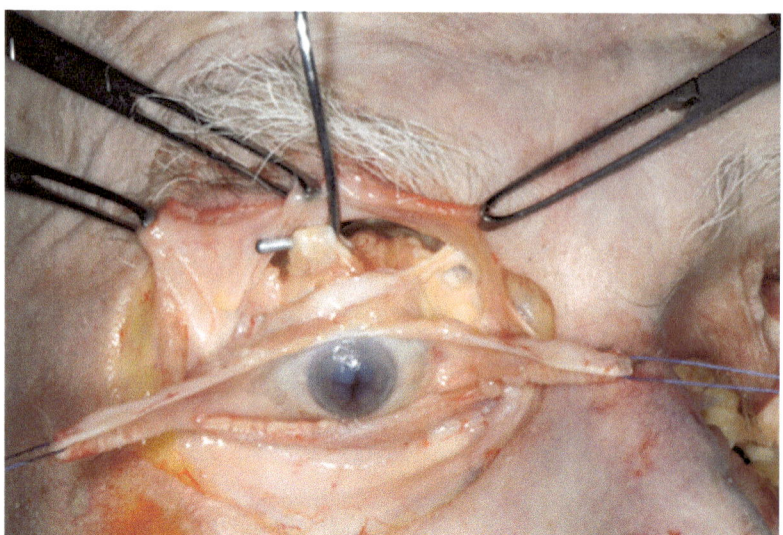

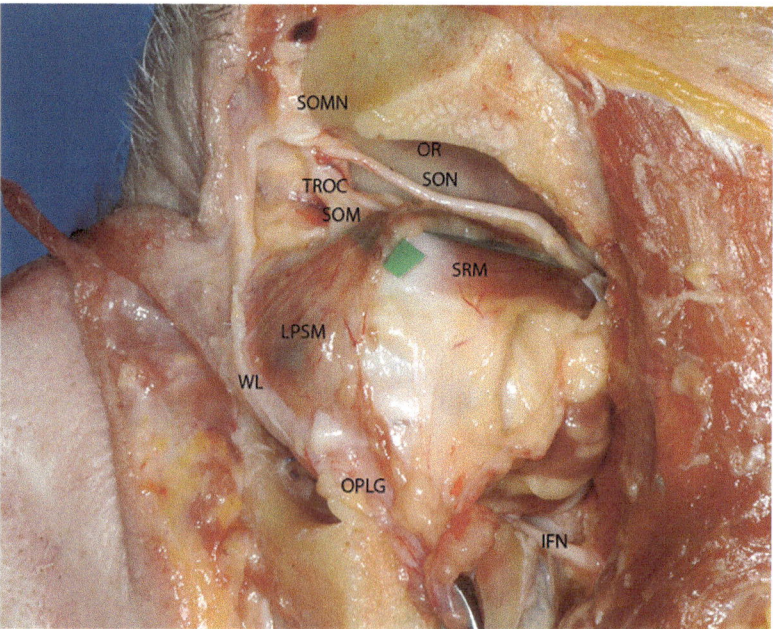

Fig. 3.5 Anatomic specimen – Lateral view of superior entrance of right orbit—after removal of lateral orbital wall. Upper eyelid stretched downward and sideways over the globe. LPSM fanning out from the junction (marked with green strip) with the superior rectus muscle (SRM) that is running backward. Whitish tendon of superior oblique muscle (SOM) exits from the trochlea and passes underneath the SRM. The white cable structure traversing directly underneath the orbital roof corresponds to the supraorbital nerve (SON). Whitnall's ligament (WL) spans along the anterior LPSM border. *IFN* infraorbital nerve, *OPLG* orbital part of lacrimal gland, *SOMN* supraorbital margin, *OR* orbital roof, *SON* supraorbital nerve, *SOM* superior oblique muscle, *LPSM* levator palpebrae superior muscle, *SRM* superior rectus muscle, *WL* Whitnall's ligament, *IFN* infraorbital nerve, *TROC* trochlea

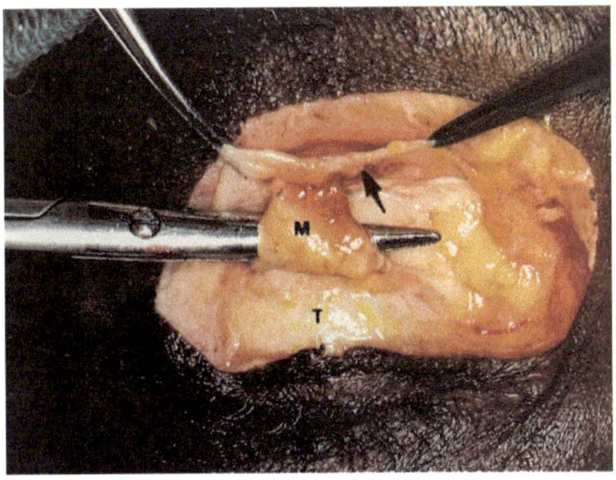

Fig. 3.6 Anatomic specimen – Upper eyelid - Deep structures exposed. The levator palpebrae superior (black arrow) held upward with tweezers, separated from the underlying musculus tarsalis superior (Müller), supported by the tip of scissors. T = tarsal plate

The band-like lateral horn of the levator divides the lacrimal gland incompletely into the orbital and palpebral lobes before inserting on Whitnall's lateral orbital tubercle as one of the components of the lateral retinaculum. The medial horn passes over the superior oblique muscle tendon, blends with the reflections of the medial canthal tendon, and joins the medial retinacular structures departing from the posterior lacrimal crest for osseous attachment. The cutaneous insertion of the aponeurosis is effected by terminal fibers traversing the orbicularis muscle and forming a distinct supratarsal fold. A subset of deeper fibers attaches to the anterior surface of the upper tarsus.

The innervation of the LPSM is by the superior CN III division.

The superior tarsal muscle of Müller's (STM) is an involuntary smooth muscle innervated by postganglionic sympathetic fibers derived from the paravertebral chain—superior cervical ganglion; the fibers run with the carotid plexus to enter the orbit via the superior orbital fissure. The sympathetic branches run mainly along the infratrochlear and lacrimal branches of the ophthalmic nerve.

STM is incorporated in the undersurface of the levator palpebrae anterior to WL, is adherent to the palpebral conjunctiva posteriorly, and inserts into the superior tarsal border (Figs. 3.1 and 3.6). It measures 8–12 mm in length and is present across nearly the whole width of the tarsus. STM works in synergy with the LPSM to raise the upper eyelid. The ligament prevents the upper lid from pulling away from the globe during elevation by the LPSM. Additional smooth muscles fibers may arise from Tenon's capsule.

The STM is separable from the LPM, whose aponeurosis continues on top of the tarsal plate (Fig. 3.6). Müller's muscle is separated from the conjunctiva and the levator aponeurosis by a thin fibrovascular tissue layer. In fact, Müller's muscle elevates the upper eyelid during sympathetic stimulation as in excitement.

The muscle of Riolan [2] consists of three subdivisions [1]. (1) The pars ciliaris/marginalis (striated muscle, the fibers of which are separate bundle from the pretarsal OOM running anterior to the tarsal plate and in parallel to the Gray line, (2) the pars subtarsalis, a smaller fibers bundle which is located posterior to the Meibomian orifices, and (3) the pars fascicularis. This third subdivision traverses the marginal surface of the tarsus and connects the former two muscle groupings.

Sensory innervation of the upper eyelid is provided by several branches of the CN V.

Clinical Implications

In patients presenting with Graves' disease (Chap. 15 and 16), retraction of the upper eyelid >> lower eyelid retraction is a common sign, which results from excessive sympathetic activity within Müller's muscle; to a lesser degree the LPSM may retract; inflammatory changes and fibrosis may also play a role in this process [11]. The involvement of the lateral horn of the LPSM in particular is supposed to be responsible for the typical aspect of the lateral flare (Fig. 3.7a–c).

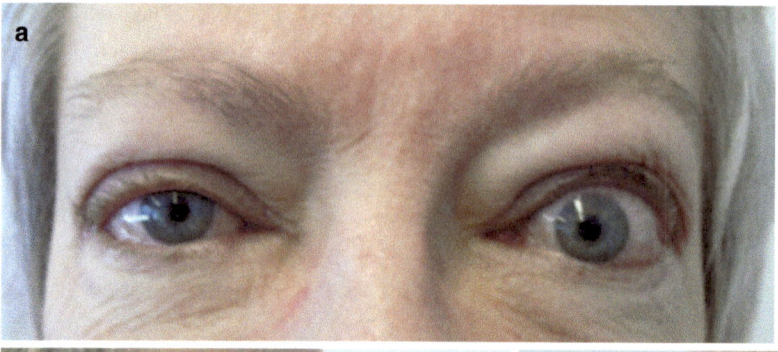

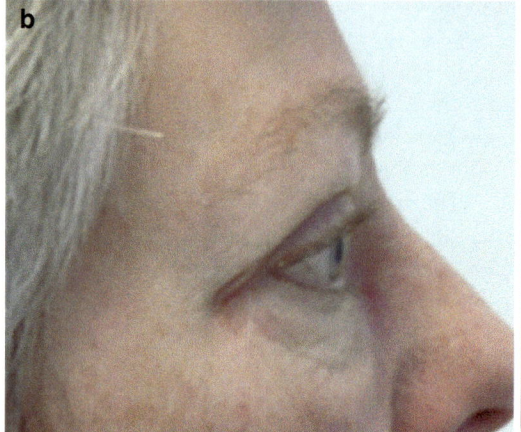
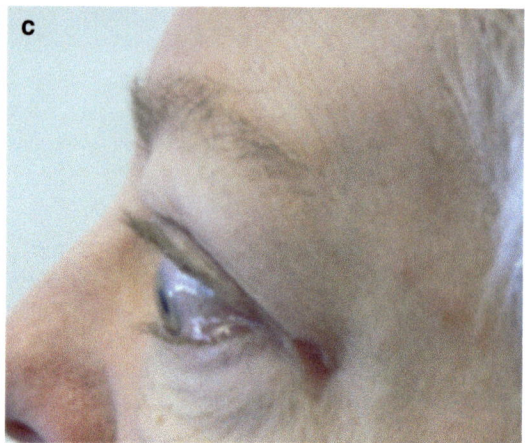

Fig. 3.7 (**a**) Frontal view of patient with unilateral Graves' orbitopathy OS. (**b**) Lateral view of patient with unilateral Graves' orbitopathy – unaffected side, (**c**) Lateral view of patient with unilateral Graves' orbitopathy – affected side, exophthalmos present

Whitnall's Ligament

Whitnall's superior suspensory ligament [8, 9] in the upper eyelid represents a portion of the globe suspensory system (Figs. 3.4 and 3.8).

Whitnall's ligament in context with the LPSM, trochlea, and lacrimal gland. The capsulopalpebral fascia located anterior to the inferior oblique muscle in the inferior orbit forms Lockwood's ligament (LL)—a kind of analogue to Whitnall's ligament. Thus WL is a supportive fascia structure of the upper eyelid and in fact a condensation of the fascial sheath of the LPSM that thickens near the level where the levator muscle blends into the aponeurosis just behind the superior orbital rim approximately 18–20 mm above the superior border of the tarsus (Fig. 3.8). Whitnall's ligament has a counterpart: the intermuscular transverse liga-

ment underlying the LPSM on top of the superior rectus muscle. Both function as a sleeve supporting the LPSM and prevent the upper eyelid from pulling away from the globe during elevation. Medially the ligament attaches onto the periosteum of the medial orbital wall and the suspensory system of the trochlea, laterally the ligament blends with the capsule of the lacrimal gland and the lateral orbital wall superior to the gland.

WL serves multiple functions: it contributes to the suspension of the lacrimal gland as well as to LPSM, superior oblique muscle, and origin of Müller's muscle.

Further contributing structures to the support system are fascial condensations, the medial and lateral check ligaments, Lockwood's inferior ligament, lacrimal ligaments, and a complex network of multiple septa.

Fig. 3.8 Shematic view - anterior orbit and tarsoligamentary apparatus—schematic view. (With permission from https://surgeryreference.aofoundation.org)

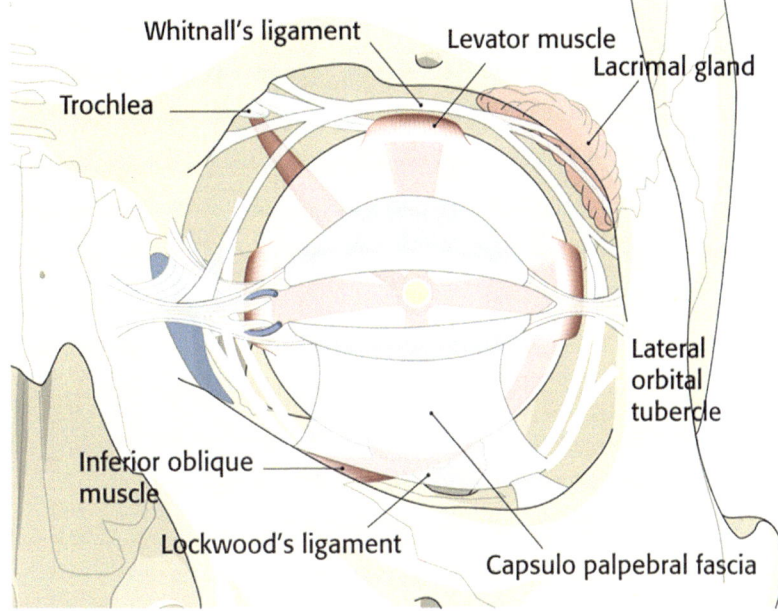

Lower Eyelid

Lid Retractors

The lower eyelid retractors consist of two principal layers, the capsulopalpebral fascia and Müller's inferior tarsal muscle. The innervation to the sympathetic inferior tarsal muscle travels along with branches of the infraorbital nerve. The capsulopalpebral fascia is a supportive structure analogous to the LPSM.

Fibrous extensions originate from the sheath of the inferior rectus muscle to compose the head of the capsulopalpebral fascia. This head transforms into an envelope around the inferior oblique muscle. Anterior to the muscle the inferior and superior fascial portions conjoin to form Lockwood's inferior suspensory ligament. The superior portion of the capsulopalpebral fascia then extends posteriorly into the inferior conjunctival fornix and converts into Tenon's capsule.

The inferior portion fuses with the orbital septum anteriorly, connects through the orbicularis oculi muscle (OOM) to the skin, and inserts more upwardly on the inferior border of the lower tarsus. The dermal attachments are the basis of the lower eyelid crease. A limited depression of the lower eyelid margin occurs in downgaze to enable an unimpaired visual field. This movement is essentially a synergic effect of the inferior rectus muscle contraction which is translated by the capsulopalpebral head and the inferior fascial portion.

The sympathetic smooth inferior tarsal muscle is an accessory lower eyelid retractor. It originates and lies posterior to the capsulopalpebral fascia, extends upward distal from Lockwood's ligament (LL), and inserts near and at the base of the inferior tarsus. LL is a band-like fascial sling, ± 45 mm in length and ± 6 mm wide. It is connected to the medial and lateral check ligaments.

It prevents downward and backward displacement of the globe and supports it as in a hammock. In its center, it is formed as a connective tissue thickening in the lower portion of Tenon's caspule. Medially, Lockwood's ligament blends with Horner's muscle and the medial check ligament. These insert onto the posterior lacrimal crest. Part of its medial head joins the medial horn of the levator aponeurosis.

Laterally the LL spreads out two heads.

The anterior lateral LL head inserts onto the inferior border of the lateral canthal ligament.

The posterior lateral LL head joins the lateral retinaculum at Whitnall's tubercle together with the orbital septum, the lateral check ligament of the lateral rectus muscle, and the lateral horn of the levator aponeurosis.

Lockwood's ligament extends as composed of two frontal layers: an anterior layer which interdigitates with the orbital septum and a posterior layer which fuses with the inferior border of the tarsal plate.

Lockwood's ligament stabilizes spatial anatomy within the orbit during function.

The cutaneous innervation of the lower eyelid [12] is provided by the infraorbital nerve (ION) and the zygomaticofacial nerve (ZFN). The nerves run within and inferior to the epimysium of the orbicularis muscle and perforate the orbicularis muscle perpendicular to distribute to the overlying skin. Terminal branches of ION are mainly distributed medial to the lateral canthus. Most terminal branches of the ZFN are distributed lateral to the lateral canthus.

Clinical Implications

In Graves' disease, **traction** on capsulopalpebral fascia of inferior rectus muscle **contracture** will result in retraction of the lower eyelid: if this results in inability to close the eyelids properly this may cause lagophthalmos, when the eyeball turns upwards while showing the white lower surface of the globe.

Several surgical approaches to the orbit, more specifically the orbital floor have been described.

All of them carry their own (dis)advantages [13–15] and vary in the extent of bony exposure.

The transconjunctival approach has fewest lower eyelid complications and reduces the prevalence of ectropion, however carries an increased risk of entropion

when compared to the subciliary approach. Lacrimal system damage can occur.

The transcutaneous subciliary—and the subtarsal approach [16, 17] reduce the prevalence of entropion, the risk of ectropion increases including a higher risk of scleral show. The infraorbital approach includes the advantage not to directly interfere with the eyelid lamellae. Transcutaneous approaches however do leave a visible scar, though when applied in an available skin crease, scar visibility may be minimal. The trans (retro)caruncular approach [18–28] is a safe and effective method for decompression of the medial orbital wall in case of Graves' exophthalmos or for reconstruction of the medial orbital wall in case of a fracture; a combined transcaruncular - transconjunctival approach may be indicated for reconstruction of large orbital wall defects involving both medial wall and orbital floor.

Canthal Ligaments

Medial Canthal Ligament (MCL)

The upper and lower tarsal plates convert into the superior and inferior crura, fibrous bands, and fuse to form the common medial ligament. Medial extensions of Whitnall's ligament attach to the common ligament and the posterior lacrimal crest as do the medial extensions of Lockwood's ligament (Fig. 3.8).

The crura are located between the OOM anteriorly and the conjunctiva posteriorly. Separate limbs originate from the common ligament and course to attach to the bone at the medial orbital rim and to the lacrimal sac. The anterior limb, approximately 10 mm in length and 4 mm in thickness inserts to the frontonasal process of the maxilla anterior and above the anterior lacrimal crest [15, 29]; as such it is the strongest component of the medial canthal tendon complex and provides the major support of the medial canthal angle. The thinner posterior limb, passing between the canaliculi, fans out and inserts to the posterior lacrimal crest. This limb provides a backward pull to maintain the eyelid in a posture tangential to the globe surface. The superior limb, connecting the anterior and posterior limb inserts onto the orbital process of the frontal bone.

The resultant vector of all the canthal limbs and attachments suggests that resuspension of the entire complex following disruption should be directed posteriorly and superiorly toward the posterior lacrimal crest.

Both the pretarsal and the preseptal OOM portions bifurcate into a separate rear (i.e. deep head) and forward-directed (i.e. superficial head) stripe just like the prongs of a pair of open scissors.

The posteriorly departing muscle pathways of the upper and lower OOM lid portions correspond to the deep head of the pretarsal (DHPT—Horner's) muscle and the deep head of the preseptal (DHPS—Jones) muscle, respectively, both fanning out in a further medial course. The anterior or superficial stripes conform with the superficial head of the pretarsal (SHPT) muscle and the preseptal (SHPS) OOM portions (Fig. 3.3a,b).

The superficial heads of the upper and lower lid pretarsal OOM (SHPT) in conjunction with the superior and inferior muscles of Riolan extend over and interdigitate anteromedially with the crura of the medial canthal ligament. They invest both lacrimal canaliculi, cover the ampullae, and insert onto the anterior limb of the canthal tendon.

Reaching the level of the common canaliculus, the medial extensions of upper and lower lid DHPT fuse and compose a prominent flat muscular bulge with a vertical height of 6 mm and a thickness of 2.5 mm, referred to as Horner's muscle. Horner's muscle continues behind the posterior limb of the medial canthal ligament to insert at the posterior

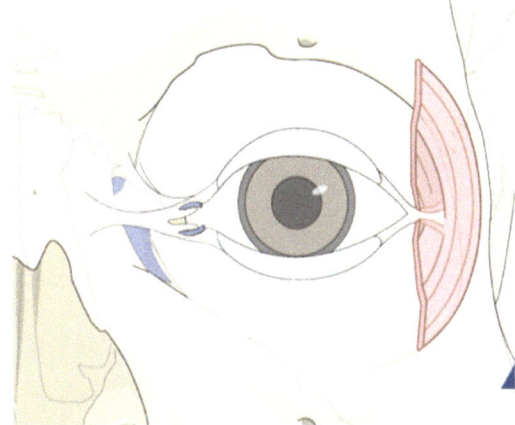

Fig. 3.9 Schematic view – Medial canthal ligament in relation to lacrimal sac/fossa forming a "roof" (With permission from https://surgeryreference.aofoundation.org)

lacrimal crest (Fig. 3.3a,b): thus forming the deepest layer of the inner canthus tissues.

The DHPS or Jones' muscle predominantly turns toward the lateral aspect of the lacrimal sac and creates a "second" layer which is somewhat overlapping the lateral part of the DHPT (Fig. 3.3a,b).

The SHPT and the SHPS remain foremost, run along the anterior surface of the lacrimal sac, and fuse with the medial canthal tendon to create a third most superficial tissue layer (Figs. 3.3b).

The posterior limb of the medial canthal ligament forming a frontal sheet to Horner's muscle is adjoined by the medial horn of the levator aponeurosis, the posterior layer of the orbital septum, and the medial check ligament altogether composing the medial retinaculum. Horner's muscle tone contributes to the apposition of the eyelid to the globe during eyelid closure.

The superior arm of the medial canthal ligament forms a kind of "roof" of the lacrimal sac/fossa and blends with the fascia of the lacrimal sac (Figs. 3.8, 3.9, and 3.10). The pulling actions of the DHPT and the DHPS are supposed to contribute to the induction of negative pressure in the tear sac supporting the lacrimal pump mechanism thereby.

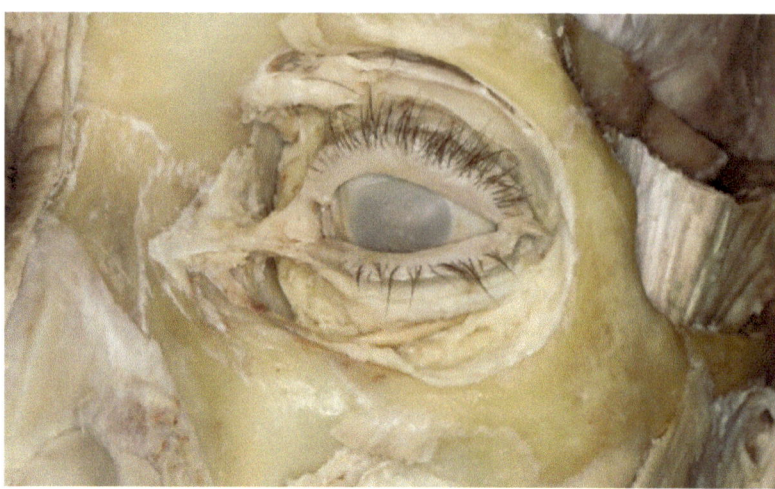

Fig. 3.10 Anatomic specimen _Medial canthal ligament, and some of the fascial structures in the anterior orbit—OOM removed. MCL components—superior and inferior crura fuse to a common tendon that attaches to the periosteum of the medial orbital rim (frontonasal maxillary process, orbital process of frontal bone, nasal and lacrimal bones) with three limbs (or arms)—anterior, superior (anterior lacrimal crest), and posterior (posterior lacrimal crest—not visible)

Lateral Canthal Ligament (LCL)

The lateral canthal ligament has features of both a ligament and a tendon.

The lateral endings of the tarsal plates pass over into the crura of the lateral canthal ligament. The crura unite resulting in the common tendon with both superficial and deep components.

The superficial flat ligamentous layer is continuous with the overlying orbital septum and interdigitations of the pretarsal OOM [30]. It measures approximately 10 mm in length/width; this portion coalesces with the periosteal surface of the lateral orbital rim and the temporalis fascia (Fig. 3.11).

The deep LL components arise from the lateral aspect of the tarsal plate to develop a cable-like common tendon (Figs. 3.11, 3.12, and 3.13). This inserts onto the lateral orbital tubercle of Whitnall. This tubercle is a small roundly protuberance of about 2–3 mm diameter located on the inner orbital plate of the zygoma (OPZ), immediately (2–4 mm) in the marginal territory posterior to the lateral orbital rim and about 10 mm beneath the frontozygomatic suture (FZS).

In summary, Whitnall's lateral orbital tubercle is important as marking the point of attachment of the:

- Lateral common canthal tendon (originally termed the "lateral palpebral ligament")
- Lateral check ligament—of the lateral rectus muscle (LRM)
- Lateral horn of the LPSM aponeurosis
- Lateral end of Lockwood's inferior suspensory ligament of the eyeball

Somewhat paradoxically the eponymous superior transverse ligament of the eye, commonly addressed as Whitnall's ligament is not listed having a specific attachment to the "tuberculum orbitale" in the original publications [8, 9]. Indeed Whitnall's ligament inserts into the periorbita of the lacrimal fossa up to 10 mm above the lateral orbital tubercle by way of the fascia of the lacrimal gland.

The attachments of the lateral canthal tendon, the lateral LRM check ligament, and the lateral horn of the LPSM aponeurosis extend along the lateral orbital wall posterior to the tubercle in a successive order from the rim before they end. The LCL continues for about 5 mm, the check ligament for approximately 6.5 mm, and the lateral aponeurosis horn roughly for 8.5 mm.

The complex arrangement of the entire ligamentous and fascial lateral extensions is summarized as the lateral retinaculum.

Fig. 3.11 Anatomic specimen – Superficial lateral canthal ligamentous layer with underling fat pad (Eisler) attaching to the periosteum of the lateral orbital rim - anatomic specimen

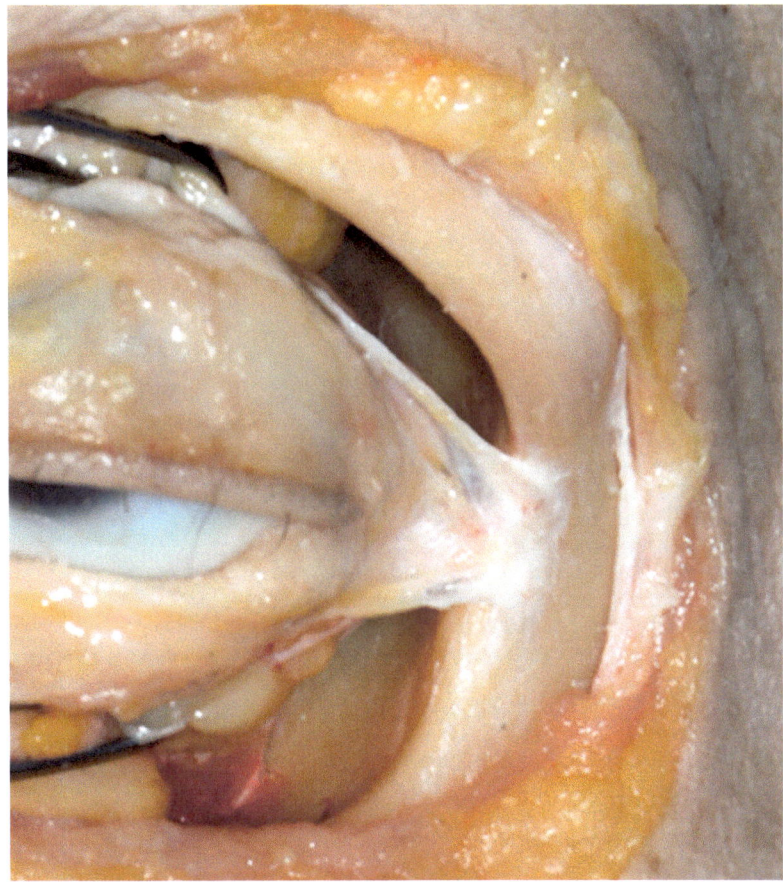

Fig. 3.12 Anatomic specimen. Deep lateral canthal tendon, converging from the crura and attaching to Whitnall's tubercle

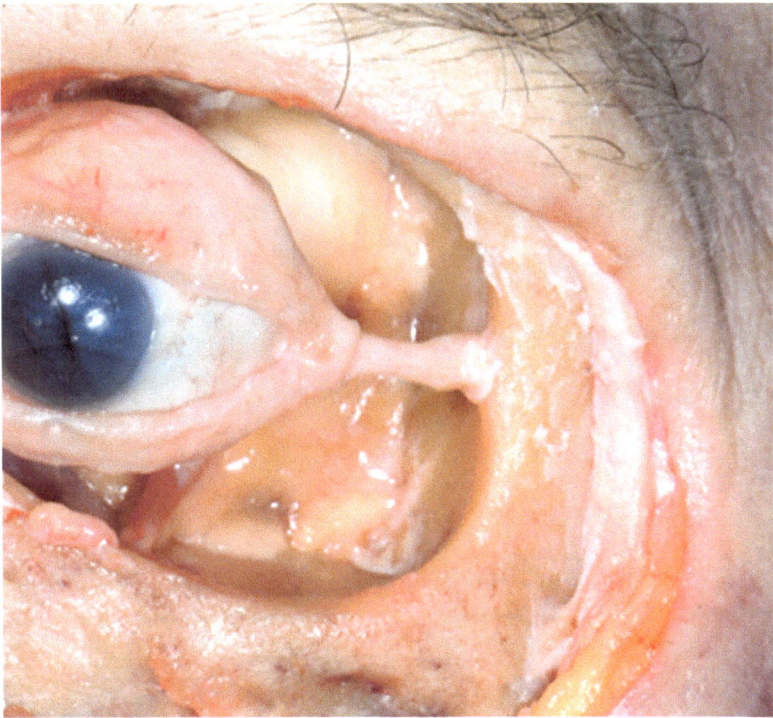

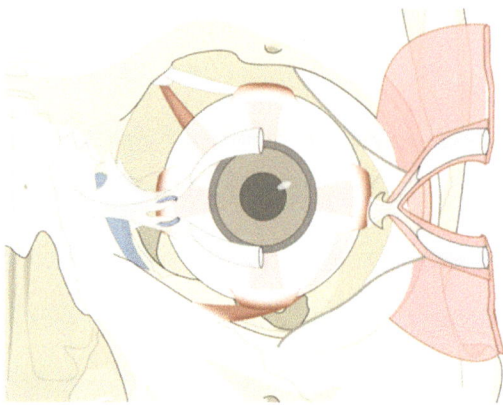

Fig. 3.13 Schemata view – : transview of tarsoligamentary apparatus and globe featuring the orbital cavity. The lateral OOM, including the lateral canthal ligament complex half reflected laterally. Superior and inferior tarsal plates continue medially into the superior and inferior crus, the crura fuse into a common ligament and split progressing further medially into an anterior and posterior arm-limb of the medial canthal ligament (fibrous connective tissue structure) attaches respectively anteriorly to the periosteum of the orbital process of the maxillary bone in front and above of the anterior lacrimal crest, posteriorly to the posterior lacrimal crest, just above the entrance of the nasolacrimal duct; the anterior and posterior arms are connected by a superior limb of the medial canthal ligament to form a horizontal raphe: "roof" of the lacrimal fossa which inserts into the periosteum of the orbital process of the frontal bone (roof = anterior–posterior webbing between the anterior and posterior arm of the medial canthal ligament) (With permission from https://surgeryreference.aofoundation.org)

The Periorbita and Orbital Septum

Periorbita

The periorbita is identical with the periosteal lining covering the internal orbit, that starts at the anterior bony aperture. While orbital fat is hold into place, the periosteal envelope serves as a covering membrane. In the anterior area, the periorbita is continuous with the orbital rim periosteum and the orbital septum. The fusion zone between periorbita and periosteum is thickened: arcus marginalis. This is the origin of the orbital septum, corresponding to the deep layer of the galea aponeurotica.

Posteriorly, in the orbital apex, the periorbita communicates with the dura mater of the middle cranial fossa and cavernous sinus through the optic canal and the SOF. Here, it also contributes to the Zinn's ring. Infero-dorso-laterally, the periorbita extends into the inferior orbital fissure (IOF) and fuses with Müller's orbital smooth muscle which covers the IOF. Inferior to the rear sinkhole of the IOF the periorbita transforms into the periosteal lining of the adjacent pterygopalatine and infratemporal fossa. At the lacrimal crest, the periorbita splits into a thin layer for the lacrimal fossa and a thick layer, the fascia lacrimalis which invests the lacrimal sac.

Orbital Septum

The orbital septum is a fibroelastic multilayer membrane. It defines the anterior boundary of the orbit as a barrier that separates facial from orbital structures.

The orbital septum originates from the arcus marginalis orbitae and consists of two layers. The outer superficial layer is formed by the deep layer of the galea (Fig. 3.1). The inner, deep layer is the anterior coursing fascia of the periorbita, which further separates into an upward layer superiorly covering the frontal bone and the inner layer becoming the orbital septum of the upper eyelid.

Within the eyelids, the orbital septum separates the anterior from the posterior lamella: the septum passes between the OOM and the fat pockets.

In the upper eyelid, the orbital septum inserts and fuses with the LSPM aponeurosis, several mm above the tarsal plate. Finally, the superficial layer of the septum continues further downward onto the anterior tarsal surface.

In the lower eyelid, the orbital septum fuses with the capsulopalpebral fascia inferior to the tarsus before they insert to the tarsal edge and the lower canthal crus together.

From the latter, as with Lockwood's ligament, the orbital septum continues to insert onto the posterior lacrimal crest. An anterior septal layer inserts to the common medial canthal ligament and to the anterior lacrimal crest finally enclosing the lower part of the lacrimal sac anteriorly.

Another intermediate septal layer passes backward around the lacrimal sac in the upper and lower eyelid. Its insertions connect to the posterior limb of the canthal tendon and to the posterior lacrimal crest, frontal to Horner's muscle. In the lower eyelid, the lower border of the layer fuses to the periorbita at the opening of the nasolacrimal duct. Laterally, the septum becomes intertwined with the superficial lateral canthal ligament and its circumferential borders insert onto the orbital margin of the zygomatic bone.

Lacrimal Functional Unit

The overall lacrimal functional unit [4, 31–35] is organized in three partitions: a glandular secretion system, a tear fluid distributional area (ocular surface, conjunctiva, cornea), and a drainage pathway, that collects, conveys, and empties the tear fluid into the lower meatus of the nasal cavity toward an orifice anteriorly underneath the inferior concha. The unit modulates lacrimation, protection of the ocular surface including immune responses from the anterior segment of the eye, and neurosensory perception (e.g., pain sensation).

Tear Producing Glands

Main and Accessory Aqueous Lacrimal Glands

The aqueous tear fluid production is effected by the main lacrimal gland [36–38] with contributions from numerous small accessory lacrimal glands. The accessory glands are located in the eyelid conjunctiva along the nonmarginal tarsal borders of the tarsi (Wolfring glands) and at the fornices—predominantly the superior fornix—(Krause glands) (Figs. 3.1 and 3.14).

The main lacrimal gland is a bilobed exocrine gland consisting of a larger superior orbital part and a smaller inferior palpebral part. The anterior horizontal lobe portions are sandwiched around the concave lateral horn of the LPSM aponeurosis, whereas the parenchyma of the gland is continuous around the posterior edge of the horn.

Both lobes are located posterior to the orbital septum. The orbital lobe is affixed to the lacrimal fossa in the anterolateral orbital roof. The fibrous periosteal attachments include Whitnall's ligament. The palpebral lobe extends laterally into the superotemporal fornix of the upper lid, where it is adherent to the conjunctiva through which it becomes visible with lid eversion. Anteriorly the palpebral lobe spreads beyond the superior orbital margin, while the orbital lobe has a posterior orientation and conforms to the space between the orbital wall and the globe surface. The position and footprint area of the overall gland may vary considerably to behind the globe, over the superior rectus muscle (SRM) and the vertical midline of the globe as far as alongside the temporal aspect to the lower border of the lateral rectus muscle (LRM).

Excretory ducts from both lobes all empty at the superotemporal conjunctival fornix. Several interlobular ducts connect the orbital lobe to the ducts of the palpebral lobe. About 10–15 other ducts from the orbital lobe route independently through the palpebral lobe. Hence resection of the palpebral lobe ablates the lacrimal flow of the entire gland.

The arterial supply of the main gland derives from the lacrimal artery, a branch of the ophthalmic artery, with medial and lateral ramifications inside the lobes, subsequent branches supplying the palpebral marginal arcades, and a terminal conjunctival network. Venous drainage is via similar tributaries to the superior ophthalmic vein.

Sensory innervation is provided by the lacrimal nerve (¬CN V1). The aqueous secretion by the lacrimal gland is subject to parasympathetic and sympathetic stimulation via the VIIth CN parasympathetic pathway and via postganglionic axons from the superior cervical ganglion (SCG) (see autonomic innervation).

The accessory lacrimal glands of Wolfring and of Krause (Fig. 3.14) basically resemble the lobe structures and histologic characteristics of the main lacrimal gland. The nodule-shaped glands are individual organs with their own interstitial connective tissue coatings. They are located in the lamina propria of the conjunctiva and their ducts open onto the conjunctival surface. Wolfring's glands are the larger type nod-

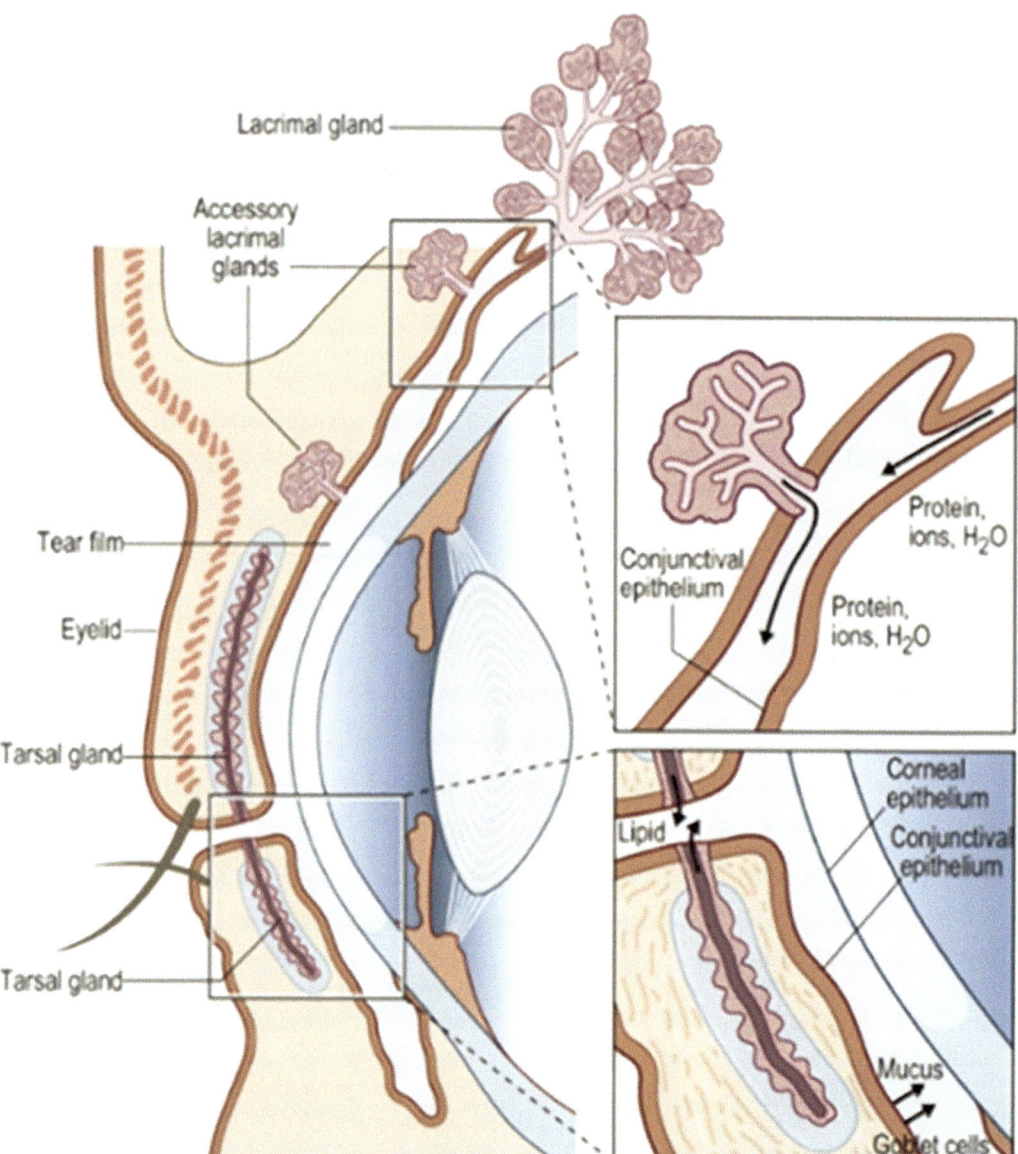

Fig. 3.14 Orbital glands that contribute to the various components (see insets) of the preocular tear film. (With permission from Grays Anatomy 42nd edition)

ules found in numbers of 2–5 above or within the upper border of the superior tarsus near its midline and in a number of 2 within the lower border of the inferior tarsus. The total more numerous Krause's glands lie in the upper fornix—about 20–40—below the palpebral lobe of the main lacrimal gland and in the lower fornix—about 6–8. Autonomic innervation in human accessory lacrimal glands has long been confirmed with parasympathetic prevailing over sympathetic formations.

Conjunctival Goblet Cell (CGC) Population

Goblet cells [39–41] of the conjunctiva are plump, rounded specialized cells that extend through the entire thickness of the stratified epithelium.

They can occur individually or within clusters; the goblet cell population is distributed throughout the palpebral and bulbar conjunctiva with the greatest density in the medial canthus/

lower fornix region and sparser dispersion over the superior and inferior bulbar regions. In regions with sparser numbers, single goblet cells exceed cell clusters.

Most importantly, conjunctival goblet cells secrete polymeric gel-forming and membrane-bound mucins onto the ocular surface. Based on their mucin secretion, several functions can be attributed to CGCs such as lubrication, maintenance of surface wetting and the tear film across the epithelium, prevention of corneal surface damage and infection.

Tarsal Glands

Processions of individual tarsal glands of Meibom (Meibomian Glands—MG) [42] are embedded in parallel arrangement perpendicular to the eyelid margins within connective tissue fibers of the tarsal plates (Figs. 3.1 and 3.14). MGs are a distinct variant of large sebaceous glands that are not associated with hair follicles.

The tarsi in the upper eyelids are fitted with between 20 and 40 MGs, the lower tarsi with between 20 and 30 MGs. The vertical measure of the meibomian glands follows the height of the tarsal plates and consequently differs according to the shape and dimensions of the upper and lower tarsi. A single MG is composed of multiple holocrine secretory acini, which are clustered in tiers around the long vertical axis of a central duct to which they link in oblique direction via short connecting ductuli. The open end of the central duct is lined by keratinized epidermal layers grown in from the free lid margin. This orifice is typically positioned at the posterior lid margin directly in front of the mucocutaneous junction and inner lid border. MGs produce a clear oily, lipid-rich secretion (i.e.,"Meibum"), that forms part of the tear film composition.

Ciliary Glands

There are two kinds of glands associated with the lashes on the eyelid margins ("eyelid cilia"), the modified apocrine sweat (sudor) glands of Moll, and the pilosebaceous glands of Zeis (Fig. 3.1).

Moll glands are tubules that form a long stretched spiral widening root in contrast to the coiled tubuloalveolar body of common sweat cells. They have a wide-lumen ampulla and open with a narrow duct into a ciliary follicle close to the surface or run into the duct of a sebaceous Zeis gland or directly onto the surface of the lid margin between the eyelashes—often with one orifice amid two lashes.

Moll glands are located anterior to the Meibomian glands and extend deep into the lid next to the tarsal plates. They are more numerous and more developed in the lower eyelid.

Zeis glands are larger than Moll glands and directly associated with each eyelash follicle. Usually two Zeiss glands exude their secretions through their excretory ducts surrounding the mid-length portion of the ciliary follicle. Each Zeis gland consists of clusters of several sac-shaped enlargements, the alveoli, which unite into a short wide duct.

Ocular Tear Film

The tear fluid is a complex solution [43–45]composed of water, enzymes, proteins, immunoglobulins, lipids, various metabolites, and exfoliated epithelial and polymorphonuclear cells. The classical concept of an interpalpebral, trilaminar, preocular tear film with a superficial lipid phase (from tarsal Meibomian glands) overlying an intermediate aqueous phase (from main and accessory lacrimal glands) and an innermost mucinous layer (from conjunctival goblet cell population) [46] is queried by novel research.

The three layered architecture is only applicable to the preocular fluid within the interpalpebral aperture, that is present in the interblink period. The modern reappraisal proposes a model that encompasses a coherent fluid extending over the whole "ocular surface (OS) system." This integrates all conjunctival and corneal epithelia into a single three dimensional sack-like territory with the retropalpebral pouches up into the fornices. As a consequence thereof, the tear fluid is distributed in three continuous compartments, the fornix conjoined with the retrotarsal space ("cul de sac"), the tear menisci and the preocular area, all subject to the condition that the eyes are open. The preocular tear film is spread over the exposed bulbar conjunctiva and the contours of the cornea. Over the corneal region, this film is extraordinarily thin with a thickness of 2–5.5 μm.

The threefold junctional interface between the occlusal surface of the lid margin, the preocular tear film, and the atmosphere gives rise to the tear menisci. Thus they fill in and run along the corner profiles of the lower and upper lid margins with a concave front and a prismatic (or wedge-shaped) cross-sectional geometry. They act as a reservoir to supply the fluid which renovates the preocular tear film at each blink. The meniscus parameters (height, width, and radius of curvature) can be used to compute the meniscus volume which is supposed to correlate with the overall tear volume.

The tear film adhesion to the surface of the conjunctival and corneal epithelium is increased by large transmembrane mucins which, as part of the glycocalyx, attach to the microplicae of the cells and extend into the preocular tear film. The preocular tear fluid corresponds to a single layer of mucoaqueous gel with a decreasing mucous density toward the surface. It is still unclarified whether this is true for all OS compartments. The mucoaqueous gel makes up the bulk of the tear film and lies beneath a very thin lipid layer (at mean 42 nm). The lipid layer forms an ever present sealant closing off the mucoaqueous gel between the eyelid margins, where it is secreted

from the Meibomian glands and spreads onto the tear film with each blink.

An intact lipid layer shielding is important in stabilizing the tear film and preventing excessive evaporation of the tear fluid. The preocular/precorneal tear film provides the primary refractive medium for light entering the visual system and provides a protective environment for all tissues involved in the OS.

Lacrimal Drainage Pathway—Nasolacrimal Sac/Duct

The meniscal stream carries the tear fluid along the upper and lower eyelid margins [46] toward the lacrimal puncta. Each punctum is a small or transversely oval-shaped orifice located on the peak of the papilla lacrimalis near the medial end of the lid margin next to the nasal canthal angle (Fig. 3.15). Both puncta face backward into the groove between the plica semilunaris and become only visible when the lids are everted. From the punctal outflow, the drainage continues sequentially via the upper and lower canaliculi into the lacrimal sac. Each canaliculus has a short vertical (2 mm) and then after a dilatation at right-angular turn ("ampulla"), a longer (8–10 mm) horizontal part. The horizontal canaliculi converge medially, the upper being shorter than the lower. Contraction of the lacrimal part of the orbicularis oculi muscle, i.e., Horner-Duverney's

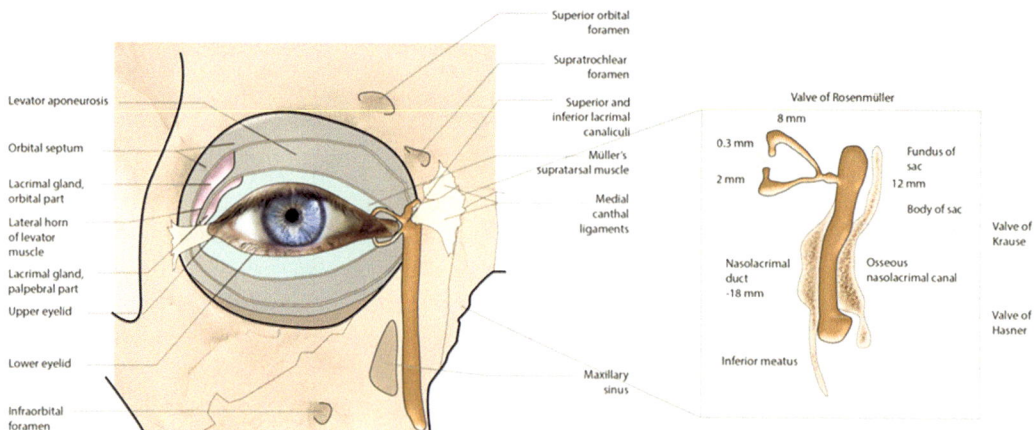

Fig. 3.15 Schematic frontal view of lacrimal drainage system: orbital septum and LSPM partially removed to expose lacrimal gland. Inset: details of lacrimal drainage system and nasolacrimal duct. (With permission from S. Steenen)

muscle causes the tear fluid in the canaliculi to be transported toward the lacrimal sac.

The horizontal canaliculi enter the lacrimal sac on the posterolateral surface either united to a short (1–2 mm) common canaliculus piercing the lacrimal fascia or independently. An in-folding of the lacrimal sac mucosal lining is supposed to have a one-way valvular function at the internal canalicular entrance (valve of Rosenmüller) that might prevent a retrograde reflux eventually emerging at the puncta.

The lacrimal sac, a membranous conduit measuring approx. 0.2–0.5 cm in diameter and approx. 1.2–1.5 cm in length lies in the lacrimal fossa. Its upper end (fornix of lacrimal sac) is closed in a dome-shaped fashion, its lower part merges into the nasolacrimal duct configuring an isthmus at the entrance level into the bony canal.

The nasolacrimal duct is approx. 1.2–2.4 cm long. From the sac, it descends following the course of the intraosseous maxillary canal at an angulation of 15° to 30° degrees in a backward and of 5° in a lateral direction; the lower meatal duct portion with its terminal aperture ("lacrimal ostium") is accommodated in the lateral wall of the inferior nasal passageway. The sagittal and vertical position as well as the shape of the ostium are most variable within limits of 2.0 –3.5 cm behind the anterior border of the nostril sill and 0.9–2.2 cm above the nasal floor in a round, punctiform, slit-like—linear, vertically or transversely oriented configuration, single or in a duplicated array.

Numerous mucosal folds in distinct places within the nasolacrimal duct have been described as valves and are still labeled with eponyms of anatomic celebrities but with no evidenced valvular function. The plica lacrimalis or valve of Hasner is the most constant variant situated on the side of inferior turbinate just prior the ostium.

The fluid dynamics within the lacrimal system are associated with pumping mechanics during eyelid closure and muscular contraction (e.g., preseptal and pretarsal OOM) involved therein.

The lower system comprises the lacrimal sac and the lacrimal duct. The lacrimal sac is anterior to the orbital septum, nestled in its own fascia in the lacrimal fossa. The lacrimal crest of the maxillary bone forms the anterior border of the lacrimal fossa.

The tendinous insertions of the OOM bind the lacrimal sac anteriorly and posteriorly, aiding in the movement of tears. Posterior to the sac, the deep heads of the pretarsal OOM insert. The deep heads of the preseptal OOM attach the lacrimal sac latero-posteriorly. Anterior to the sac, the superficial heads of the pretarsal and preseptal orbicularis muscles insert onto the anterior crest of the lacrimal fossa. Together, these insertions help squeeze the lacrimal sac to move tears forward through the system.

Plica Semilunaris

The plica semilunaris is a crescent-shaped conjunctival fold lateral to the caruncle conjoining to the bulbar conjunctiva and lacrimal portion of the eyelids. It surrounds the medial limbus from the superior to the inferior fornix. The plical fold functions in the distribution of the tear fluids, maintains the lacrimal lake, and keeps the lacrimal puncta in contact with the lake. Though the plica is not very prominent, it provides enough resiliency to the movements of the globe and the eyelids. It unfolds and flattens or is retracted on abduction or adduction, respectively, what allows the lacrimal puncta to hold up their position.

Lacrimal Caruncle

The lacrimal caruncle is a soft, pink bump with ovoid shape found in the tear lake at the inner canthus inferomedial to the plica semilunaris. It is oriented obliquely with an inferolateral angulation, while its superior border is at level with the inferior lid margin. The surface is lined with a nonkeratinizing epithelium giving it the look of conjunctival tissue. In fact, the caruncules feature skin appendages (pilosebaceous units, sweat gland, hair follicles, connective tissue, fat) (Fig. 3.16), since they derive from the lower eyelid margin developmentally. Equally the caruncles have direct connections to the lower lid retractors and to the MRM. Congenital supernumerary caruncles are extremely rare.

Fig. 3.16 Histologic section _ Caruncle overlapped by plica semilunaris—horizontal histological preparation. A goblet-cell-rich epithelial surface, a pilosebaceous Zeis gland emptying into a hair follicle (partially sectioned) and the tubuloalveolar body of a sweat gland are proof of the skin appendage character

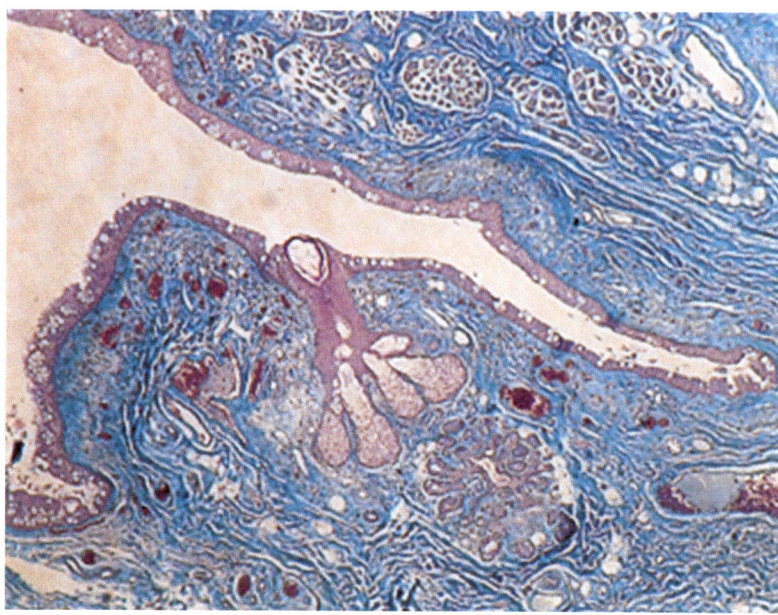

Clinical Implications

A trans- or preferably retrocaruncular approach [19, 20] provides direct and reliable adequate access to the medial aspect of the orbital floor and medial wall of the orbit posteriorly toward the orbital apex. It is a safe, rapid, and cosmetically pleasing surgical approach. The approach can be combined with a transconjunctival access and a lateral canthotomy to gain overall access to the medial orbital wall, the orbital floor, and the lateral orbital wall. The retrocaruncular approach benefits the transcaruncular insofar that there is less risk of tarsus exposure and thus a decreased risk of lid complications. As the transcaruncular approach in fact consists of an inferior transconjunctival approach extending medially and superiorly, it divides the cutaneous-tissue caruncle. Horner's muscle fixates the caruncle to the posterior lacrimal crest and helps to support the medial eyelid; as the retrocaruncular incison is performed lateral to the caruncle, the risk of damage to Horner's muscle is minimized. Also some lower eyelid retractors attach to the caruncle, if division of the caruncle is performed, this may result in an inferolateral shift of the lateral aspect of the caruncle.

The Ocular Globe

The Globe—Gross Anatomical Outline/Overview

The ocular globe or eyeball has a volume of about 6–8 cm^3, a vertical diameter of 23.5 mm, and an anterior to posterior diameter, called the eye/optical axis of 23 mm (Fig. 3.17).

The anatomic equator divides the globe topographically into two unequal halves: the anterior and the posterior hemispheres; it is a virtual line which corresponds to the greatest circumference of the eyeball (Fig. 3.17b, c). The cornea constitutes the anterior segment and occupies 1/6th of the surface area, the larger posterior segment is composed of the scleral shell and has a larger radius of curvature than the anterior segment.

The eyeball is surrounded by Tenon's capsule a fascial bulbar sheath that runs from the optic nerve nearly to the corneal limbus or sclerocorneal junction, 2 mm dorsal to the corneal limbus. It is firmly adherent to the episclera. The white outer layer of the eyeball is called sclera, which is continuous with the transparent cornea at the front of the eye. The sclera is covered by the bulbar conjunctiva, the cornea lacks such conjunctival coverage. Both fibrous layers maintain the shape and size of the eyeball. The larger part, the firm sclera protects the inner contents of the

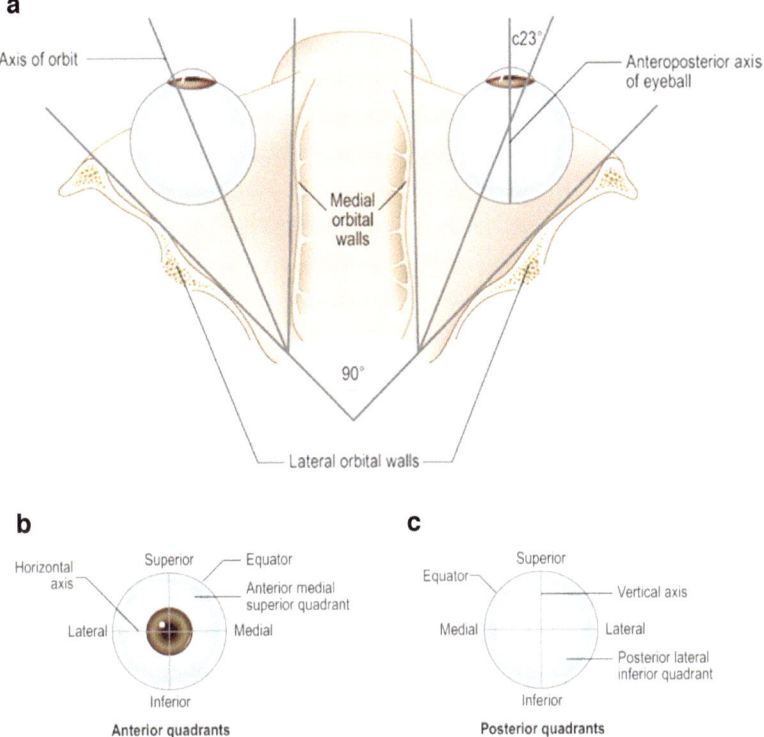

Fig. 3.17 Schematic view. Optic and orbital axis. (**a**) The orbits are pyramidal shaped: the apex = posteriorly, the base = anteriorly. The orbital axis runs from posterior to anterior in the bony orbit: it is a line that passes through the apex of the orbit and the center of the opening of the orbit Fig. 3.17a). The optic-ocular axis however: from posterior globe to anterior globe does not run parallel to the orbital axis, when there is a forward gaze. It is a straight line passing through the geometrical center of the lens. (**b**) coronal view of the anterior globe surface half up to the equator level with division into 4 anterior quadrants. (**c**) coronal view of the posterior globe surface half up to the equator level with division into 4 posterior quadrants. (With permission from: Grays Anatomy 42nd edition)

globe; in addition, it provides attachment to the extra ocular muscles. As such, Tenon's capsule, a dense elastic connective tissue membrane, is perforated to allow for adequate functional insertion of the EOM tendons.

The transparent cornea has a refractive power of 40–45 diopters and together with the lens (refractive power of 20–22 diopters) it assures that light beams are focused on the retina. The cornea has its thinnest aspect in the center. Its outer surface is covered by a multilayer and has an epithelium which is continuous with the adjacent bulbar conjunctiva, the structure that covers the anterior part of the globe, i.e., sclera and the backside of the eyelids.

For descriptive purposes, the conjunctiva may be divided into three subdivisions:

1. Tarsal or palpebral lining the eyelids

2. Forniceal, lining the upper and lower fornices

3. Bulbar, overlying the sclera on the anterior portion of the globe (Fig. 3.1)

While the cornea is mostly avascular besides vascular supply at the region of the limbus, many sensory nerves derived from the ciliary nerves reach the cornea and make it a highly sensitive area. At the posterior pole of the globe, the sclera is perforated by fibers of the optic nerve at the lamina cribrosa.

The limbus is created by the sclerocorneal junction.

The canal of Schlemm, a vascular scleral venous sinus is located within the posterior part of the sclerocorneal junction. The canal collects the aqueous humor from the anterior chamber and drains into the veins of the eyeball thereby

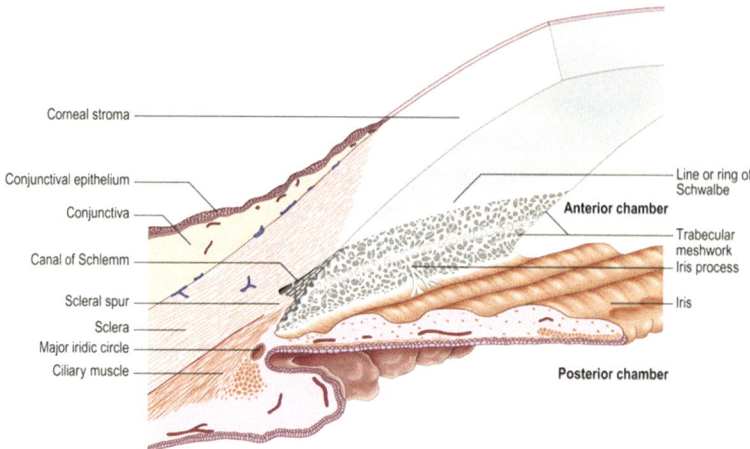

Fig. 3.18 Schematic view – Anterior chamber angle; the trabecular meshwork is an annulus of tissue spanning the angle; its meshes are shown in transverse section opposite the canal of Schlemm and partly attached to the scleral spur. A single iris process bridges the angle, connecting the trabecular meshwork to the anterior tissues of the iris. (with permission from: Grays Anatomy 42nd edition)

maintaining fluid homeostasis. The adjacent trabecular meshwork is a specialized tissue, draining aqueous humor from the anterior chamber and as such responsible for controlling the intraocular pressure.

There are in fact two fluid (aqueous humor)-filled chambers in the eye: an anterior chamber and posterior chamber. The anterior chamber including the trabecular meshwork is located between the cornea and the iris, the posterior chamber between the backside of the iris, the lens capsule, and the ciliary processes or corona ciliaris, which is part of the ciliary body (Fig. 3.18). The aqueous humor is produced by the ciliary body and flows into the anterior chamber through the narrow slit communication between the iris and the lens.

Centrally, behind the lens, the vitreous body (corpus vitreum) is located against the inner surface of the retina.

The uvea, or uveal tract is composed of three parts: iris, ciliary body, and choroid.

The iris, from anterior to posterior consists of an anterior border layer, a stroma including the sphincter pupillae muscle, an anterior pigment epithelium including the dilatator muscle and a posterior pigmented layer. The ciliary body, contiguous with the iris anteriorly and with the choroid posteriorly contains the ciliary muscle and the ciliary processes.

The largest part of the uvea is the choroid, the vascular layer of the eye ball, which extends from the ora serrata and is responsible for the nourishment of the outer half of the retina.

The iris, the diaphragm of the eye, a circular pigmented organ, the part of the uvea located anteriorly, conveys the pupil, a central hole looking black. This opening is regulated by the activity of the sphincter and dilatator pupillae muscle; the sphincter muscle being strongest.

The ciliary body also inhabits the smooth ciliary muscle and the ciliary processes which suspend the lens by the ciliary zonulae of Zinn: suspensory ligaments closely related to the ciliary processes and connecting the capsule of the lens to the ciliary body.

The lens itself, positioned behind the iris is situated on the anterior surface of the vitreous body. The ciliary muscle changes the curvature of the lens, necessary for accommodation, i.e., focusing of the eye. The central portion of the anterior lens is related to the pupil.

Contraction of ciliary muscle relieves tension upon the lens resulting in more convex form, which is made possible by the outermost layer of the lens, the highly elastic capsule. Increase of lens curvature resulting after contraction of ciliary muscle allows for near vision.

The choroid, supplied by the posterior ciliary arteries lies between the sclera and retina but is absent in the region of the lamina cribrosa to allow the optic fibers to exit to form the optic nerve. The choroid is built up of four layers: from outward, larger vessels inward containing smaller vessels up to capillaries and an inner layer, Bruch's membrane, the lamina basalis or lamina vitrea.

The serrated junction between the retina and the ciliary body is called ora serrata: the termination of the retina which corresponds with the acute transition between the nonphotosensitive, nonfunctional area of the ciliary body and the complex multilayered photosensitive region of the retina. The inner surface of the choroid is covered by a pigmented single cell layer (stratum pigmenti) which continues around the whole inner surface of the eye.

The innermost layer, the so called neural tunic or retina consists of ten layers, from outward inwards the most fundamental being the layer of rods and cones (photoreceptors), the layer of bipolar cells, and a layer of ganglionic stratum of the retina. The innermost layer, the stratum internum borders the aqueous humor from the retina. External to the outer layer of rods and cones is the stratum pigmenti, which finally absorbs the light which has been transmitted to all the other inward directed layers. Located in the posterior retina is the macula lutea (central portion of vision) with the fovea centralis, measuring 1 mm². The fovea (centralis), a depression within the macula is located in a straight line behind the lens. Here the retina is markedly thin because of the absence of ganglionic cells; the rods (night vision) gradually disappear from the periphery to the center and are replaced by cones (day and color vision). The fovea centralis contains only cones.

During daylight, visual acuity is high including color vison, cones are stimulated more so at the fovea centralis. At night, mainly rods, which are present more abundantly in the macula than in the fovea centralis are activated and visual acuity is low. In the macula, around the fovea, the ganglionic cell layer of the optic nerve is thick. There is a point where to the nerve fibers converge to result in the optic disk: optic papilla. Here, no other layers are present: the disc is insensitive to light: the blind spot in the visual field. From the optic disc, the nerve fibers continue, after passing through the lamina cribrosa to form the myelinated optic nerve.

The retina is mainly supplied by the central retinal artery. A second (choroidal) circulation comes via the short and long posterior ciliary arteries. The central artery of the retina passes through the lamina cribrosa adjacent to the optic nerve at the location of the optic disk. The artery is accompanied by the central vein which drains in the cavernous sinus.

The globe is surrounded extensively by connective tissue septa including the extraocular muscles (EOM) connecting them all to the periorbita. It consists of a complicated network which maintains the spatial relationship and supports the coordinated globe position during ocular movement. It also serves as a protection system, especially in coadjuvancy with the abundant presence of orbital fat.

The optical axis extends from the anterior to the posterior pole of the eye and is defined by the straight line going through the geometrical center of the lens and the cornea.

The visual axis represents the direction of gaze; an axis from a fixation point toward the center of the pupillary entrance subsequently reaching the fovea.

As mentioned above the orbital axis is the line from the center of the optic foramen in the apex extending anteriorly, laterally, and inferiorly to the middle of the orbital aperture.

The optical axis and orbital axis do not align when looking straight forward (Fig. 3.17). The angle between the optical axis and the visual axis is called angle alpha.

Clinical Implications

An A-P diameter of 24 mm is seen in emmetropic (normal) eyes. An axis of > 24 mm is called myopia, of < 23 mm hypermetropia (Fig. 3.19).

Myopia (Nearsightedness)—difficulty focusing on faraway objects because of an abnormally elongated eyeball. Light rays come into focus before they reach the retina and begin to diverge again by the time they fall on it. Corrected with concave lenses, which cause light rays to diverge slightly before entering the eye.

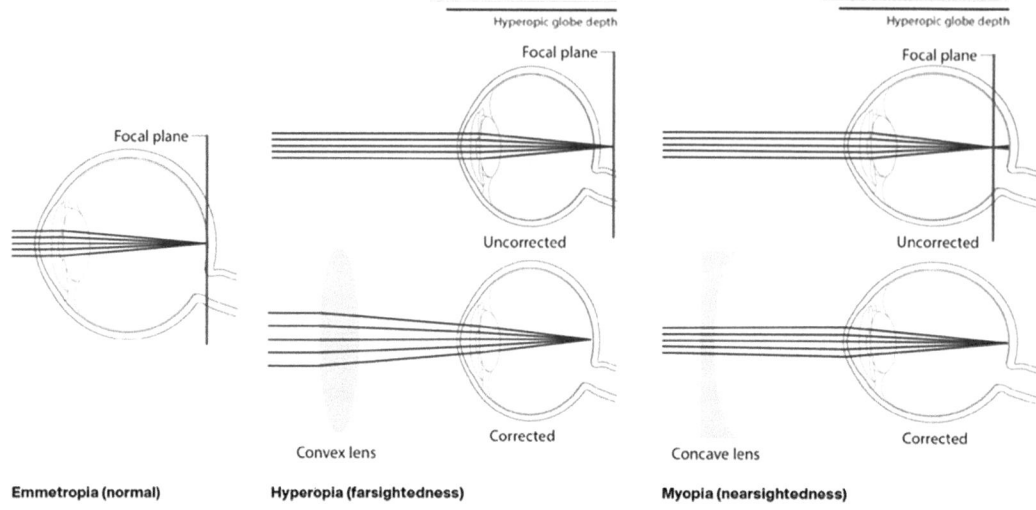

Fig. 3.19 Normal emmetropic eye in contrast to hyperopia and myopia and the effects of corrective lenses. (With permission from S. Steenen)

Hyperopia (Farsightedness)—difficulty focusing on nearby objects because of an abnormally short eyeball. The retina lies in front of the focal point of the lens, and the light rays have not yet come into focus when they reach the retina. Corrected with convex lenses, which cause light rays to converge slightly before entering the eye.

Presbyopia—Declining ability to focus on nearby objects as one ages. An effect of declining elasticity of the aging lens, often first noticed around age 40–45. Results in difficulty in reading and doing close handwork. Corrected with reading glasses or bifocal lenses.

Astigmatism—Inability to simultaneously focus light rays that enter the eye on different planes. Focusing on vertical lines, such as the edge of a door, may cause horizontal lines, such as a tabletop, to go out of focus. Caused by a deviation in the shape of the cornea so that it is shaped like the back of a spoon rather than like part of a sphere. Corrected with "cylindrical" lenses, which refract light more in one plane than another.

Zinn's Ring and Extraocular Muscle System

Zinn's Ring

The common annular tendon is composed of fibrous tissue condensations from the periorbita within the orbital apex, the dura lining the SOF, and optic canal and the optic nerve sheath. The apical orbital area is located posteriorly in the orbit, exact definitions however vary (Chap. 2). The rectus extraocular muscles (EOM) originate from the ring, while the LPSM, the superior oblique muscle (SOM), and the inferior oblique muscle (IOM) outside the Zinn's ring (Figs. 3.20, 3.21a, b, and 3.22). The rectus muscles diverge anteriorly to insert onto the sclera (Figs. 3.22 and 3.23).

Zinn described the orbital fibrotendinous ring already in 1755 (J. G. Zinn 1727–1769 German Botanist and Anatomist Göttingen) (Fig. 3.24a–e). In the classic description, Zinn's ring is the common origin of the rectus musculature (Figs. 3.20 and 3.24).

As the term apex already suggests, the orbital apical part is the narrowest compartment of all (Chap. 2). The bony boundaries of the apex con-

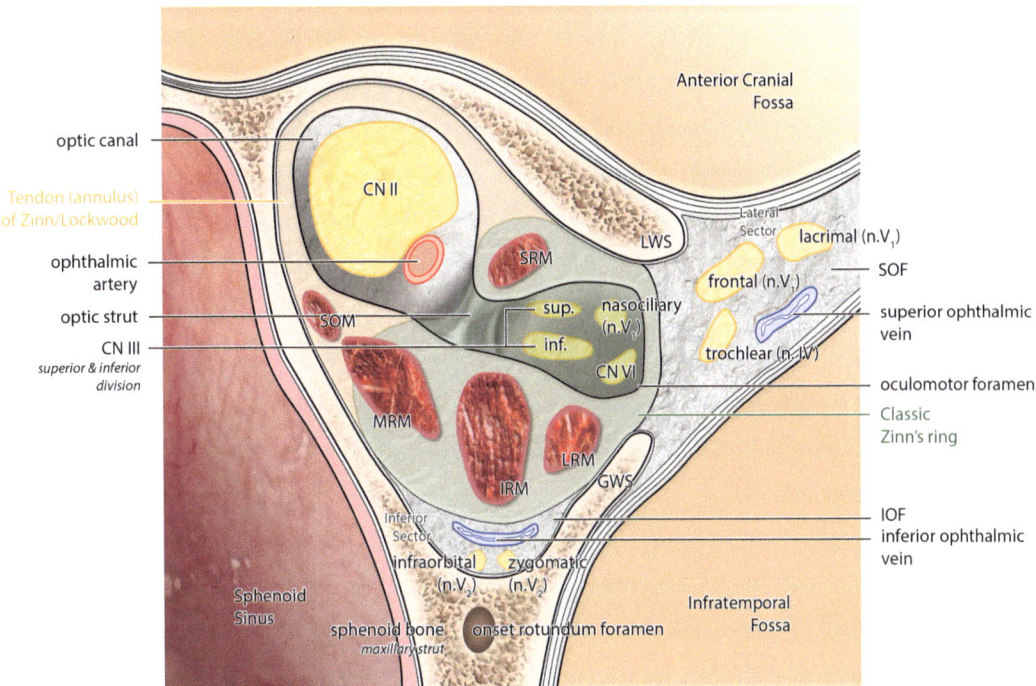

Fig. 3.20 Schematic view – Zinn's ring. Classic Zinn's ring is outlined: the annulus tendon of Zinn which is the common origin of the four rectus muscles. The entire circumference of Zinn's ring includes the opening of the optic canal (OC) and the central portion of the superior orbital fissure (SOF). The optic strut is located inferolateral to the optic canal and divides the ring into the superomedial annular foramen which transmits the optic nerve (CN II) and ophthalmic artery (OA) and the superolateral foramen, i.e. the intra-annular central portion of the superior orbital fissure (SOF) which allows passage of the oculomotor nerve (CN III) inferior and superior division, abducens nerve (CN VI), and the nasociliary nerves. There are two extra-annular compartments -superolateral portion of the SOF and the inferiorly located posterior portion of the inferior orbital fissure (IOF). *CN II* optic nerve, *CN III* oculo-motor nerve, *CN VI* abducens nerve, *SRM* superior rectus muscle, *SOM* superior oblique muscle, *MRM* medial rectus muscle, *IRM* inferior rectus muscle, *LRM* lateral rectus muscle, *SOF* superior orbital fissure, *IOF* inferior orbital fissure, *LWS* lesser wing sphenoid, *GWS* greater wing sphenoid. (With permission from S. Steenen)

sist of the lesser wing of the sphenoid: the roof, the ethmoidal sinus: the medial wall, the greater wing of the sphenoid: the lateral wall and the maxillary strut, the bony extension of the greater wing of the sphenoid connecting to the lesser wing: the floor of the apex. In fact once we cross the maxillary strut posteriorly, we enter the apex orbita [15, 17, 47] (Chap. 2).

Zinn's ring begins at the orbital openings of the optic canal and SOF and is continuous with the dura mater, cavernous sinus, the optic canal, and the optic nerve sheath. It extends posteriorly along the optic strut coursing intracranially, originating from the lateral wall of the sphenoid body. The tissues are blended in a funnel, circular array structure extending backward rather than a ring, though for reasons of simplicity they are still portrayed as a cylindrical ring. The terminal posterior tip of the tendinous funnel inserts at the infraoptic tubercle or a canalicular depression which is located below and posterior to the optic strut.

The central SOF compartment or superolateral foramen, accommodates the oculomotor (CN III), and abducens (CN VI), and nasociliary nerves

(branch of CN V1) as well as the sensory and sympathetic roots of the ciliary ganglion. The CN III divides into its superior and inferior division before entering the ring. Since the superolateral

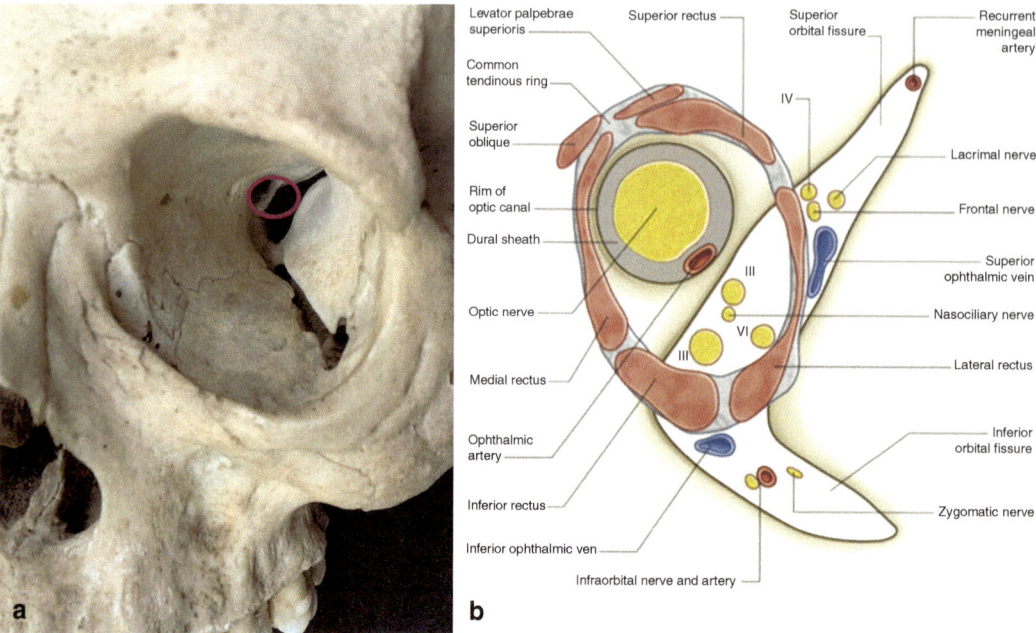

Fig. 3.21 (**a**) Anatomic specimen – Skull with purple outline of the position of Zinn's ring in the posterior orbit, i.e., within the apex of the orbit (**b**) Schematic view of the common tendinous ring with regard to the location of superior orbital fissure (SOF) and inferior orbital fissure (IOF); muscle origins have been superimposed: the attachment of the four rectus muscles at the ring is depicted as well as the position of the LPSM, SOM and IOM just outside of the ring. The relative positions of the nerves and vessels that enter the orbital cavity by passing through the superior orbital fissure, optic canal, and oculomotor foramen are shown. The levator palpebrae superioris and superior

oblique lie external to the common tendinous ring to it. The ophthalmic veins frequently pass through the ring. A recurrent meningeal artery may run from the orbit into the cranial cavity through the upper lateral extension of the SOF instead of a separate crania orbital foramen (COF). The inferolateral component of the annulus of Zinn divides the SOF into an intra-annular compartment corresponding to the so-called oculomotor foramen and two extra-annular compartments - the superolateral SOF and the posterior OIF inferiorly. *CN III* oculomotor nerve, *CN IV* trochlear nerve, *CN VI* abducens nerve. (With permission from: Grays Anatomy 42nd edition)

Fig. 3.22 Anatomic specimen – Rectus muscles originating from Zinn's ring. Note that the four rectus muscles project forward to enclose the intraconal space while the SOM and the IOM are positioned outside the tendinous ring and thus are located in the extraconal space. *TROC* trochlea, *SOM* superior oblique muscle, *LPSM* levator palpebrae superior muscle, *OPLG* orbital part lacrimal gland, *SRM* superior rectus muscle, *ON* optic nerve, *MRM* medial rectus muscle, *LRM* lateral rectus muscle, *IRM* inferior rectus muscle, *IOM* inferior oblique muscle, *CN III* oculomotor nerve

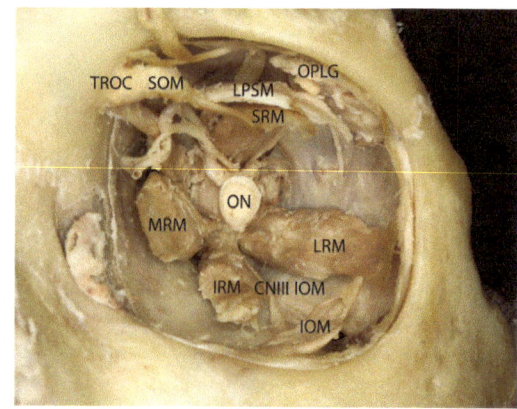

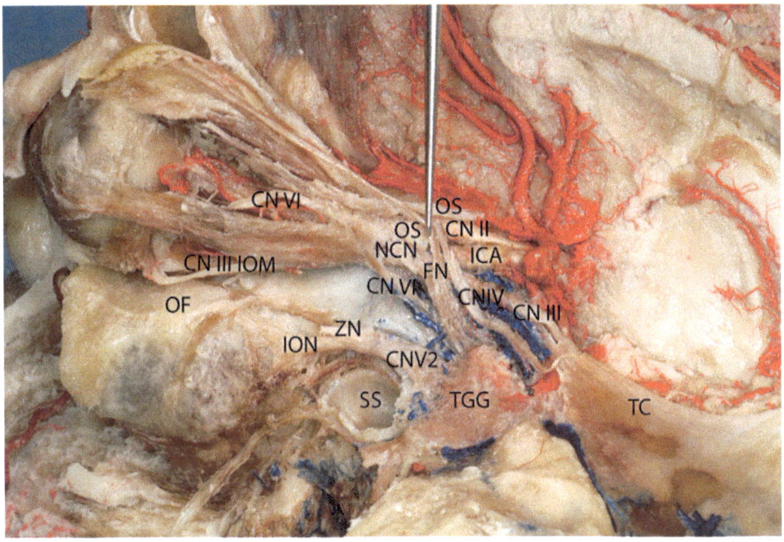

Fig. 3.23 Anatomic specimen – Contents of posterior orbit with focus on oculomotor foramen. Optic strut (OS) partially removed and oculomotor foramen opened from lateral side—frontal nerve (FN) passing through the lateral part of the oculomotor foramen retracted upward to expose nasociliary nerve (NCN) oculomotor (CN III) and abducens nerve (CN VI). FN and trochlear nerve (CN IV) are displaced superomedially on top of the inferior end of the optic strut (next to metal probe). CN VI enters the ocular surface of the lateral rectus muscle. CN III running medially to the trochlear and abducens nerve. *CN II* optic nerve, *CN III—IOM* oculomotor branch to inferior oblique muscle, *OS* optic strut, *CN IV* trochlear nerve, *FN* frontal nerve, *CN VI* abducens nerve, *CN V2* trigeminal nerve— maxillary division, *ION* infraorbital nerve, *ICA* internal carotid artery, *OF* orbital floor, *SS* sphenoid sinus, *TC* tentorium cerebelli, *TGG* trigeminal ganglion, *NCN* nasociliary nerve, *ZN* zygomatic nerve

foramen contains two of the optomotor nerves (CN III and CN VI), it is termed the oculomotor foramen (Figs. 3.20 and 3.21b).

The superior extra-annular compartment of the SOF transmits the frontal and lacrimal nerves (branches of CN V1), the trochlear nerve (CN IV), and the superior ophthalmic vein (Figs. 3.20 and 3.21b). The inferior extra-annular compartment contains the inferior ophthalmic vein and sympathetic fibers which accompany the internal carotid plexus eventually reaching the ciliary ganglion as well as the infraorbital and zygomatic nerves embedded in extraconal fat.

Summarizing, the annulus of Zinn surrounds the optic foramen and oculomotor foramen which encloses the central 1/3rd of the superior orbital fissure (Figs. 3.20 and 3.21a, b).

The extraocular muscle (EOM) system generates the coordinated voluntary ocular movements via six extraocular muscles, striated skeletal four rectus and two oblique muscles (Fig. 3.22).

The lateral, inferior, and medial rectus muscles (LRM, IRM, MRM) originate from the inferior tendon of Zinn's ring, which is located inferior to the optic foramen; the superior rectus muscle (SRM) originates from the superior margin of the optic foramen and completes the upper aspect of the ring (Fig. 3.21b). Altogether, the rectus muscles arrange in a cone shape when they run anteriorly and form a "sleeve"-like structure surrounding and bounding a separate compartment behind the ocular globe (Fig. 3.22). This results in the differentiation into an intra- and extraconal space, the intraconal space located within a musculofascial cone with the base anterior, formed by the posterior half of the globe which converges posteriorly on the tendinous ring at the orbital apex where the rectus muscles insert (Figs. 3.22, 3.24, and 3.25)—see

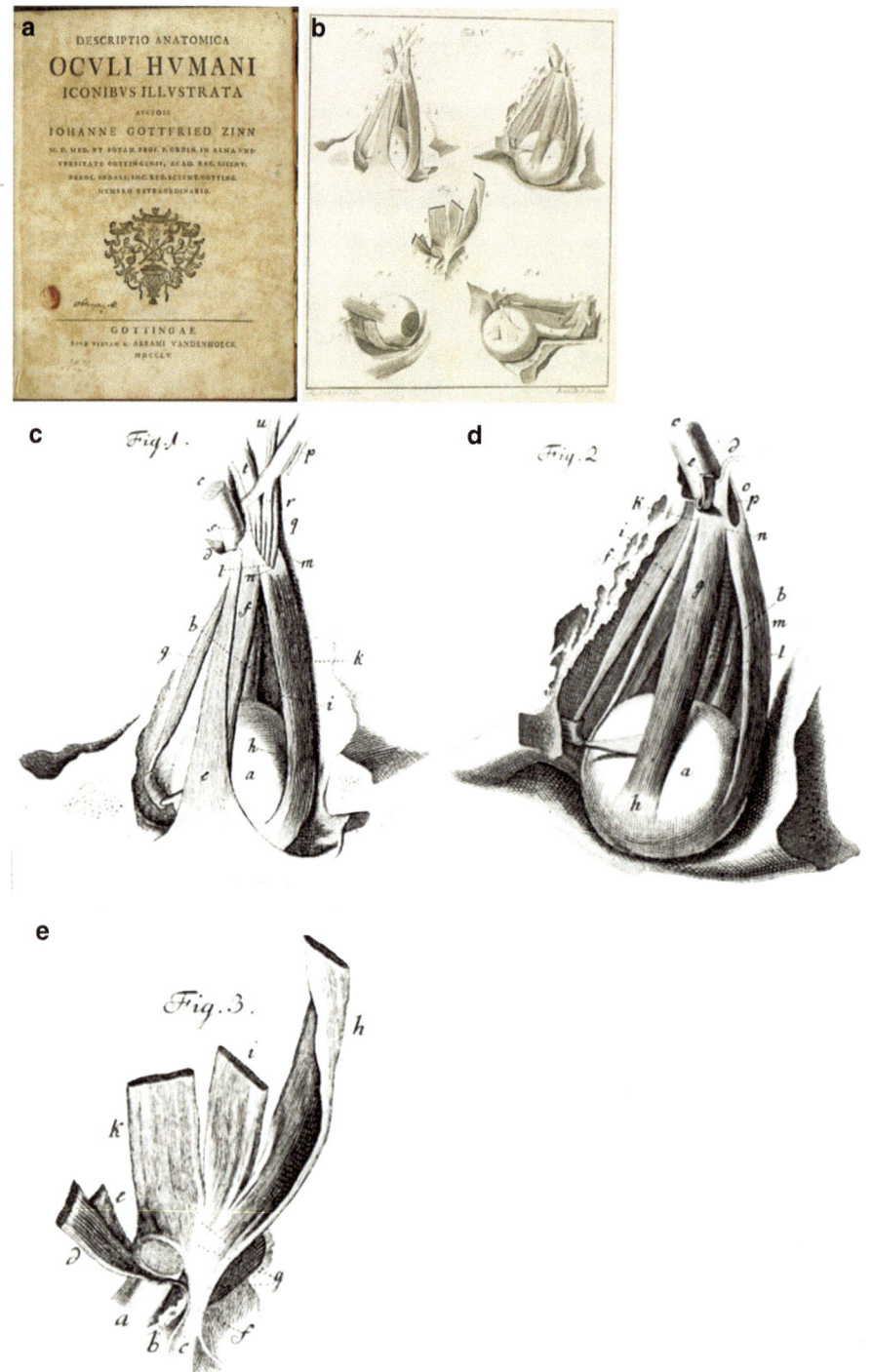

Fig. 3.24 (**a**) Cover title of Johann Gottfried Zinn's Book—Descriptio anatomica oculi humani iconibus illustrata, Gottingae: Viduam B. Abrami Vandenhoeck, 1755. (**b**) Original drawings of Table V titled "Musculi bulbi oculi," from J.G. Zinn's book and its figures 1 - 5 showing EOM and the 'common' annular tendon (Zinn's ring) from different perspectives. (**c**) Fig. 1 from Table V enlarged - superior view of left orbit with LPSM, SRM, SOM, LRM and the 'common' (= 'Ligamento communi' in original terminology) annular tendon (**d**) Fig. 2 from Table V enlarged – same view as in Fig. 1 after removal of the LPSM, ethmoidal air cells opened. (**e**) Fig. 3 from Table V enlarged - inferior frontal view of left annular/'common' tendon encompassing the optic nerve a (cross section anterior) superomedially and with the EOM (from left to right) MRM, SOM in the background, LPSM, SOM and LRM

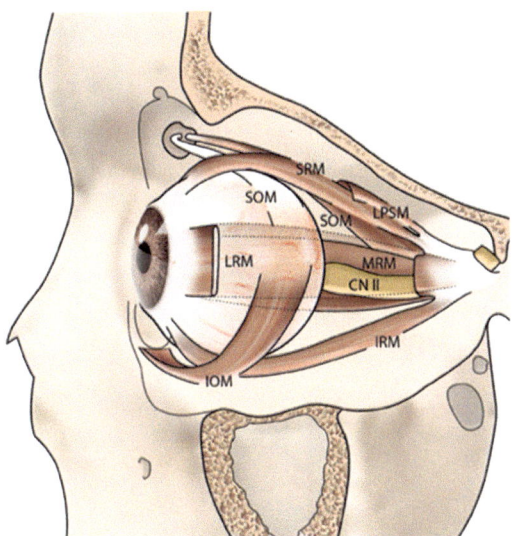

Fig. 3.25 Schematic lateral view—EOM: conal arrangement between Zinn's ring and scleral/globe attachments. Note SOM and IOM attachments posterior to the globe—other than rectus muscles: lateral rectus muscle anterior attachment anterior to the globe equator. *LRM* lateral rectus muscle, *SOM* superior oblique muscle, *SRM* superior rectus muscle, *LPSM* levator palpebrae superior muscle, *MRM* medial rectus muscle, *IRM* inferior rectus muscle, *CN II* optic nerve. (With permission from S. Steenen)

below chapter Topographic spaces of the orbit. Inferiorly, connective tissue extends from the annulus (Zinn's ring) to blend with Müller's orbital smooth muscle.

The LPSM arises from the periorbita outside Zinn' ring over the lesser wing of the sphenoid just medial of the SRM. It runs in close contact just above and medially to the SRM until the muscles diverge at the level of the SRM scleral insertion.

The origin of the SOM is assigned to superomedial portion of the common annular tendon (Zinn's ring) deriving from the periorbita covering the sphenoid body. The IOM does not originate from the annulus, but arises from the periosteum of the inferior orbital rim immediately lateral to the opening of the bony lacrimal duct.

As the CN IV emerges the annular tendon, it innervates the SOM from the extraconal surface, this is in contrast to the innervation on the intraconal (ocular) side for all other EOM.

Approximately 10 mm anterior to the optic strut, the EOM (LRM, IRM, MRM, SRM) and the SOM continue anteriorly as separate structures parallel to the orbital walls until they insert

into the sclera anterior or posterior (SOM) to the equator of the globe (Fig. 3.23).

On their way, the MRM [48] and SRM show some origin from the adjacent dura covering the optic nerve. Lattice-like tissue septa separate the origin of the rectus muscles while radial septa connect the muscle to the adjacent orbital wall [49, 50]. These connective tissue septa are highly organized and form an accessory locomotor framework with structural organization and constant pattern; the soft tissue system, apart from connective tissue contains vessels, nerves, and smooth musculature. The relation of the eye musculature, the orbital walls, and the globe with these connective tissue structures allows for smooth normal eye movement [49, 50].

Before entering Zinn's ring, the intracranial neurovascular structures run via the optic canal, pass the optic foramen to enter the orbital apex [47]: the optic nerve (ON), the ophthalmic artery (OA), and the postganglionic sympathetic nerves that accompany the carotid plexus.

Tenons capsule invests the anterior projection of the EOM. The muscle fibers are separated from each other by a surrounding endomysium. Halfway through forward, the muscle cone is centered within the orbit, ± 8 mm from the adjacent orbital walls. When the muscles enter Tenons capsule at the posterior surface of the globe, fascial strands from the muscle sheaths are interconnecting with Tenons capsule to contribute to the pulley—suspensory system.

A complex system of thin fibrous sheaths and fat tissue is present between the muscles and this functions together with connective tissue structures as a pulley through which the muscle moves and determines the direction of the muscle pull.

Once Tenons capsule is entered, the rectus muscles arc over the globe, flatten once across a model of orbital mechanics based the equator, and insert onto the sclera via a tendinous ligamentary band, approximately 10 mm in width (Fig. 3.26).

A fascial condensation from the sheath of the medial rectus muscle contributes to the medial check ligament and this, in conjunction with the medial horn of the LPSM, attaches behind the posterior lacrimal crest to the orbital septum and to the caruncle and plica semilunaris. The lateral check ligament is a fascial condensation from the lateral rectus sheath to Whitnall's tubercle; addi-

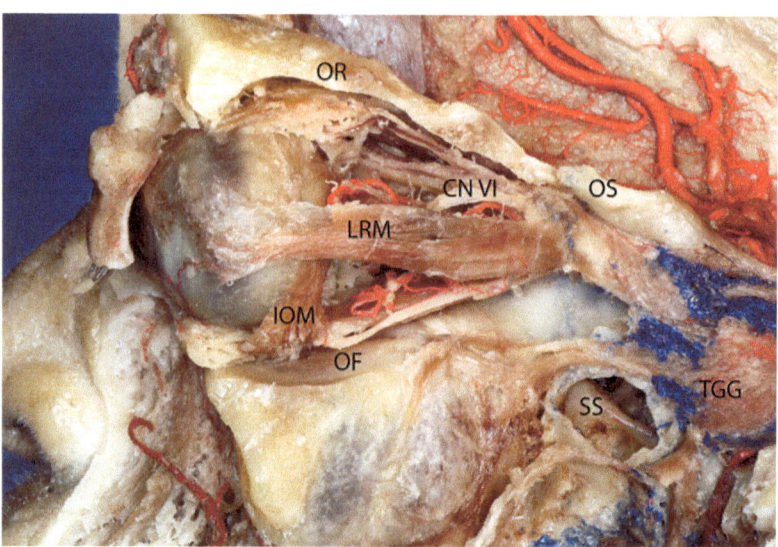

Fig. 3.26 Anatomic specimen – EOM cone—lateral view, attachments of LRM and IOM to the sclera anterior and posterior to the equator of the globe, respectively; insertion of LRM at the tendinous ring of Zinn. IOM runs inferior and medial to IRM. Relation to orbital roof (OR) and orbital floor (OF). Neighborhood of sinus sphenoidalis and trigeminal ganglion. *LRM* lateral rectus muscle, *OR* orbital roof, *IOM* inferior oblique muscle, *OF* orbital floor, *CN VI* abducens nerve, *OS* optic strut, *SS* sphenoid sinus, *TGG* trigeminal ganglion

tional attachments are present to the orbital septum and fornix of the conjunctiva.

The SOM runs forward and becomes invested by collagen with elastin fibers. It is intimately involved in the connective supportive system of the globe.

Unlike the rectus and IOM pulleys, the SOM has a rigid pulley in the superonasal orbit, the trochlea, that redirects the intermediate SOM tendon and consecutive muscle portion to the postero-supero-lateral globe quadrant.

The trochlea is a circular fibrocartilaginous structure, about 5 mm, firmly adherent to a bony fovea—occasionally to a spine of the frontal bone at the fovea trochlearis (Fig. 3.22); it consists of fibrocartilaginous elements making up an overall cylindrical shape with a characteristic posterolateral flange at the anterior end:

- A biconcave cartilage "saddle" of limited size (ca. 4 × 4 × 5.5 mm—width/height)
- An inner-tube fibrovascular or trabecular sheath providing the outer laminae for a fluid-filled bursa-like paratendinous space
- An outermost layer of dense fibrous condensations with a bony securement folding

Presumptively, the SOM tendon moves within the trochlea by staggered interfiber sliding actions in concert with telescoping of the entire tendon bundle.

The SOM then finally inserts on the sclera on the posterotemporal globe surface, which explains its function: it will rotate the globe: infraduction (depression), incycloduction (internal rotation), and slight abduction.

The IOM is unique because of its origin far apart from Zinn's ring.

It arises from the periosteum of the medially located orbital surface of the infraorbital rim, 1.5 mm lateral to the entrance of the nasolacrimal canal. Its course is posteriorly, lateralward and upward at an angulation of about 50° to the medial orbital wall, bends around the lateral globe curvature toward its final insertion to the sclera at the postero-inferior-lateral aspect of the globe. Along its course, it passes inferior to the IRM where both enveloping sheaths of the IOM and IRM fuse. Here, thickening of Tenons capsule is present contributing to Lockwood's ligament. Its function is extorsion (external rotation), elevation, and abduction: it rotates the eye and moves it upward and outward.

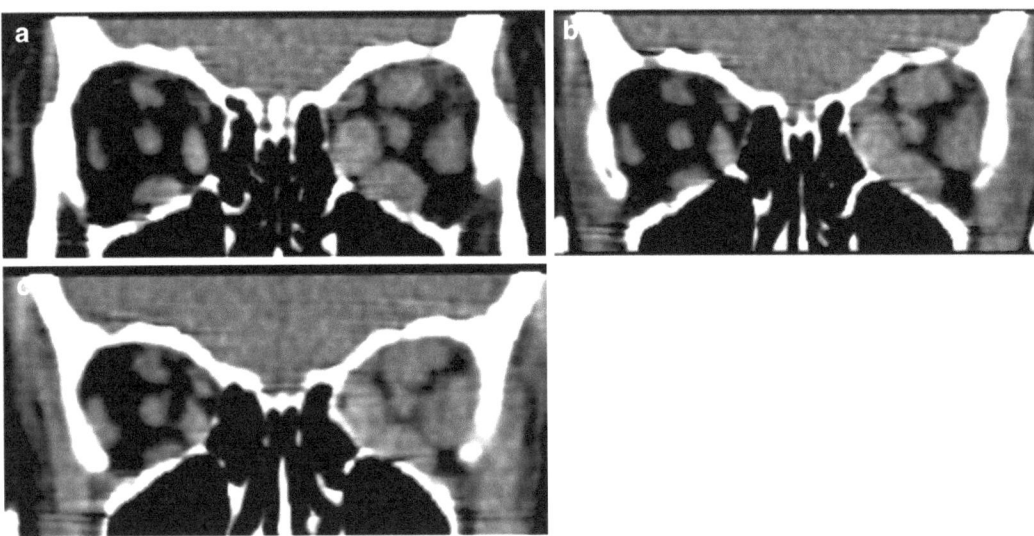

Fig. 3.27 (a–c) Coronal CT scan slides of the orbit: A = anterior; B = midway; C = posterior. Progressive orbital apical crowding in a female patient with unilateral Graves' orbitopathy. Note the increase in volume of the rectus musculature in the left orbit

Clinical Implications

In case of Graves' orbitopathy (Chaps. 15 and 16), enlargements of the EOM can be seen. They concern the muscle-bellies themseselves, not the tendons. Enlargement can reach 2–3 times the normal volume. Although all EOM can be involved, the inferior and medial rectus are involved most frequently.

The enlargement is caused by inflammatory changes including accumulation of glycoprotein, mucopolysaccharides as well as GlycosAminoGlycans (GAG) [11]. Enlargement can be massive causing apical compression.

The volume changes are responsible for the proptosis. In extreme volume increase of EOM, apical crowding can result which may lead to progressive pressure on the optic nerve: malignant Graves' orbitopathy: urgent apical decompression is indicated in such cases (Fig. 3.27a–c).

Eventually, inflammatory contracture will often result in motility disturbances and subsequently diplopia.

The anterior herniation of extraconal fat in patients with Graves' disease affects the upper eyelid.

Traumatic rupture of the superior oblique tendon in approximation of the trochlea is a rare cause of acquired Brown's syndrome: restricted elevation in adduction: upward gaze limitation.

Brown in 1949 [51–53] described a shortening of the tendon sheath of the superior oblique tendon. The origin of this entity however is still subject of debate.

Surgical reinsertion is highly favorable for the outcome and to prevent malfunction of the SOM.

As the trochlea is located very anteriorly and as such likely to be affected in medial orbital injury, direct trauma to the trochlea—pulley and/or superior oblique muscle tendon may result: acquired Brown's syndrome.

In case of an orbital roof fracture (Chap. 11), also restriction of the superior oblique muscle tendon may occur [53] Acquired Brown's syndrome [54] can be treated by superior oblique and trochlear luxation.

Once the connective tissue framework within the orbit is ruptured, motility disturbances of the globe may result [49, 50].

Topographic Spaces of the Orbit

The orbital cavity itself is divided into several spaces [14]: the subperiorbital or subperiosteal space, the extraconal and intraconal spaces, and the lacrimal space. The subperiorbital or subperiosteal space is a potential space. The cone-shaped array of the EOM and the connective septa system separate the extraconal from the intraconal space. Thus the extraconal space corresponds to the compartment outside the musculofascial cone. It is bounded by the periorbita externally and by the orbital septum anteriorly. The inner side of this space is formed by the fascial EOM envelopes.

The intraconal space is retrobulbar and lies within the musculofascial cone; so its anterior base is the posterior sphere of the globe while its sides are formed by the extraocular muscle contiguous to Zinn's Ring within the orbital apex.

The intraconal space contains the CN II, the superior & inferior division CN III (including the parasympathetic motor root to the ciliary ganglion and the ganglion itself), nasociliary branches of CN V1 including the long and short ciliary nerves and CN V1, and branches of the ophthalmic artery including the central retinal artery.

The extraconal space contains the lacrimal and frontal and extraconal branches of the nasociliary branches: ethmoidal and infratrochlear nerve (CN V1), zygomatic and infraorbital branches (CN V2), CN IV, branches of the ophthalmic artery and vein, and the lacrimal gland.

Both compartments contain abundant intraconal and extraconal fat which protects and facilitates motility. The lacrimal gland fossa can be regarded as a distinct compartment accommodating the orbital lobe of the lacrimal gland.

Adipose Body of the Orbit (ABO)

Fat, the adipose body of the orbit (ABO) is abundant within the confines of the orbital cavity [5] and around the outer orbital area. The ABO fills all spaces in the orbit left empty by the periorbita, connective tissue septa system, globe, muscles, neurovascular, lacrimal and glandular structures and represents almost half of the total orbital volume. It is not consistently unraveled yet whether the ABO is a continuous entity with ramified projections or if it is partitioned into single zones. This seems of interest because fat compartments might behave differently with aging or a disease process. The retrobulbar intraconal and extraconal part of the ABO accounts for most of the intraorbital fat.

The extraconal collection of fat continues anteriorly and distributes circumferentially around the globe (Fig. 3.28a, b). This peribulbar ABO area appears to furnish fine extensions next to the fornix margins into the upper and lower eyelids where they merge into two or three fat compartments, respectively.

Recently the lower, potentially also the upper, eyelid fat pads were reported as discrete compartments being separated from the posterior intraorbital ABO by kind of a fascial membrane, called the circumferential intraorbital retaining ligament. This ligamentous barrier is ascribed to the connective tissue system and the EOM pulleys. It attaches alongside the equator of the globe and along the perimeter of the orbital wall facing to it.

The quality of the ABO fat lobules, i.e., their size and shape, differs corresponding to their topography. The anterior peripheral orbital zones show a small packed and rather fibrous lobulation, while the retrobulbar central zone back into the apex presents large, egg-shaped lobules woven into a few thin fibrous septa.

Besides a cushioning function, a model of orbital mechanics based upon Finite-Element Analysis (FEA) confirms the theory that the supporting action of the orbital fat contributes to the suspension of the eyeball and to the stabilization of the rectus EOM gliding tracks.

Distinct preaponeurotic fat pockets are located directly deep to the septum.

The fat pockets are anterior extensions of extraconal orbital fat.

In the upper eyelid, two fat pockets are located anterior to the LPSM, divided in a medial and a larger central location and covered by a thin capsule. Interlobular septa are abundantly present. The lateral compartment is occupied by the lacrimal gland.

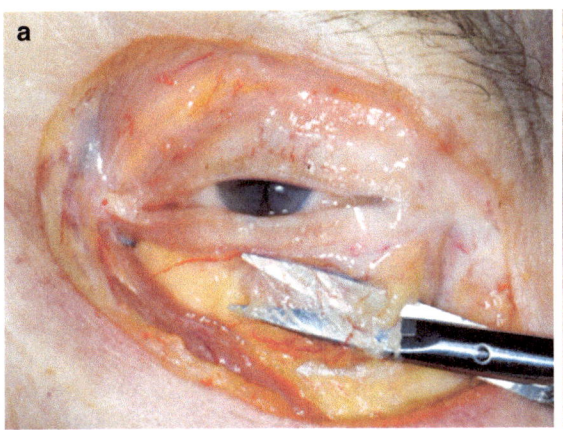
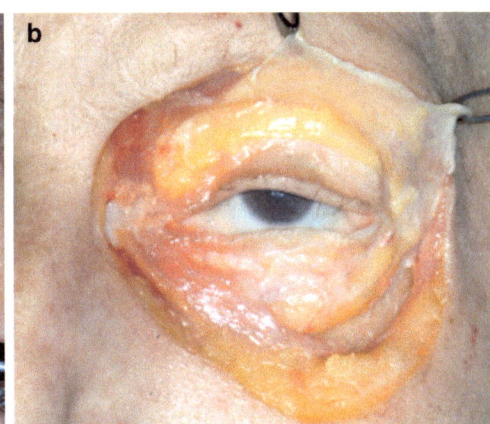

Fig. 3.28 Anatomic specimens _ (**a**) Orbital septum elevated (scissors) over fat compartments in the lower eyelid (**b**) Fat compartments of the eyelids. Upper eyelid: two retroseptal or preaponeurotic fat pads plus the lacrimal gland underneath the elevated septum. Lower eyelid: three fat pads still contiguous and covered by orbital septum—

Cranial to the orbicularis muscle, the frontalis muscle is separated from the underlying periosteum by a fat pocket: the superior Retro-Orbicularis-Oculi Fat pocket (ROOF). This fat tissue extends from the supraorbital notch to the temporal ligament laterally. It is buried in a split of the deep galea layer. The fat pocket can descend into the upper eyelid through the orbicularis retaining ligament, projecting downward just anterior to the orbital septum and so behind the orbicularis muscle.

In the lower eyelid, there is a medial, central, and lateral fat pocket.

The arcuate expansion of Lockwood's ligament spreads in between the central and lateral (temporal) fat pocket. The lateral fat pocket can be multilocular. Eisler's "fat" pocket is a minor accessory fat pad located superficial to and immediately above Whitnall's tubercle. The central and medial fat pockets are separated by the inferior oblique muscle (IOM).

Along the face, a system of several layers of fat pads is present. Superficially, there is the subcutaneous fat layer which is absent in the eyelids.

In the malar region, the Sub-Orbicularis-Oculi Fat (SOOF) pocket is located inferior to the inferior orbital rim. A medial component along the orbital rim, the medial SOOF is distinguished from a lateral SOOF compartment extending over the malar prominence. The lateral SOOF appears to connect by means of a temporal tunnel to the inferior temporal compartment. Superiorly it is bordered by the orbicularis retaining ligament, which separates it from the eyelid.

Neurovascular Anatomy of the Orbit

Optic Nerve (CN II)

It is essential to understand that the so called N. opticus is a white matter protrusion from the diencephalon which according to the current state of research is unlike other cranial or peripheral nerves incapable to regenerate in terms of a restitutio ad integrum. Even though a label such as "fasciculus opticus cerebralis" would appear more meaningful, the expression optic nerve is the conventional nomenclature. Each CN II is part of the afferent visual pathway (Functional Category: Special Somatic Afferent—SSA) composed of approximately 1.2 million retinal ganglion cell axons which are myelinated by oligodendrocytes posterior to the lamina cribrosa. CN II passes from the globe and the orbit via the optic canal to the optic chiasm with a total length of 45–50 mm and can be divided into four zones: intraocular, intraorbital, intracanalicular, and intracranial.

The intraocular component or optic nerve head (optic papilla) is only 1 mm thick. The axons leaving the retina to form the optic nerve posteriorly exit through the perforations in the sclera (lamina cribrosa) and become myelinated.

The intraorbital CN II portion or retrobulbar segment is 3–4 mm in diameter and 25–30 mm long. This portion runs a serptine course to cover the 20 mm spatial distance between the optic foramen and the posterior pole of the eyeball. This reserve length allows for movement and a certain limit of distension. Before entry into the canal, CN II passes through the superomedial annular foramen of Zinn's ring. Meningeal layers (pia mater, arachnoidea, dura) envelope the CN II along its intraorbital and intracanalicular course. The subarachnoidal space around CN II contains cerebrospinal fluid and communicates with the intracranial subarachnoidal space (chiasmal cistern).The central retinal artery, originating from the ophthalmic artery, and an accompanying vein pierce the dural nerve sheath from inferomedially some 5–15 mm posterior to the globe, advance within the subarachnoid space, and penetrate the nerve stroma in a vertical direction on their way to the retina.

The intracanalicular CN II inside the LWS (optic canal—Table 3.1) is 5–8 mm long and supplied with pial branches of the ophthalmic artery which runs along the inferolateral surface of the nerve. Within the canal, the optic nerve is vulnerable to compression.

The two intracranial CN II parts, running along the medial aspect of the anterior clinoid process travel toward the optic chiasm (Fig. 3.29). Approximately 10 mm long, they lie in the subarachnoid cistern of the optic chiasm directly beneath the frontal lobes and with the internal carotid at their lateral surfaces.

The CN II contains the afferent, sensory limb for the light reflex. Anatomically, the afferent limb consists of the retina—optic nerve—pretectal nucleus in the midbrain (level superior colliculus). The efferent limb has nerve fibers running along the CN III.

Oculomotor Nerve (CN III)

The oculomotor nucleus is located in the midbrain in the gray substance of the floor of the cerebral aqueduct; the (accesory parasympathetic) nucleus Edinger-Westphal is located dorsal to the oculomotor nucleus and provides the autonomic function of the oculomotor nerve to the intrinsic eye-musculature. The somatic portion of the oculomotor nucleus of the midbrain contains a paired topographic individual ocular muscle localization of motor neurons.

CN III exits ventrally from the brain stem in front of the pons (interpeduncular space) and once it runs lateral to the posterior clinoid process, it pierces the dura mater at the top of the clivus and enters the lateral roof of the cavernous sinus lateral to the abducens nerve. It then runs through the lateral wall of the cavernous sinus just above the trochlear nerve, lateral to the intracavernous ICA (Fig. 3.30). Anteriorly in the cavernous sinus, the oculomotor nerves receive sympathetic fibers from the superior cervical sympathetic ganglion via the ICA plexus; there is no direct functionality with the CN III, however small branches communicate with the CN III as it passes through the cavernous sinus, finally reaching the end organ: the superior tarsal muscle, i.e., Müller's muscle.

It enters the orbit through the central part of the SOF (Table 3.1), the superolateral or oculomotor foramen of Zinn's ring close to the lateral surface of the optic strut (Figs. 3.20, 3.21b, and 3.23).

The nerve splits into its two divisions within the SOF or sometimes in the anterior cavernous sinus already. The CN III innervates all EOM as well as the LPSM: functional category – General Somatic Efferent (GSE). The smaller superior division passes the oculomotor foramen next to the tendinous attachment of the superior rectus muscle and sends branches to the ocular (inferior) surfaces in the posterior third of the SRM and LPSM. The nerve fibers pass medially around the superior rectus muscle to insert into the LPSM.

Table 3.1 Bony openings related to neuro and vascular anatomy

	Acronym	Location	Nerve(s)	Artery(ies)	Vein(s)	Further Structure(s)
Supraorbital Notch / Foramen	SONF	supraorbital margin	sensory branch of CN V1: • frontal nerve – supraorbital nerve	supraorbital artery	supraorbital vein	
Frontal Notch/ Foramen	FONF	supraorbital margin	sensory branch of CN V1: • frontal nerve – supratrochlear nerve	supratrochlear artery	supratrochlear vein	
Infraorbital Groove/ Canal	IOG	orbital plate of maxilla/ orbital floor	sensory branch of CN V2: • maxillary nerve – infraorbital nerve	infraorbital artery (+ perforators)	infraorbital vein	
Infraorbital Foramen	IOFN	facial antral wall	sensory branch of CN V2 (maxillary nerve): infraorbital nerve	infraorbital artery	infraorbital vein	
Inferior Orbital Foramen	IOF	GWS, ZB, MB, PB	sensory branches of CN V2: • zygomatic nerve • infraorbital nerve parasympathetic branches from pterygopalatine ganglion	infraorbital artery	inferior ophthalmic vein/ communication with pterygoid plexus	periorbita orbital muscle - Müller extension of buccal fat pad
Inferior Orbital Fissure – Rear Sinkhole / Pterygopalatine Fossa	IOFRS	GWS, MB, PB	CN V2 – maxillary nerve via foramen rotundum • nerve of pterygoid canal (= deep and greater petrosal nerve) via pterygoid canal (Vidian)			
Anterior Ethmoidal Foramen / Canal	AEF	anterior fronto-ethmoidal junction	• anterior ethmoidal nerve	anterior ethmoidal artery	anterior ethmoidal vein	
Posterior Ethmoidal Foramen / Canal	PEF	posterior fronto-ethmoidal junction	• posterior ethmoidal nerve	posterior ethmoidal artery	posterior ethmoidal vein	
Nasolacrimal Groove/ Fossa/ Canal	NLC	frontal corner of inferomedial orbita quadrant				nasolacrimal sac and duct
Zygomaticoorbital Foramen/ Canal	ZOF	OPZ	sensory branch of CN V2: • maxillary nerve – zygomatic nerve	collateral arterioles	collateral venules	
Zygomaticofacial Foramen/ Canal	ZFF	facial surface of zygomatic body	branch of ZOF: • zygomatico facial nerve	collateral arterioles	collateral venules	
Zygomaticotemporal Foramen/ Canal	ZTF	internal/posterior surface of zygomatic body and /or OPZ	branch of ZOF: • zygomatico temporal nerve	collateral arterioles	collateral venules	
Superior Orbital Fissure – Above Zinn's Ring / Annular Tendon / Lateral Superior SOF Sector			CN V1 branch: • lacrimal nerve CN V1 branch: • frontal nerve • CN IV = • trochlear nerve	[variant: recurrent meningo-lacrimal artery – middle meningeal artery]	superior ophthalmic vein	
Superior Orbital Fissure – Through Zinn's Ring / Annular Tendon / Central SOF Sector	SOF	LWS and GWS	CN V1 branch: • nasociliary nerve CN III: • superior and • inferior division of oculomotor nerve • CN VI = • abducens nerve sympathetic root of ciliary ganglion			
Superior Orbital Fissure – Below Zinn's Ring / Annular Tendon / Inferior Medial SOF Sector					inferior ophthalmic vein	
[Cranio orbital Foramen]	COF	between sphenoid body and two roots of LWS		[recurrent meningo-lacrimal artery – middle meningeal artery]		
Optic Foramen/ Canal	OFC		• CN II = • optic nerve sympathetic fibers from internal carotid plexus (ICP)	ophthalmic artery		meningeal sheaths

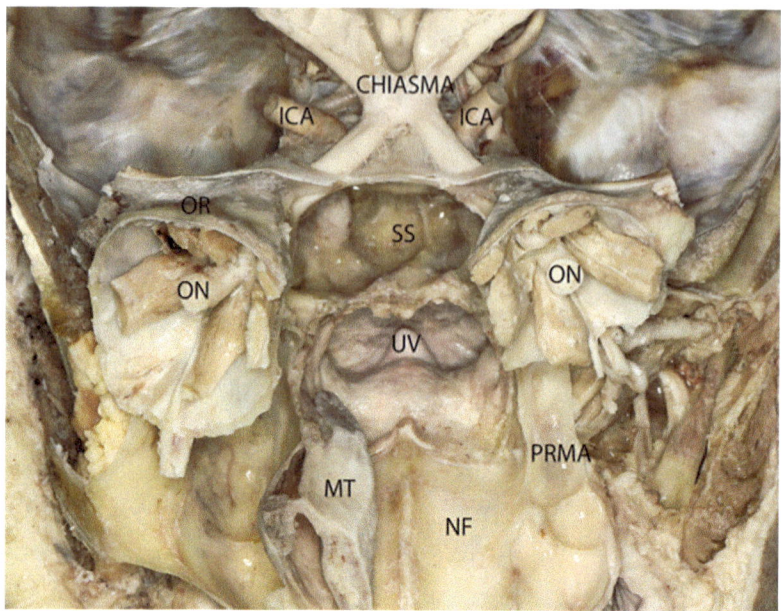

Fig. 3.29 Anatomic specimen – Coronal section through the midorbits posterior to the eyeball, midfacial skeleton, skullbase, and frontal bone around the orbits removed to show contents of orbital apices and retrobulbar optic nerve segments (intraorbital, interacanlicular, intracra- nial) leading to the chiasma opticum. *MT* middle turbi- nate, *NF* nasal floor, *PRMA* posterior recess of maxillary antrum, *UV* uvula, *ON* optic nerve, *ICA* internal carotid artery, *SS* sphenoid sinus, *OR* orbital roof

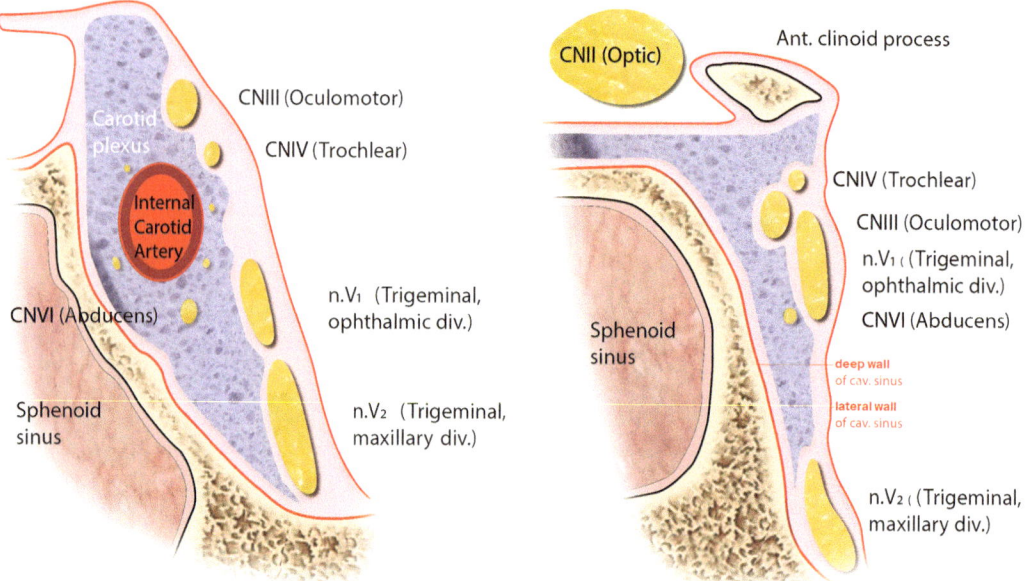

Fig. 3.30 Schematic coronal view of cavernous sinus; note that the CN VI abducent nerve is located within the sinus, near the ICA, in contrast with the CN III, IV, and CNV1 and CNV2, which are located within the sinus wall. (With permission from S. Steenen)

The larger inferior division, as it enters the intraconal space divides in the orbital apex into the following three subdivisions: medial, central, and lateral, which initially all run lateral to the optic nerve. The medial branch crosses beneath the optic nerve to reach the ocular (lateral) surface of the MRM in its posterior third. The central branch runs anteriorly to innervate the IRM from its ocular (superior) surface, again posteriorly. The lateral branch is the longest and travels anteriorly along the lateral border of the IRM to enter the IOM near its midpoint (Fig. 3.23). The latter branch carries the parasympathetic fibers (functional category: General Visceral Efferent—GVE) which, after synapsing in the ciliary ganglion, continue within the short ciliary nerves to:

- The involuntary musculus sphincter pupillae (pupillary constriction)
- The involuntary ciliary muscle (accommodation control)
- Regulate the flow of aqueous humor into Schlemm's canal
- Carry sensory proprioceptive neurons from the IOM which run with the lateral branch of the inferior CN III division

The CN III is responsible for the efferent limb of the light reflex: pretectal nucleus—Edinger Westphal (midbrain)—preganglionic parasympathetic fibers—ciliary ganglion (synapse)—postganglionic fibers—ciliary sphincter: control of the diameter of the pupil to regulate the light intensity entering the eye.

Clinical Implications

Interference of the oculomotor nerve results in a downward and outward deviation of the globe (functioning superior oblique and lateral rectus muscle) including ipsilateral ptosis of the upper eyelid. The pupil is dilated without reaction (direct and consensual) to light (musculus sphincter-constrictor pupillae) or accommodation (ciliary muscle).

The branch to the inferior oblique muscle is just underneath the periorbit and at risk for lesions by compression when the soft tissues are elevated e.g., lacerations of the periorbita during exploration of the orbital floor. The intraoperative mydriasis is not necessarily a direct effect from irritation of the ciliary ganglion but also from the compression of its parasympathetic root which derives from the CN III branch to the inferior oblique muscle. This parasympathetic root is located more anteriorly than the ciliary ganglion (compare Ciliary ganglion section).

Trochlear Nerve (CN IV)

CN IV is the single cranial nerve emerging from the dorsal aspect of the brain stem. It has the smallest caliber of all CNs and the longest intracranial course of approximately 40 mm; it is a somatic efferent nerve (functional category - GSE) that innervates the SOM. From the inferior tectal lamina, CN IV is curving around the cerebral peduncle above the pons and along the free edge of the cerebellar tentorium. The dura is penetrated inferior and lateral to the entry point of CN III into the cavernous sinus. Due to this long pathway [55], CN IV appears particularly vulnerable to injury from blunt head trauma.

Within the lateral sinus wall, CN IV moves up continuously on its way forward to finally cross over CN III before entrance into the SOF (Fig. 3.23). It runs outside Zinn's ring, in classical terms through the superolateral narrow portion of the SOF (Table 3.1), accompanied by the frontal and lacrimal branches from the ophthalmic trigeminal division, overlaying it at first (Fig. 3.21b). Along its intraorbital course, CN IV remains entirely extraconal. The frontal and lacrimal nerves move to the lateral side of the CN IV. From a superomedial position in the orbital apex, CN IV turns medially across or rarely piercing the LPSM to run on the orbital (superolateral) SOM surface. The neuromuscular junction is commonly situated on the extraconal IOM surface in the posterior or middle one-third of the muscle. CN IV may separate consecutively in up to four branches with two branches in the majority. A recent study indicates that CN IV is not always a pure motor nerve but occasionally has a mixed nerve fiber composition conveying motor and sensory (nociceptive) information.

Clinical Implications
As CN IV partly runs adjacent to the medial bony wall, the nerve is prone for injury in case of midfacial trauma.

Trigeminal Nerve (CN V)

Somatosensory Innervation of the Orbit

The trigeminal nerve CN **V** represents the major sensory nerve of the face. Its sensory ganglion (synonym: Gasserian or semilunar ganglion) is seated in an impression near the apex of the petrous bone in the middle cranial fossa.

"Meckel's cave" is a dural pocket that contains the sensory and motor roots of the trigeminal nerve, the trigeminal ganglion, and the trigeminal cistern.

The sensory nerves of the orbit come from the ophthalmic trigeminal division (**V**1) for the main part. The maxillary division (**V**2) contributes the zygomatic nerve and the infraorbital nerve, which pass through the inferior and lateral orbit.

Trigeminal Ophthalmic Division (CN V1)—Ophthalmic Nerve

The trigeminal ophthalmic division (CN **V**1) is the smallest of the three peripheral trigeminal divisions.

CN **V**1 is responsible for the sensory innervation of the eye [56], orbit, ocular adnexa, and forehead above the palpebral fissure, nasal dorsum and sidewalls, upper nasal cavity including the septum, ethmoid, sphenoid, frontal sinuses, and anterior cranial fossa. Of particular note is the cornea which is the most densely innervated tissue in the human body.

Within the lateral wall of the cavernous sinus, the ophthalmic nerve branches into the lacrimal, frontal, and nasociliary nerves which enter the orbit through the SOF. The lacrimal and frontal nerves run outside Zinn's ring, the nasociliary nerve passes through the superolateral foramen of the ring.

The nasociliary nerve, inside the apex sends out a communicating branch, the sensory root as well as sympathetic fibers, to the ciliary ganglion. The sympathetic axons travel anteriorly in the orbit via the nasociliary and lacrimal nerves to innervate the sympathetic eyelid muscles; sensory fibers are carried from the iris, ciliary body, and cornea.

To arrive in the superomedial orbit, the nasociliary nerve exits the muscle cone in the interspace between the MRM below and the SOM and SRM above. Joined with branches of the ophthalmic artery, the anterior and posterior ethmoidal nasociliary nerve branches leave the orbit through the same named foramina. The terminal branch of the nasociliary is the infratrochlear nerve (ITN). This advances inferior to the SOM and trochlea to the medial palpebral commissure, where it splits into multiple terminal ramifications to the eyelid, lacrimal sac, caruncule, and the external nose. It exchanges fibers with the supratrochlear nerve (STN), the terminal branch of the frontal nerve.

The frontal nerve (FN), the largest CN **V**1 branch, runs outside Zinn's ring (Fig. 3.21b) and passes through the narrow superior lateral SOF sector on the medial side of the lacrimal nerve, below the trochlear \nerve (CN IV) and above the superior ophthalmic vein. The frontal nerve runs forward on top of the LPSM surface covered by the periorbita just below the longitudinal midline of the orbital roof (Fig. 3.31). At this level, the nerve bifurcates into the supraorbital and supratrochlear nerves. The supraorbital nerve maintains the forwards course of the frontal nerve now on top of the LPSM and exits the orbit at the supraorbital notch or foramen. Its terminal branches, a singular lateral deep branch and several superficial branches fanning out medially, bundle sensory fibers from the middle upper eyelid, the conjunctiva, eyebrows, forehead, and scalp as far back as to the vertex. Compared with the supraorbital, the supratrochlear nerve has a smaller diameter. It turns medially and passes above the trochlea to leave the supraorbital rim medially of the supraorbital nerve either through a foramen or notch. It receives fibers from

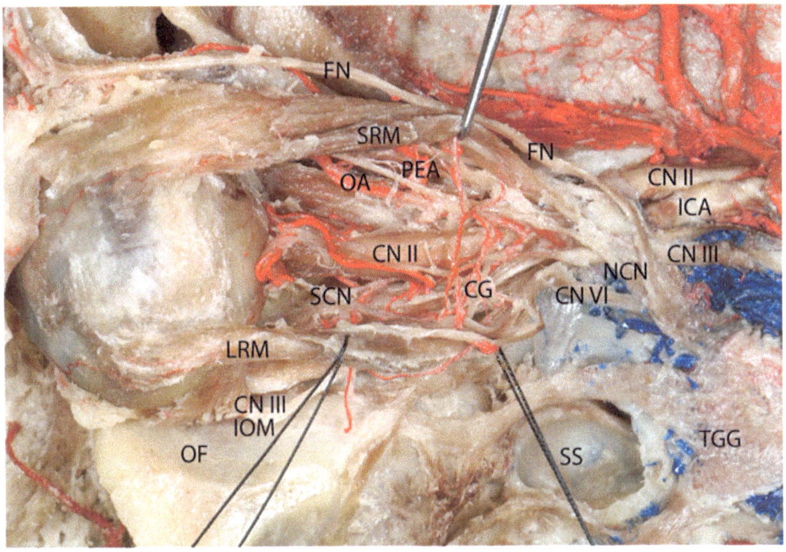

Fig. 3.31 Anatomic specimen – Lateral view of the orbital contents after opening the oculomotor foramen between the origin superior rectus muscle (SRM) and the lateral rectus muscle (LRM). LRM (black sutures) and SRM (hook) retracted to expose the ciliary ganglion (CG). CG is located between the CN II and LRM, medially it is related to the ophthalmic artery. *OA* ophthalmic artery, *PEA* posterior ethmoidal artery, *FN* frontal nerve, *SRM* superior rectus muscle, *CN II* optic nerve, *CG* ciliary ganglion, *ICA* internal carotid artery, *CN III* oculomotor nerve, *NCN* nasociliary nerve, *CN VI* abducens nerve, *LRM* lateral rectus muscle, *SCN* short ciliairy nerve, *OF* orbital floor, *IOM* inferior oblique muscle, *SS* sphenoid sinus, *TGG* trigeminal ganglion

the medial upper eyelid and brow, lateral glabellar region, and the lower forehead up to the midline. The main branches of the FN and SON run deep to the orbicularis oculi muscle, from the supraorbital notch to the superior border of the tarsal plate. The STN and ITN run between the orbicularis muscle and the overlying skin. The lacrimal nerve continues to run forward between the orbicularis oculi muscle and the tarsal plate. The upper eyelid is primarily supplied by the SON and the FN while the medial extension is supplied by the STN and ITN and the lateral extension mainly by the lacrimal nerve.

The intraorbital course, the occurrence, form, and position of the nerve exit points as well as the extraorbital distribution of both the surpraorbital and supratrochlear nerves and their branches/ subbranches are highly diverse and the subject of numerous recent studies in the context of corneal neurotization, migraine surgery, and cosmetic filler injections.

The lacrimal nerve (LN), the smallest branch of CN V1 runs outside Zinn's ring and passes the SOF on the lateral side of the frontal nerve. Its course in the superolateral orbital quadrant fol-lows the superior border of the LRM. It provides the sensory innervation to the lacrimal gland, conjunctiva, and is variably involved in the parasympathetic secretomotor supply of the gland (see later). After switchover in the pterygopalatine ganglion, the postsynaptic parasympathetic fibers are carried within the zygomatic (= CN **V**2 branch) and zygomaticotemporal (ZTN) nerves to join the lacrimal sensory nerve across a communicating branch before they enter the gland. Different from this, least frequently occurring in classic textbook descriptions, a direct entry of the ZTN without a preceeding communication to the lacrimal nerve appears as the most commonly found variant. A dual connection via provision of a communicating branch to the lacrimal nerve and a direct ZTN contact to the gland takes the second place in frequency.

Trigeminal Maxillary Division (CN V2)—Maxillary Nerve

The maxillary nerve carries sensory information from the lower eyelid, the cheek, upper lip, anterior upper gingiva, nasal, palatal, pharyngeal mucosa, and maxillary, ethmoid, and sphenoid sinus.

After giving off the middle meningeal nerve once it leaves the Gasserian ganglion, the CN V2 enters the pterygopalatine fossa through the foramen rotundum and splits into three major branches: sphenopalatine nerve, the infraorbital nerve, and the zygomatic nerve; the zygomatic nerve may also divert from the infraorbital nerve during its course in the pterygoid fossa on its way toward the IOF.

The sphenopalatine (pterygopalatine) nerve courses straight downward. Some branches pass through the pterygopalatine ganglion without synapses, the main part continues below the ganglion and enters the pterygopalatine canal.

The infraorbital nerve, having reached the pterygopalatine fossa, turns anterolaterally to pass through the IOF sinkhole or the infratemporal fossa, respectively to ascend immediately toward the posterior end of the infraorbital groove. The opening to the groove is situated just below the bone level of the orbital floor. Running anteriorly, in the midorbit the infraorbital groove becomes roofed and converts into the infraorbital canal. The canal is minimally longer (average 14 mm; range 5–22 mm) than the groove (average 13 mm; range 7–22 mm). On its way anteriorly, the canal descends from the orbital floor progressively into the maxillary sinus. The common axis of the groove and canal follows a slight forward curvature with a posterolateral start from just anterior of the IOF isthmus directed to the infraorbital foramen as an anteromedial emerging point. As mentioned alongside this passage, the infraorbital nerve is accompanied by the infraorbital artery giving off multiple small branches (perforator vessels to IRM) from the open upside of the groove (Fig. 3.32). Inside the canal portion, the nerve delivers the middle and anterior superior alveolar rami. Outside the infraorbital foramen, the infraorbital nerve splits into a fan of branches, the inferior palpebral, the superior labial, and the internal and external nasal rami (Fig. 3.33).

Parasympathetic and sympathetic fibers are associated with the course of the infraorbital nerve.

The arterial strand divides into several smaller branches which perforate to the periorbita and the rectus muscle along the top side of the infraorbital groove. These branches must be meticulously cauterized during subperiorbital dissection away from the orbital floor.

If not handled with great care, these perforater branches can be responsible for the occurrence of a retrobulbar hematoma, i.e., orbital compartment syndrome (Chap. 13).

The zygomatic nerve is a branch of the maxillary nerve (CN V2). It may appear to be a branch of the infraorbital nerve, exiting either just before or while the infraorbital nerve traverses the IOF.

The zygomatic nerve deviates laterally already beneath the IOF and periorbita level to turn toward the inside of the lateral orbital wall, entering the zygomatico-orbital foramen where it may subdivide into the zygomaticofacial (ZFN) and zygomaticotemporal (ZTN) nerves, departing the orbit (Table 3.1). The ZFN and ZTN may differ in their course and mutual conversion during the depart and exit between varying constellations and numbers of foramina inside and outside the orbital cavity. According to their openings at the facial and temporal surface of the zygoma, the nerves are sensory for a skin territory in the upper cheek, lateral lower eyelid, and lower temple.

Clinical Implications

In a superior orbital fissure (SOF) syndrome [57, 58], the neural structures passing through the fissure may be involved to a different degree. In the complete clinical picture, the combination of motor impairment due to CN III–CN IV–CN VI lesions results in ophthalmoplegia and subsequently proptosis due to decreased tension of the EOM. Ptosis will occur due to loss of function of the LPSM & loss of sympathetic input to Müllers superior tarsal muscle (long ciliary nerves). Focus impairment (loss of accommodation) and a fixed, dilated pupil (loss of parasympathetic innervation by ciliary nerves—inferior div. CN III to

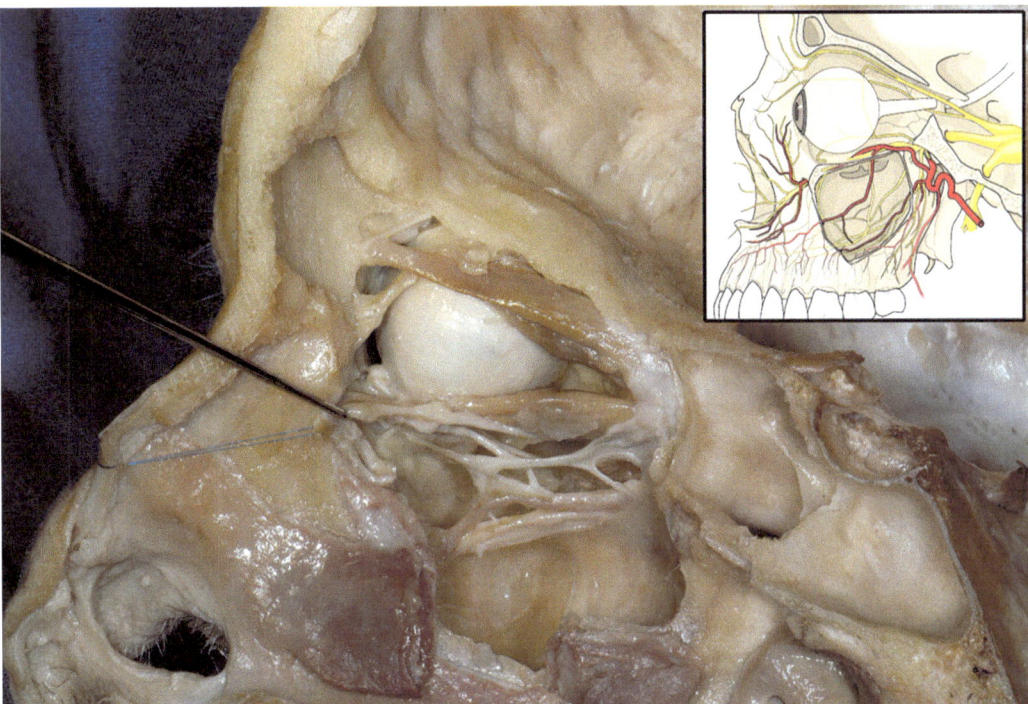

Fig. 3.32 Anatomic specimen – Orbital contents—medial approach with infraorbital neurovascular bundle in the center—inferior oblique muscle and oculomotor nerve inferior branch hook retracted—anatomic specimen. Nerve branches entangled with numerous small arterial branches exiting from a larger arterial strand. The small arteries correspond to perforators entering the periorbita on their way to the inferior rectus muscle undersurface,

Inset: Schematic view – midsagittal plane through left orbit and maxillary antrum—schematic view. CN **V**2 and infraorbital neurovascular bundle with entrance of the artery into the orbital floor via the infraorbital groove. The perforators (see text) derive from this part of the infraorbital artery (With permission from https://surgeryreference.aofoundation.org).

ciliary and sphincter pupillae muscle) are present. The corneal reflex is absent due to loss of afferent input (long-short ciliary nerves) from the CN **V**1 (ophthalmic division).There is paresthesia of the upper eyelid and lacrimal hyposecretion (CN V1 frontal, lacrimal, and nasociliary branches) (**Fracture sites compare**—Fig. 3.34a, b). Loss of vision indicates involvement of the orbital apex (superomedial foramen of Zinn's ring) and the condition is called orbital apex syndrome.

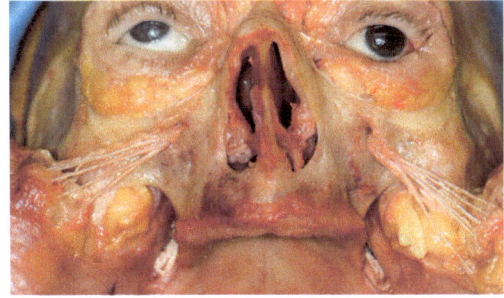

Fig. 3.33 Anatomic specimen – Fanning of infraorbital artery and nerve fascicles from infraorbital foramina. The innervation pattern provides sensory supply to the peeled off soft tissues all around the lower circumference of the bony orbit, including the lower eyelids

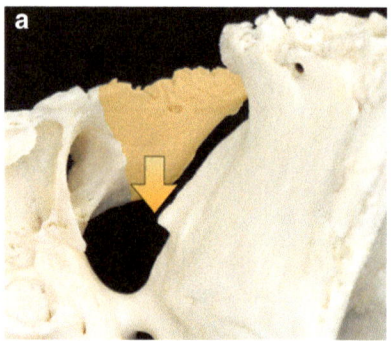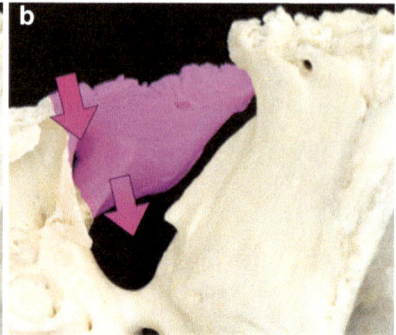

Fig. 3.34 Illustration – Superior orbital fissure syndrome and trauma mechanism responsible for differences in clinical symptomatology (**a**) Fracture site in the lateral part of the lesser wing of sphenoid (LWS) and downward displace-ment - optic canal intact CN II function most likely preserved. (**b**) Fracture location in the medial lesser wing sphenoid including the superomedial foramen (optic canal) and downward displacement – full blown orbital apex syndrome.

Abducens Nerve (CN VI)

The abducens nerve exits the brain stem ventrally at the pontomedullary junction. The neurons arise in the paired motor nuclei in the pons. After an ascending prepontine intracranial course, it pierces the dura of the posterior cranial fossa on the clivus. CN VI advances from its petroclival entrance point surrounded by a dural sleeve which is situated in between the sphenopetroclival venous confluence and the lateral wall of the cavernous sinus. CN VI runs rather in the body then in the lateral wall of the cavernous sinus (Fig. 3.30). The sleeve represents a short osteofibrous tube named Dorello's canal, which passes below the butterfly-shaped petrosphenoid ligament. The vulnerability of CN VI in severe cranial and cervical trauma or by increased intracranial pressure cerebral herniation and downward migration of the brainstem is attributed to its firm rostral tethering within Dorello's canal. Within the cavernous sinus, CN VI continues forward on the lateral side of the posterior vertical segment of the internal carotid artery and medial to the ophthalmic nerve (CN **V**1) (Fig. 3.30). The SOF is passed inside the superolateral part of Zinn's ring inferior to the superior CN III divison, inferolateral to the nasociliary nerve, and lateral to the inferior CN III division (Table 3.1). In the orbital apex, CN VI shifts laterally to spread several subbranches along the ocular (medial) LRM aspect (Figs. 3.23, 3.26, 3.31). The neuromuscular junction is located in the posterior or middle one-third of the muscle. Sympathetic fibers are carried along with CN VI.

Just as CN III and CN IV, the abducens carries sensory fibers from proprioceptive receptors in the EOM.

Clinical Implications

Because of its long and tortuous course along the cranial base, the CN VI is vulnerable especially for compressive and/or traumatic injuries, intracranial tumor lesions, and infections [59].

Since CN VI does not lie within the lateral wall of the sinus as do the oculomotor and trochlear nerves but runs in the sinus itself lateral to the internal carotid artery, it is generally affected first in case of increased pressure within the cavernous sinus (Fig. 3.30). Sympathetic fibers from the carotid plexus to the dilator pupillae muscle run with the abducens for a short distance: thus a fixed pupillary constriction in combination with abducens palsy may occur.

Autonomic Nervous System

The autonomic nervous system of the orbit and eye has many functional tasks, which are mediated via parasympathetic and sympathetic pathways and their interrelated or antagonistic action:

• Pupil dilatation/constriction
• Control of ocular accommodation

- Influence on light reflex circuits and convergence
- Width control of the palpebral aperture (STM and ITM)
- Regulation of intraocular pressure—aqueous humor homeostasis
- Modulation of lacrimal secretion—tear flow production
- Control of ocular blood flow

Parasympathetic Pathways

The head has four parasympathetic ganglia: the ciliary-, the pterygopalatine-, the otic ganglion and the submandibular ganglion. The orbita related ciliary- and pterygopalatine ganglion will be discussed.

Parasympathetic innervation of the eyes commences from two origins:

- Edinger Westphal Nucleus (EWN):
  EWN is the accessory subdivision of CN III nucleus lying in the rostral mesencephalon.
- Superior Salivatory Nucleus (SSN):
  SSN is the parasympathetic (lacrimal) constituent of CN VII nucleus lying in the medulla oblongata.

The preganglionic neurons originating in EWN gather with the oculomotor nerve, to proceed along the "IIIrd CN Parasympathetic Pathway," to the ciliary ganglion (CG) and synapse there to their postganglionic projections via short ciliary nerves.

The preganglionic neurons from the SSN course within the intermediate nerve (parasympathetic and sensory part of CN VII) to the geniculate ganglion and along the greater petrosal nerve following the "VIIth CN Parasympathetic Pathway" to synapse in the pterygopalatine ganglion (PPG). The greater petrosal nerve unites with the deep petrosal nerve, which consists of postganglionic sympathetic fibers from the ICA plexus, to form the pterygoid nerve (Vidian) inside the same-named canal. On this way, the fibers for the parasympathetic root as well as for the sympathetic root are conveyed to the PPG.

The sensory fibers of the maxillary nerve directly pass the PPG without synaptic contacts except for some short fibers which constitute the sensory root of the PPG. Vasodilator and secretomotor fibers from PPG are distributed conjoining with the sensory branches of the maxillary nerve (zygomatic nerve [− occasionally communicating with lacrimal nerve], greater and lesser palatine nerves, the nasopalatine, and the pharyngeal nerve) to the lacrimal gland and mucous membranes of the nose, hard palate, upper lip and gums, soft palate, tonsils, and upper part of the pharynx, respectively.

The IOF provides the intraorbital entrance for the "VIIth CN parasympathetic pathway" and postganglionic sympathetic fibers.

Sympathetic Pathways

The sympathetic innervation starts from the posterolateral hypothalamus and descends inferomedial cell column of the gray substance extending within the C8–T2 segments of the spinal cord. The neurons give rise to preganglionic fibers that connect to the paravertebral sympathetic chain ganglia and continue to ascend in the sympathetic trunk to the superior cervical ganglion (SCG) where they synapse. The complete sympathetic supply of the head emerges from SCG.

The final postganglionic sympathetic pathways follow the ICA through the foramen lacerum/carotid canal into the cavernous sinus, the departure region for the terminal passage into the orbit [60] is both through the optic canal and the SOF.

Still proximal to Zinn's ring, a plexus of sympathetic nerve surrounds the ophthalmic artery. The main sympathetic access route into the orbit is by means of the sensory ophthalmic nerve branches (CN V1)—frontal, lacrimal, and nasociliary, while the extraocular motor nerves—oculomotor (CN III), trochlear (CN IV), abducens (CN VI), and the ophthalmic artery plexus are only partly involved in the sympathetic supply. Orbital arteries are closely associated with orbital sensory nerves.

The nasociliary nerve trunk gives off small sympathetic ramifications traveling anteriorly toward the ophthalmic artery, the extraocular motor nerves, and the CG.

The tunica media muscle layers of the ophthalmic artery and all branches including the central retinal artery, frontal artery, and lacrimal artery, are innervated by sympathetic axons in contrast to the venous drainage system (superior ophthalmic vein). This suggests sympathetic control of the arterial flow within the eye and orbit.

The common path of extraocular motor nerves and sympathetic fibers may be explainable by the provision of sympathetic input for the regulation of the MRM and LRM muscle pulleys.

The long ciliary nerves (from the nasociliary nerve) carry sympathetic axons anteriorly through the suprachoroidal space to innervate the ciliary body, iris (dilator pupillae), and trabecular meshwork.

Apart from sympathetic nasociliary branches, the CG receives sympathetic fibers directly from extensions of the ICA plexus and the ophthalmic artery.

The delicate fascicles of the short ciliary nerves arise from the CG and transport these sympathetic nerve fibers as well as parasympathetic fibers. While the parasympathetic fibers innervate the ciliary body (i.e., ciliary muscle) and iris (sphincter pupillae), the sympathetic fiber subset is nonspecific and provides vasoconstriction for the uveal blood vessels. The ocular adnexal structures such as the tarsal muscles and lacrimal gland are further destinations of sympathetic efferents in the orbit.

The sympathetic innervation to the superior (Müller) and inferior tarsal muscles [61] derives most likely from the lacrimal and infratrochlear branches of the ophthalmic nerve. The long ciliary nerves bring in additional sympathetic supply to Müllers superior tarsal muscle. The perivascular sympathetic plexus around the lacrimal artery continues with the vasculature into the lacrimal gland and appears to modulate its secretory function by regulation of the blood flow and direct innervation of the acini. Moreover some sympathetic fibers pass along the external carotid artery, through the pterygopalatine fossa to enter the orbit with the maxillary artery and the infraorbital nerve.

Sympathetic innervation of the inferior orbitalis muscle of Müller, the smooth muscle across the IOF and adding to the inferior tarsal muscle is conveyed from the PPG and travels along branches of the infraorbital nerve, such as the zygomatic nerve.

Clinical Implications

Horner's syndrome [59, 62, 63] is caused by an interruption of the oculosympathetic pathway (pointed out above). Depending on the extra-, intracranial or intraorbital location of the "sympathetic disruption," it may result in ipsilateral miosis (pupillary constriction/anisocoria due to paresis of dilatator muscle) and mild ptosis (due to paresis of Müller's superior tarsal muscle), both contributing to pseudoenophthalmus (sinking back of the eye in the orbit). Ipsilateral impairment of sweating and vasoconstriction in the face (facial anhidrosis) ensues from damage at the thoracic/cervical level. In contrast, lesions at the height above the superior cervical ganglion—after the sudomotor and vasomotor fibers have branched off—show only limited involvement of the face.

Important note: Careful observation will reveal that the reaction of the pupils to direct and consensual light and to accommodation is preserved since these functional circuitries do not rely on sympathetic nerve action.

Ciliary Ganglion (CG)

The CG represents one of the four (submandibular, otic, pterygoplatatine and ciliary ganglia) parasympathetic cranial ganglia, and is associated with the ophthalmic division of the CN **V** (nasociliary) and the inferior division of CN III.

It is the pre-postganglionic relay center of the "IIIrd CN Parasympathetic Pathway."

CG lies embedded in fat at the orbital apex, between the lateral aspect of CN II and the LRM most often in the midhalf between Zinn's ring and the back of the eyeball, 1 cm anterior to the SOF (Fig. 3.31).

There are numerous variations in size, shape, number, and location of CG [64].

So it can have a round, ovoid or an irregular, star-like flattened shape. The mean size approximates to 2.5 mm in horizontal diameter, 1.5 mm in vertical height, and 0.5–1 mm in thickness.

Rarely the CG may be located in close contact with the inferior division of the oculomotor nerve (Fig. 3.35). It is most often located lateral to the CN II, variable in the area between retrobulbar and ventral to the maxillary strut, i.e., apex orbitae (Fig. 3.36).

The CG has inputs by sensory, parasympathetic, and sympathetic fiber projections, which join it from the posterior aspect and why they are named roots; it is associated with roots from the nasociliary (CN V1) and oculomotor nerve (CN III) as well as with direct sympathetic rami from the internal carotid plexus (Fig. 3.37).

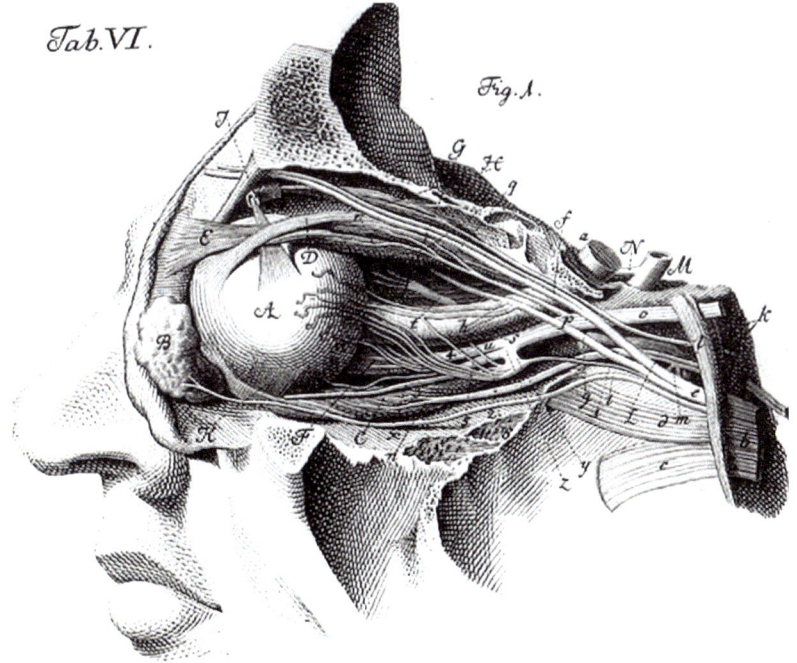

Fig. 3.35 Lateral view of the orbital contents. Original drawing Fig. 1 from Table VI titled "Nervi & musculorum bulbi oculi", y = ciliary ganglion. From Johann Gottfried Zinn's Book—Descriptio anatomica oculi humani iconibus illustrata, Gottingae: Viduam B. Abrami Vandenhoeck, 1755.

Fig. 3.36 Schematic Drawing - Location of the ciliary ganglion (CG). Frequency of distribution (percentages) between the posterior aspect eyeball and Zinn's ring. 1 = CG; 2 = long branch of inferior division of CN III to IOM. (With permission from S. Steenen)

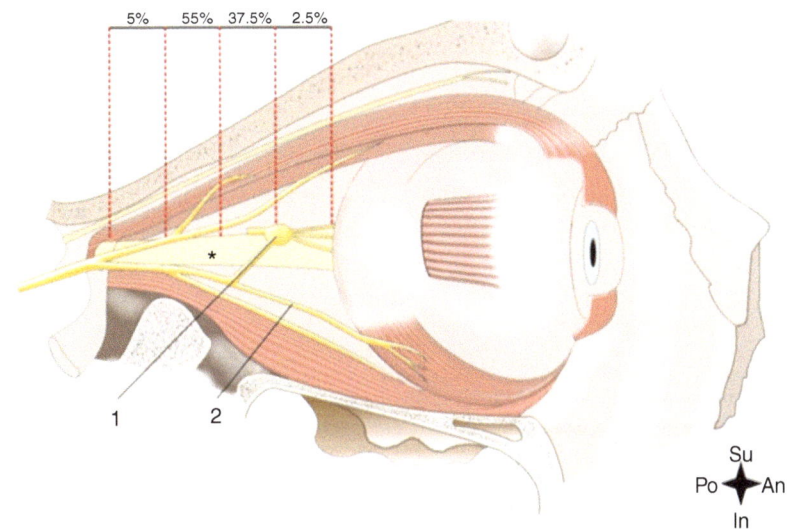

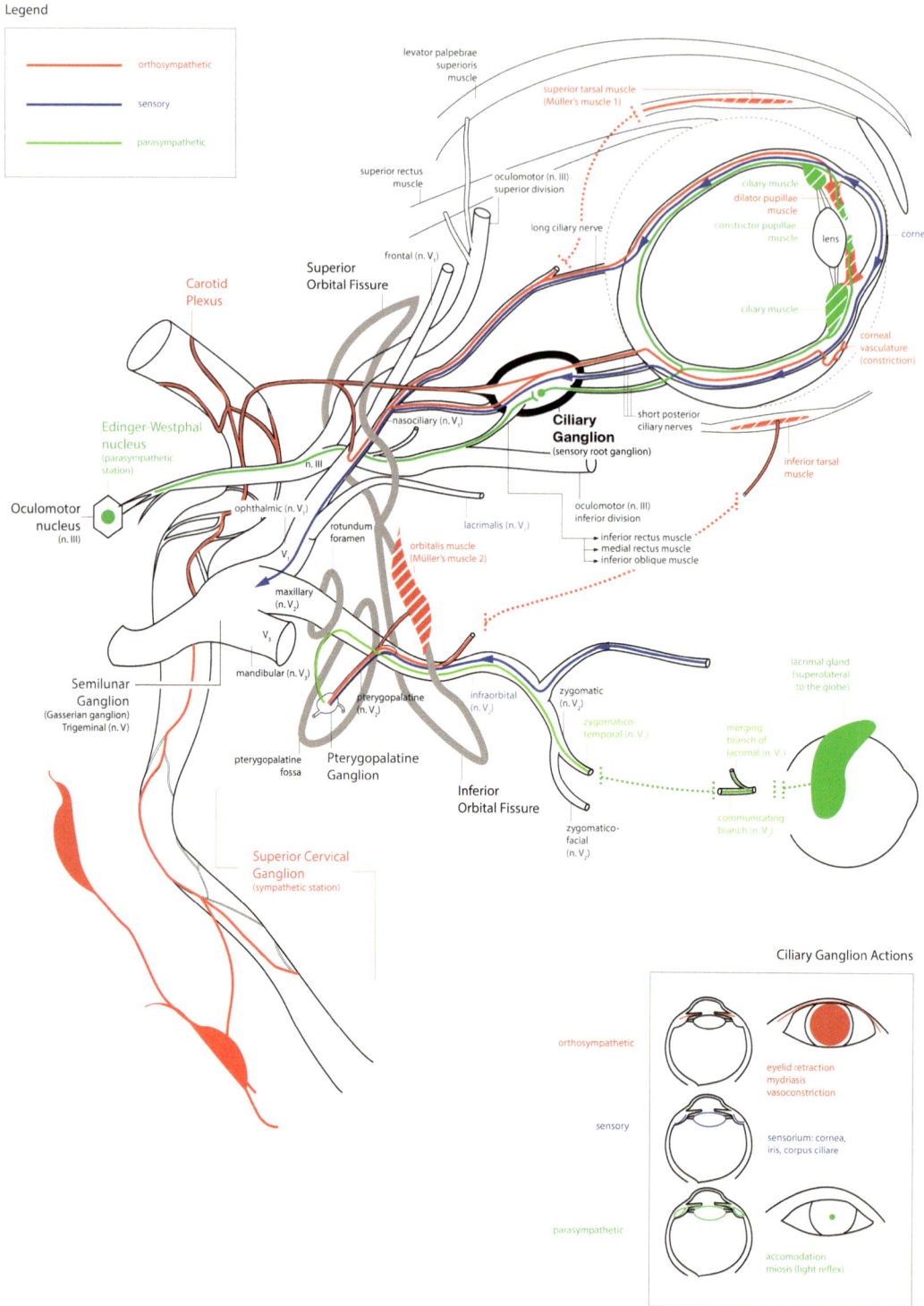

Fig. 3.37 Synoptic illustration of cranial autonomic nerve system and its function ciliary ganglion and course of (para) sympathetic nerves; overview of general somato-sensory afferent (GSA) and general visceral efferent (GVE) pathways via the CG and PPG into the orbit—eye. (for a detailsed description – see text) Red line = (ortho-) sympathetic; blue line = sensory; green line = parasympathetic. (With permission from S. Steenen)

The sensory CG root is provided by a single branch diverting from the nasociliary nerve (CN V1) proximal to Zinn's ring. It carries sensory fibers from the globe as short posterior ciliary nerves, which convey sensation from the eyeball and cornea, enters the ciliary ganglion, and without synapse transmits to the nasociliary nerve. Some sensory fibers directly enter the nasociliary nerve within the long ciliary nerves.

The parasympathetic or so called motor CG root consists of a commonly single fiber strand leaving from the inferior division of the oculomotor nerve (CN III) or more specifically from its longest branch supplying the inferior oblique muscle.

Within the CG, the preganglionic parasympathetic fibers synapse (Fig. 3.37). The postganglionic parasympathetic output passes into the short ciliary nerves numbering up to 20 branches. In their initial course, the short ciliary nerves stay in a lateral position to CN II; finally they pierce the sclera at the posterior globe close to the entrance of the optic nerve. Within the globe, they run anteriorly in the suprachoroidal space, innervat-

ing the ciliary muscle (regulation of ocular accommodation) while a minor (3–5%) portion innervates the sphincter pupillae muscle (regulation of pupil constriction).

The sympathetic CG root fibers arise from the carotid plexus, pass the SOF, and travel through the ciliary ganglion without synapsing to assort within the short posterior ciliary nerves. They are in charge of vasoconstriction of the uveal vasculature. Some other sympathetic fibers travel with the nasociliary nerve initially and proceed as long posterior ciliary nerves to the superior tarsal muscle (Müller) and dilatator pupillae muscle, as mentioned above (Fig. 3.37).

Pterygopalatine Ganglion (PPG)

The PPG also known as sphenopalatine, nasal or Meckel's ganglion is the largest cranial parasympathetic ganglion and the pre-postganglionic switchover of the "VIIth CN Parasympathetic Pathway" (Figs. 3.37, 3.38, and 3.39). It receives

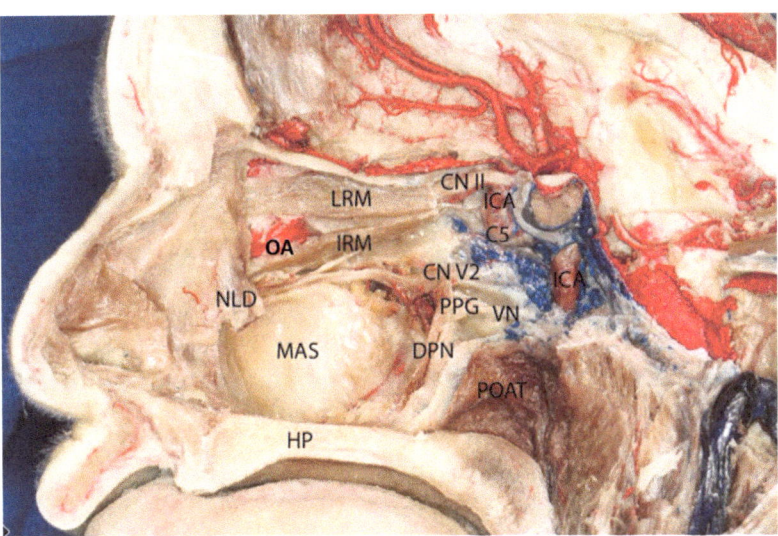

Fig. 3.38 Anatomic specimen – Paramedian sagittal section on left side of a head: EOM muscles from the medial side, Zinn's ring removed and sphenoid sinus walls removed to show the course of CN II and the ophthalmic artery arising from the ICA. Below the posterior orbit, the pterygopalatine ganglion (PPG) and some of its connections are visible. ICA—C5 or the clinoid segment, is a short segment of the internal carotid that begins after the artery exits the cavernous sinus at the proximal dural ring and extends distally to the distal dural ring, after which the carotid artery is considered "intradural" having entered the subarachnoid space. *VN* vidian nerve, *CS* cavernous sinus, *HP* hard palate, *MAS* maxillary antrum/sinus, *DPN* descending palatine nerves, *NLD* nasolacrimal duct (right), *POAT* pharyngeal opening of auditory tube, OA ophthalmic artery, *LRM* lateral rectus muscle, *IRM* inferior rectus muscle, *CN II* optic nerve, *ICA* internal carotid artery, *CN V2* trigeminal nerve—maxillary division, *C5* clinoid segment, *PPG* pterygopalatine ganglion

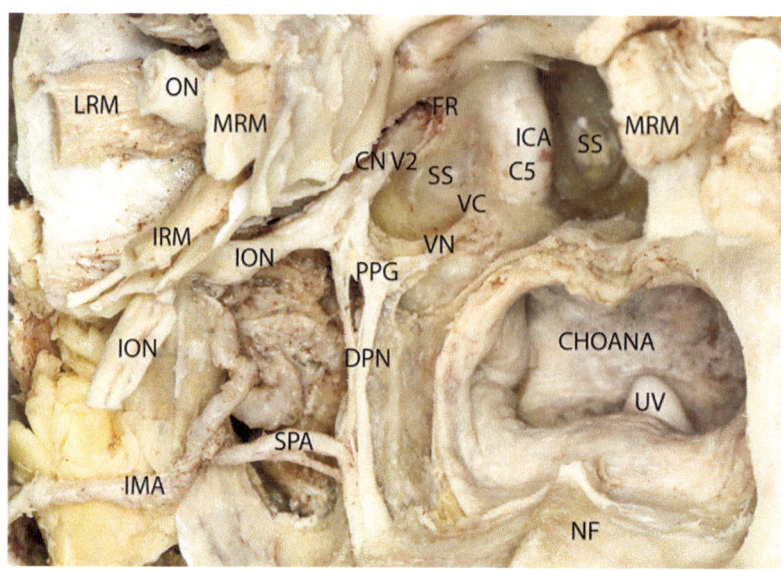

Fig. 3.39 Anatomic specimen – Pterygopalatine ganglion (PPG) in its fossa (PPF). All vertical walls of the maxillary antrum removed including perpendicular plate of the palatine bone. Orbital floor and medial orbital wall superomedially encompassing the orbital apex with cut ends of EOM and CN II, optic nerve ON. Medial pterygoid plate exposed next to choanal aperture on medial aspect. Sphenoid sinus wide open with ICA prominence (C5 segment) and opticocarotid recess. Internal maxillary artery (IMA) pulled inferolaterally out of the PPF. Sphenopalatine artery (SPA) branching off and joining the decending bundle of palatine nerves (DPN) toward palatine foramina. PPG in the center of the photograph with its three major connections—maxillary nerve (CN

V2), Vidian nerve (VN), and descending palatine nerve (DPN). Maxillary nerve coming from foramen rotundum (FR) (partially drilled away) delivers multiple small rami to PPG before transforming into the infraorbital nerve (ION) that enters the orbital floor. *IMA* internal maxillary artery, *VC* vidian canal, *ON* optic nerve, *LRM* lateral rectus muscle, *MRM* medial rectus muscle, *IRM* inferior rectus muscle, *ION* infraorbital nerve, *CN V2* trigeminal nerve—maxillary division, *FR* foramen rotundum, *SPA* sphenopalatine artery, *DPN* descending palatine nerve, *PPG* pterygopalatine ganglion, *VN* vidian nerve, *ICA* internal carotid artery, *C5* clinoid segment, SS sphenoid sinus, *UV* uvula, *NF* nasal floor

input from a sensory, parasympathetic, and sympathetic roots, which reach it via the maxillary nerve branches and the pterygoid (Vidian) nerve. It is located in the pterygopalatine fossa [65], where it lies lateral to the sphenopalatine foramen, medial to the foramen rotundum and anterior to the aperture of the pterygoid (Vidian) canal.

Looking from the front to the bony orbit, the PPG projects medio-posteriorly underneath the orbital process of the palatine bone where it clings along the transition from the perpendicular plate.

It can be of oval, triangular or square shape with a mean size of 5 mm in horizontal and 6.5 mm in vertical dimension. PPG may be bipartite and there may be a number of associated microganglia.

The sensory PPG root connections descend from the maxillary division of the trigeminal

nerve (CN V2 and rr. ganglionares n. maxillaris) after entering the pterygopalatine fossa (PPF) through the foramen rotundum (Fig. 3.39). These maxillary rami pass the PPG and do not synapse. In the PPF, they emerge as nasopalatine, pharyngeal, greater and lesser palatine nerves. Together these nerves conduct sensation (General Somatosensory Afferents—GSA) from the mucosa of the hard palate, upper gums, soft palate, tonsils, and the naso- and upper orpharyngeal walls.

The parasympathetic PPG root is brought about by the nerve of the pterygoid canal (Vidian). The Vidian nerve enters the PPF posteriorly and conglomerates with the PPG (Figs. 3.38 and 3.39). Within the PPG, the preganglionic parasympathetic fibers originating from the greater petrosal partition of the Vidian nerve synapse with postganglionic parasympathetic secretomotor neurons, which meet with the maxillary and ophthalmic divisions of the trigeminal nerve (CN V1 and V2) to travel to their destinations in the mucosa of the nasal cavity, naso- and oropharynx, and the upper oral cavity as well as to the lacrimal gland. The parasympathetic supply to the lacrimal gland is presently debated and may be reached within the zygomatic nerve, the ZTN, and possibly the lacrimal nerve itself.

The sympathetic PPG root corresponds to fibers from the deep petrosal nerve having made their way within the Vidian nerve. These postganglionic sympathetic fibers traverse the PPG interruptedly. They are mainly distributed to the nasal and pharyngeal mucosa with a few fibers reaching the lacrimal gland.

Sympathetic innervation to Müller's inferior orbital muscle and to the inferior tarsal muscle derives from the PPG via the infraorbital nerve CN V2. Among the numerous PPG outlets, there are delicate sensory orbital rami which are supplemented by sympathetic fibers from the ICP. This doublet leaves into the orbital apex and forms the retroorbital plexus, which again send out fine rami eventually targeting for the lacrimal gland.

Arterial Supply to the Orbit

The arterial system supplying the orbit, eyeball, and adnexa features an extensive collateral circulation system interconnecting the branches of the ophthalmic artery (OA) originating from the internal carotid artery (ICA) and linking with the external carotid artery (ECA) [66].

The ICA-/ECA-derived network is set up in between the:

- Frontal branches of superficial temporal artery—supraorbital/supratrochlear arteries
- Facial/angular artery—medial palpebral/dosal nasal arteries
- Maxillary/sphenopalatine artery—ethmoid arteries
- Transverse facial and deep temporal artery—lacrimal artery

Clinical Implications

The knowledge of the anatomic functional basis of pupillary control is important in orbital surgery. During intraorbital surgery, too much pressure may be exerted within the orbital apex, i.e., next to the ciliary ganglion.

Parasympathetic fibers synapse in the ciliary ganglion and continue within the short ciliary nerves to innervate the sphincter pupillae. The sympathetic fibers cross the ciliary ganglion uninterruptedly and continue as well in the short ciliary nerves but have no relevance for pupillary function.

Increased pressure during intraorbital procedures within the orbital apex next to the ciliary ganglion may lead to interferences with the synapting parasympathetic CG fibers. It is assumed that the parasympathetic synapses within the CG are particularly susceptible to pressure, what may result in pupillary dilatation.

To distinguish whether the function of the CN II is affected or not in case of intraoperative mydriasis, the swinging flashlight test can be carried out. When a flashlight is shown onto the eye with the dilated pupil, pupillary constriction will result in the opposite eye provided that CN II function is intact. When the flashlight is swung over to the undisturbed contralateral eye, this will react with pupillary constriction, while the pupil of the affected eye will remain wide. This confirms an undisturbed bilateral CN II Function and impairment of the CG or CN III as the cause of the mydriasis.

Intraoperative mydriasis may also be due to compression of the parasympathetic CG root which derives from the CN III branch to the inferior oblique muscle (Figs. 3.31 and 3.37). This parasympathetic root is located more anteriorly than the CG.

CN III lesions proximal of the CG, involvement of the CG synaptic sites or compression of the distal parasympathetic CG root cannot be differed intraoperatively.

Intraoperative pupillary dilatation must be appreciated as a specific warning sign of too much and/or too long pressure on the delicate neurovascular orbital tissue portions. If this pressure continues, it may eventually result in a dilated pupil (sphincter pupillae muscle disturbance) and may occur in combination with a loss of the accommodation reflex owing to CN III paresis or ciliary muscle disturbance, what constitutes a serious permanent complication.

Adie's myotonic pupil (Holmes-Adie Syndrome) is the result of postganglionic parasympathetic denervation in the CG. The reaction of the abnormally mid-dilated pupil to light is sluggish and poor and much less than the response to accommodation. The pupil remains constricted and will dilate very slowly only.

So there is a light-near dissociation, though the accommodation is impaired (tonic near constriction and hyperopia). Other features may consist of decreased deep tendon reflexes secondary to degeneration of dorsal root ganglia in the spinal cord.

External Carotid Artery (ECA)

The infraorbital artery, one of the terminal branches of the maxillary artery represents the only direct supply of the ECA system to the orbit. It leaves from the pterygopalatine fossa and enters the orbit through the IOF (Table 3.1) to follow the same course as the infraorbital nerve (CN V2). During passage through the infraorbital groove, it gives off small branches (perforators) extending into the fat and muscles (IRM and IOM) of the lower orbit (Fig. 3.32a, b).

The middle meningeal artery (MMA), another terminal branch of the maxillary artery, that enters the middle cranial fossa through the foramen spinosum may connect via a recurrent meningeal branch of the lacrimal artery (LA) or give off the OA. The recurrent vessel connection is oriented backward and passes through the outermost part of the superior lateral SOF sector or a distinctive foramen in the greater wing of the sphenoid (GWS) lying anterolaterally to the upper end of the SOF, which is called the meningo-orbital or cranio-orbital foramen (COF). The direction of flow for an aberrant OA is downstream from the MMA, of course.

Internal Carotid Artery (ICA)

Ophthalmic Artery (OA)

The OA is the primary source of arterial blood supply to the orbit. In general, it is the first intracranial branch of the ICA and exits off from superomedial convexity of the supraclinoid/ophthalmic ICA segment (C 6) in the subarachnoid space above the cavernous sinus. At this origin, the OA lies medial to the anterior clinoid process [67] and on the inferior side of CN II.

Within the optic canal (Table 3.1), the artery is fixed to the dural sheath and still kept in an inferolateral position. The orbit is entered through the supermedial foramen of Zinn's ring. Different from this typical textbook arrangement, some variants can be encountered in terms of the OA following a separate course through an additional proximal access foramen to the optic canal or through a duplicate bony passage below the canal. As a special feature the OA and or the lacrimal artery (LA) can arise from the middle meningeal artery (MMA) [68] (Fig. 3.40a–c).

Or the OA and LA may loop between the ICA and the MMA (Fig. 3.40c) as a baseline for all other branches. The OA may also be duplicated. In as much as the OA exclusively exits from the MMA, the entire arterial supply of the orbit will depend on the ECA. Therefore an arterial branch traversing a GWS foramen or the upper lateral end of the SOF occurring during a deep lateral periorbital dissection should be preserved unless its provenance is clarified beyond any possible doubt.

The vessel variations [69, 70] are reflected in diversified courses through the bony gateways (superior lateral SOF sector and/or COF). In the

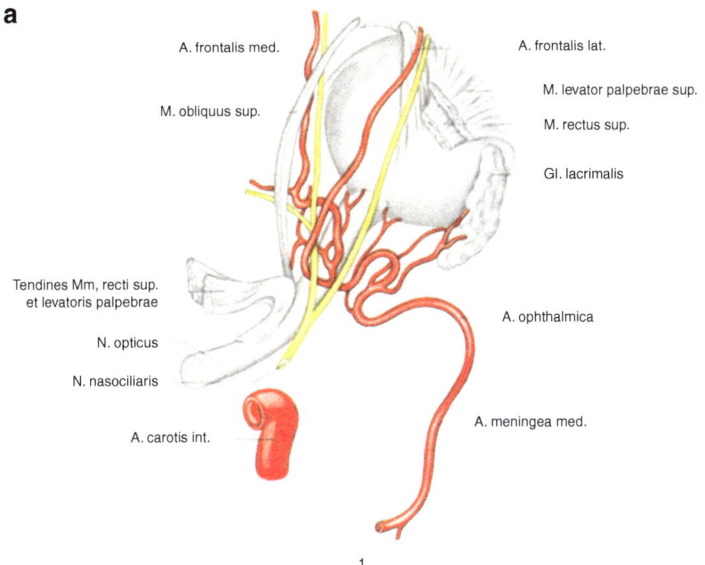

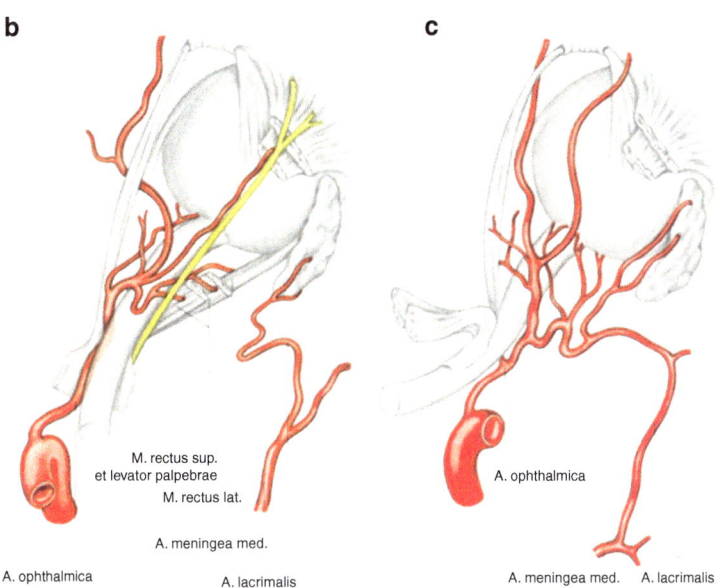

Fig. 3.40 Schematic views – Variations in origin and course of ophthalmic (OA) and lacrimal artery (LA). (**a**) OA originating from medial, meningeal artery (MMA). (**b**) LA originating from medial, meningeal artery (MMA). (**c**) Loop between OA and MMA via LA giving off several branches. (With permission from: Krmpotic-Nemanic J, Draf W, Helms J. Chirurgische Anatomie des Kopf-Hals Bereichs. Springer Verlag, Berlin Heidelberg 1985)

rare case that the MMA originates from the OA instead of the maxillary artery, the foramen spinosum will not exist.

Once inside the orbital apex, the OA typically crosses over the CN II from a posterolateral to an anteromedial location between the MRM and the SOM.

The intraorbital course of the OA is characterized by tortuosity and extensive arborization (Figs. 3.31 and 3.41).

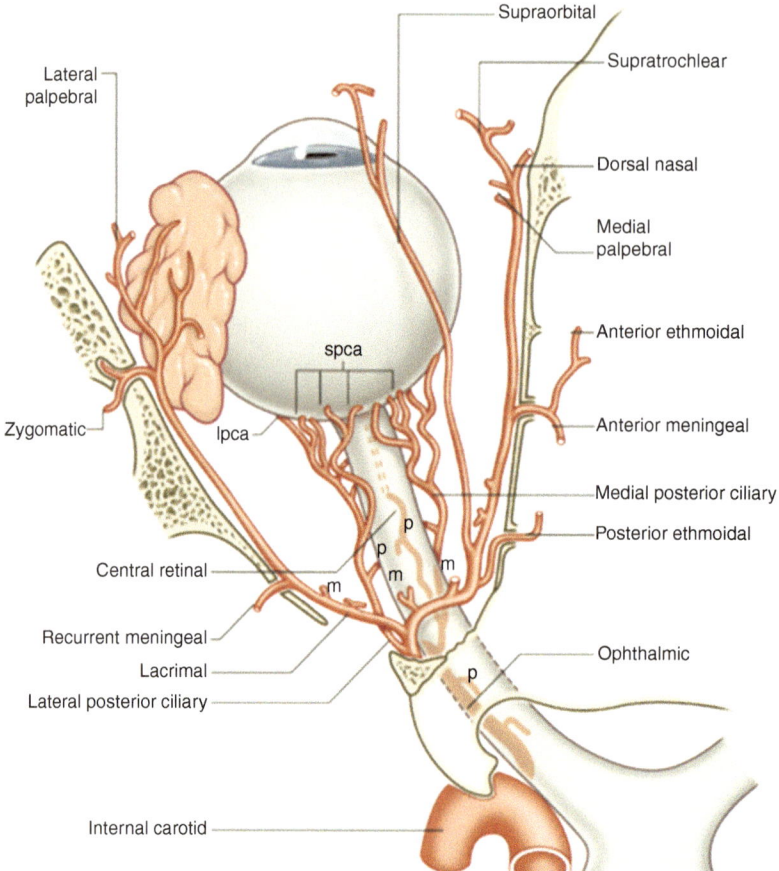

Fig. 3.41 Orbit from superior view exhibiting Orbital distribution of the ophthalmic artery (OA) and its branches. *g* globe, *lpca* long posterior ciliary artery, *m* muscular arteries, *on* optic nerve, *p* pial arteries, *spca* short posterior ciliary arteries. (With permission from: Gray's Anatomy 42nd edition)

The branches and subbranches can be grouped according to their topography (orbital, ocular, extraorbital) and supply area, based on the site of origin from the stem vessel (medial, lateral, etc.) or in line with their sequence of origin (orbital apex toward aditus/rims, initial to terminal branches). In a posterior–anterior direction, the OA distributes the following branches (Fig. 3.41).

- Central Retinal Artery (CRA): is the first ocular branch from the OA originating at the ramification of the OA and the lacrimal artery near the anterior orbital apex. The CRA turns inferiorly to pierce the optic nerve sheath from its lower-medial aspect.

The distance from the CN II/eyeball junction posteriorly to the CRA entry point into the CN II sheath is between 7 and 10.5 mm. The CRA travels then a short distance in the subarachnoid space and zigzags vertically into the nerve stroma and forward again to the optic disc.

- Posterior Ciliary Arteries: two or infrequently three lateral and medial long posterior ciliary arteries (lpca) originate from the lacrimal artery and/or from the OA itself. These vessels pass forward next to CN II, enter the sclera, and course between choroid and sclera around the lateral and medial orbital circumference to reach and penetrate the ciliary muscle. By anastomoses among one another and with the anterior ciliary arteries, they form the greater arterial circle of the iris. A bundle of 15–20 short posterior ciliary arteries (spca) emanates from the long posterior ciliary nerves near the posterior globe and pierces the sclera in a ring around CN II (= Zinn-Haller ring) and branches to supply the choroid and the optic nerve.

- Muscular Arterial Branches to Rectus EOMs and Anterior Ciliary Arteries: the muscle arteries (m) for the rectus musculature depart in the proximity of the OA and lacrimal artery divergence. They run forward along the ocular surfaces or inside the muscular substance to enter the globe at the scleral tendinous insertions. Here the muscular branches subdivide into pairs of anterior ciliary branches except for the LRM, which carries only one anterior ciliary artery. The Lacrimal Artery (LA): an early larger branch diverges laterally into the superotemporal orbit at the OA crossing site over CN II. LA first runs intraconally, then accompanies the lacrimal nerve (CN V1) along the superior LRM border and distributes to the lacrimal gland, and to the conjunctiva and lids in the lateral corner of the eye via the superior and inferior lateral palpebral arteries. Proximal to the lacrimal gland, already a descending branch divides into the zygomaticotemporal and zygomaticofacial arteries. These run along the lateral orbital wall and exit the orbit together with the homonymous nerves (ZTN and ZFN) through the respective foramina.

The lateral long posterior ciliary artery branches off medially to parallel CN II together with the medial long posterior ciliary artery.

The recurrent meningeal artery branch returns laterally to communicate with a frontal branch of the middle meningeal artery (MMA) via the SOF or the COF.

- Supraorbital artery (SA): is given off from the OA at the crossroads over CN II. It courses with the supraorbital nerve (CN V1) on top of the LPSM underneath the periorbita to leave the orbit through the supraorbital notch or foramen to the skin of the eyebrows and forehead.

- Ethmoidal Arteries: the posterior ethmoidal artery (PEA) and the anterior ethmoidal artery (AEA) rise from the OA along its anterior course vis-a-vis the medial orbital wall in transverse direction with a given distance in between them. The vessels penetrate the periorbita to enter their foramina at the frontoethmoidal suture together with the homonymous nasociliary nerve branches (CN V1). PEA usually passes over the SOM or between the MRM and SOM. It is variable in size, topographic location, and may even be absent.

The AEA is a more constant and somewhat larger vessel. It runs between the MRM and SOM. The AEA is the first transverse element to be encountered in a subperiorbital dissection of the superonasal quadrant.

Inside the ethmoid, it ascends to enter the upper nasal cavity and to penetrate the roof to direct across the anterior cranial fossa to the cribriform plate.

Both the AEA and PEA (Fig. 3.31) are supplemented with anastomoses via the sphenopalatine artery as well as the angular/facial artery provide the blood supply of a widespread area, the entire ethmoidal sinuses, the infundibulum of the frontal sinus, the upper nasal cavity including the septum and the skin over the cartilaginous nasal vault with additional descending branches.

- Medial Palpebral Arteries: the superior and inferior medial palpebral arteries arise from the OA below the trochlea and descend to the lacrimal sac and the upper and lower eyelids, where they form rows of marginal and peripheral arcades connecting with the lateral palpebral arteries of the LA.

- Frontonasal artery: corresponds to the distal OA end that splits into the following terminal branches.

- Supratrochlear artery: it runs above the trochlea and over the medial upper orbital rim to the paramedian forehead (formerly known as medial frontal artery).

- Dorsal nasal artery: it comes to lie midway in between the trochlea and the medial canthal tendon and continues anteriorly for the supply of the lacrimal sac and the skin of the nasal sidewall and dorsum.

Venous Outflow of the Orbit

The venous drainage from the orbit [71, 72] can be organized into three routes—upstream return into the cavernous sinus, downward return into the pterygoid plexus, and anterior leave by communication with the venous system of the face. The veins of the orbit are suggested to be valveless. Other than the veins elsewhere in the body, the orbital veins are not as closely associated with their arterial counterparts.

Two principal venous systems and their tributaries deal with the outflow from distinct distribu-

tions: the superior ophthalmic vein (SOV) drains the superomedial orbit, whereas the inferior ophthalmic vein (IOV) drains the inferolateral orbit (Fig. 3.42).

Nasofrontal Vein: corresponds to the confluence of the supratrochlear and angular veins and their union to the SOV.

Superior Ophthalmic Vein (SOV): begins medially above the anterior eyeball in continuation of the nasofrontal, passes the SOF (Table 3.1), and empties into the cavernous sinus. In the anterior section, the course of the SOV coincides with parts of the OA.

The SOV passes backward alongside the trochlea and the SOM. It crosses above CN II toward the lateral SRM border to bend down into the lateral orbital apex, which it leaves between the heads of the SRM and LRM for the SOF. Usually the SOF runs through the narrow superior lateral sector outside Zinn's ring. At the SOF level, the SOV meets with the IOV. Fused to

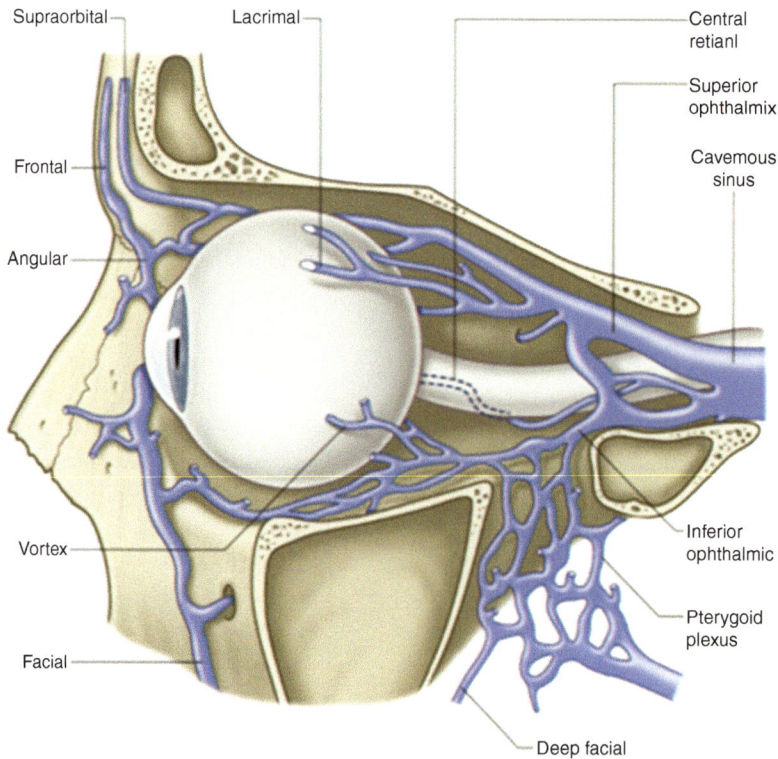

Fig. 3.42 Illustration - Orbit with globe from lateral side with a simplified overview of the principal vessels and tributaries of orbital venous drainage system. (With permission from: Gray's Anatomy 42nd edition)

a common conduit, they merge with the anteroinferior extensions of the cavernous sinus.

Ethmoidal Veins: most often there are two veins paralleling their corresponding arteries over some part. The anterior ethmoidal vein drains into the SOV, the posterior ethmoidal vein joins a venous webbing under the orbital roof.

Muscular Venous Branches from the EOMs: The venous branches from SOM and SRM consistently drain into the SOV. The MRM and LRM branches may drain into the SOV, IOV or lacrimal vein (LV).The IRM and IOM branches connect to the venous plexus relating to the orbital floor.

Vorticose/Vortex veins (VVs): a set of commonly four, occasionally up to eight VVs drain the choroid. Each eyeball quadrant has at least one VV emanating posterior of the prime meridian. The superior medial VV empties either into the SOV or the lacrimal vein LA. The inferior medial and lateral VVs feed into the SOV or IOV either separately or joined together.

Lacrimal Vein: accompanies the LA and usually opens into the SOV from the lateral side further posteriorly than the other tributaries. It may be large enough to rank as a third main orbital vein, in particular if it enters the cavernous sinus separately. It is usually joined by muscular branches from the SRM and LRM and by a vortex vein.

Collateral Veins: the upper and lower orbital venous divisions communicate via several collateral veins. They all run a course from the floor to the roof. With regard to their topographical position inside the orbit, they are called anterior, medial, lateral, and posterior collateral veins.

Ciliary Veins: accompany their corresponding arteries. The anterior ciliary veins drain the ciliary body and greater circle of the iris into the muscular branches.

Central Retinal Vein: leaves the meningeal optic nerve sheath nearer behind the globe than the central retinal artery and passes into the orbital apex to open directly into the cavernous sinus or less frequently into the SOV, IOV or posterior collateral vein.

Inferior Ophthalmic Vein (IOV): develops from the posterior end of a venous plexus which relates to the orbital floor by covering around the IRM and underneath CN II in the midorbit. The anterior part of the plexus receives the VVs from the lower globe hemisphere and inferior muscular branches, the posterior part connects with the collateral veins.

The IOV runs posteriorly in parallel to the CN III branch to the IOM and to the IRM, respectively before it may divide into two or three branches. A consistent superior branch passes into the cavernous sinus in common with the SOV. Another direct branch connects to the cavernous sinus through the inferior SOF sector below Zinn's ring. A variable third lower branch may communicate with the pterygoid plexus through the IOF.

The infraorbital vein accompanies the infraorbital artery and nerve. It passes along the canal/grove and through the IOF (Table 3.1) to drain into the pterygoid plexus. Its tributaries emerge from tissues close to the orbital floor and communication with the IOV.

Cavernous Sinus

The cavernous sinus (CS) (Figs. 3.30 and 3.42) or the "lateral sellar compartment" belongs to the cranial dural venous sinuses [72, 74, 75]. The paired CS expand on each side of the sella turcica, the pituitary gland, and the lateral aspect of the sphenoid body including the adjacent part above the petrous apex. The architecture of the CS walls is composed of meningeal and periosteal/periorbital layers [76]. The CS walls surround a system of three longitudinal venous axes and the C4 segment of the ICA with a sympathetic nerve plexus around it. The oculomotor (CN III), trochlear (CN IV), and ophthalmic nerves (CN V1) course in the lateral wall (Fig. 3.30). The abducens nerve (CN VI) passes forward within the sinus in between the medial side of CN V1 and the lateral side the ICA. The longitudinal "intermediate venous axis" laying between the ICA and cranial nerves contributes to the venous drainage from the directly adjacent orbit [77]. The sinus is larger posteriorly and gets more narrow anteriorly, where it reaches into the orbital apex at the SOF [78]. The lower

SOF margin, made up by the junction of the sphenoid body and GWS, is located at the same level as the lower acute angled CS edge. The medial and lateral CS walls meet along the lower edge, so that the sinus obtains a triangular cross section. The lower CS edge extends from across the base of the GWS to the lateral border of the dorsum sellae leaning against the lateral margin of the sella turcica. The transitional region between the orbital apex and the lateral sellar compartment via the SOF has been conceptualized in terms of a "lateral sellar orbital junction" (LSOJ) which is situated in the wide inferomedial SOF portion. According to the LSOJ concept, the SOF is not a mere portal but a three-dimensional structure contributing to a sequence of compartments from the cone of the orbital apex over a neural compartment into the intracranial venous compartment corresponding to the cavernous sinus. The PPF has an extension via the IOF and the SOF. Though it is addressed as the pterygopalatine compartment subjacent to the LSOJ, it is not counted as a constituent of it.

Lymphatic Drainage

In the eyelids, lymphatic drainage [79, 80] is restricted to the region anterior to the orbital septum and provided by a superficial pretarsal and a deep posttarsal system.

The superficial system drains the lymph fluid from the OOM and the overlying skin.

The deep system drains the tarsus and conjunctiva. Drainage is into preauricular and submandibular nodes. The retaining ligament at the infraorbital rim separates lymphatic drainage locations: lower eyelid edema stops "acutely" at the rim.

Orbital edema resolves via the conjunctival lymphatic system. Edema in the posterior orbit drains into the cavernous sinus, also carrying the risk of infectious spread intracranially and the risk of cavernous sinus thrombosis.

The retina and cornea lack lymphatic vessel systems.

Periorbital Dissection—Deep Orbit Versus Rule of Halves: 24 to 12 to 6

The anterior IOF loop has been proposed as an anatomic landmark to define the point of entrance into the deep orbit in post-traumatic surgical dissection and repair, with the two arguments that the frontal orbital cross section begins to taper backward continuously from there on, making a subperiosteal (periorbital) dissection progressively difficult and with the high frequency at which it is involved by relevant traumatic defects.

Clinical Implications

The IOV and/or SOV are the terminations of the angular vein plexus from the facial vein system; since the majority of these veins are valveless permitting flow in both directions, it has long been assumed that dentofacial infections may easily spread along these veins in a retrograde fashion into the cavernous sinus carrying the risk of a cavernous sinus thrombosis. Recently, it has been shown that various vascular mechanisms in the head & neck region may be involved in the context of inadvertent spread of cosmetic facial fillers and embolic material into the orbit [73]. There is a rich interconnection of anatomic territories (angiosomes) which are linked together by functional anastomotic ("choke") vessels. So links between the ophthalmic-, facial-, superficial temporal-, and maxillary arteries are present. A plexus of a large caliber facial venous network drains into the orbits along the inner canthus concurrently with arteriovenous shunting and true anastomoses between ophthalmic artery—angular facial artery. According to current hypothesis, different vascular pathways—separately or in combination—may contribute to the spread of infectious emboli into the cavernous sinus and dural veins.

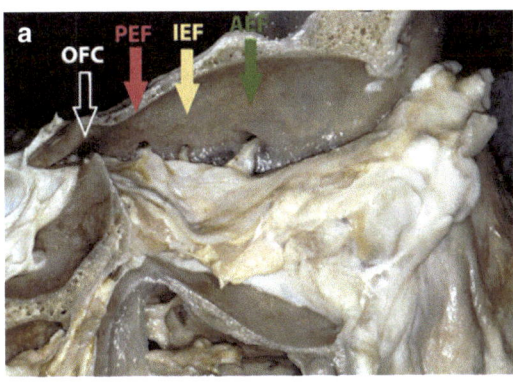
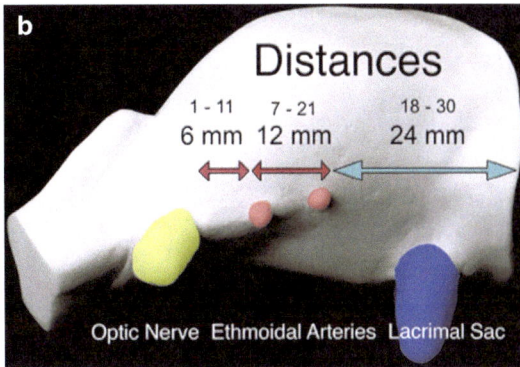

Fig. 3.43 Rule of halves explained. (**a**) Anatomic specimen – Look into the medial orbital wall. Threefold variation of ethmoidal foramina—periorbital sac emptied to show the sleeve-like extensions to AEF, IEF variant, PEF, and optic foramen/canal (OFC). *AEF* anterior ethmoidal foramen, *PEF* posterior ethmoidal foramen, *IEF* intermediate ethmoidal foramen, *OFC* optic foramen/canal. (**b**) Medial surface view on orbital contents model -with the distances 24 mm to 12 mm to 6 mm—in detail: average distance of 24 mm from anterior lacrimal crest to anterior ethmoidal foramen, 12 mm between anterior ethmoidal formanen and posterior ethmoidal foramen, and another 6 mm span from posterior ethmoidal foramen to optic foramen/canal

The concept of the deep orbit [81] suggests using the following four hard and soft tissue structures for orientation and pathfinding during dissection of the inferior and lateral wall: infraorbital nerve (canal groove/sulcus)/IOF / GWS (central trigone part/sphenotemporal buttress) / upper plate of the orbital process of the palatine bone.

These structures concenter at the confluence of the orbit, which corresponds to the area in direct proximity of the IOF isthmus.

Often, orbital depth gauging data derived from anthropometric studies cannot provide appropriate guidelines for safe distance dissection because of variation, particularly in trauma due to severe multifragmentation, displacement, soft tissue disruption, and fat herniation. So, the preference is to dissection in a subperiosteal (subperiorbital) plane to identify the leading structures [82, 83]. Apart from the EF, the medial orbit does not feature any reliable orientation aid to prevent interference with the optic foramen and the optic nerve [84].

Despite the well-known uncertainties in terms of number and zonal location of the EF, their potential distances in relation to the lacrimal crest, among each other, and to the optic foramen (Fig. 3.43a) are referred to by a well-known mnemonic, the rule of halves: 24 mm to 12 mm to 6 mm by Rontal [85] brings one to remind the distances (Fig. 3.43b) [81].

Conclusion

Anatomy has and will always be one of the most important and challenging fundaments of Medicine. Its contribution is indispensable.

When surgery is performed, knowledge of basic anatomy is a conditio sine qua non.

We realize that this chapter on soft tissue in and around the orbit is far from complete. The different types of tissues involved in the orbital contents are highly complex. Even today, despite the tremendous amount of existing information accumulated by well-respected medical researchers in the past, new insights are still continuously coming. We have presented a brief compilation of some anatomic issues here and hope that it will supply the reader with useful information to facilitate surgical interventions within the orbit.

Acknowledgements Special thanks to Dr. Serge Steenen who has taken care of drawing several illustrations in this Chap. 3, in which he tried to reflect anatomical reality in a truthful manner. Big thanks go to Gerhard Poetzel and Rudolf Herzig, the photographers at the OMFS Departments, LMU Munich, for their excellent support in editing the anatomic images.

References

1. Lipham WJ, Tawfik HA, Dutton JJ. A histologic analysis and three-dimensional reconstruction of the muscle of Riolan. Ophthalmic Plast Reconstr Surg. 2002;18(2):93–8. https://doi.org/10.1097/00002341-200,203,000-00002.
2. Riolan J. Anthropographia et osteologia. Paris: Moreau; 1626.
3. Hwang K, Wu XJ, Kim H, Kim DJ. Sensory innervation of the upper eyelid. J Craniofac Surg. 2018;29(2):514–7. https://doi.org/10.1097/SCS.0000000000004155.
4. Hollinshead WH. Anatomy for surgeons: the head and neck. New York: Harper & Row Publishers; 1982.
5. Kels BD, Grzybowski A, Grant-Kels JM. Human ocular anatomy. Clin Dermatol. 2015;33(2):140–6. https://doi.org/10.1016/j.clindermatol.2014.10.006.
6. Zide BM. Surgical anatomy around the orbit. The system of zones. Philadelphia, PA: Lippincott Williams and Wilkins; 2006.
7. Lemke BN, Stasior OG, Rosenberg PN. The surgical relations of the levator palpebrae superioris muscle. Ophthalmic Plast Reconstr Surg. 1988;4(1):25–30. https://doi.org/10.1097/00002341-198,801,130-00004.
8. Whitnall S. On a ligament acting as a check to the action of the levator palpebrae superioris muscle. J Anat Physiol Lond. 1910;45(Pt 2):131–9.
9. Whitnall SE. Anatomy of the human orbit and accessory organs of vision, Facsimile Reprint Original Edition 1921 ed. New York: Robert E Krieger Publishing Company; 1921.
10. Lim HW, Paik DJ, Lee YJ. A cadaveric anatomical study of the levator aponeurosis and Whitnall's ligament. Korean J Ophthalmol. 2009;23(3):183–7. https://doi.org/10.3341/kjo.2009.23.3.183.
11. Wiersinga WM, Kahaly GJ, editors. Graves' orbitopathy. A multidisciplinary approach – Questions and answers, 3rd, revised and expanded edition ed. Basel: Medical and Scientific Publishers; 2017.
12. Hwang K, Nam YS, Choi HG, Han SH, Hwang SH. Cutaneous innervation of lower eyelid. J Craniofac Surg. 2008;19(6):1675–7. https://doi.org/10.1097/SCS.0b013e31818c0495.
13. Al-Moraissi E, Elsharkawy A, Al-Tairi N, et al. What surgical approach has the lowest risk of the lower lid complications in the treatment of orbital floor and periorbital fractures? A frequentist network meta-analy-

sis. J Craniomaxillofac Surg. 2018;46(12):2164–75. https://doi.org/10.1016/j.jcms.2018.09.001.
14. Cornelius CP, Mayer P, Ehrenfeld M, Metzger MC. The orbits-anatomical features in view of innovative surgical methods. Facial Plast Surg. 2014;30(5):487–508. https://doi.org/10.1055/s-0034-1,394,303.
15. Dutton JJ, Waldrop TG. Atlas of clinical and surgical orbital anatomy. 2nd ed. Elsevier Saunders; 2011.
16. Al-Moraissi EA, Thaller SR, Ellis E. Subciliary vs. transconjunctival approach for the management of orbital floor and periorbital fractures: A systematic review and meta-analysis. J Craniomaxillofac Surg. 2017;45(10):1647–54. https://doi.org/10.1016/j.jcms.2017.07.004.
17. Rootman J. Orbital surgery—a conceptual approach. 2nd ed. Philadelphia, PA: Lippincott Williams & Wilkins; 2014.
18. Garcia GH, Goldberg RA, Shorr N. The transcaruncular approach in repair of orbital fractures: a retrospective study. J Craniomaxillofac Trauma. 1998;4(1):7–12.
19. Goldberg RA, Mancini R, Demer JL. The transcaruncular approach: surgical anatomy and technique. Arch Facial Plast Surg. 2007;9(6):443–7. https://doi.org/10.1001/archfaci.9.6.443.
20. Graham SM, Thomas RD, Carter KD, Nerad JA. The transcaruncular approach to the medial orbital wall. Laryngoscope. 2002;112(6):986–9. https://doi.org/10.1097/00005537-200,206,000-00009.
21. Hayek G, Mercier P, Fournier HD. Anatomy of the orbit and its surgical approach. Adv Tech Stand Neurosurg. 2006;31:35–71. https://doi.org/10.1007/3-211-32234-5_2.
22. Janfaza P, Nadol JBJ, Galla RJ, Fabian RL, Montgomery WW. Surgical anatomy of the head and neck, first Harvard University Edition. Harvard University Press; 2001.
23. Kempton SJ, Cho DC, Thimmappa B, Martin MC. Benefits of the retrocaruncular approach to the medial orbit: A clinical and anatomic study. Ann Plast Surg. 2016;76(3):295–300. https://doi.org/10.1097/SAP.0000000000000531.
24. Nguyen DC, Shahzad F, Snyder-Warwick A, Patel KB, Woo AS. Transcaruncular approach for treatment of medial wall and large orbital blowout fractures. Craniomaxillofac Trauma Reconstr. 2016;9(1):46–54. https://doi.org/10.1055/s-0035-1,563,390.
25. Pausch NC, Sirintawat N, Wagner R, Halama D, Dhanuthai K. Lower eyelid complications associated with transconjunctival versus subciliary approaches to orbital floor fractures. Oral Maxillofac Surg. 2016;20(1):51–5. https://doi.org/10.1007/s10006-015-0526-1.
26. Rontal E, Rontal M, Guilford FT. Surgical anatomy of the orbit. Ann Otol Rhinol Laryngol. 1979;88(3 Pt 1):382–6. https://doi.org/10.1177/000348947908800315.
27. Shen YD, Paskowitz D, Merbs SL, Grant MP. Retrocaruncular approach for the repair of medial orbital wall fractures: an anatomical and clinical study.

Craniomaxillofac Trauma Reconstr. 2015;8(2):100–4. https://doi.org/10.1055/s-0034-1,375,168.

28. Shorr N, Baylis HI, Goldberg RA, Perry JD. Transcaruncular approach to the medial orbit and orbital apex. Ophthalmology. 2000;107(8):1459–63. https://doi.org/10.1016/s0161-6420(00)00241-4.

29. Zide BM, McCarthy JG. The medial canthus revisited—an anatomical basis for canthopexy. Ann Plast Surg. 1983;11(1):1–9. https://doi.org/10.1097/00000637-198,307,000-00001.

30. Jordan DR, Gupta S, Hwang I. The superior and inferior components of Whitnall's ligament. Ophthalmic Surg Lasers. 2001;32(2):173–4.

31. Ali MJ, Zetzsche M, Scholz M, Hahn D, Gaffling S, Heichel J, Hammer CM, Bräuer L, Paulsen F. New insights into the lacrimal pump. The Ocular Surfaec. 2020;18:689–98.

32. Maliborski A, Różycki R. Diagnostic imaging of the nasolacrimal drainage system. Part I. Radiological anatomy of lacrimal pathways. Physiology of tear secretion and tear outflow. Med Sci Monit. 2014;20:–628, 638. https://doi.org/10.12659/msm.890098.

33. Scott G, Balsiger H, Kluckman M, Fan J, Gest T. Patterns of innervation of the lacrimal gland with clinical application. Clin Anat. 2014;27(8):1174–7. https://doi.org/10.1002/ca.22447.

34. Seifert P, Spitznas M. Demonstration of nerve fibers in human accessory lacrimal glands. Graefes Arch Clin Exp Ophthalmol. 1994;232(2):107–14. https://doi.org/10.1007/bf00171672.

35. Zoumalan CI, Joseph JM, Lelli GJ Jr, et al. Evaluation of the canalicular entrance into the lacrimal sac: an anatomical study. Ophthalmic Plast Reconstr Surg. 2011;27(4):298–303. https://doi.org/10.1097/IOP.0b013e31820d1f7b.

36. Lorber M. Gross characteristics of normal human lacrimal glands. Ocul Surf. 2007;5(1):13–22. https://doi.org/10.1016/s1542-0124(12)70049-6.

37. Lorber M, Vidić B. Measurements of lacrimal glands from cadavers, with descriptions of typical glands and three gross variants. Orbit. 2009;28(2–3):137–46. https://doi.org/10.1080/01676830902766014.

38. Seifert P, Spitznas M, Koch F, Cusumano A. The architecture of human accessory lacrimal glands. Ger J Ophthalmol. 1993;2(6):444–54.

39. Gipson IK. Goblet cells of the conjunctiva: A review of recent findings. Prog Retin Eye Res. 2016;54:49–63. https://doi.org/10.1016/j.preteyeres.2016.04.005.

40. Jakobiec FA, Eagle RC Jr, Selig M, Ma L, Shields C. Clinical implications of goblet cells in dacryoadenosis and normal human lacrimal glands. Am J Ophthalmol. 2020;213:267–82. https://doi.org/10.1016/j.ajo.2020.01.029.

41. Kessing SV. Mucous gland system of the conjunctiva. A quantitative normal anatomical study. Acta Ophthalmol (Copenh). 1968;Suppl 95:91.

42. Knop E, Knop N, Millar T, Obata H, Sullivan DA. The international workshop on meibomian gland dysfunction: report of the subcommittee on anatomy, physiology, and pathophysiology of the meibomian gland. Invest Ophthalmol Vis Sci. 2011a;52(4):1938–78. https://doi.org/10.1167/iovs.10-6997c.

43. Cher I. Fluids of the ocular surface: concepts, functions and physics. Clin Exp Ophthalmol. 2012;40(6):634–43. https://doi.org/10.1111/j.1442-9071.2012.02758.x.

44. Willcox MDP, Argüeso P, Georgiev GA, et al. TFOS DEWS II tear film report. Ocul Surf. 2017;15(3):366–403. https://doi.org/10.1016/j.jtos.2017.03.006.

45. Yokoi N, Bron AJ, Tiffany JM, Maruyama K, Komuro A, Kinoshita S. Relationship between tear volume and tear meniscus curvature. Arch Ophthalmol. 2004;122(9):1265–9. https://doi.org/10.1001/archopht.122.9.1265.

46. Wolff E. The mucocutaneous junction of the lid-margin and the distribution of the tear fluid. Trans Ophthalmol Soc UK. 1946;66:291–308.

47. Engin Ö, Adriaensen G, Hoefnagels FWA, Saeed P. A systematic review of the surgical anatomy of the orbital apex. Surg Radiol Anat. 2021;43(2):169–78. https://doi.org/10.1007/s00276-020-02573-w.

48. Kakizaki H, Zako M, Nakano T, Asamoto K, Miyaishi O, Iwaki M. Direct insertion of the medial rectus capsulopalpebral fascia to the tarsus. Ophthalmic Plast Reconstr Surg. 2008;24(2):126–30. https://doi.org/10.1097/IOP.0b013e3181647cb2.

49. Koornneef L. New insights in the human orbital connective tissue. Result of a new anatomical approach. Arch Ophthalmol. 1977;95(7):1269–73. https://doi.org/10.1001/archopht.1977.04450070167018.

50. Koornneef L. Orbital septa: anatomy and function. Ophthalmology. 1979;86(5):876–80. https://doi.org/10.1016/s0161-6420(79)35444-6.

51. Fu L, Malik J. Brown syndrome. Treasure Island, FL: StatPearls; 2021.

52. Lauer SA, Sauer H, Pak SM. Brown's syndrome diagnosed following repair of an orbital roof fracture: a case report. J Craniomaxillofac Trauma. 1998;4(4):20–2.

53. Laure B, Arsene S, Santallier M, Cottier JP, Sury F, Goga D. Desinsertion post-traumatique de la poulie du muscle oblique supérieur. [Post-traumatic disinsertion of the superior oblique muscle trochlea]. Rev. Stomatol Chir Maxillofac. 2007;108(6):551–4. https://doi.org/10.1016/j.stomax.2007.08.003.

54. Warrier S, Wells J, Prabhakaran VC, Selva D. Traumatic rupture of the superior oblique muscle tendon resulting in acquired Brown's syndrome. J Pediatr Ophthalmol Strabismus. 2010;47(3):168–70. https://doi.org/10.3928/01913913-20,100,505-08.

55. Takezawa K, Townsend G, Manavis J, Ghabriel M. Aberrant distribution of the trochlear nerve: A cadaveric study supported by immunohistochemistry. Ann Anat. 2017;213:1–7. https://doi.org/10.1016/j.aanat.2017.04.001.

56. Krause W. Termination of the nerves in the conjunctiva. J Anat Physiol Lond. 1867;1:346.

57. Reymond J, Kwiatkowski J, Wysocki J. Clinical anatomy of the superior orbital fissure and the orbital

apex. J Craniomaxillofac Surg. 2008;36(6):346–53. https://doi.org/10.1016/j.jcms.2008.02.004.

58. Rohrich RJ, Hackney FL, Parikh RS. Superior orbital fissure syndrome: current management concepts. J Craniomaxillofac Trauma. 1995;1(2):44–8.

59. Striph GG, Burde RM. Abducens nerve palsy and Horner's syndrome revisited. J Clin Neuroophthalmol. 1988;8(1):13–7.

60. Thakker MM, Huang J, Possin DE, et al. Human orbital sympathetic nerve pathways. Ophthalmic Plast Reconstr Surg. 2008;24(5):360–6. https://doi.org/10.1097/IOP.0b013e3181837a11.

61. Manson PN, Lazarus RB, Morgan R, Iliff N. Pathways of sympathetic innervation to the superior and inferior (Muller's) tarsal muscles. Plast Reconstr Surg. 1986;78(1):33–40. https://doi.org/10.1097/00006534-198,607,000-00004.

62. Hartmann B, Kremer I, Gutman I, Krakowski D, Kam J. Cavernous sinus infection manifested by Horner's syndrome and ipsilateral sixth nerve palsy. J Clin Neuroophthalmol. 1987;7(4):223–6. https://doi.org/10.3109/01658108709007456.

63. Martin TJ. Horner syndrome: A clinical review. ACS Chem Neurosci. 2018;9(2):177–86. https://doi.org/10.1021/acschemneuro.7b00405.

64. Haladaj R. Anatomical variations of the ciliary ganglion with an emphasis on the location in the orbit. Anat Sci Int. 2020;95(2):258–64. https://doi.org/10.1007/s12565-019-00518-x.

65. Lundy JA, McNary T. Neuroanatomy, pterygopalatine ganglion. Treasure Island, FL: StatPearls; 2021.

66. Ducasse A, Segal A, Delattre JF, Burette A, Flament JB. Participation of the external carotid artery in orbital vascularization. J Fr Ophtalmol. 1985;8(4):333–9.

67. Seoane E, Rhoton AL Jr, de Oliveira E. Microsurgical anatomy of the dural collar (carotid collar) and rings around the clinoid segment of the internal carotid artery. Neurosurgery. 1998;42(4):869–84.; discussion 884–866. https://doi.org/10.1097/00006123-199,804,000-00108.

68. Abed SF, Shams P, Shen S, Adds PJ, Uddin JM, Manisali M. A cadaveric study of the cranio-orbital foramen and its significance in orbital surgery. Plast Reconstr Surg. 2012;129(2):307e–11e. https://doi.org/10.1097/PRS.0b013e31821b6382.

69. Bertelli E, Regoli M, Bracco S. An update on the variations of the orbital blood supply and hemodynamic. Surg Radiol Anat. 2017;39(5):485–96. https://doi.org/10.1007/s00276-016-1776-9.

70. Hayreh SS. Orbital vascular anatomy. Eye (Lond). 2006;20(10):1130–44. https://doi.org/10.1038/sj.eye.6702377.

71. Cheung N, McNab AA. Venous anatomy of the orbit. Invest Ophthalmol Vis Sci. 2003;44(3):988–95. https://doi.org/10.1167/iovs.02-0865.

72. Spektor S, Piontek E, Umansky F. Orbital venous drainage into the anterior cavernous sinus space: microanatomic relationships. Neurosurgery. 1997;40(3):532–9; discussion 539–540

73. Taylor GI, Shoukath S, Gascoigne A, Corlett RJ, Ashton MW. The Functional anatomy of the ophthalmic angiosome and its implications in blindness as a complication of cosmetic facial filler procedures. Plast Reconstr Surg. 2020;146(4):745. https://doi.org/10.1097/PRS.0000000000007155.

74. Campero A, Campero AA, Martins C, Yasuda A, Rhoton AL Jr. Surgical anatomy of the dural walls of the cavernous sinus. J Clin Neurosci. 2010;17(6):746–50. https://doi.org/10.1016/j.jocn.2009.10.015.

75. Rhoton AL. The middle cranial base and cavernous sinus. In: Dolenc VV, Rogers L, editors. Cavernous sinus—developments and future perspectives. Wien: Springer Verlag; 2009. p. 3–26.

76. Kawase T, van Loveren H, Keller JT, Tew JM. Meningeal architecture of the cavernous sinus: clinical and surgical implications. Neurosurgery. 1996;39(3):527–34; discussion 534–526. https://doi.org/10.1097/00006123-199,609,000-00019.

77. Mitsuhashi Y, Hayasaki K, Kawakami T, et al. Dural venous system in the cavernous sinus: a literature review and embryological, functional, and endovascular clinical considerations. Neurol Med Chir (Tokyo). 2016;56(6):326–39. https://doi.org/10.2176/nmc.ra.2015-0346.

78. Froelich S, Abdel Aziz KM, van Loveren HR, Keller JT. The transition between the cavernous sinus and orbit. In: Dolenc VV, Rogers L, editors. Cavernous sinus—developments and future perspectives. New York: Wien Springer Verlag; 2009. p. 27–33.

79. Dickinson AJ, Gausas RE. Orbital lymphatics: do they exist? Eye (Lond). 2006;20(10):1145–8. https://doi.org/10.1038/sj.eye.6702378.

80. Gausas RE, Gonnering RS, Lemke BN, Dortzbach RK, Sherman DD. Identification of human orbital lymphatics. Ophthalmic Plast Reconstr Surg. 1999;15(4):252–9. https://doi.org/10.1097/00002341-199,907,000-00006.

81. Evans BT, Webb AA. Post-traumatic orbital reconstruction: anatomical landmarks and the concept of the deep orbit. Br J Oral Maxillofac Surg. 2007;45(3):183–9. https://doi.org/10.1016/j.bjoms.2006.08.003.

82. Cornelius CP, Stiebler T, Mayer P, Smolka W, Kunz C, Hammer B, Jaquiéry C, Buitrago-Téllez C, Leiggener CS, Metzger MC, Wilde F, Audigé L, Probst M, Strong EB, Castelletti N, Prein J, Probst FA. Prediction of surface area size in orbital floor and medial wall fractures based on topographical subregions. J. CranioMaxillofac Surg. 2021a;49(7):598–612.

83. Cornelius CP, Probst F, Metzger MC, Gooris PJJ. Skeletal features and some notes on the peri-

orbital lining. Atlas Oral Maxillofac Surg Clin. 2021b;29(1):1–18.

84. Abuzayed B, Tanriover N, Gazioglu N, Eraslan BS, Akar Z. Endoscopic endonasal approach to the orbital apex and medial orbital wall: anatomic study and clinical applications. J Craniofac Surg.

2009;20(5):1594–600. https://doi.org/10.1097/SCS.0b013e3181b0dc23.

85. Rontal E, Rontal M, Guilford FT. Surgical anatomy of the orbit. Ann OtoloRhino Laryngol. 1979;88(3PtI):382–6. https://doi.org/10.1177/000348947908800315.

Open Access This chapter is licensed under the terms of the Creative Commons Attribution 4.0 International License (http://creativecommons.org/licenses/by/4.0/), which permits use, sharing, adaptation, distribution and reproduction in any medium or format, as long as you give appropriate credit to the original author(s) and the source, provide a link to the Creative Commons license and indicate if changes were made.

The images or other third party material in this chapter are included in the chapter's Creative Commons license, unless indicated otherwise in a credit line to the material. If material is not included in the chapter's Creative Commons license and your intended use is not permitted by statutory regulation or exceeds the permitted use, you will need to obtain permission directly from the copyright holder.

Imaging of the Orbit: "Current Concepts"

4

Maartje M. L. de Win

Learning Objectives
- To gain knowledge about state-of-the-art orbital imaging techniques and when to use them in clinical practice.
- To review CT and MRI scans of the orbit in a systematic way in order to identify relevant findings.
- To become familiar with radiological appearances of common and more uncommon orbital pathology.

Introduction

Many different conditions can affect the small anatomical space of the orbit, and imaging with computer tomography (CT) or magnetic resonance imaging (MRI) can be essential in the evaluation of orbital disease to make a correct diagnosis and guide and evaluate medical or surgical treatment. Common orbital diseases that may require imaging are orbital trauma, Graves' orbitopathy (GO), orbital infection, and orbital soft tissue lesions like intraorbital vascular lesions or inflammation or lesions of the bony

orbit. A structured way of reviewing the images and knowledge of orbital anatomy, imaging techniques, indications for orbital imaging, and the clinical and radiological presentation of orbital pathology are crucial for optimal evaluation.

Continuing technical developments created ample improvement in the imaging possibilities over the last decades. CT high-resolution images of the orbit, with the ability to reconstruct images in different planes, allow excellent depiction of the bony structures but also very reasonable delineation of soft tissues. However, MRI provides optimal soft-tissue contrast allowing for accurate depiction of the extension of pathology. Besides the standard anatomical imaging with MRI, it is also possible to look at biological phenomena of orbital pathology with functional MRI techniques like diffusion-weighted imaging (DWI) and perfusion-weighted imaging (PWI). Combining parameters from different imaging techniques in so-called multiparametric imaging creates additional value especially in the differential diagnosis of orbital soft-tissue masses. Moreover, ongoing developments in postprocessing tools strongly improve the usefulness of imaging, for example, in the presurgical planning and in the preoperative navigation during orbital reconstruction surgery. In the near future, also artificial intelligence (AI) will be applied to further improve orbital imaging and decision-making.

With the continuously improving possibilities in diagnostic imaging, evoking amplification of imaging indications, and the amount of orbital

M. M. L. de Win (✉)
Department of Radiology and Nuclear Medicine, Amsterdam University Medical Centers, Amsterdam, The Netherlands
e-mail: m.m.dewin@amsterdamumc.nl

© The Author(s) 2023
P. J. J. Gooris et al. (eds.), *Surgery in and around the Orbit*,
https://doi.org/10.1007/978-3-031-40697-3_4

121

scans, the challenge is to keep considering principles like radiation protection and cost-effectiveness.

In this chapter, we give an overview of the current orbital imaging techniques, their indications and usefulness in clinical practice, some guidance to a structured approach of the images, and radiological features of some common and less common orbital conditions.

Imaging Techniques

CT and MRI are the main imaging modalities to evaluate orbital pathology. Imaging of the bony orbit through conventional radiographs has become obsolete and has been replaced by the much more sensitive CT scans. The only indication for a conventional radiograph of the orbit nowadays is to exclude metallic foreign bodies in patients who need to undergo an MRI scan and have an increased risk for orbital metallic foreign bodies. Ultrasound can be useful in preseptal and ocular lesions, but is of limited value for retrobulbar lesions. Ultrasound may provide information about tumor shape, internal reflectivity, and vascular flow, but is operator-dependent, less suited for follow-up, and has a low specificity identifying malignant tumors [1]. Nuclear medicine techniques like FDG-PET-CT have limited value in assessing orbital disease due to high FDG uptake in extraocular muscles and typical small volume of orbital disease. FDG-PET, however, might be useful in screening for systemic lymphoma or detecting the primary tumor in case of orbital metastasis. The use of ^{68}Ga-DOTATATE PET-CT might be helpful to confirm the diagnosis of optic nerve sheet meningioma in selected cases when the diagnosis is not completely sure from the MRI [2].

CT

CT uses ionizing radiation (x-rays) to make cross-sectional images of the body. The X-rays are detected by a detector on the opposite side of the body. From the raw data, images are reconstructed from measurements of attenuation coefficients of the x-rays through the body. The reconstructed images reflect different densities of the body tissues, represented by Hounsfield units (HU) [3].

The advantages of CT are clear, it is a very fast and widely available imaging technique that gives excellent images of the bony orbit, but also information of the soft tissues. With the current multidetector CT (MDCT) machines, it is possible to scan the orbit with sub-millimetric and near isotropic (identical in all directions) resolution in only a few seconds, which makes it also suitable for restless patients in the emergency department [4].

Useful scan parameters for an orbital CT protocol on a state-of-the-art MDCT scanner are 120 kV, 90mAs, collimation 128–192 mm with 0.6 mm slice thickness, and pitch of 0.8. Parameters should be chosen with the ALARA principle (as low as reasonable achievable) in mind to keep the radiation dose as low as possible while maintaining diagnostic accuracy.

Because of the near isotropic resolution, the raw images from a single scan can be reconstructed in additional coronal or sagittal planes through multiplanar reconstructions (MPR). Moreover, different reconstruction filters can be applied to get optimal images from the orbital bone and from the soft tissues. Preferably, 1–2 mm reconstructions should be made for the bone setting and 1.5–3 mm for the soft-tissue images. Axial MPRs should be reconstructed when the original scan is not symmetric. Coronal MPRs should be made perpendicular to the orbital axis. Sagittal MPRs can be made on indication although for the assessment of orbital fractures these reconstructions are paramount. Metal artifact reduction (MAR) algorithms can be applied to improve image quality in patients with metal implants [5]. 3D reconstructions can be beneficial for surgical planning as they provide insight into the fracture malalignment prior to reconstructions (Fig. 4.1). In Part 5 of this book, treatment planning after trauma, including the use of CT-based reconstruction and planning software, will be discussed in more detail.

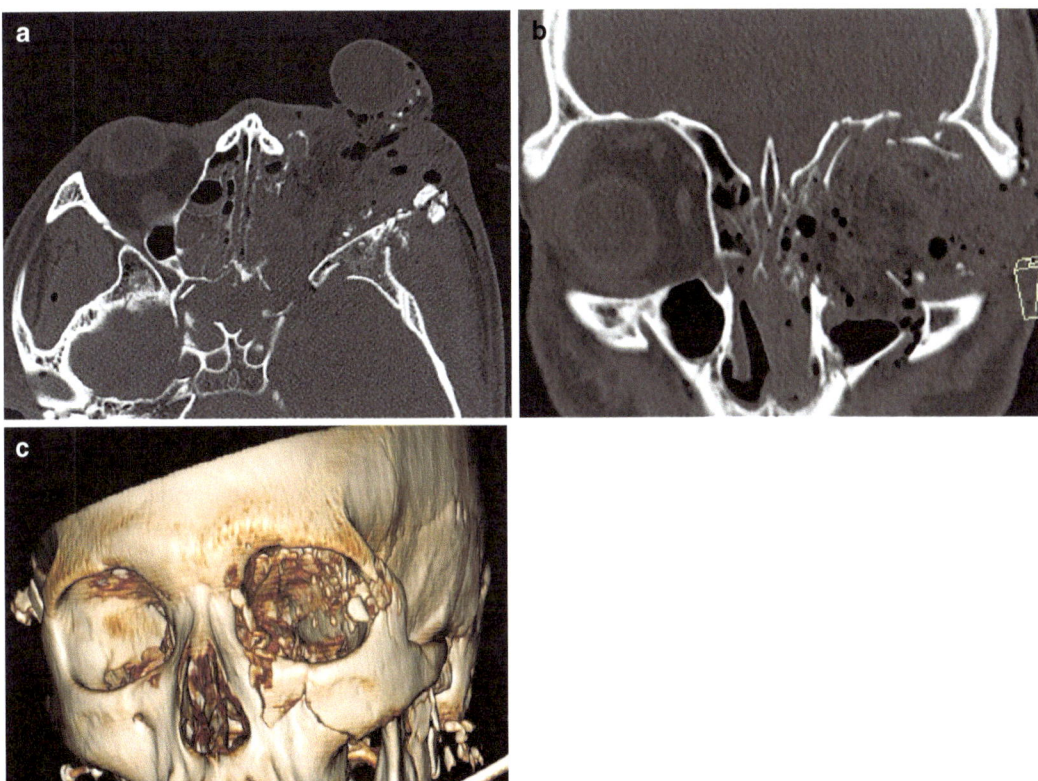

Fig. 4.1 Axial (**a**), coronal (**b**), and 3D volumetric bone reconstructions (**c**) of the CT from a patient after extensive orbital trauma due to a shooting accident. Note the luxation of the left globe and fractures of all the orbital walls. 3D reconstructions can be beneficial for surgical planning of complex fractures as they provide insight into the fracture malalignment prior to reconstructions

Intravenous iodinated contrast can be used for optimizing delineation of normal and pathological structures. In orbital imaging, the main indication for CT with intravenous contrast is evaluation of orbital infection. For tumor imaging, MRI is imaging method of first choice. In patients with impaired renal function or a history of severe allergy to iodinated contrast agents, a CT without contrast or an MRI scan should be considered. When this is not possible, the patient should be prepared with medication in order to prevent a severe allergic reaction or with prehydration and posthydration to prevent postcontrast acute kidney injury (PC-AKI) [6, 7]. Every hospital should have a (local) protocol on prevention and management of adverse events in patients at risk receiving iodinated contrast.

CT should be the preferential imaging technique when evaluation of the bony orbit is of main concern, like in trauma. Also in other urgent situations, such as evaluation of the extent of orbital cellulitis or postoperative complications, CT will be the imaging method of first choice in most hospitals. Moreover, CT is the preferred imaging method to evaluate orbital complications of chronic sinonasal disease, like silent sinus or sinonasal mucoceles involving the orbital wall [8].

Cone Beam CT

Cone beam CT (CBCT) is a relative recent CT imaging technique providing low-dose and high-spatial resolution visualization of high-contrast structures. In the head and neck, it is mainly used for dental and mastoid imaging. Although it gives excellent spatial resolution of the bony structures, the soft-tissues of the orbit cannot reliably be evaluated with CBCT because of its limited contrast resolution [9].

MRI

The advantage of orbital imaging with MRI, when compared to CT, is the excellent soft-tissue contrast, without the use of ionizing radiation. The latter makes it more suitable for children and young adults and for patient who need repeated orbital imaging. The downside, besides the higher costs, is the longer scanning time; different series and planes have to be scanned separately and take on average 2–4 min, making the average scanning time for an orbital MRI about 10–15 min for a standard protocol without contrast (i.e., for GO) up to 30 min for a more extensive tumor imaging protocol. Most children below the age of 6–7 will therefore need sedation or general anesthesia with adequate patient monitoring to undergo MRI scanning.

The superior visualization of soft-tissues makes MRI imaging the preferred method for imaging orbital soft-tissue masses and other more complex orbital disease. Orbital MRI scans can be obtained on both 1.5 T or 3 T MRI scanners and are usually performed with a head coil; only for superficial or globe lesions, a surface coil could have additional value. 3 T has the advantage of a higher signal-to-noise ratio, which can be translated into a better spatial resolution in the same acquisition time, but it also has more challenges regarding susceptibility artifacts.

A routine orbital MRI protocol includes a precontrast turbo spin-echo (TSE) T1- and T2-weighted imaging (WI) without fat suppression and a series with fat suppression (FS) that is relatively independent of field inhomogeneity like STIR (short-TI inversion recovery), T2-weighted TIRM (turbo inversion recovery), or Dixon technique. Images should be acquired with high in-plane spatial resolution and slice thickness not exceeding 3 mm. With the state-of-

the-art MRI scanners, it is also possible to acquire volumetric scans in 3D, from which the other planes can be reconstructed.

A time of flight MR angiography (TOF MRA) without the need of IV contrast or a dynamic MRA after IV contrast could be added as an additional series when (high-flow) vascular lesions, such orbital arteriovenous malformation (AVM) or (dural) carotid cavernous fistula, are suspected.

The T1-WI and T2-WI without fat suppression give excellent anatomical detail and contrast between the orbital fat that is very bright and the other soft tissues, like the extraocular muscles (EOMs), the lacrimal gland and the optic nerve sheet complex. This facilitates assessing the extension of orbital disease. With fat-suppression techniques (i.e. STIR, TIRM, Dixon), edema is accentuated, facilitating the detection of active inflammation/disease. The fat suppression of the STIR technique is superior to most other fat-suppression techniques especially near metallic foreign bodies and near tissue interfaces with high susceptibility differences (like interface between the orbit, skull base, and paranasal sinus); moreover, it has the additional feature of additive T1 and T2 contrast [10]. The Dixon technique has the advantage to provide images with and without fat suppression from a single acquisition [11].

MRI protocol without intravenous contrast is generally adequate for evaluation of GO and follow-up of known benign orbital diseases like venous malformations. For more complex orbital pathology, like tumors, infectious or inflammatory disease, additional postcontrast series after injection of gadolinium-based contrast are indicated. The postcontrast T1-WI should be made with fat suppression; otherwise, it might be difficult to recognize enhancement within the bright orbital fat.

Functional/Advanced MRI Techniques

Besides the routine anatomical MRI series, more advanced techniques such as diffusion-weighted imagine (DWI) and perfusion-weighted imaging

(PWI) allow studying of functional aspects of orbital lesions. This can help in further characterizing orbital pathology and a more accurate differentiation between malignant and benign disease.

Diffusion-Weighted Imaging (DWI)

The principle of DWI relies on the Brownian motion of water molecules in a voxel of tissue, i.e., within the intracellular and extracellular fluid of the voxel. In simplified terms, the higher the cellularity or cellular swelling, the more restricted the motion will be [12]. From the DWI, apparent diffusion coefficients (ADC) values are calculated and represented in the ADC maps. In active inflammation like GO, ADC values will be increased because of increased interstitial edema (increased diffusion), while in highly cellular tumors, like lymphoma or in purulent abscesses with highly packed inflammatory cells, ADC values will be low (restricted diffusion). DWI is therefore one of the techniques that can be used in further characterizing orbital lesions, especially in differentiating benign from malignant disease. The most frequently used DWI technique, especially in brain imaging, is the echo-planar imaging (EPI)-based sequence. The advantage is that it is a fast technique, but the imaging quality of this sequence in the orbit can be limited by the inhomogeneity artifacts caused by the interfaces between air, bone, and soft tissue in the orbit. When metallic reconstruction material is applied in the orbit, EPI-DWI sequences will be noninterpretable in most cases because of the susceptibility of artifacts resulting in severe distortions. Non-EPI-DWI like the turbo spin-echo (TSE) DWI sequence is an alternate technique with less susceptibility of artifacts and image distortion than on EPI-DWI scans. Although this sequence in general has a longer acquisition time, the non-EPI technique is preferable to the EPI-DWI technique when scanning the head and neck, including the orbit because of better image quality, both at 1.5 T and 3 T [13, 14]. Advisable is to use at least 3 b values, e.g., $b = 0$, $b = 500$ and $b = 1000$ s/mm^2 to calculate reliable ADC values.

Perfusion-Weighted Imaging (PWI)

PWI is a technique that gives insight into the perfusion of tissues by blood. The main technique used in head and neck imaging is the dynamic contrast-enhanced (DCE) perfusion. It relies upon the T1 shortening induced by an intravenously injected gadolinium-based contrast bolus passing through tissue. Before (at baseline) and after injection of the contrast, fast repeated T1 images with a temporal resolution of about 3–4 s are obtained during 4–5 mins and the increase of the signal is measured. The increase of the signal (enhancement) is mainly caused by the accumulation of gadolinium in the extravascular extracellular space. This is dependent on the microvascular characteristics of the tissue, like blood volume, blood flow, permeability, and the availability of the extravascular space. In characterization of an orbital lesion, this can provide more information about angiogenesis and capillary permeability, features related to aggressive behavior, tumor grade, and prognosis [15]. With postprocessing software perfusion characteristics can be analyzed in a specific region of interest (ROI), for example, in an orbital mass. Cystic and hemorrhagic areas should be excluded. The scan can be analyzed in a qualitative way by looking at time-intensity curves (TICs), from which parameters like TTP (time-to-peak), iAUC (initial area under the curve), and washout rates can be calculated. In orbital imaging, three main TIC curves are usually recognized: persistent uptake pattern (mostly benign), plateau pattern (indeterminate), and washout pattern (mostly malignant); the latter can be subdivided into rapid uptake with rapid washout or with slow washout [15, 16]. This simple way of reading the DCE images is informative and easily applied in clinical practice.

Multiparametric Imaging and Future Directions

In multiparametric imaging, the outcomes of different imaging techniques, i.e., CT, anatomic and functional MRI techniques, are combined. The combining of functional characteristics of DWI and PWI additional to pure morphologic features can improve diagnostic accuracy, especially in the evaluation of orbital soft-tissue masses like vascular pathology, inflammatory disease, and neoplasms [16–18].

In recent years, an increasing amount of studies were performed on more quantitative analyses of MRI images. Examples of quantitative MRI include T2 mapping of the spin–spin relaxation time and quantifying the amount of edema and fatty infiltration using fat–water imaging with Dixon techniques that can be used to quantify clinical activity in GO [19]. However, the development in quantification of MRI is mainly focused on DWI and PWI parameters.

The frequently used ADC value derived from DWI is a quantitative parameter. However, the mean ADC value obtained from a manually drawn ROI cannot represent the heterogeneity of the whole lesion. Whole-tumor histogram analysis of the ADC maps can generate several diffusion parameters and could be valuable in differentiating orbital tumors [20, 21]. Moreover, ADC values calculated by a conventional mono-exponential model cannot separate the pure water molecular diffusion from the water molecular diffusion in capillaries, and this can affect the measurements. Intravoxel incoherent motion (IVIM), based on a biexponential model by using multiple b values, enables quantification of both diffusion and perfusion in a single acquisition without the use of a contrast agent. Recent studies already showed that the true diffusion coefficient (D) and the perfusion fraction (f) might help to improve the characterization of orbital lesions [22, 23].

Quantitative perfusion parameters of PWI can be obtained through pharmacokinetic modeling with the modified Tofts model after assessing the arterial input function (AIF). With this model, perfusion parameters including tumor blood flow (TBF), k^{trans} (volume transfer constant between the plasma and the extracellular extravascular space), v_e (extravascular extracellular volume fraction), and tumor flow residence time τ can be calculated from the DCE series [24–26]. However, DCE has the disadvantage that intravenous contrast agent is indispensable to obtain the parameters. Arterial spin labeling (ASL) is an emerging perfusion-weighted MRI technique that allows for absolute tissue perfusion quantification without the use of extrinsic contrast agents [27].

Besides quantification of imaging parameters, the use of postprocessing and planning software will take a leap forward in clinical practice. This will be more extensively discussed in the other chapters. Another promising future direction in diagnostic imaging is the development of AI that probably will play an important role in further improving orbital imaging and decision-making in next decades. An initial study already showed that an AI framework could accurately differentiate orbital cavernous venous malformation from schwannomas [28].

Structural Review of Orbital Imaging

Many different diseases can affect the orbit, and a systematic way to analyze an orbital CT or MRI scan will improve the quality of the review and narrow the differential diagnosis. Orbital lesions should be evaluated in multiple planes, at least in the axial and coronal planes. For some lesions, sagittal planes should be obtained for an optimal evaluation.

A structured review based on the compartment model can be a useful strategy to analyze orbital scans [29]. The first step is to localize the origin of the orbital pathology in one of the orbital compartments, i.e., paranasal sinus, bone, extraconal space, muscle cone, intraconal space, optic nerve, globe or lacrimal fossa. This allows significant reduction in the number of differential diagnoses as these compartments contain different tissues. However, there are also diseases that can present in multiple compartments.

Very important in further narrowing the differential diagnosis is to combine the imaging parameter with the demographic and clinical features of a patients. The quality of the radiological report highly depends on the quality of the clinical information and the questioning of the referring physician [30].

Structured reporting can be useful to improve the communication between radiologist and clinician and to improve the clinical significance of the radiological report [31]. The structured reporting should be tailored to the clinical question, a checklist can be helpful, and radiologist and clinicians could discuss together what is important to be mentioned in the radiological report.

Table 4.1 gives an overview of the preferred imaging techniques, the desired series, planes and a review checklist for most common orbital imaging indications.

Table 4.1 Preferred imaging technique, series, planes, and review checklist for most common orbital imaging indications [1, 2]

Indication	Preferred imaging technique	IV contrast	Series/ reconstructions	Planes/ MPRS	Review checklist
Orbital trauma	CT (additional MRI in complicated cases)	No	CT: bone and soft tissue	Axial and coronal	Foreign bodies
					Orbital and maxillofacial fractures
					Orbital soft tissues, muscle entrapment
					Associated injuries
Graves' orbitopathy	MRI > CT	No	MRI: T1, T2, STIR, (DWI) CT: bone and soft tissue	Axial and coronal	Proptosis
					EOM enlargement and edema
					Orbital fat volume and edema
					Lacrimal gland enlargement and edema
					Optic nerve stretching
					Apical crowding
Orbital infection	CT = MRI	Yes	CT: postcontrast, bone, and soft tissue MRI: T1, T2, STIR, DWI, Postcontrast T1-FS	Axial and coronal	Preseptal vs postseptal extension
					Edema, subperiosteal empyema, true intraorbital abscess
					Thrombosis VOS
					Extraorbital complications: cavernous sinus thrombosis, intracranial empyema/abscess
					Cause: sinusitis
Orbital soft-tissue masses	MRI (additional CT when bone involvement)	Yes	MRI: T1, T2, STIR, DWI, Postcontrast T1-FS, (DCE)	Axial, coronal (and sagittal)	Delineation: well-defined vs infiltrative
					Extension: orbital compartments, soft tissues, bone, periorbital extension
					Signal characteristics (T1, T2, diffusion)
					Enhancement: pattern and TIC
Lesions of the bony orbit	CT (additional MRI when soft-tissue involvement)	Depends	CT: bone and soft tissue MRI: T1, T2, STIR, DWI, Postcontrast T1-FS, (DCE)	Axial, coronal	Aspect: lytic, cystic, sclerotic, mixed
					Delineation (well-defined, infiltrating, remodeling, permeative)
					Extension: what bones, intraorbital and periorbital extension
					Involvement orbital apex and optic canal
					Enhancement
					Signal characteristics on MRI (T1, T2, diffusion)

Orbital Trauma

Imaging plays a key role in assessing the orbit after facial trauma, evaluating occult orbital injuries, the bony orbit and deep orbital structures, and identifying foreign bodies. High-resolution CT is imaging method of first choice in trauma, and the scan should be reconstructed in different planes and in both bone and soft-tissue reconstructions. Usually, in high-impact trauma, a CT scan of the whole skull is made from which separate brain and facial/orbital reconstructions can be made. On indication, CTA can be performed to asses for dissection, traumatic pseudoaneurysm, or carotid cavernous fistula. Plain radiography has very limited sensitivity in detecting orbital fractures and cannot asses the intraorbital soft tissues. MRI has limited value in the setting of acute trauma, but could be of additional value in complex cases or later in follow-up, especially when traumatic injury of the optic nerve or other soft tissues is suspected.

The first step in reviewing orbital trauma CT is to exclude foreign bodies. Metallic and glass foreign bodies have an increased attenuation and are best delineated on CT. Wooden foreign bodies can have the same attenuation as air, mimicking orbital gas, but within days this attenuation may increase.

When the orbital walls are fractured, the review of the bony structures should address the displaced fragments, the involvement of the orbital apex, and the infraorbital and the optic canal. The soft tissues of the orbit should be evaluated for intraorbital hematoma, fat stranding, and whether there is entrapment of the inferior and/or medial rectus muscles and fat. Also, the form and position of the globe should be evaluated, as well as the position of the lens. Signs of globe rupture include loss of volume and contour and intraocular gas. Retinal detachment is shown as a V-shaped hyperdense configuration, because the retina is relatively fixed at the level of the optic disc on the posterior side and at the ora serrata anteriorly. In choroidal detachment, lentiform hyperdense lesions are seen, not limited anteriorly by the ora serrata.

It is also important to analyze if fractures are limited to the orbit or part of more extensive maxillofacial fractures. Several classification systems and recognizable patterns exist for the imaging evaluation of maxillofacial fractures [32]. Knowledge of these classification systems can be important in effectively describing the fractures and aiding in clinical decision-making. Besides identifying all of the fractures and using the appropriate classification system, the radiologist also needs to recognize injuries that may be associated.

Isolated fractures to the orbital floor and orbital wall are often referred to as blowout fractures (BOF). This mainly involves the medial and/or inferior walls of the orbit and results in displacement of the fracture fragments into the ethmoid sinus, nasal cavity, or maxillary sinus. Indication for surgery is discussed in Chap. 10, but parameters that are taken into account for decision-making are size of the fracture fragments, displacement of orbital contents, like hypoglobus or enophthalmos, and entrapment of the extraocular muscles. On CT, kinking of the medial or inferior rectus muscle is a sign of entrapment. A specific type of BOF is the trapdoor fracture, where the bony fragment recoils back to its original position and entraps the muscle outside the orbital wall (Fig. 4.2). It is mainly seen in children and young adults due to the elasticity of the orbital floor and is considered a surgical emergency as the muscle can become ischemic. "Blow-up" fractures are a rare type of BOF that involve the orbital roof, and they are associated with traumatic intracranial injury. In "blow-in" fractures, the frontal bone is displaced inferiorly, resulting in decreased orbital volume.

The orbital fractures can also be part of naso-orbital-ethmoid fractures (NOE), zygomatico-maxillary complex fractures (ZMC), transfacial fractures (Le Fort type 2 or 3), or complex midfacial fracture (CMF). NOE fractures involve the nasal bones, the ethmoidal bones, and the medial orbital wall. They can be classified according to Markowitz into type I, with a single fracture fragment; type II, comminuted; and type III, comminuted with the medial canthal tendon disrupted from the bone [33].

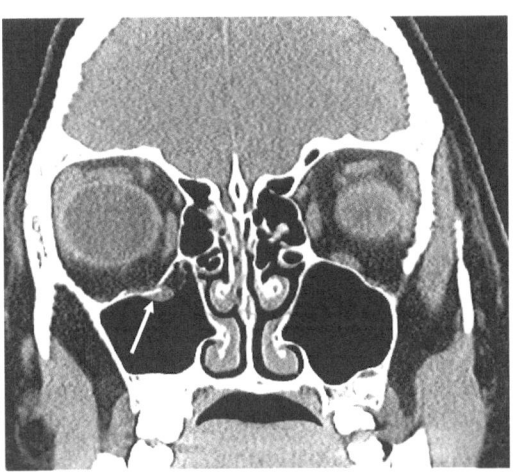

Fig. 4.2 Trapdoor fracture of the orbital floor on the right side. Due to repositioning of the fracture fragment in its original position, the inferior rectus muscle is entrapped outside the orbit in the maxillary sinus. This special type of blow-out fracture is mainly seen in children and young adults due to the elasticity of the orbital floor and is considered a surgical emergency as the muscle can become ischemic. Entrapment of the muscle is best seen in the soft-tissue reconstruction

In ZMC fractures, there is an isolated fracture of any part of the zygomatic bone, which includes the lateral and inferior orbital rims, the internal lateral orbital wall, the zygomaticomaxillary buttress, and the zygomatic arch [34]. In ZMC fractures, it is important to check what parts of the zygoma are involved, if fracture parts are displaced and if there are associated intraorbital injuries.

The characterizing features of Le Fort fractures are involvement of the maxilla and fracture of the pterygoid plates [35]. In type II and III, the orbit is involved. The type II Le Fort, there is a fracture of the pterygoid plates together with fractures of the inferior orbital rim, the medial orbital rim, and the nasal bone or nasofrontal suture. In this type, the maxilla and nasal regions are mobile from the rest of the face. A type III Le Fort fracture is a fracture of the pterygoid plates plus a fracture of the zygomatic arch, lateral and medial orbital rims, and nasal bones or nasofrontal sutures. In a type III Le Fort fracture, the entire midface is mobile.

Graves Orbitopathy (GO)

Besides the clinical parameters combined in the clinical activity score (CAS), imaging has important additional value in diagnosing and monitoring GO (also referred to as thyroid-associated orbitopathy or thyroid-associated ophthalmopathy) [36, 37]. Imaging also plays a role in the differential diagnosis, when the clinical presentation is atypical. The typical radiological presentation of GO is a bilateral proptosis due to bilateral enlargement of the extraocular muscles, with sparing of the tendons, as well as an increase in the orbital fat volume. Involvement of the extraocular muscles in decreasing order of frequency are the levator palpebrae superioris, the inferior rectus, the medial rectus, the superior rectus, the lateral rectus, and oblique muscles. The order can be memorized by the mnemonic I'M SLOW (except for the levator muscle). The most important differential diagnosis is inflammatory myositis, which often involves the tendinous insertion.

CT has been the preferred imaging method for GO for years. It can measure proptosis, detect muscle enlargement, increased orbital fat volume and lacrimal gland volumes, optic nerve stretching, apical crowding, and remodeling of the bony orbit [38–40]. A positive correlation can be seen between active inflammation and the volume and density of the intraorbital fat, EOMs, and lacrimal gland [41, 42]. The measurement of the orbital volume apex crowding index (ratio between soft tissue and orbital fat volume of the apex) and thickening of the medial rectus muscle could help to predict dysthyroid optic neuropathy (DON), a serious complication of GO [43, 44]. It is also shown that it is possible to reliable evaluate these CT characteristics of GO with low-dose CT with an iterative model reconstruction (IMR) algorithm to reduce radiation dose to the ocular lens [45].

However, CT only provides limited information about disease activity and has limited correlation with the CAS. With MRI, it is possible to gain additional information about disease activity through its ability to reflect the edematous

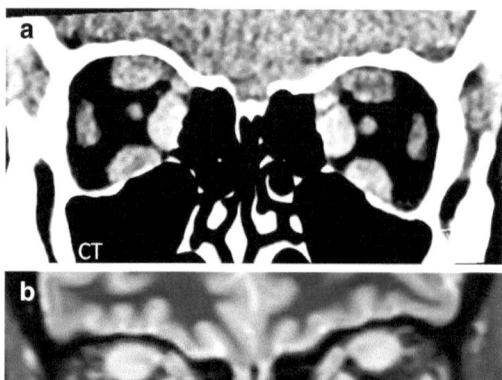
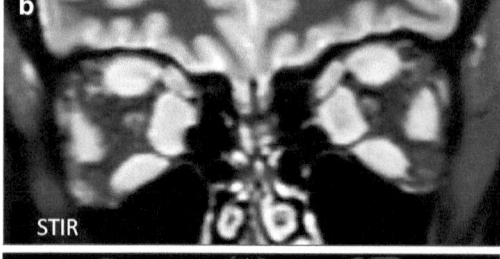
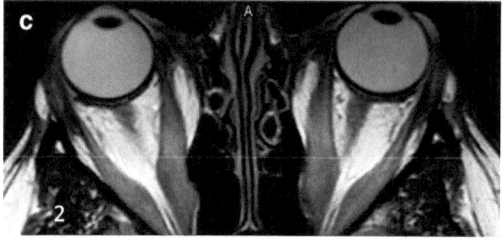

Fig. 4.3 Coronal CT with soft-tissue reconstruction (**a**) coronal STIR MRI (**b**) and axial T2-wighted MRI (**c**) in a patient with Graves' orbitopathy. On CT, the bilateral thickening of the EOMs is appreciated, and the hyperintense signal on STIR and T2 in the EOMs and orbital fat reflects edema as a sign of active inflammation, which can be relevant additional information. Note the marked proptosis, and the sparing of the tendons, which is typical for GO

changes in the orbital soft tissues (Fig. 4.3). The STIR technique or T2-WI with fat suppression (i.e., TIRM, Dixon) is most reliable to show soft-tissue edema. The increased signal intensity (ratio) in the EOMs, and to a lesser extent in the orbital fat and lacrimal glands, reflecting edema and inflammation, positively correlate with the CAS and can be used to evaluate medicamentous treatment or to predict worsening of GO [19, 46–49].

Besides signal intensity on the fat-suppressed series, DWI can be useful. An increased ADC value, reflecting increased diffusion is seen in the EOMs and lacrimal glands of active GO compared to inactive disease and healthy subjects [50, 51], and this can be detected even in an early stage of disease [52, 53]. Also, this technique can be helpful in monitoring treatment response [54].

A structural review of an orbital CT or MRI in GO patients should include measurement of the degree of proptosis, of the thickness of the EOMs, the orbital fat and lacrimal gland volumes and protrusion (through volumetric measurement or visual inspection), the degree of optic nerve stretching, and apical crowding. For MRI, additional evaluation of the edema of the EOMs and lacrimal gland should be performed, visually or more quantitative by measuring signal intensity ratios (SIRs) compared to the temporal muscle or with measurement of ADC values.

Orbital Infections

Preseptal cellulitis is mainly a clinical diagnosis that does not need additional imaging. In postseptal orbital cellulitis, however, imaging is often indicated to determine the extent of the disease and to make the decision between conservative and operative management. Imaging can be performed both with postcontrast CT of with MRI. The CT has the advantage that it is very fast and widely available, also besides office hours, thereby making it the preferential imaging method in most cases. CT can also reveal possible underlying sinusitis as the most common cause of orbital cellulitis and guide orbital or sinonasal surgery.

In children and young adults, MRI is the preferred imaging technique when available in orbital cellulitis because there is no ionizing radiation involved. In young kids, this advantage should be weighed against the use of general anesthesia often needed in children below 7 years old. Also, when limited availability of MRI could lead to significant delay, CT should be considered. An alternative strategy could be the use of a recently proposed rapid MRI in children without the use of intravenous contrast or general anesthesia [55].

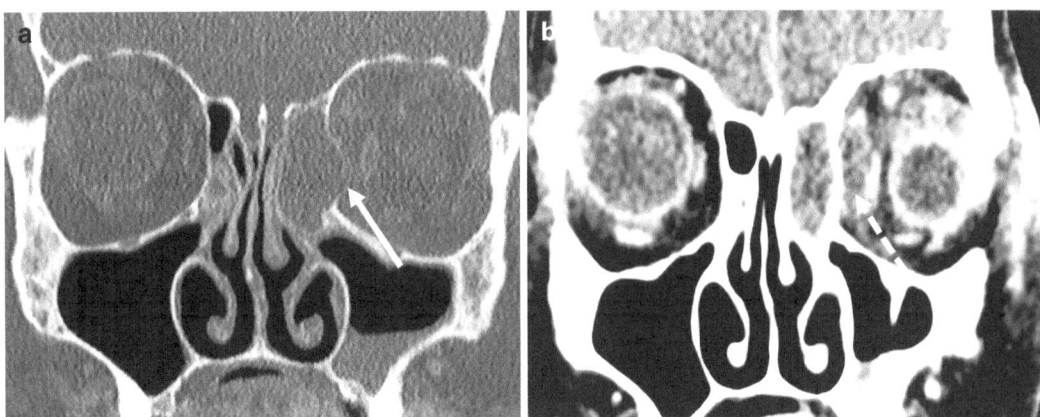

Fig. 4.4 Subperiosteal empyema due to infected mucoceles in the left ethmoid in a patient presenting with proptosis. Note the remodeling of the lamina papyracea due to the mucocele (arrow), best recognized on the bone reconstructions (**a**); the subperiosteal empyema is best visualized on the soft-tissue reconstructions after IV contrast (dashed arrow) (**b**)

Because of the superior soft-tissue contrast, it is easier to differentiate suppurative collections from orbital fat infiltration with MRI compared to CT. Especially, the differentiation between true intraorbital abscess from subperiosteal empyema (Fig. 4.4) or intraorbital fat infiltration is important because the first needs surgical intervention, while the other conditions can often be managed with conservative treatment [56]. Adding DWI to the MRI protocol further improves the diagnostic confidence in cases of subperiosteal empyema or orbital abscess, even without the use of intravenous contrast, because the purulent material will show diffusion restriction [57].

MRI is also more sensitive in detecting possible complications, like superior orbital vein or intracranial (cavernous) sinus thrombosis, epidural empyema, or intracranial abscess [58, 59]. Therefore, in case of clinical presentation with a progressive orbital apex syndrome or neurological impairment, imaging with MRI is preferred. This also applies for rapidly progressive orbital infections in immunocompromised or diabetic patients when invasive fungal infections like angioinvasive aspergillus of rhino-orbital cerebral mucormycosis should be conceded. In these cases, urgent MRI is preferred to CT to precisely determine the extent of infection in surrounding tissues, because of the bad prognosis without aggressive treatment [60, 61].

Orbital Soft-Tissue Lesions

The most important role of imaging in orbital soft-tissue lesions or masses is to appoint the extent of disease and to make a correct (differential) diagnosis. The most common orbital soft-tissue masses include orbital inflammation, vascular or lymphatic malformations (of which the cavernous venous malformation is most common), developmental cysts, benign tumors like schwannoma, optic nerve sheet meningioma, solitary fibrous tumors (mostly benign, but up to 20% can be malignant) or pleomorphic adenoma of the lacrimal gland and malignant tumors like lymphoma, metastasis, and malignant tumors of the lacrimal gland. In children, most common orbital masses are infantile hemangioma, developmental cysts, optic nerve glioma, retinoblastoma, and rhabdomyosarcoma. For the differential diagnosis, radiological findings should be combined with clinical information, laboratory data, and information about other organs that might be involved.

Making the right diagnosis on orbital masses can be challenging, and it was even reported that ophthalmologists and radiologists could only give a correct diagnosis in less than 50% of the cases when compared to histology [62]. Orbital inflammation, for example, has many appearances mimicking other diseases like infection, Graves' orbitopathy, optic nerve sheet meningioma, lymphoma, or other malignant tumors [63, 64]. Moreover, orbital inflammatory disease can be idiopathic (IOI) or non-specific orbital inflammation (NSOI) or can be related to a systemic inflammatory or granulomatous disease, like rheumatoid arthritis, seronegative spondyloarthropathies, sarcoidosis, IgG4-related ophthalmic disease (IgG4-ROD), or granulomatosis with polyangiitis (GPA). Although there are some distinctive imaging characteristics, it is often not possible to distinguish these different types of inflammation solely based on imaging. With MRI, the extent of the lesion can be determined and this helps to classify orbital inflammation according to the location including diffuse, focal mass-like, anterior, posterior and apical, or more specific if certain orbital structures are involved, like myositis, dacryoadenitis, periscleritis or perineuritis [65].

When the lesion is easily approached, a biopsy can be taken for histology, but it can be essential to have a radiological diagnosis before planning the biopsy or surgery. For apical lesions, it is hard to obtain a biopsy safely, so ophthalmologists could decide to start treatment based on the clinical and radiological diagnosis alone. Moreover, for lesions like pleomorphic adenomas or dermoid cysts, it is essential to know the diagnosis before surgical intervention. Biopsy can be avoided in most of these cases and the lesions should be removed in toto with their capsules to prevent recurrence or chemical inflammation.

CT, anatomic, and functional MRI techniques are often complementary in evaluating an intraorbital mass lesion. Combining different imaging techniques, in so-called multiparametric imaging (Fig. 4.5), showed to improve the diagnostic accuracy of orbital soft-tissue masses [16–18].

Involvement of the bony orbit is best-evaluated with CT. When the orbital bone is involved, it is important to discriminate bone erosion or destruction from bone remodeling. Destruction is mainly seen in aggressive lesions like malignancy or osteomyelitis, while remodeling is mainly seen in slow-growing lesions like mucoceles, cavernous venous malformations, pleomorphic adenomas, or schwannomas. Osseous

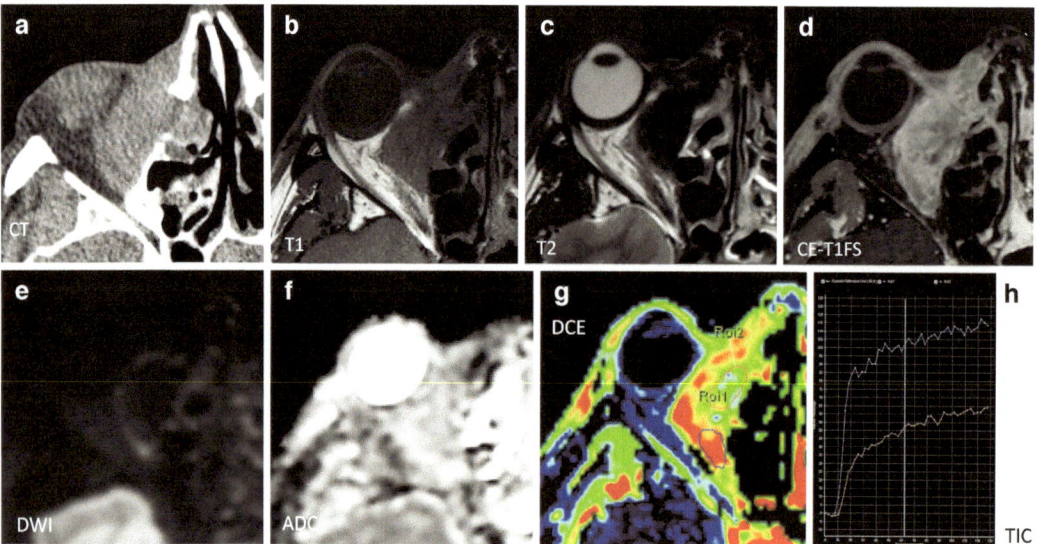

Fig. 4.5 Multiparametric imaging of a lesion medial in the right orbit, combining parameters from CT (**a**), T1WI (**b**), T2WI (**c**), contrast-enhanced T1 with fat suppression (**d**), DWI (**e**) with the ADC map (**f**), and DCE (**g**) with the time-intensity curve (**h**). This lesion shows bone destruction, sclerosis with low signal on T2, homogeneous enhancement, no diffusion restriction (ADC = 1.1×10^{-3} mm²/s), and a persistent uptake on the TIC. This is suggestive of an benign mass, most likely a sclerosing form of orbital inflammation, which was proven by biopsy (histology: sclerosing inflammation, not otherwise specified)

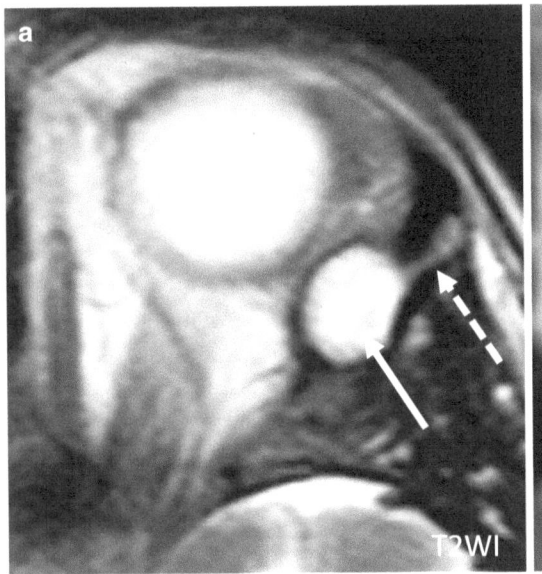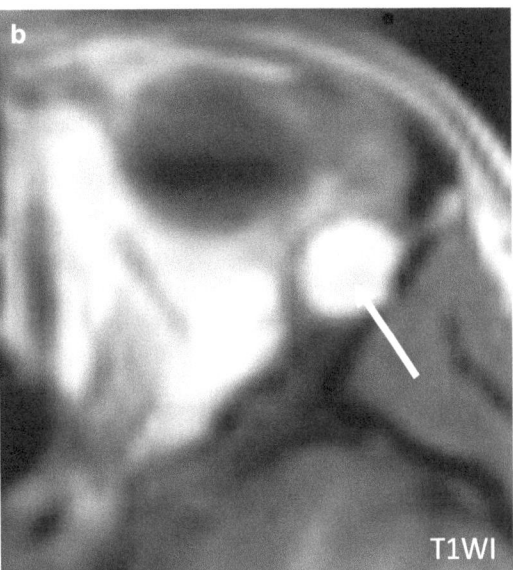

Fig. 4.6 Dermoid cyst lateral in the left orbit of a child. Note the hyperintense signal on both T2WI (**a**) and T1WI (**b**) due to fatty content (arrows) and the extension into the widened frontozygomatic suture (dashed arrow). Due to the typical appearance on MRI, presurgical biopsy can be avoided in most of these cases and the lesions should be removed in toto with their capsules to prevent recurrence or chemical inflammation

destruction is also frequently seen in GPA and is caused by necrosis. With CT, it also possible to appreciate calcifications within a lesion, for example, phleboliths, which are characteristic for low-flow venous malformations. Often, an CT scan is needed for per operative navigation.

The T1-WI, T2-WI, and T1-FS after contrast are used to evaluate the extent of the lesion in six directions and the involvement of the different orbital compartments and structures. Some lesions are characterized by a specific location like optic nerve gliomas originating from the optic nerve, optic nerve sheets meningiomas appearing with the typical 'tram-track sign" surrounding the nonenhancing optic nerve, pleomorphic adenomas originating from the lacrimal gland, or myositis involving the EOMs. Other lesions, like most inflammatory and vascular of malignant lesions are not restricted to a specific location. Besides the origin and extent of the lesion, also potential extraorbital manifestation, like perineural spread or intracranial extension, should be evaluated with MRI. Further, lesion

boundaries are appreciated, discerning well-circumscribed lesions from more infiltrative lesions, as well as potential perilesional edema.

Most of the orbital lesions are T1 iso-intense or hypointense, like in orbital inflammation and orbital tumors. Only lesions that contain fat, such as dermoid cysts (Fig. 4.6), melanin in case of melanotic melanomas, or that contains hemorrhagic components can show T1 hyperintense components.

On T2-WI, the signal intensity depends on components like edema and fibrosis; edema in case of active inflammation being hyperintense, while fibrosis in case of sclerosing orbital inflammation appearing very hypointense. Inflammatory lesions that show very low signal on T2-WI, implicating extensive fibrosis, will be less responsive to steroid treatment than lesions that show hyperintense T2 signal due to active inflammation. Very low T2 signal can also be seen in IgG4-related orbital inflammation. Fluid-containing lesions like cysts and venous or lymphatic malformations are often very bright on T2, and this is exaggerated when applying fat saturation. Cavernous venous malformations can be recognized by the T2 hypointense pseudo-capsule.

A next step in the approach is to evaluate the enhancement pattern. In general, enhancement will be more homogeneous in case of inflammation and lymphoma than in case of venous malformations, infection, or other malignancies. In well-circumscribed ovoid orbital masses, the enhancement spread pattern over time can be used to differentiate orbital schwannomas, which show start of contrast-enhancement from a wide area, from cavernous venous malformations that demonstrate a heterogeneous moderate enhancement starting from one point or portion within the lesion [66].

DWI helps to distinguish benign from malignant orbital lesions, especially in differentiating orbital inflammatory disease from lymphoma. In general, inflammatory lesions have increased diffusion with high ADC values because of freely diffusible water molecules in the edema, while malignant lesions show more restricted diffusion with lower ADC values because of higher cellular content. Because lymphomas are very cellular tumors, they have low ADC values. A threshold value for ADC of 1.15×10^{-3} mm^2/s was proposed to differentiate benign from malignant orbital masses with an accuracy of more than 90% [67]. Another study showed that lesions with an ADC of less than 0.93×10^{-3} mm^2/s are likely to be malignant with a 90% probability, while lesions with an ADC value more than 1.35×10^{-3} mm^2/s are likely to be benign with more than 90% probability [68].

The addition of PWI to the MRI protocol might further improve the lesion characterization, although there is less evidence for the impact of DCE on the diagnostic accuracy when compared to DWI [17]. While persistent uptake on the time-intensity curve is typical for benign lesions and washout is highly suspicious for malignant lesions, the curve quite often demonstrates a plateau, which is undetermined and not contributing to the differential diagnosis [69]. At least, PWI has proved to be useful in selected

cases, i.e., to differentiate the rare orbital solitary fibrous tumors from other well-defined orbital lesions like schwannomas or cavernous vascular malformations due to their typical washout TIC pattern [70, 71].

Lesions of the Bony Orbit

The most common lesions of the orbital walls are related to trauma (discussed in previous paragraph) and related surgery. Other more common lesions are often related to sinonasal disease like mucoceles (Fig. 4.4) and long-standing polyposis with remodeling of the medial orbital wall leading to proptosis or the silent sinus syndrome with depression of the orbital floor leading to enophthalmos due to negative pressure in the maxillary sinus (Fig. 4.7). Also, sinonasal squamous cell carcinoma or esthesioneuroblastoma can involve the orbital walls. Primary lesions of the orbital bone can be classified into congenital lesions (i.e., dermoid, epidermoid), fibro-osseous lesions (i.e., fibrous dysplasia, osteoma, ossifying fibroma), benign tumors (i.e., meningioma,

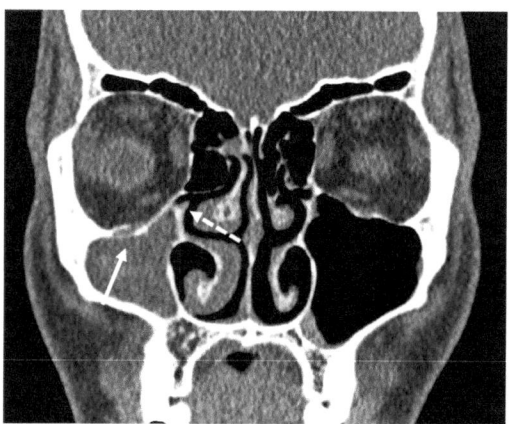

Fig. 4.7 Silent sinus syndrome in a patient presenting with painless enophthalmos on the right side due to depression of the orbital floor (arrow). This is caused by a atelectasis of the maxillary sinus walls due to negative sinus pressure in a chronic opacified maxillary sinus. There is lateral displacement of the uncinate process in contact with the lamina papyracea(dashed arrow)

Langerhans cell histiocytosis, hemangioma of the bone, aneurysmal bone cyst, mucoceles), malignant primary (i.e., Ewing or osteosarcoma), and other malignant tumors (i.e., plasmacytoma, lymphoma, metastasis).

CT is the preferential imaging method in evaluating lesions of the bony orbit, but additional MRI can be desirable in complex cases to evaluate the bone marrow or when adjacent soft tissues are involved. A systematic approach is essential. The first step is to evaluate whether the lesion originates from the bone or whether the lesion arises from the soft tissues with secondary involvement of the bone. When the lesion originates from the bone, the next step is to analyze that the morphology is the lesion well-define or ill-defined, osteolytic, or sclerotic. Other clues are periosteal reaction, bone remodeling, cortical destruction, matrix calcification, and whether lesions are monostotic or polyostotic. Destruction is mainly seen in aggressive lesions like malignancy or osteomyelitis, while remodeling is mainly seen in slow-growing lesions like mucoceles, bone cysts, or fibrous dysplasia. Also, in bone lesions the clinical history and especially the age of the patient are very important for the differential diagnosis. A well-defined lytic lesion in a young child is suggestive of an (epi)dermoid cyst or Langerhans cell histiocytosis, while an ill-defined lytic lesion is suspicious of osteomyelitis, Ewing sarcoma, osteosarcoma, or leukemia. In elderly patients, a lytic lesion is most likely to be metastasis or plasmacytoma/ multiple myeloma.

A lesion where both CT and MRI are indicated is the spheno-orbital or sphenoid wing meningioma. They can be recognized by their typical appearance, centered within the greater sphenoid wing with reactive hyperostosis of the bone, best appreciated with CT, and a soft-tissue component with intraorbital and intracranial extension along the dura, best delineated with MRI (Fig. 4.8). These meningiomas often cause narrowing of the orbital apex due to the hyperostosis or due to the soft-tissue component with involvement of the superior orbital fissure and less frequently the optic nerve canal [72].

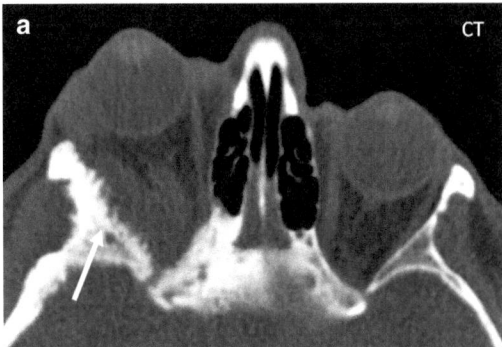
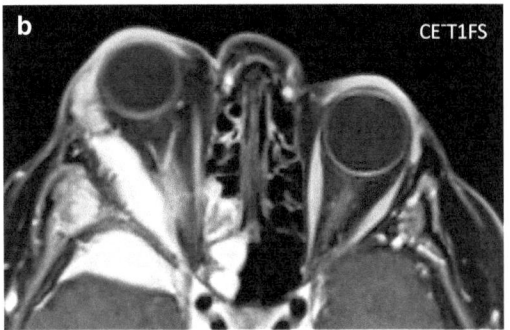

Fig. 4.8 Typical example of a spheno-orbital or sphenoid wing meningioma in a middle-aged female, presenting with proptosis of the right eye. Note the additional value of both CT and MRI: the reactive hyperostosis of the sphenoid bone (arrow) is best appreciated with CT (**a**), while the soft-tissue component with intraorbital and intracranial extension is best reviewed on MRI (**b**). Narrowing of the orbital apex and optic nerve compression due to the combination of hyperostosis and soft tissue component

Conclusions

Imaging using CT or MRI will provide crucial additional information in the evaluation of many orbital conditions. Because of its superior bony characterization and fast acquisition, CT is the imaging method of first choice in urgent situations like trauma, infection, and when the lesions arise from the orbital wall. CT is also useful in preoperative planning and preoperative navigation. For evaluating complex orbital disease, MRI is the preferred modality. With its superior soft-tissue differentiation, MRI is ideal for determining extent of orbital lesions, including inflammatory disease, orbital vascular lesions, and tumors. By adding functional MRI techniques, like DWI and PWI, and by combining the parameters from different imaging techniques in so-called multiparametric imaging, it is possible to further improve biological characterization of these lesions. A structured way of reviewing the orbital images, knowledge of radiological appearance of most common orbital pathology and combining it with important clinical information is essential to create added value of orbital imaging for patient care.

References

1. Lanni V, Iuliano A, Fossataro F, Russo C, Uccello G, Tranfa F, et al. The role of ultrasonography in differential diagnosis of orbital lesions. J Ultrasound. 2021;24(1):35–40.
2. Al Feghali KA, Yeboa DN, Chasen B, Gule MK, Johnson JM, Chung C. The use of (68)Ga-DOTATATE PET/CT in the non-invasive diagnosis of optic nerve sheath meningioma: a case report. Front Oncol. 2018;8:454.
3. Ambrose J, Hounsfield G. Computerized transverse axial tomography. Br J Radiol. 1973;46(542):148–9.
4. Prokop M. New challenges in MDCT. Eur Radiol. 2005;15(Suppl 5):E35–45.
5. Katsura M, Sato J, Akahane M, Kunimatsu A, Abe O. Current and novel techniques for metal artifact reduction at CT: practical guide for radiologists. Radiographics. 2018;38(2):450–61.
6. Rosado Ingelmo A, Dona Diaz I, Cabanas Moreno R, Moya Quesada MC, Garcia-Aviles C, Garcia Nunez I, et al. Clinical practice guidelines for diagnosis and management of hypersensitivity reactions to contrast media. J Investig Allergol Clin Immunol. 2016;26(3):144–55. quiz 2 p following 55
7. van der Molen AJ, Reimer P, Dekkers IA, Bongartz G, Bellin MF, Bertolotto M, et al. Post-contrast acute kidney injury. Part 2: risk stratification, role of hydration and other prophylactic measures, patients taking metformin and chronic dialysis patients: recommendations for updated ESUR contrast medium safety committee guidelines. Eur Radiol. 2018;28(7):2856–69.
8. Albadr FB. Silent sinus syndrome: interesting computed tomography a and magnetic resonance imaging findings. J Clin Imaging Sci. 2020;10:38.
9. Miracle AC, Mukherji SK. Conebeam CT of the head and neck, part 1: physical principles. AJNR Am J Neuroradiol. 2009;30(6):1088–95.
10. Krinsky G, Rofsky NM, Weinreb JC. Nonspecificity of short inversion time inversion recovery (STIR) as a technique of fat suppression: pitfalls in image interpretation. AJR Am J Roentgenol. 1996;166(3):523–6.
11. Simon J, Szumowski J, Totterman S, Kido D, Ekholm S, Wicks A, et al. Fat-suppression MR imaging of the orbit. AJNR Am J Neuroradiol. 1988;9(5):961–8.
12. Hagmann P, Jonasson L, Maeder P, Thiran JP, Wedeen VJ, Meuli R. Understanding diffusion MR imaging techniques: from scalar diffusion-weighted imaging to diffusion tensor imaging and beyond. Radiographics. 2006;26(Suppl 1):S205–23.
13. Hirata K, Nakaura T, Okuaki T, Kidoh M, Oda S, Utsunomiya D, et al. Comparison of the image quality of turbo spin echo- and echo-planar diffusion-weighted images of the oral cavity. Medicine (Baltimore). 2018;97(19):e0447.
14. Feeney C, Lingam RK, Lee V, Rahman F, Nagendran S. Non-EPI-DWI for detection, disease monitoring, and clinical decision-making in thyroid eye disease. AJNR Am J Neuroradiol. 2020;41(8):1466–72.
15. Jittapiromsak N, Hou P, Liu HL, Sun J, Schiffman JS, Chi TL. Dynamic contrast-enhanced MRI of orbital and anterior visual pathway lesions. Magn R eson Imaging. 2018;51:44–50.
16. Xu XQ, Hu H, Liu H, Wu JF, Cao P, Shi HB, et al. Benign and malignant orbital lymphoproliferative disorders: differentiating using multiparametric MRI at 3.0T. J Magn Reson Imaging. 2017;45(1):167–76.
17. Russo C, Strianese D, Perrotta M, Iuliano A, Bernardo R, Romeo V, et al. Multi-parametric magnetic resonance imaging characterization of orbital lesions: a triple blind study. Semin Ophthalmol. 2020;35(2):95–102.
18. Ro SR, Asbach P, Siebert E, Bertelmann E, Hamm B, Erb-Eigner K. Characterization of orbital masses by multiparametric MRI. Eur J Radiol. 2016;85(2):324–36.
19. Das T, Roos JCP, Patterson AJ, Graves MJ, Murthy R. T2-relaxation mapping and fat fraction assessment to objectively quantify clinical activity in thyroid eye disease: an initial feasibility study. Eye (Lond). 2019;33(2):235–43.
20. Ren J, Yuan Y, Wu Y, Tao X. Differentiation of orbital lymphoma and idiopathic orbital inflammatory pseudotumor: combined diagnostic value of conventional MRI and histogram analysis of ADC maps. BMC Med Imaging. 2018;18(1):6.
21. Xu XQ, Hu H, Su GY, Liu H, Hong XN, Shi HB, et al. Utility of histogram analysis of ADC maps for differentiating orbital tumors. Diagn Interv Radiol. 2016;22(2):161–7.
22. Jiang H, Wang S, Li Z, Xie L, Wei W, Ma J, et al. Improving diagnostic performance of differentiating ocular adnexal lymphoma and idiopathic orbital inflammation using intravoxel incoherent motion diffusion-weighted MRI. Eur J Radiol. 2020;130:109191.
23. Lecler A, Duron L, Zmuda M, Zuber K, Berges O, Putterman M, et al. Intravoxel incoherent motion (IVIM) 3 T MRI for orbital lesion characterization. Eur Radiol. 2021;31(1):14–23.
24. Khalifa F, Soliman A, El-Baz A, Abou El-Ghar M, El-Diasty T, Gimel'farb G, et al. Models and meth-

ods for analyzing DCE-MRI: a review. Med Phys. 2014;41(12):124301.

25. Erb-Eigner K, Asbach P, Ro SR, Haas M, Bertelmann E, Pietsch H, et al. DCE-MR imaging of orbital lesions: diagnostic performance of the tumor flow residence time tau calculated by a multi-compartmental pharmacokinetic tumor model based on individual factors. Acta Radiol. 2019;60(5):643–52.

26. Tofts PS, Brix G, Buckley DL, Evelhoch JL, Henderson E, Knopp MV, et al. Estimating kinetic parameters from dynamic contrast-enhanced T(1)-weighted MRI of a diffusable tracer: standardized quantities and symbols. J Magn Reson Imaging. 1999;10(3):223–32.

27. Eissa L, Abdel Razek AAK, Helmy E. Arterial spin labeling and diffusion-weighted MR imaging: utility in differentiating idiopathic orbital inflammatory pseudotumor from orbital lymphoma. Clin Imaging. 2020;71:63–8.

28. Bi S, Chen R, Zhang K, Xiang Y, Wang R, Lin H, et al. Differentiate cavernous hemangioma from schwannoma with artificial intelligence (AI). Ann Transl Med. 2020;8(11):710.

29. Goh PS, Gi MT, Charlton A, Tan C, Gangadhara Sundar JK, Amrith S. Review of orbital imaging. Eur J Radiol. 2008;66(3):387–95.

30. Castillo C, Steffens T, Sim L, Caffery L. The effect of clinical information on radiology reporting: a systematic review. J Med Radiat Sci. 2021;68(1):60–74.

31. Ganeshan D, Duong PT, Probyn L, Lenchik L, McArthur TA, Retrouvey M, et al. Structured reporting in radiology. Acad Radiol. 2018;25(1):66–73.

32. Patel R, Reid RR, Poon CS. Multidetector computed tomography of maxillofacial fractures: the key to high-impact radiological reporting. Semin Ultrasound CT MR. 2012;33(5):410–7.

33. Markowitz BL, Manson PN, Sargent L, Vander Kolk CA, Yaremchuk M, Glassman D, et al. Management of the medial canthal tendon in nasoethmoid orbital fractures: the importance of the central fragment in classification and treatment. Plast Reconstr Surg. 1991;87(5):843–53.

34. Kelley P, Hopper R, Gruss J. Evaluation and treatment of zygomatic fractures. Plast Reconstr Surg. 2007;120(7 Suppl 2):5S–15S.

35. Rhea JT, Novelline RA. How to simplify the CT diagnosis of Le fort fractures. AJR Am J Roentgenol. 2005;184(5):1700–5.

36. Mourits MP, Koornneef L, Wiersinga WM, Prummel MF, Berghout A, van der Gaag R. Clinical criteria for the assessment of disease activity in Graves' ophthalmopathy: a novel approach. Br J Ophthalmol. 1989;73(8):639–44.

37. Lo C, Ugradar S, Rootman D. Management of graves myopathy: orbital imaging in thyroid-related orbitopathy. J AAPOS. 2018;22(4):256 e1–9.

38. Nugent RA, Belkin RI, Neigel JM, Rootman J, Robertson WD, Spinelli J, et al. Graves orbitopathy: correlation of CT and clinical findings. Radiology. 1990;177(3):675–82.

39. Harris MA, Realini T, Hogg JP, Sivak-Callcott JA. CT dimensions of the lacrimal gland in Graves orbitopathy. Ophthalmic Plast Reconstr Surg. 2012;28(1):69–72.

40. Tan NYQ, Leong YY, Lang SS, Htoon ZM, Young SM, Sundar G. Radiologic parameters of orbital bone remodeling in thyroid eye disease. Invest Ophthalmol Vis Sci. 2017;58(5):2527–33.

41. Le Moli R, Pluchino A, Muscia V, Regalbuto C, Luciani B, Squatrito S, et al. Graves' orbitopathy: extraocular muscle/total orbit area ratio is positively related to the clinical activity score. Eur J Ophthalmol. 2012;22(3):301–8.

42. Byun JS, Moon NJ, Lee JK. Quantitative analysis of orbital soft tissues on computed tomography to assess the activity of thyroid-associated orbitopathy. Graefes Arch Clin Exp Ophthalmol. 2017;255(2):413–20.

43. Goncalves AC, Silva LN, Gebrim EM, Matayoshi S, Monteiro ML. Predicting dysthyroid optic neuropathy using computed tomography volumetric analyses of orbital structures. Clinics (Sao Paulo). 2012;67(8):891–6.

44. Weis E, Heran MK, Jhamb A, Chan AK, Chiu JP, Hurley MC, et al. Quantitative computed tomographic predictors of compressive optic neuropathy in patients with thyroid orbitopathy: a volumetric analysis. Ophthalmology. 2012;119(10):2174–8.

45. Lee HJ, Kim J, Kim KW, Lee SK, Yoon JS. Feasibility of a low-dose orbital CT protocol with a knowledge-based iterative model reconstruction algorithm for evaluating Graves' orbitopathy. Clin Imaging. 2018;51:327–31.

46. Higashiyama T, Nishida Y, Morino K, Ugi S, Nishio Y, Maegawa H, et al. Use of MRI signal intensity of extraocular muscles to evaluate methylprednisolone pulse therapy in thyroid-associated ophthalmopathy. Jpn J Ophthalmol. 2015;59(2):124–30.

47. Higashiyama T, Iwasa M, Ohji M. Quantitative analysis of inflammation in orbital fat of thyroid-associated ophthalmopathy using MRI signal intensity. Sci Rep. 2017;7(1):16874.

48. Chen W, Hu H, Chen HH, Su GY, Yang T, Xu XQ, et al. Utility of T2 mapping in the staging of thyroid-associated ophthalmopathy: efficiency of region of interest selection methods. Acta Radiol. 2020;61(11):1512–9.

49. Tachibana S, Murakami T, Noguchi H, Noguchi Y, Nakashima A, Ohyabu Y, et al. Orbital magnetic resonance imaging combined with clinical activity score can improve the sensitivity of detection of disease activity and prediction of response to immunosuppressive therapy for Graves' ophthalmopathy. Endocr J. 2010;57(10):853–61.

50. Abdel Razek AA, El-Hadidy M, Moawad ME, El-Metwaly N, El-Said AA. Performance of apparent diffusion coefficient of medial and lateral rectus muscles in Graves' orbitopathy. Neuroradiol J. 2017;30(3):230–4.

51. Razek AA, El-Hadidy EM, Moawad ME, El-Metwaly N, El-Said AAE. Assessment of lacrimal glands in

thyroid eye disease with diffusion-weighted magnetic resonance imaging. Pol J Radiol. 2019;84:e142–e6.

52. Kilicarslan R, Alkan A, Ilhan MM, Yetis H, Aralasmak A, Tasan E. Graves' ophthalmopathy: the role of diffusion-weighted imaging in detecting involvement of extraocular muscles in early period of disease. Br J Radiol. 2015;88(1047):20140677.

53. Politi LS, Godi C, Cammarata G, Ambrosi A, Iadanza A, Lanzi R, et al. Magnetic resonance imaging with diffusion-weighted imaging in the evaluation of thyroid-associated orbitopathy: getting below the tip of the iceberg. Eur Radiol. 2014;24(5):1118–26.

54. Lingam RK, Mundada P, Lee V. Novel use of non-echo-planar diffusion weighted MRI in monitoring disease activity and treatment response in active Grave's orbitopathy: an initial observational cohort study. Orbit. 2018;37(5):325–30.

55. Jain SF, Ishihara R, Wheelock L, Love T, Wang J, Deegan T, et al. Feasibility of rapid magnetic resonance imaging (rMRI) for the emergency evaluation of suspected pediatric orbital cellulitis. J AAPOS. 2020;24(5):289e1–4.

56. Van der Veer EG, van der Poel NA, de Win MM, Kloos RJ, Saeed P, Mourits MP. True abscess formation is rare in bacterial orbital cellulitis; consequences for treatment. Am J Otolaryngol. 2017;38(2):130–4.

57. Sepahdari AR, Aakalu VK, Kapur R, Michals EA, Saran N, French A, et al. MRI of orbital cellulitis and orbital abscess: the role of diffusion-weighted imaging. AJR Am J Roentgenol. 2009;193(3):W244–50.

58. Branson SV, McClintic E, Yeatts RP. Septic cavernous sinus thrombosis associated with orbital cellulitis: a report of 6 cases and review of literature. Ophthalmic Plast Reconstr Surg. 2019;35(3):272–80.

59. Cumurcu T, Demirel S, Keser S, Bulut T, Cavdar M, Dogan M, et al. Superior ophthalmic vein thrombosis developed after orbital cellulitis. Semin Ophthalmol. 2013;28(2):58–60.

60. Raab P, Sedlacek L, Buchholz S, Stolle S, Lanfermann H. Imaging patterns of rhino-orbital-cerebral mucormycosis in immunocompromised patients : when to suspect complicated mucormycosis. Clin Neuroradiol. 2017;27(4):469–75.

61. Thurtell MJ, Chiu AL, Goold LA, Akdal G, Crompton JL, Ahmed R, et al. Neuro-ophthalmology of invasive fungal sinusitis: 14 consecutive patients and a review of the literature. Clin Exp Ophthalmol. 2013;41(6):567–76.

62. Koukkoulli A, Pilling JD, Patatas K, El-Hindy N, Chang B, Kalantzis G. How accurate is the clinical and radiological evaluation of orbital lesions in comparison to surgical orbital biopsy? Eye (Lond). 2018;32(8):1329–33.

63. Ferreira TA, Saraiva P, Genders SW, Buchem MV, Luyten GPM, Beenakker JW. CT and MR imaging of orbital inflammation. Neuroradiology. 2018;60(12):1253–66.

64. Gordon LK. Orbital inflammatory disease: a diagnostic and therapeutic challenge. Eye (Lond). 2006;20(10):1196–206.

65. Yesiltas YS, Gunduz AK. Idiopathic orbital inflammation: review of literature and new advances. Middle East Afr J Ophthalmol. 2018;25(2):71–80.

66. Tanaka A, Mihara F, Yoshiura T, Togao O, Kuwabara Y, Natori Y, et al. Differentiation of cavernous hemangioma from schwannoma of the orbit: a dynamic MRI study. AJR Am J Roentgenol. 2004;183(6):1799–804.

67. Razek AA, Elkhamary S, Mousa A. Differentiation between benign and malignant orbital tumors at 3-T diffusion MR-imaging. Neuroradiology. 2011;53(7):517–22.

68. Sepahdari AR, Politi LS, Aakalu VK, Kim HJ, Razek AA. Diffusion-weighted imaging of orbital masses: multi-institutional data support a 2-ADC threshold model to categorize lesions as benign, malignant, or indeterminate. AJNR Am J Neuroradiol. 2014;35(1):170–5.

69. Yuan Y, Kuai XP, Chen XS, Tao XF. Assessment of dynamic contrast-enhanced magnetic resonance imaging in the differentiation of malignant from benign orbital masses. Eur J Radiol. 2013;82(9):1506–11.

70. Yang BT, Wang YZ, Dong JY, Wang XY, Wang ZC. MRI study of solitary fibrous tumor in the orbit. AJR Am J Roentgenol. 2012;199(4):W506–11.

71. Zhang Z, Shi J, Guo J, Yan F, Fu L, Xian J. Value of MR imaging in differentiation between solitary fibrous tumor and schwannoma in the orbit. AJNR Am J Neuroradiol. 2013;34(5):1067–71.

72. Kirollos RW. Hyperostosing sphenoid wing meningiomas. Handb Clin Neurol. 2020;170:45–63.

Open Access   This chapter is licensed under the terms of the Creative Commons Attribution 4.0 International License (http://creativecommons.org/licenses/by/4.0/), which permits use, sharing, adaptation, distribution and reproduction in any medium or format, as long as you give appropriate credit to the original author(s) and the source, provide a link to the Creative Commons license and indicate if changes were made.

The images or other third party material in this chapter are included in the chapter's Creative Commons license, unless indicated otherwise in a credit line to the material. If material is not included in the chapter's Creative Commons license and your intended use is not permitted by statutory regulation or exceeds the permitted use, you will need to obtain permission directly from the copyright holder.

Part III

Functional Aspect of the Globe

Vision

5

Maarten P. Mourits

Learning Objectives

- Seeing is not just receiving visual information.
- Seeing involves a process in which images are modified and interpreted by complex retinal and cerebral activities.
- Watching is a conscious and attentive form of seeing; e.g., the viewer knows what he sees.
- Seeing must be learned.
- Stereoscopic binocular single vision is the highest level of seeing.

Introduction

Aristotle (384–323 BCE) stated that of all our senses the eye is the most important. The recent observation that "about one-third of the human brain is dedicated to the mission of vision" [1] seems to confirm his statement after almost 2500 years. Until the year 1600, the so-called emission theory to explain how the eyes work prevailed. Perhaps based on the observation that cats show bright yellow-green eyes in the dark, Greek scientists and many others after them believed that eyes emitted beams that were constantly scanning the surroundings. Around 1600, it was Keppler who discovered that it is the other way around: Light beams enter the eye (immission theory), and the world is depicted on our retina. However, the difference between the meaning of the words seeing and watching illustrates that passive and active aspects are involved. Images projected on the retina need to enter numerous neural networks before they get a meaning. Already Descartes argued that visual images are processed as codes by our central nervous system [2].

Over the centuries, vision, as compared to other senses, only gained importance. At present, most human beings spend perhaps more than half the time they are awake looking at some kind of display: working, studying, relaxing, shopping, gathering information, exchanging personal information, reading, and anything; all can be done with smart devices and our eyes (and fingers). My personal screen time this week was 4.5 h a day, and this concerned just my desktop, not my other devices.

What Is Seeing?

Seeing is not only the ability to distinguish details, which allows us to read and write; it is also the perception of a three-dimensional field that offers

The original version of the chapter has been revised. A correction to this chapter can be found at https://doi.org/10.1007/978-3-031-40697-3_23

M. P. Mourits (✉)
Amsterdam University Medical Centers, Location AMC, Amsterdam, The Netherlands

© The Author(s) 2023, corrected publication 2024
P. J. J. Gooris et al. (eds.), *Surgery in and around the Orbit*,
https://doi.org/10.1007/978-3-031-40697-3_5

us a position in spatial planning. With extremely fast eye movements (up to 700° per second), we scan our surroundings. Less than 2% of what we see is imaged in detail. Anything seen beyond this area is blurred. These parts of our visual field are being filled in with (visual) information by our brain based on previous experiences. The brain compares new images with those it has assembled during our life and focuses on what is new or different. In this way, our visual apparatus is a very economic and fast-responding system: Our attention is constantly drawn to those details that are new or do not fit in our previous perceptions [1].

The subjective aspect of seeing explains why sometimes two individuals see different things, although they look at the same thing. Even more fascinating is that a healthy, normal seeing individual alternatingly can see two different images when looking at one object. It appears impossible to see both images at the same time. Look at Fig. 5.1, what do you see: a duck or a rabbit?

What we actually see are differences in black and white, lines, and shapes resulting in figures that are saved in our visual memory system. These resulting figures can be interpreted as either a duck or a rabbit. Different neural networks compete for what appears in our awareness: the duck or the rabbit (Fig. 5.1).

Visual Pathways

Light beams are firstly refracted on our cornea, the window of our eye, secondly by our lens and next focused on our retina. Photons then travel through the transparent layers of the retina to the second deepest, the layer of the rods and cones. Here, they evoke an electrical signal, which is processed and sent backward to the second superficial layer, the nerve fiber layer, which eventually forms the optic nerve.

The visual information travels along the optic nerve to the optic chiasm, the lateral geniculate nuclei, and finally along the optic radiation to the primary visual cortex in the occipital lobes of our brain. Here, the images from both eyes are merged (if there is sufficient similarity) into a single, stereoscopic, three-dimensional percept. The globes or eyeballs can be regarded as a during the evolution pushed-forward part of our brain.

The visual system not only comprises afferent pathways as described above. The geniculate nuclei and the primary visual cortex connect with a broad network of cerebral regions, which in turn connect with the superior colliculus in the brainstem. From here, efferent fibers run to the dilatator and sphincter muscles of the iris, which results in a pupil size inversely proportional to the amount of light that falls on the eye. This reflex can be simply tested with a penlight in a semi-dark environment, with the person being examined looking into the distance. Because of mutual connections, not only the illuminated eye shows narrowing of the pupil, also the pupil of the fellow eye shows an identical contraction, which is called consensual reaction. A relative afferent pupil block (RAPD) indicates that something is wrong along this reflex arc, e.g., a tumor compressing the optic nerve or any other part of the optic pathways. In anterior uveitis (or iridocyclitis), the pupil cannot dilate as the iris is "glued" to the lens.

From nuclei in the midbrain, three oculomotor nerves (cranial nerves III, IV, and VI) innervate six eye muscles (per eye): four straight and two oblique muscles. A miraculously refined coordination system allows simultaneous movements of our eyes in all directions to a limit of about 50°. This coordination system is easily disturbed: Consuming a little too much alcohol already can result in blurred vision.

Fig. 5.1 Duck or rabbit? (Wikipedia—this image is in the public domain)

The retina is composed of ten layers. The most posterior is the pigment epithelium. Anteriorly lies the layer of the rods and cones. The human eye contains approximately 120 million rods and 7 million cones. The bipolar cells transmit the signals from the rods and cones to the more anteriorly located ganglion cells. A single bipolar cell may receive signals from several rods, which helps to intensify the light and movement sensitivity. In the center of the retina, called the macula lutea, the red- and green-sensitive cones are one-to-one connected to ganglion cells, which helps to create sharp vision. Amacrine and horizontal cells are dispersed between the bipolar cells and increase signal contrast. Horizontal cells transmit signals from cones and rods to other cones and rods and to several ganglion cells. Amacrine cells disperse signals from a bipolar cell to several ganglion cells. The axons of the ganglion cells transport the modified signals to the brain. Finally, Müller's cells compose a framework for the other cells.

The cones and rods contain light-absorbing pigments, for which composition vitamin A plays an important role. Vitamin A deficiency leads to hemeralopia: night blindness [2]. The rods contain the pigment-containing protein rhodopsin, which is sensitive to most forms of visible light. Rhodopsin cannot distinguish between different colors, which explain why we cannot see colors at night time when there is insufficient light to stimulate the cones. The rods of patients suffering from complete color blindness (or day blindness) are overstimulated during daytime with light. The rods are completely saturated and cannot see differences in brightness anymore. Night blindness is not always easily recognized. Day and night blindness are extremely rare.

Seeing Must Be Learned

Similar to playing the violin or golf, seeing must be learned. This occurs during the first seven years of life, provided that a clear image can be projected on the retina. When the pupil is covered by the eyelid (blepharoptosis), when the optical media (cornea, lens, vitreous) are not clear, or when the eyes are not straight, this learning process is impeded and the eye becomes amblyopic (i.e., lazy eye). In contrast to other learning processes, the ability to learn seeing comes to a complete halt at the age of about seven. Moreover, during those seven years, what has been learned already can be lost again if something happens that prevents the projection of a clear image on the retina. The earlier in life such an event takes place, the worse the visual outcome. The fact that input of the whole body is required for the development of proper seeing is demonstrated in the next experiment. Inside a cylinder with vertical black and white bars, one kitten was allowed to walk, whereas the other kitten was carried around in a box. Both kittens received exactly the same visual input, but only the kitten that was allowed to walk by itself developed normal vision [3].

Visual Acuity

Usually, the first thing that will be measured when you visit an optometrist or ophthalmologist with the complaint that you "see less" is visual acuity. This is a function of the macula lutea, the center of the retina in which the concentration of cones is maximal and, therefore, where the ability to see images in great detail is optimal. The outcome of this examination depends on the quality of your macula, but also of the refractive abilities of your eye. It was the Dutch ophthalmologist Frans Cornelis Donders (1818–1889), whose studies of pathology and physiology established the base for the correction of nearsightedness, farsightedness, and astigmatism. In nearsightedness (or: myopia), the focal point of the reflected image of the outside world lies in front of the retina, which explains why the image is not sharp. Myopic patients are corrected with concave lenses. In patients with farsightedness or hypermetropia, the focal point lies behind the retina. Without help, a young hypermetropic patient can create a sharp image by accommodating, e.g., by making his lens rounder. However, constant accommodation—for seeing nearby and in the distance—imposes an extra effort, and this may lead to fatigue and/or burning eyes or other nonspecific complaints. With aging, the ability of the human lens to become rounder decreases and nearby objects cannot be seen sharp anymore. This is called presbyopia. Presbyopia is superimposed on myopia and hypermetropia. Hypermetropic or presbyopia patients are corrected with convex lenses. People who are not myopic or hypermetropic are called emmetropic. The aging process of the human lens causes emmetropic people to look for reading glasses around the age of 45. Astigmatism, finally, is an imperfection of the curvatures of the cornea or (less frequently) of the lens. It can be corrected by aspheric, toric lenses.

Visual acuity is tested using derivatives of the Snellen optotypes. Herman Snellen (1834–1908) was Donder's successor as director of the "Ooglijders Gasthuis" in Utrecht. Compare an O and a C, depicted in black on a white background. When one is able to see the opening in the C as a distinct area (and assuming one is familiar with the Latin alphabet), one can conclude to see a C. This is the principle of Snellen's optotypes. With the Snellen chart, showing optotypes in decreasing size, the visual acuity (or the resolving power of the eye) is measured (Fig. 5.2).

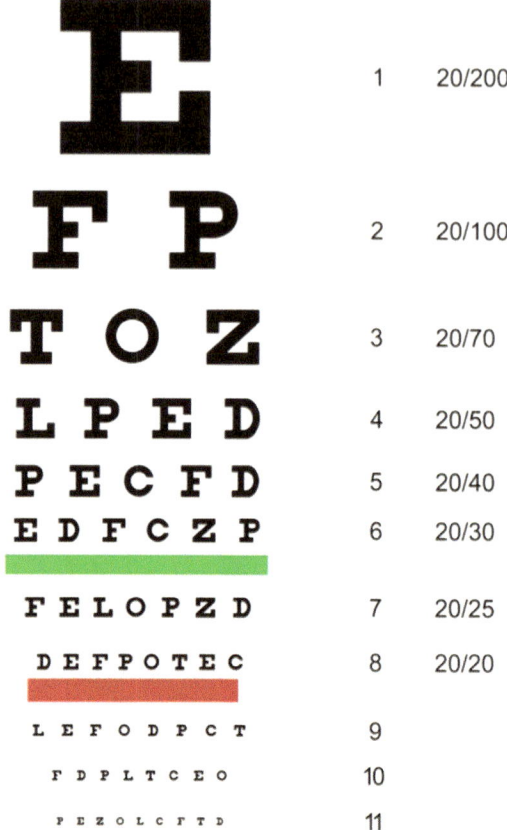

Fig. 5.2 Snellen chart (Wikipedia—Creative Commons Attribution-Share Alike 3.0 Unported license)

Color Vision

Light is defined as the part of electromagnetic radiation that is visible to the human eye and covers wavelengths between 400 nm (violet) and 750 nm (red). Objects that completely absorb light of all visible wavelengths appear as black, whereas objects that almost completely reflect light beams of all wavelengths are experienced as white. Strictly speaking, therefore, color is no intrinsic physical characteristic of objects. The sensation of color is created by our eyes and brain to make a distinction between objects possible. Color facilitates detecting borders of objects. While rods are 1000 times more sensitive to light than cones, they do not discern colors as the pigment-containing protein rhodopsin cannot discriminate between different wavelengths. In contrast, cones contain different pigment-containing proteins that transform light into elec-

trical signals. There are three types of human cones, responding to different wavelengths: blue-sensitive (short wavelengths), green-sensitive (medium wavelengths), and red-sensitive (long wavelength). Sixty-five percent of our cones are red-sensitive, 33% are green-sensitive, and only 2% are blue-sensitive. Already located in the retina, circuits of cells make mixing of colors possible, which eventually leads to the ability to distinguish more than 100 different colors (if circumstances are optimal). More than 99% of all colorblind people are suffering from a red–green color vision deficiency. The X-linked recessive red–green deficiency affects up to 8% of males.

In most orbital diseases, color vision hardly plays a role. The exception is Graves' disease. Loss of color vision, especially over the blue axis, is an early manifestation of dysthyroid optic neuropathy (DON) [4]. Color vision is mostly tested with Ishihara's red–green pseudo-isochromatic color plates. Although Ishihara's test is theoretically not the most sensitive test for patients with developing DON, it was found that almost all patients suspected of having DON responded abnormally to this test [5].

Seeing in Darkness

The human eye is able to see objects that are hardly illuminated and also things that receive more than one billion times more light (from the sun) [2]. However, it cannot see such objects at the same time. The human eye is gifted with a high potential of contrast sensitivity within a narrow range of light intensities [2]. In order to achieve this, the eye is able to adapt itself to the brightness of the light. Seeing under low-intensity light circumstances is called scotopic vision. Photopic vision is seeing under daylight circumstances. Cones are only stimulated when the light intensity surpasses a certain minimum and react immediately. Rods are far more light-sensitive, but need an adaptation time of minutes when light intensity decreases suddenly, e.g., when entering a dark room. Rods produce differences in gray, but no colors. Sharp vision disappears at night time.

Visual Field

The visual field is the total area that can be seen without moving one's head and eyes. It is grossly oval with its horizontal diameter larger than its vertical. The visual field of one eye covers that of the other eye for a large part. Only the center of the visual field allows binocular vision; in other parts, there is just monocular vision. The visual field warns of fixed or moving objects in the periphery and is thus of paramount importance in traffic circumstances. When fixating an object, we can distinguish a left and a right part of the visual field. The right part of the visual field of the right eye is projected on the nasal part of the retina, crosses the chiasm, and is then projected on the left hemisphere. The visual field is usually tested, each eye separately, with a Goldman (Chap. 6) or Humphrey perimeter. The fully developed classical visual field defects, such as homonym, heteronym, or bitemporal hemianopsia, are rare in orbital diseases. Instead, we find enlargement of the blind spot or cecocentral and paracentral defects, for instance, in patients with an optic nerve sheath meningioma or DON [6]. Typical eye diseases that are related to visual field defects are glaucoma and the family of inherited disorders of the photoreceptors called retinitis pigmentosa.

Stereoscopic Vision

When we fixate an object with both eyes, each eye sees that object from a different angle. In people with straight eyes, the two images are projected on corresponding retina points and are fused into one single image in the brain. We call this sensory fusion. The result is a sense of depth perception. A point in space further away from the fixated point is depicted nasally from the fovea (binasal disparity) and is therefore experienced as further away from the eye. A point closer to the eye than the fixated point is depicted temporarily from the fovea (bitemporal disparity) and is therefore appreciated as closer to the eye (Fig.5.3). Hence, stereoscopic seeing requires a geometrical construction, built up by two pupils and a fixation point. Prior to sensory fusion, the eyes have to be directed at the focus of attention. This is called motor fusion. Sensory and motor fusion are complementary. Binocular single vision and diplopia are extensively discussed in Chap. 6. Stereoscopic seeing enables evaluation of distances between two objects. Stereoscopic seeing, however, is limited. Beyond a distance of approximately five meters, this system adds little to monocular seeing because the differences in the angles between the two eyes become too small to be detected by our brain.

With only one eye, a limited form of stereoscopy is possible. This is called psychophysic stereoscopy. It is based on the (known) size of objects, the parallax, and accommodation state of our eyes.

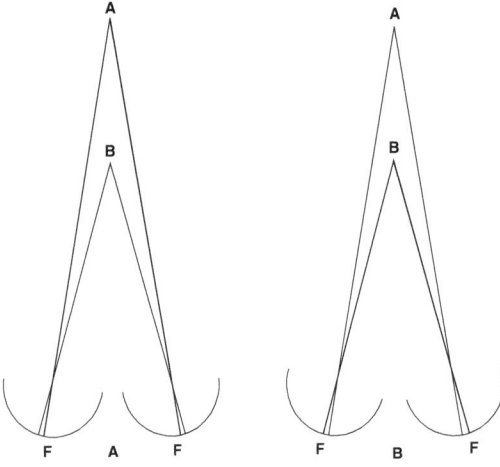

Fig. 5.3 Stereoscopic vision. On the left side, point A is being fixated. Point B, closer by, is being depicted on the retina with temporal disparity. On the right side, point B is being fixated. Point A, at a more remote distance, is being depicted with nasal disparity

For most daily activities, psychophysic stereoscopy suffices. For instance, individuals who have only one functional eye are allowed by law to drive a passenger car. In contrast, cataract surgery requires double-eyed stereoscopic vision. Stereoscopic depth perception is considered the ultimate level of seeing. Not every individual reaches this level. As mentioned before, seeing must be learned and ocular developmental disorders (e.g., squint) may prevent an individual from reaching the highest level of seeing. However, many individuals are not aware that they have no binocular single vision, nor are they aware that they have one amblyopic eye, and, nevertheless, they live happily.

Facial Recognition

Facial recognition is an extremely important characteristic of animals that live together in groups, such as primates. Humans have developed a refined facial recognition system, which allows a young baby already to recognize its mother's face. The center of the retina (e.g., the macula lutea) and occipital brain center(s) are essential for facial recognition. Patients with advanced stages of macular degeneration (the most frequent cause of impaired vision in the Western world) lose the ability to recognize faces, which in turn may lead to social isolation.

References

1. Eagleman D. The brain. The story of you. Edinburgh: Canongate Books; 2016. p. 42.
2. Crone RA. Licht, kleur, ruimte. De leer van het zien in historisch perspectief. Houten: Bohn Stafleu Van Loghem; 1992.
3. Held R, Hein A. Movement-produced stimulation in the development of visually guided behavior. J Comp Physiol Psychol. 1963;56:872–6.
4. Neigel JM, Rootman J, Belkin RI, Nugent RA, Drance SM, Beattie CW, Spinelli JA. Dysthyroid optic neuropathy: The crowded orbital apex syndrome. Ophthalmology. 1988;95:1515–21.
5. McKeag D, Lane C, Lazarus JH, Baldeschi L, Boboridis K, Dickinson AJ, et al. Clinical features of dysthyroid optic neuropathy: a European group on graves' orbitopathy (EUGOGO) survey. Br J Ophthalmol. 2007;91:455–8.
6. Newman SA. Cecocentral scotoma: a neuro-ophthalmic revisionist approach. IOVS. 2015;56:2605.

Open Access This chapter is licensed under the terms of the Creative Commons Attribution 4.0 International License (http://creativecommons.org/licenses/by/4.0/), which permits use, sharing, adaptation, distribution and reproduction in any medium or format, as long as you give appropriate credit to the original author(s) and the source, provide a link to the Creative Commons license and indicate if changes were made.

The images or other third party material in this chapter are included in the chapter's Creative Commons license, unless indicated otherwise in a credit line to the material. If material is not included in the chapter's Creative Commons license and your intended use is not permitted by statutory regulation or exceeds the permitted use, you will need to obtain permission directly from the copyright holder.

Diplopia

6

Yvette Braaksma-Besselink
and Hinke Marijke Jellema

Learning Objectives
- Diplopia can be either monocular, binocular, or gaze dependent.
- Be aware of the possibility of pre-existing strabismus.
- The absence of diplopia in an individual does not automatically imply that this person has binocular single vision.
- Careful examination of the ocular movements is necessary to differentiate neurogenic from mechanic causes of diplopia and to diagnose any pre-existing strabismus.
- During follow-up, patients may be helped with some form of occlusion or Fresnel prisms to prevent diplopia.
- Different surgical procedures are available to create a useful field of binocular single vision.
- If a significant amount of cyclotorsion is measured, a poorer prognosis for both adjusting to the Fresnel prism as well as for the outcome of strabismus surgery is given.

- A full field of binocular single vision may not be reached despite surgical treatment in patients with orbital pathology.

Introduction

Each of us has double vision. Our eyes allow us to see the world as it is; our minds allow us to see the world as it can be.—Patti Dawn Swansson

The perception of depth through the use of two eyes is a naturally occurring visual process, which is often taken for granted. The importance of binocular single vision becomes painfully apparent once it is lost. Once a diagnosis is made, requiring quick surgical or medicinal intervention and especially in case of multidisciplinary approach, the complexity of diplopia often fades into the background. However, the basis of the binocular system plays an important role during the course of treatment, especially when different medical specialists are to decide on the treatment, sometimes based upon the presence or absence of diplopia. Extricating the true cause of diplopia may be difficult. The orthoptist, however, is trained in examining ocular movement and binocular single vision. The orthoptist may play a crucial role in the early stages of diagnosing in orbital disorders and certainly in the follow-up through the different stages of treatment, whether conservative or surgical.

Y. Braaksma-Besselink · H. M. Jellema (✉)
Department of Ophthalmology, Amsterdam
University Medical Centers, Location AMC,
Amsterdam, The Netherlands
e-mail: y.c.besselink@amc.uva.nl;
h.m.jellema@amsterdamumc.nl

© The Author(s) 2023
P. J. J. Gooris et al. (eds.), *Surgery in and around the Orbit*,
https://doi.org/10.1007/978-3-031-40697-3_6

Levels of Binocular Single Vision and Diplopia

Physiology Binocular Vision: The Cyclopean Eye

Binocular single vision is the simultaneous use of both eyes to give a single mental impression in normal visual conditions. Normal binocular vision is possible when the left and right fovea fixate on the same point and there is no manifest deviation (Fig. 6.1a). The two images of both fovea are perceived as one: The brain combines the images of both eyes to a single three-dimensional percept. This combining of the two images from each fovea is called the cyclopean eye (Fig. 6.1b). Double vision occurs when either two eyes perceive two separate images or when the perceived images are of dissimilar quality (Fig. 6.1c, d). Binocular double vision or diplopia is an indication that at least both eyes perceive an image.

Different Levels of Binocular Vision

Binocular single vision is the result of a complicated system that consists of several levels. The first level is simultaneous perception, i.e. the ability to perceive an image with both eyes at the same time. The second level is fusion, which is the ability to appreciate two images that are alike and to interpret them as one. The highest grade of binocular single vision is stereoscopic vision, which is the perception of relative depth of objects in space. This can be tested in a clinical setting through the use of stereoscopic tests, such as the stereofly test or the TNO stereopsis test (Fig. 6.2).

Diplopia in Childhood

In order to appreciate double vision, there are several conditions that have to be met. Binocular vision must be present since the early years of childhood. If adults have strabismus since an early age, the binocular system has adjusted to this binocular abnormality. This means that the brain is used to an ocular deviation and that the visual cortex develops an adaptation called suppression. In case of suppression, an eye is able to see in monocular condition. In binocular condition, however, the suppressed eye does not consciously perceive an image.

In some cases, this suppression also leads to amblyopia, i.e. a lower visual acuity caused by a continued state of suppression, in which the visual system delays or even stops its visual development.

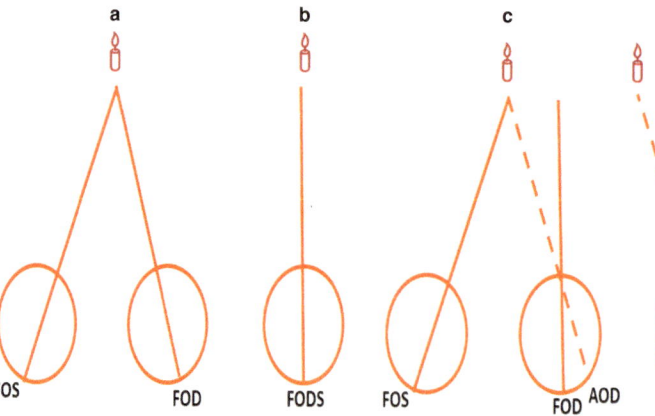

Fig. 6.1 (a–d) Normal binocular single vision when the candle is perceived on the fovea of the right and left eye (a) shown as one candle in the cyclopean eye (b). Diplopia when the candle is perceived on the fovea of the left eye (FOS) and point A of the right eye (AOD) (c) showing two candles in the cyclopean eye (d). *OS* left eye, *OD* right eye, *FOD* fovea right eye, *AOD* other point of right eye, *FOS* fovea left eye

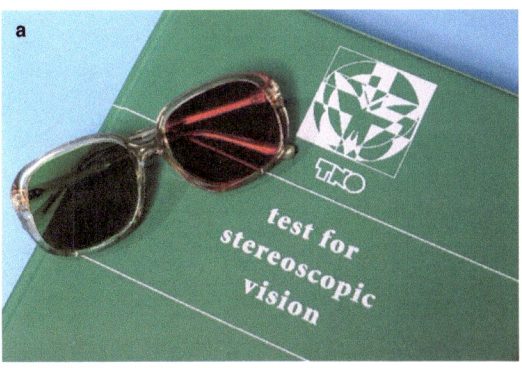 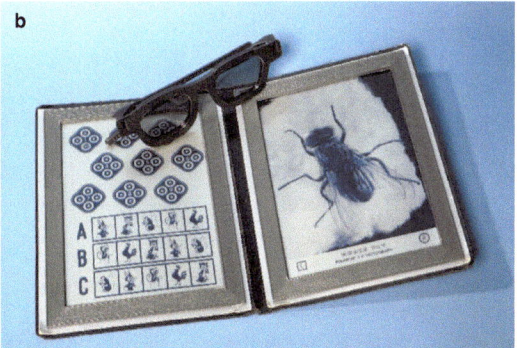

Fig. 6.2 TNO stereopsis test (**a**) and Titmus stereofly test (**b**)

If there is a sudden change in eye position in children, they might perceive a double image very shortly. The plasticity of the young brain will almost immediately adapt to this new situation, in which case suppression and amblyopia arise. Objective examination is therefore crucial in young children.

Strabismus But No Diplopia

In orbital diseases, loss of binocular single vision resulting in diplopia is a common phenomenon, but loss of binocular single vision does not always result in diplopia. Examples of the absence of diplopia in combination with absent or abnormal binocular single vision are:

1. Pre-existing strabismus since childhood, in which case the image of the deviating eye is suppressed.
2. Amblyopia in one eye.
3. Significantly reduced visual acuity due to a coexisting eye disease.
4. Significant visual field defect due to a coexisting eye disease.
5. An ocular torticollis, i.e. patients adapt an abnormal head posture to avoid double vision.

Monocular or Binocular Diplopia

Diplopia must first be sorted in either binocular or monocular diplopia. Binocular diplopia means that there is a change in eye position of one eye relative to the other, a condition to which the patient is not used to. With one eye closed, the diplopia disappears. Monocular diplopia is a form of blurred vision, but usually considered as double vision by the patient. Monocular diplopia is in almost all causes produced by abnormalities of the ocular globe, such as the ocular surface or the ocular lens.

Acquired or Long-Standing?

Once it has been established that there is indeed binocular double vision, orthoptic examination can then clarify whether double vision is caused by long-standing or recently acquired strabismus. Strabismus that has been present since childhood is most often concomitant, i.e. an ocular deviation that remains unchanged, regardless the direction of gaze. In case of incomitant strabismus, the ocular deviation changes depending on the direction of gaze (Fig. 6.3). Examining the ocular movements will distinguish between both types of strabismus (see Eye movement). It does not mean that incomitant strabismus is always acquired. If incomitant strabismus is present without diplopia in the absence of ophthalmological abnormalities, it may be assumed to be long-standing. Whether the long-standing strabismus is concomitant or incomitant, the binocular system has adjusted to this situation, as is seen in the presence of suppression fitting the previously existing deviation.

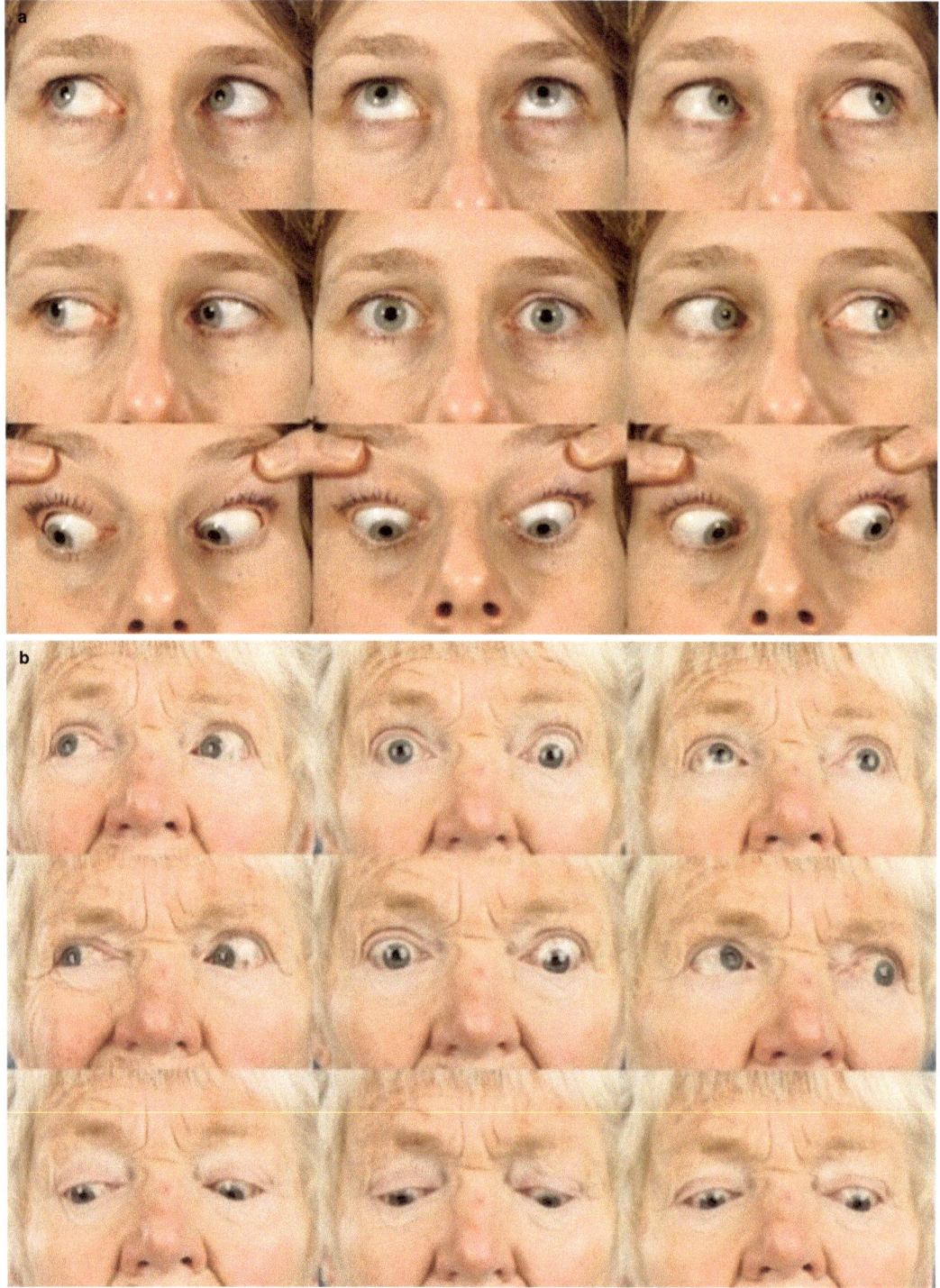

Fig. 6.3 Straight eye position with concomitant eye movements (**a**) and left hypotropia with incomitant eye movements (**b**)

Pre-existing strabismus may be disrupted in the presence or absence of mechanical or neurogenic damage and may consequently cause diplopia. Once an eye position is changed, a patient may experience double vision just as is seen in patients with acquired strabismus without an orthoptic history. This can be explained by a previously adapted binocular system or suppression that is not sufficient anymore. Patients may not even be aware of long-standing strabismus and report it as recently acquired.

In case of acquired strabismus, the type of diplopia (horizontal, vertical, cyclotorsion, or combined) will give an indication of the involved ocular muscles. Through orthoptic examination, it is possible to specify the type of diplopia as well as to determine whether ocular motility is concomitant or incomitant. Most often it is possible to define whether motility disorders are long-standing or acquired, mechanical, or neurogenic in nature. In follow-up consultations, it is possible to record whether ocular motility is stable or changing over time.

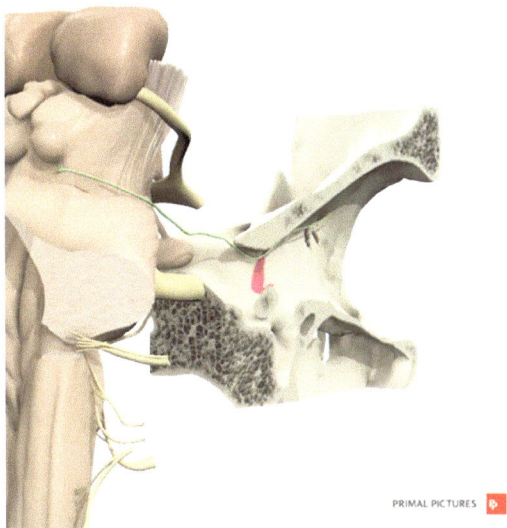

PRIMAL PICTURES

Fig. 6.4 The trochlear nerve (green) passes from the midbrain onto the lateral surface of the crus of the cerebral peduncle. It runs through the lateral dural wall of the cavernous sinus, then crosses the oculomotor nerve and enters the orbit through the superior orbital fissure, above the common tendinous ring of the recti muscles. Here, it lies above the levator palpebrae superioris muscle and medial to the frontal and lacrimal nerves [1]

Causes of Diplopia

The different types of diplopia may have several causes. Before orbital and often mechanic causes are assumed in orbital abnormalities, one also must consider a neurological cause of diplopia.

Diplopia Due to Head Trauma

Head trauma may lead to ocular nerve damage which is expressed as ocular muscle paralysis, such as lateral rectus paralysis in case of a nuclear lesion of the VIth cranial nerve. Even on a more peripheral level, one can find an ocular muscle paresis. The trochlear nerve (IVth cranial nerve) is most vulnerable to damage due to its long trajectory from the brainstem to the superior oblique muscle (Fig. 6.4). Either internuclear nerve failure or peripheral nerve damage will lead to an ocular motility disorder with diplopia. Nerve damage may be caused by disruption of the nerve in case of direct damage by, for example, bone

fragments, bruising of a nerve by brain movement, interrupted blood supply, and/or hemorrhage or compression from or within the nerve.

Diplopia is also seen in patients with superior orbital fissure syndrome, in which case the IIIrd, IVth, VIth, and Vth cranial nerves are involved in the event of bone accident in this region. In case of intraorbital damage, one may also find diplopia caused by nerve damage when near the apex of the orbit.

Of course, diplopia may be caused by direct damage to the ocular muscles as is seen in facial trauma or may be iatrogenic in nature after orbital surgery. Muscles may be contused in case of shifting bony fragments, swelling due to intramuscular hemorrhage, muscle damage due to laceration, or changed muscle mechanics due to muscle displacement. In some cases, however rare, there may even be muscle entrapment in an orbital fracture. In case of mechanical damage after an orbital blow-out fracture, the mechanism of damage leading to diplopia is quite similar to that caused by an orbital decompression operation. In case of a fracture, damage to the muscles is rarely the cause of diplopia. It may partly be due to the herniation of orbital fat and connective tissue into the surrounding sinus with subsequent traction on the muscle sheaths. In case of an orbital decompression, diplopia may be induced by means of change in the support system of the orbital content after removal of orbital walls.

Diplopia Due to Abnormal Structure

Another major cause of diplopia is the presence of an abnormal lesion in the orbit. Any abnormal volume in the region of the muscle or of the muscle itself may limit both the contraction and the relaxation of that specific movement (Fig. 6.5). Ocular muscles that have been affected by orbital pathology—such as orbital myositis, intraorbital space-occupying lesions, or inflammatory conditions such as Graves' orbitopathy (GO)—show a change in function depending on the level of involvement. In case of mild inflammation, ocular function may not be hampered at all. However, as may be seen in GO, some patients show edema of the orbital tissue including the ocular muscles, in which case there is a structural change of muscular tissue and, hence, function and mechanics.

Depending on the amount of swelling, one will find impaired muscle movement caused by tightness or contraction of the muscle. The typically enlarged eye muscles cause a limitation of eye movement in the opposite direction due to their inability to relax.

Orthoptic Investigative Procedures

In order to determine the presence and the degree of strabismus, ocular motility disorders, and/or the level of binocular single vision, the orthoptist has an array of investigative procedures to select from. At the start of the investigation, the presence of an abnormal head posture to compensate for gaze-dependent strabismus must be noted. All observations and measurements are performed with the patient assuming the primary sitting position (sitting upright, shoulders back and head upright).

Eye Position

The corneal light reflex informs about the position of the globe. Using the light reflex, an impression of the angle of deviation (strabismus) can be achieved (Fig. 6.6).

- A light reflex on the border of the pupil indicates strabismus of roughly 15°.
- A light reflex between the border of the pupil and the limbus (corneoscleral transition) indicates strabismus of about 30°.
- A light reflex on the limbus means a strabismus of about 45°.

This estimation of the angle of deviation is called the 'Hirschberg method'.

An asymmetric corneal reflex is indicative of strabismus and may be accompanied by diplopia (Fig. 6.7). We can distinguish four different types of strabismus:

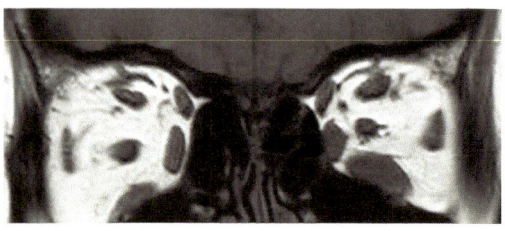

Fig. 6.5 MRI scan of a patient showing enlargement especially of the inferior rectus muscle of the left eye

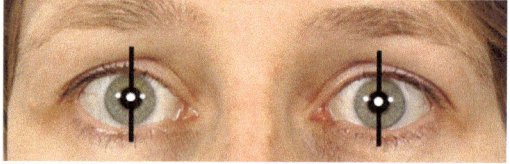

Fig. 6.6 Symmetrical light reflex with the potential of binocular single vision

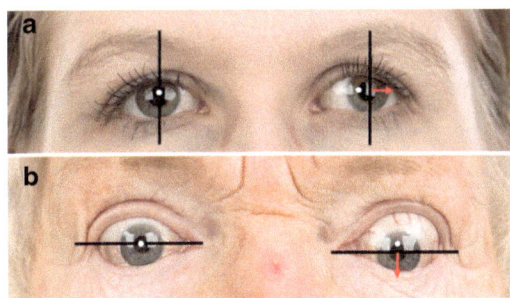

Fig. 6.7 Asymmetrical light reflex with an exotropia (**a**) and hypotropia (**b**) of the left eye

- Esotropia = inward displacement of one eye.
- Exotropia = outward displacement of one eye.
- Hypertropia = upward displacement of one eye.
- Hypotropia = downward displacement of one eye.

Measurement of Strabismus

More precisely than with the Hirschberg method, the amount of deviation can be captured through the use of prism bars at several distances or even in different directions of gaze. When measuring strabismus, the prism bar is used to divert incoming light rays of a fixation light to meet the amount of ocular deviation. When combined with the cover test, the examiner can objectify the degree of strabismus. The patient is asked to fixate on a light, while the eyes are alternately covered (Fig. 6.8). If strabismus is present, an eye will have to make a corrective movement after it is uncovered. If the amount of strabismus is equal to the strength of the prism used, the movement of the eyes is neutralized. If the strength is equal to the amount of deviation, light will fall on to the fovea and an eye does not have to adjust its position upon fixation.

Binocular Single Vision

Once strabismus has been established, one can examine the ability of binocular single vision by

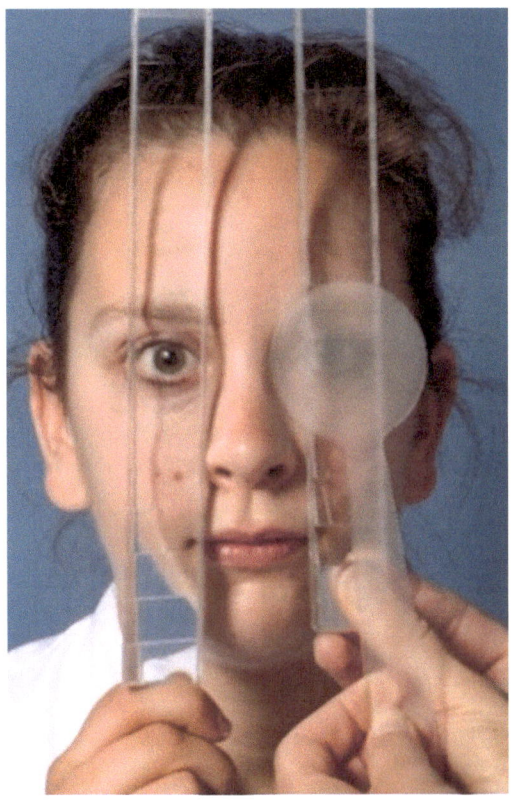

Fig. 6.8 Alternating cover test in combination with one prism basing temporal (in front of right eye) and the other prism basing down (in front of left eye) measuring and esodeviation and left hyperphoria or left hypertropia

means of adjusting the incoming image through a prism bar. This prism bar, adjusted in the specific amount that meets the ocular deviation, will cause an image to be interpreted as one (Fig. 6.9). Once fusion has been established, one is then able to report on the ability to hold on to the single image, while changing the amount of horizontal and vertical prisms. This will, then, give rise to a horizontal and vertical fusional range. The size of this range can clarify complaints. A high range may conceal ocular deviations, whereas a small range may give rise to complaints that do not seem fitting in case of a small deviation. This fusion range helps to decide whether prisms are beneficial in alleviating diplopia during daily life (see conservative treatment).

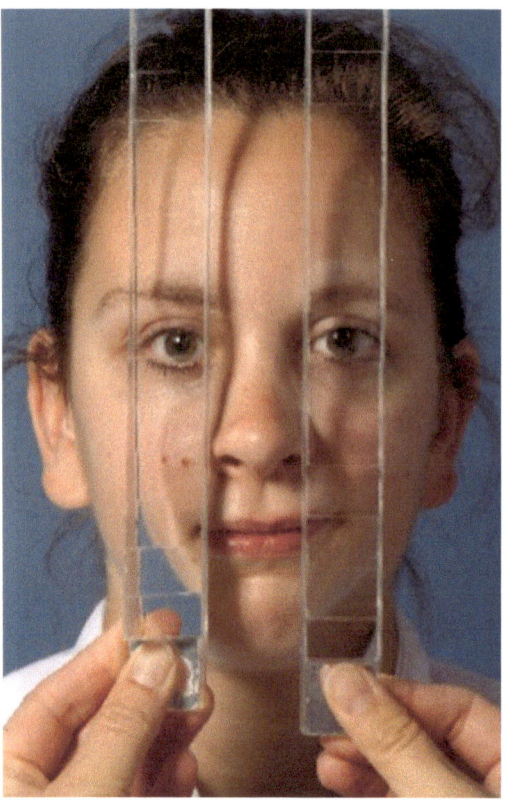

Fig. 6.9 Prism bars in front of the patient's eye to perceive binocular single vision

Fig. 6.10 The Goldman perimeter

Field of Binocular Single Vision

In acquired strabismus, diplopia is often only present in some directions of gaze. In other directions of gaze, binocular single vision exists. The localization and the extent of the area of diplopia determines the inconvenience in daily life. For instance, a patient with an orbital floor fracture may have diplopia only if looking up. As the need for looking up in everyday life circumstances is limited, especially for tall people, the impact of the diplopia will be acceptable. Generally, single vision in primary (gaze straight ahead) and reading position will often be acceptable for the patient and is often the best to be obtained, i.e. in patients with severe GO.

In case a patient reports diplopia as well as binocular single vision, a field of binocular single vision can be determined and this is very

useful in decision making and follow-up. A field of binocular single vision can be attained and quantified in several manners. In literature, Goldman perimetry is considered to be the golden standard (Fig. 6.10). One must keep in mind, however, that this device does not depict natural viewing conditions because of limited fusion. The measurement performed with the Harmswand or Maddox tangent screen (Fig. 6.11) results in a field of binocular single vision in more natural conditions.

In either approach, the patient is asked to report at what point he/she notices diplopia. In case of Goldman perimetry, the patient is asked to follow a light from a point of single vision with his/her head fixated. In the other two approaches, the patient will be asked to move his/her head while focusing on a fixated light. The field of binocular single vision is a means of quantifying double vision. Scoring a field of binocular single vision may be done with the Sullivan score [2] (Fig. 6.12). The quantitative score ranges from 0

Fig. 6.11 The Maddox tangent screen

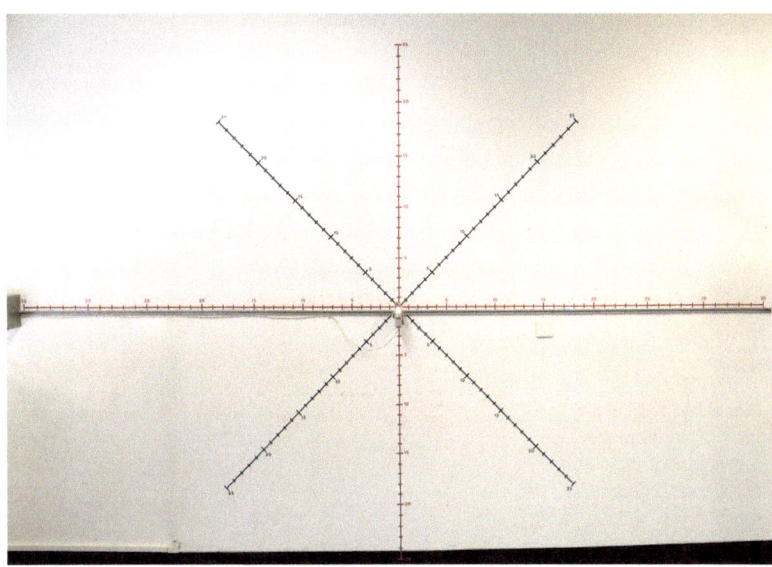

Fig. 6.12 Score field according to Sullivan et al. (1994) [2] for quantifying the field of binocular single vision ranging from 0 (no binocular single vision) to 100 (no double vision)

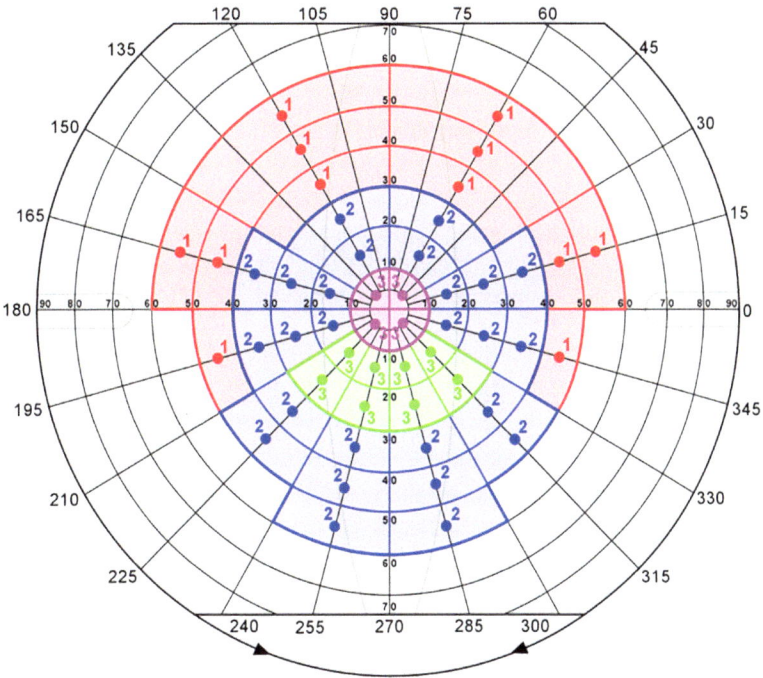

(no binocular single vision) to 100 (no double vision). This scoring system easily clarifies the progression of improvement (Fig. 6.13), worsening or change after surgery. The use of the Sullivan score field in combination with the field of binocular single vision simplifies progress in treatment and allows comparison between study groups accessible [3].

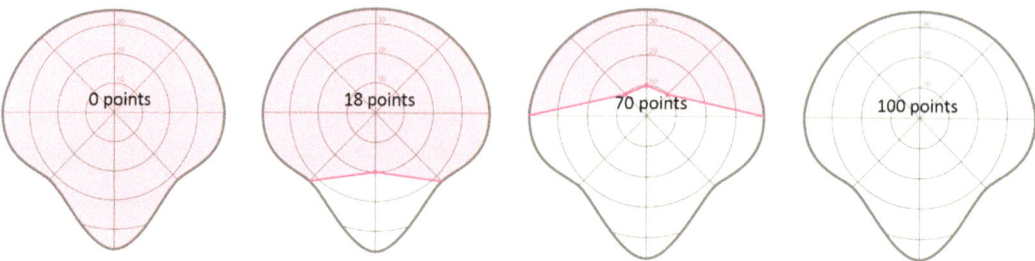

Fig. 6.13 Progress of the field of binocular single vision during follow-up measurements. White = binocular single vision; pink = diplopia

Fig. 6.14 Schematic view of horizontal eye movement of the medial and lateral rectus muscles [4]

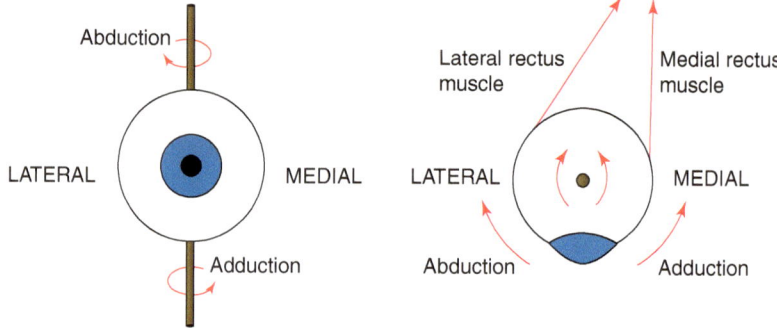

Eye Movement

Different types of eye movements can be distinguished:

- Versions: binocular and symmetrical eye movements. Both eyes look in a particular direction and move in a synchronized manner.
- Ductions: monocular eye movement of one eye in a specific direction:
  - Abduction: temporal movement.
  - Adduction: nasal movement.
  - Elevation: upward movement.
  - Depression: downward movement.
- Concomitant: No change in ocular deviation despite change in direction (Fig. 6.3a).
- Incomitant: A change of ocular deviation per direction; one eye is restricted in its movements and the other will overcompensate while moving in the same direction (Fig. 6.3b).

To determine whether strabismus is concomitant or incomitant, the ocular deviation must be judged in the nine positions of gaze. Each ocular muscle

has a different function per gaze. The axis of rotation of the normal functioning eye muscles depends on the muscle insertion to the globe. The horizontal muscles, the medial and lateral rectus, all have an almost purely horizontal action (Fig. 6.14).

The vertical and oblique muscles, however, have a more complicated action of movement due to their oblique position on the globe; not only moving the eye up and down, but also causing movement around the visual axis of the eye, which is called cyclotorsion. The visual axis is the line that connects a point in the outside world through the center of the pupil to the fovea centralis (center) of the retina.

The vertical and oblique ocular muscles have a primary, secondary, and tertiary action (Table 6.1). Depending on the position of the ocular globe, the position of the muscle will change relative to the visual axis leading to a different action. The superior oblique muscle/tendon is at an angle of 51° with the visual axis (Fig. 6.15). This means that if the eye is adducted, the visual axis is perfect aligned with the superior oblique muscle and will have a vertical action (depression).

Table 6.1 Actions of external eye muscles in primary position [5]

Muscle	Primary action	Secondary action	Tertiary action
Inferior rectus	Depression	Excycloduction	Adduction
Superior rectus	Depression	Incycloduction	Adduction
Medial rectus	Adduction	–	–
Lateral rectus	Abduction	–	–
Superior oblique	Incycloduction	Depression	Abduction
Inferior oblique	Excycloduction	Elevation	Abduction

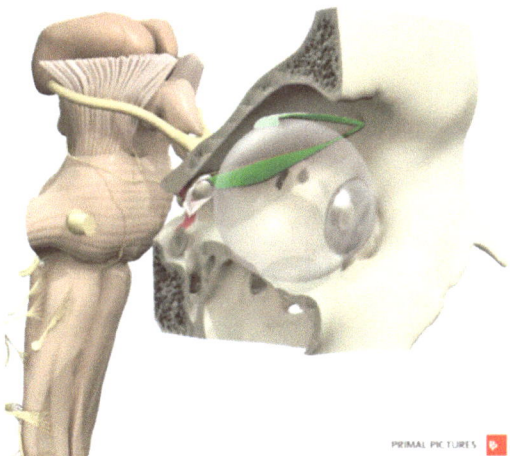

Fig. 6.15 The anatomical position of the superior oblique muscle [6]

Measurement of Incomitant Strabismus

Once it has been established that there is a form of incomitant strabismus due to a motility disorder, ocular movements will have to be documented more precisely. There are three essential components of this documentation: ductions, eye position in nine directions of gaze, and cyclotorsion. These three components are necessary for a proper diagnosis as well as essential in decision making with regard to further treatment.

Ductions

Ductions are a measurement of monocular ocular movement, more specifically of the horizontal and vertical eye muscles. Muscle function can be evaluated by means of a grading system. A grading scale from −4 to +4 is often used to quantify

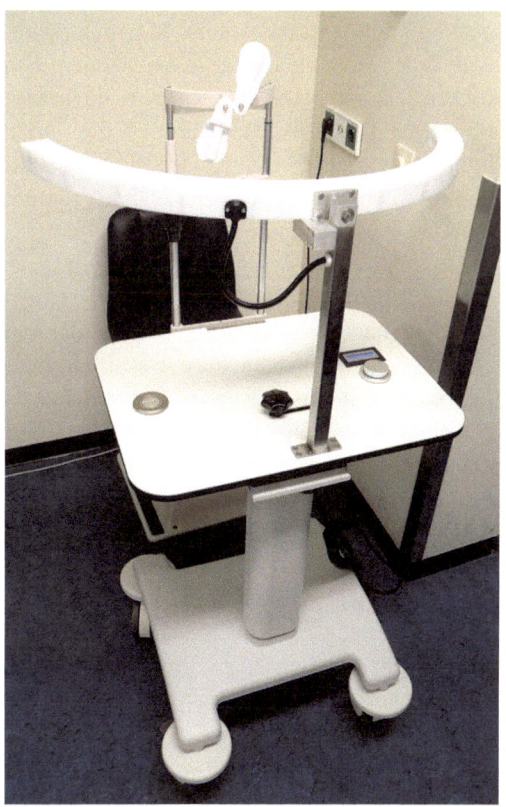

Fig. 6.16 Motility meter developed by Mourits

muscle function (+ is degree of overaction, and − is degree of underaction). In this manner, oblique muscle function can also be roughly classified. This grading system is completely subjective—and therefore observer dependent—and much less precise to a deviometer such as the Goldman perimeter or a motility meter as developed by Mourits (Fig. 6.16), although the latter cannot evaluate or grade oblique muscle function. The patient follows a moving light on an

Table 6.2 Normal values of ductions [7]; OD is oculus dexter, i.e. the right eye: OS is oculus sinister, i.e. the left eye

Ductions	OD	OS
Abduction	46	46
Adduction	48	48
Elevation	34	34
Depression	58	58

arc, while the examiner pays attention to the corneal light reflex of the patient. Maximal duction is recorded from the digital screen as seen in the picture. Though much more precise, the grading is derived from an observation of the corneal light reflex and, therefore, also subjective and observer dependent. Normal values of the ductions are shown in Table 6.2.

Nine Positions of Gaze

Once the function or limitation of an ocular muscle has been documented, the eye position should be measured in nine positions of gaze. This can be achieved through the use of a Maddox rod in combination with a tangent screen at 2.5 m (Figs. 6.11 and 6.17), a Hess motility screen (50 cm) or a Lancaster screen (1 m). Each will give the same result, although measured at different distances. Overaction and underaction of muscle function can be schematically depicted and followed over time (Fig. 6.18).

Cyclotorsion

Another function of ocular motility that is essential in diagnosis and therapeutic deliberation is cyclotorsion. Cyclotorsion is a rotation of the eye around the visual axis. The majority of patients will not spontaneously report abnormal cyclotorsion, unless specifically asked for.

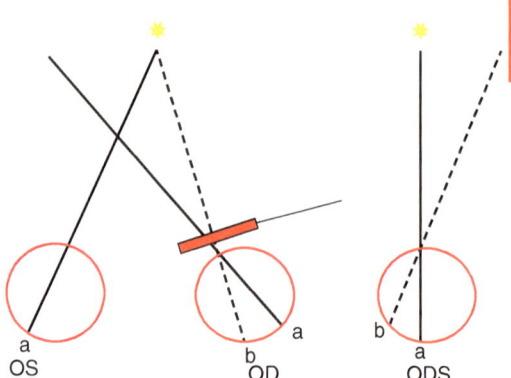

Fig. 6.17 Maddox rod in front of the right eye. The patient perceives a red line through the red lens of the Maddox rod. The red line is perceived on the right side of the light on the tangent screen at 2.5m in case of this esodeviation. The red line is projected onto the tangent screen at 2.5m, the distance between the light and the line can then be recorded in degrees

However, cyclotorsion is often indispensable in diagnosing and plans for surgical treatment especially in acquired strabismus. Measurement of torsion is helpful in identifying oblique muscle weakness or overaction, such as can be found in acquired fourth nerve palsy in which case excyclotorsion will be found. In case of congenital strabismus, cyclotorsion will not be noticed by the patient. The presence or absence of cyclotorsion can be crucial in differentiating between congenital or acquired lesions and the need for further neurological examination. Incyclotorsion, on the other hand, is often found secondary to orbital injury or orbital surgery [8]. Measurement of cyclotorsion can be performed in different positions of gaze using the Harms tangent screen or the cycloforometer of Franceschetti (Fig. 6.19).

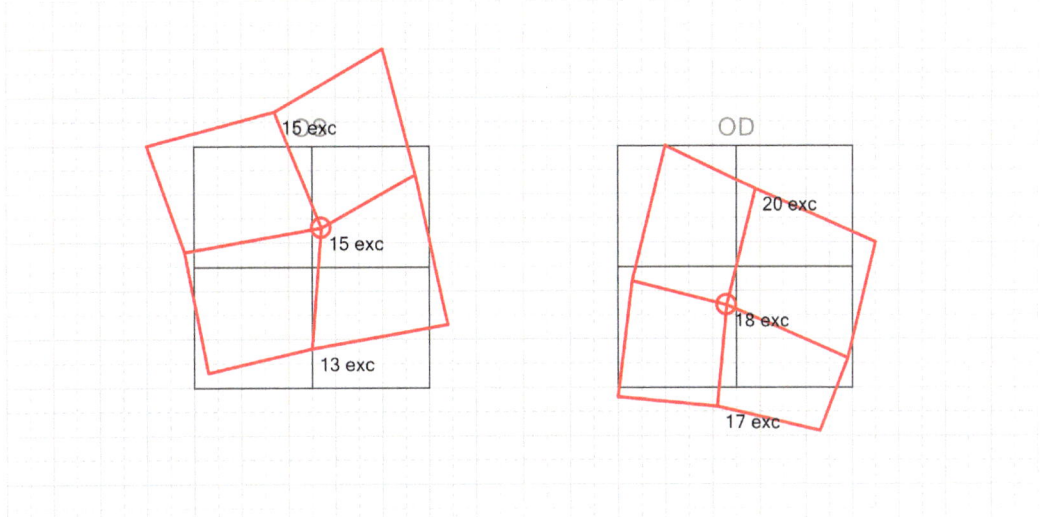

Fig. 6.18 Amsterdam motility diagram: The red lines are a depiction of the points of each reported placement of the red line on the tangent screen at 2.5m in the nine directions of gaze. The red lines show a right hypotropia (the circle in the middle of the square on the right side of the diagram is lower in ratio to the black lines which would be normal) in a patients with Graves' orbitopathy caused by a restricted elevation. On right gaze the inferior oblique muscle overacts (red line formed from the circle in the middle on left side (OS) which is deviated to the right upper corner and shows the inferior oblique function), which results in an increased right hypotropia. Cyclotorsion is recorded in three directions of gaze next to the diagram. *Exc* excyclotorsion, *OS* left eye, *OD* right eye

Fig. 6.19 The cycloforometer of Franceschetti

Evaluation

After orthoptic assessment, diagnosis, and multidisciplinary consultation, it is important to evaluate these findings in light of the patient's experience. A patient with a field of binocular single vision of 80 points can be fully incapacitated when he/she is a plasterer and uses the remaining 20% of upper field of gaze for more than 70% of the day, while a similar patient working an administrative job will report no complaints at all, although presenting with equal measurements. Several quality-of-life questionnaires are available to quantify this.

Treatment of Diplopia

The treatment of diplopia can be either conservative or surgical.

Conservative Treatment

While waiting for spontaneous improvement, nonsurgical treatment can start once a patient is presented to the orthoptist. Primarily, three options may be considered: abnormal head posture, prisms, and occlusion.

A fourth form of treatment must be contemplated in case of orbital fractures: eye movement exercises. No clinical studies have been performed to analyze the effect of monocular eye muscle exercises in patients with orbital fracture.

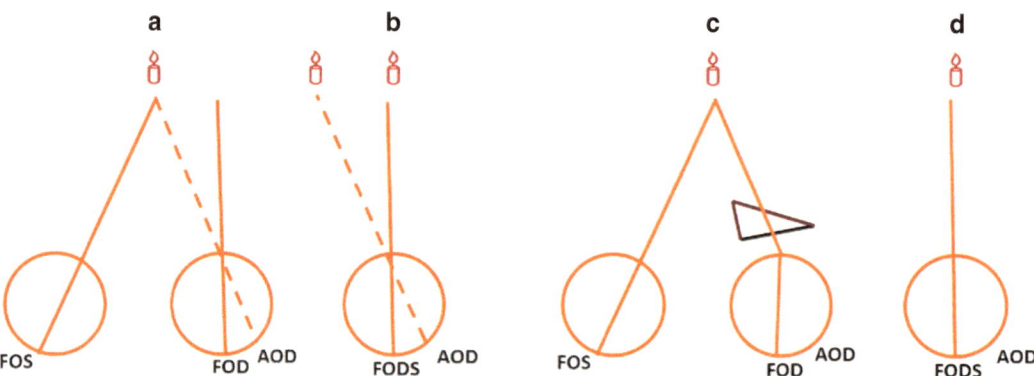

Fig. 6.20 Exotropia of the right eye (**a**) with diplopia shown in the cyclopean eye (**b**). Prism base nasal in front of deviated eye (**c**) gives single vision for the patient a shown in the cyclopean eye (**d**)

It is, however, the golden standard and the first line of treatment in these patients. During the first few weeks after trauma, the affected eye has to be actively moved in all gaze directions with the sound eye covered.

Abnormal Head Posture

Most patients with acquired strabismus and some form of binocular single vision will naturally adapt and assume an abnormal head posture to avoid double vision. Some patients, however, must be actively pointed to this possibility.

Prisms

In case of strabismus with diplopia, to achieve binocular single vision, a patient sometimes can be helped with prisms (Fig. 6.20). By means of a prism bar, ocular deviation can be neutralized and the particular prism strength can be prescribed. With changing deviation, Fresnel prisms (a ribbed piece of silicone foil that can be applied to the spectacle lens) can be used (Fig. 6.21). Fresnel prisms are available at a range from low to high power, both for horizontal and vertical diplopia (not for torsional diplopia).

The foil is prescribed for the eye with the lowest vision and/or the most restricted eye movements. Prism foils will always lead to some amount of reduced vision due to the line pattern

Fig. 6.21 Fresnel prism foil on the right spectacle lens to correct horizontal diplopia

and hence if applied in front of the eye with the lowest vision will limit complaints. It is imperative to inform the patient about this. Even when ocular movement shows an incomitant pattern, prisms are often well accepted [9]. In theory, incomitant strabismus would require different prism strength, depending on the direction of gaze. The advantage of a Fresnel prism is that binocular single vision is restored. Once single vision is achieved, most patients can make use of their innate fusional strength to maintain this single vision in most directions. During follow-up, the strength of the Fresnel prism can be adjusted, according to the eye position. The prism can eventually be grinded into the lenses as a permanent solution.

If binocular single vision cannot be achieved by means of prisms, one must be aware of cyclotorsion hampering fusion.

Occlusion

In some unfortunate cases, no acceptable field of binocular single vision can be gained with prisms or after strabismus surgery (see explanation below). Some patients can adjust to this situation and actively ignore the second image. It may help if the vision of one eye is reduced. Diplopia can be very frustrating and extremely debilitating and, therefore, may lead to the extreme choice of permanent occlusion. This can be achieved through the use of an occlusion patch or by means of fully or partially occluding a spectacle lens (Fig. 6.22) by means of a matted foil or grinded lens. If occlusion of a spectacle lens is an unacceptable option for the patient, a painted contact lens (Fig. 6.23) or even an intraocular occlusion lens is a more sophisticated solution. Explanation

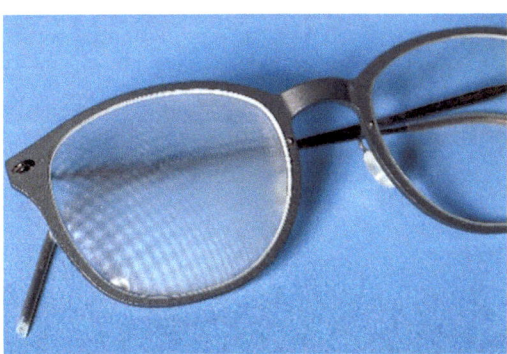

Fig. 6.22 Occlusion foil on the right spectacle glass

Fig. 6.23 Occlusion contact lens

has to be given that the visual field is significantly narrowed especially in case of a (intraocular) lens.

Surgical Treatment

It is important that repeated examinations show a stable outcome before considering any surgical treatment to resolve diplopia. However, this waiting-for-stability period can be extremely frustrating from the patient's perspective. Care professionals should be aware of this and are urged to address the reason of postponing surgery. The goal of this waiting period is to make an optimal surgical plan with a more predictable outcome based upon a stable condition. In general, at least two similar orthoptic examinations over a period of several months are necessary to decide upon which strabismus surgery is most suitable. Most authors suggest a period of 6 months of stability, this however depends on the cause of strabismus, i.e. in case of GO a stable period of 3 months can be maintained [10].

Treatment of Patients with Graves' Orbitopathy

One of the hallmarks of GO is swelling and inflammation of one or more extraocular muscles. This often results in fibrosis and loss of power of relaxation of the involved muscles.

In GO, the severity of incomitance of ocular motility results in different treatment strategies. Recession of the most affected muscle (i.e. reinsertion of the tendon of that muscle posterior to its original insertion) is the preferred choice, rather than resection (i.e. surgically shortening of the muscle) of an unaffected muscle. Depending on the degree of angular deviation, one or two muscles are recessed. The recession of the muscle will decrease its original function, yet it will increase the opposite action. For instance, a swollen medial rectus muscle leads to an abduction deficiency. Through recession this muscle's ability to abduct will increase, but its opposite function, adduction, will be diminished and although the horizontal duction range remains unchanged, ocular deviation will have shifted [10]. In all cases, resection of the muscle has to be avoided, since this would lower the total duction range and increase the incomitancy.

In severe cases, in which normal recessions (maximum of 5 mm) are not sufficient to correct the large angle of squint, elongation material (such as fascia lata or Tutopatch® implants) can be used to obtain the same effect as a recession without excessive reduction of a muscle's action [11, 12].

In approximately 70% of the GO patients that need surgery to correct their diplopia, one surgical intervention is sufficient to reach a functional field of binocular single vision [13]. However, especially in the more complex cases, a minimum of two subsequent sessions of strabismus surgery is often necessary to achieve a comfortable field of binocular single vision, e.g. in primary gaze, downgaze, and side gaze. The patient has to be informed that a full field of binocular single vision cannot be achieved.

We can distinguish a few typical ocular motility schemes in GO patients:

- GO patients with severe esotropia (esotropia of 24°) due to extensive fibrosis of both medial recti muscles: The large esodeviation is due to enlarged medial rectus muscles with limited abduction of both eyes. In the ocular motility screen, one will see the missing secondary gaze directions due to impaired ductions of both eyes (Fig. 6.24; diagram a.). Surgical plan: Recession of both medial rectus muscles is the surgery of choice [10]. Diagram 6.24b. shows the post operative result with significant improvement of the field of BSV Fig 6.24c and d.

- GO patients with a large concomitant vertical deviation (left hypertropia of 25°) due to extensive fibrosis of the inferior rectus muscle of the right eye with limited elevation and fibrosis of the superior rectus muscle of the left eye with limited depression: There is also a small esodeviation (Fig. 6.25). Surgical plan: Recession of the enlarged inferior rectus of the right eye and the superior rectus of the left eye. Diagram a of Fig. 6.25 shows the preoperative ocular deviation, diagram b shows the postoperative result.

- GO patients with abnormal head posture (chin up) due to an incomitant vertical deviation caused by asymmetric fibrotic inferior rectus muscles (Fig. 6.26): The left inferior rectus muscle is more severely affected than the right. In primary position, the patient has a left hypotropia of 9° and in downgaze no deviation. This patient perceives single vision in downgaze. Surgical plan: Asymmetrical inferior rectus muscle recession with more recession of the vertically lowered eye (left). When recession of both inferior rectus muscles is considered, one has to be aware of an increase of in cyclotorsion [14].

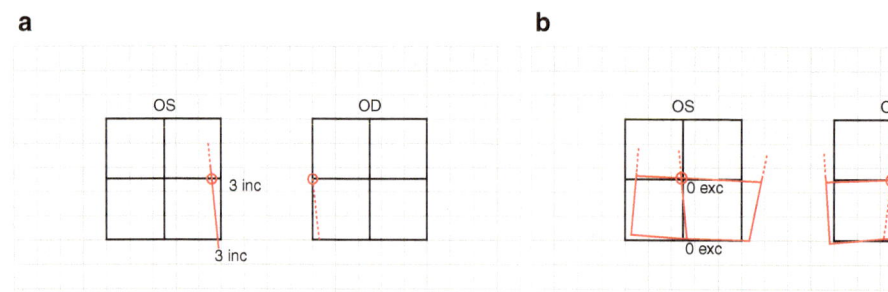

a   **b**

Fig. 6.24 Amsterdam motility diagram preoperative (**a**) and postoperative (**b**) recession of the medial rectus muscles. Field of binocular single vision before (**c**) and after

(**d**) strabismus surgery. *OS* left eye, *OD* right eye, *inc* incyclotorsion, *exc* excyclotorsion

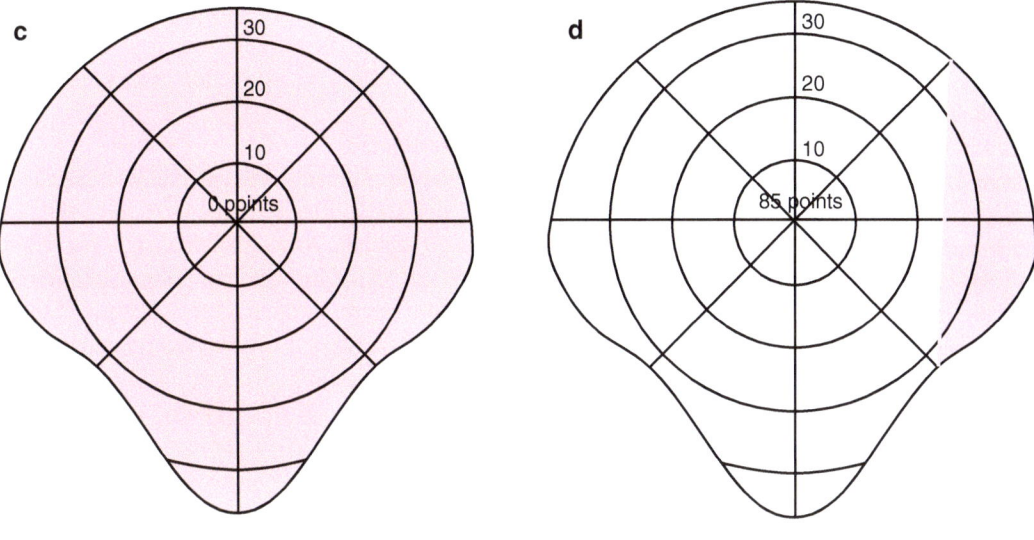

Fig. 6.24 (continued)

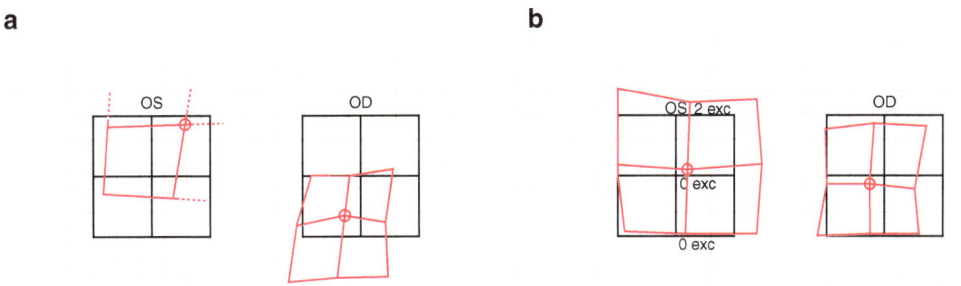

Fig. 6.25 Amsterdam motility diagram preoperative (**a**) and postoperative (**b**) inferior recession of the right eye and superior recession of the left eye. *OS* left eye, *OD* right eye, *inc* incyclotorsion, *exc* excyclotorsion

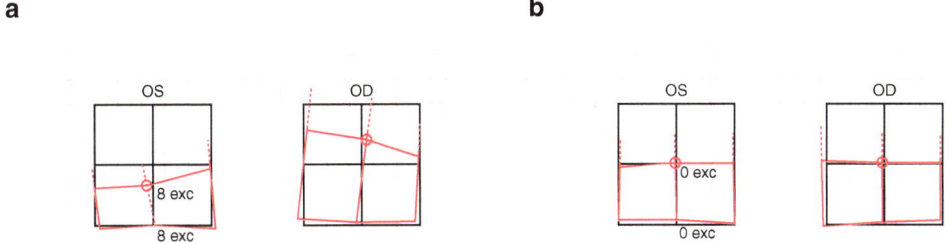

Fig. 6.26 Amsterdam motility diagram preoperative (**a**) and postoperative (**b**) asymmetrical rectus inferior recession of both eyes. *OS* left eye, *OD* right eye, *inc* incyclotorsion, *exc* excyclotorsion

Treatment of Strabismus and Diplopia in Orbital Fractures

Diplopia treatment of patients with an orbital fracture differs in several ways from patients suffering from GO. A major cause of diplopia in case of orbital trauma is edema within the orbit or within the extraocular muscles. Edema will eventually disappear. Hence, an initial wait-and-see policy is highly recommended [15, 16]. While waiting for spontaneous recovery, as mentioned earlier in this chapter, ocular eye movement exercises are key.

In case of stable orthoptic measurements, surgery can be considered. However, diplopia treatment has to follow orbital treatment. Any orbital repair can change ocular mechanics and diplopia, and this means that after orbital treatment one must again wait for orthoptic stability before deciding on further treatment of diplopia [17]. When surgical treatment is addressed, only a few options are available. In case of secondary orbital revision, any improvement of the motility should not be expected by removing or revising an implant [18]. Also, no improvement of the ductions is expected when adhesiolysis of the affected tissue is performed.

Mechanically injured, or even paretic, muscles cannot be repaired. To obtain more concomitant ocular movement and, hence, an improved field of binocular single vision, the healthy eye is the one to receive surgical treatment. In general, the overacting muscle of the other eye is recessed and/or a posterior fixation suture is placed 10–12 mm behind the insertion of the muscle (Fig. 6.27). A posterior fixation suture (e.g. Faden operation) does not affect muscle movement in primary position, but it limits an overacting muscle in the direction of its maximum action. The suture acts as a brake on the action of the healthy muscle, once the eye moves past the field of action in which the overaction takes place.

Case Report

A 68-year-old man with a medial wall fracture of the right orbit after head trauma is presented. He has no diplopia. As a child, he underwent strabismus surgery of his left eye. Large exotropia and small hypertropia of the left eye were observed (Fig. 6.28). The left eye showed impaired adduction (Table. 6.3). Visual acuity was 1.6 on the right versus 1.0 on the left, slightly amblyopic eye. Orthoptic diagnosis is a consecutive (preoperative esotropia) exotropia with adduction impairment in the left eye due to strabismus surgery with suppression. The impaired adduction and strabismus are not related to the orbital wall fracture on the right side. The patient renounced surgery as he, because of the lack of diplopia, did not consider his exotropia to be (cosmetically) disturbing.

Treatment of Strabismus Causes by Other Orbital Conditions

There are numerous other orbital conditions that may affect the mechanics of ocular movement or the ocular deviation in primary position. In general, the golden rule in decision making is to reach orthoptic stability before deciding on surgical treatment to alleviate diplopia. It is important to determine the cause of the motility disorder. Treatment can then be adapted according to the principles as mentioned in the section on GO if muscles have become fibrotic. Alternatively, treatment can be adjusted in a manner as men-

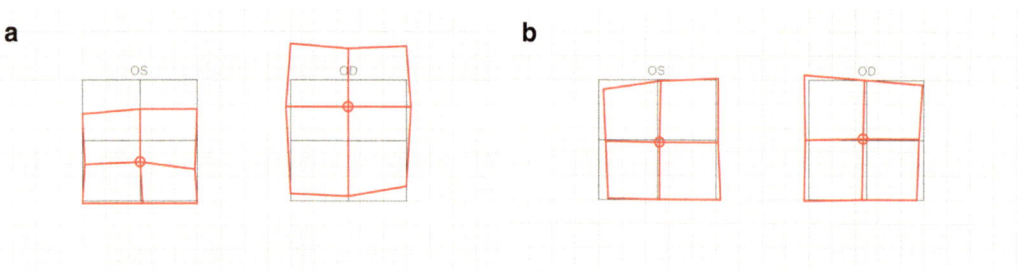

Fig. 6.27 Amsterdam motility diagram before (**a**) and after (**b**) posterior fixation suture of the superior rectus of the right eye

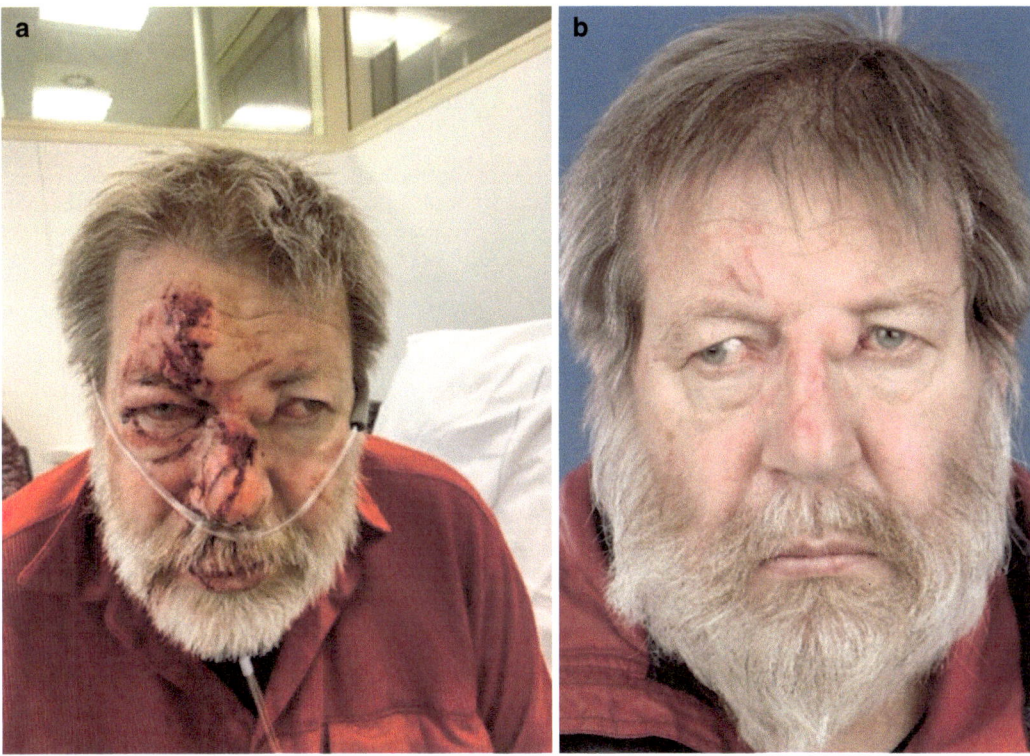

Fig. 6.28 Patient photographed 1 day (**a**) and 3 months (**b**) after trauma showing an alternating exotropia and a limitation of adduction of the left eye

Table 6.3 Ductions of patient shown in Fig. 6.28. *OD* right eye, *OS* left eye

Ductions	OD	OS
Abduction	48	48
Adduction	45	32
Elevation	32	37
Depression	54	60

tioned in the example of orbital fracture. There are numerous types of strabismus surgery that can improve symmetry in ocular movement, of which only a few are mentioned in this chapter.

Conclusion

Orthoptic examination is diverse and comprises of several components. Examination may lead to a different conclusion regarding the cause of diplopia as would be expected by only external impression and/or subjective experience of the patient. Although orthoptic measurement may not be leading in initial therapeutic decisions, it can be very

useful in case of doubt regarding the cause, type or even the absence of diplopia in orbital pathology.

One must be aware of the complexity of binocular single vision and consider possible pre-existing strabismus, complex cyclodeviations, and mechanical abnormalities as well as neurogenic disruption of the oculomotor system once ophthalmological pathologies are eliminated as causative factors.

Orthoptists can play an important supportive role in case of extensive treatment to alleviate binocular complaints. Stability of ocular movement can only be captured through orthoptic follow-up. Orthoptic conclusions are eminent in decision making for strabismus surgery.

In light of all orthoptic measurements and therapeutic options, whether conservative or surgical, one must always remember that objective findings may be miles away from subjective complaints. Double vision may be objectified in a protocolled manner, but it should be emphasized that each patient is different in mindset and his/her meaning of life.

Acknowledgments We thank professor Maarten Mourits, Linda Groenveld, Elly Merckel-Timmer, Sebastiaan de Beer, and Nynke Budding for their contributions.

All recognizable pictures are shown with written consent of the patient/volunteer.

References

1. Head and Neck, IVth nerve. Primal Pictures, Informa PLC. 2014. http://www.anatomy.tv/anatomytv/html5ui_2018/#/product/har_head_2014/release/2018_04/type/Views/id/42948/layer/3/angle/14/structureID/1093824]. Accessed 8 March 2021.
2. Sullivan TJ, Kraft SP, Burack C, O'Reilly C. A functional scoring method for the field of bincular single vision. Ophthalmology. 1992;99:575–81.
3. Jellema HM, Braaksma-Besselink YC, Limpens CEJM, Von Arx G, Wiersinga WM, Mourits MP. Proposal of success criteria voor strabismus surgery in patients with Graves' orbitopathy based on a systematic literature review. Acta Ophthalmol. 2015;93:601–7.
4. Head and Neck, Eye Movement. Primal Picture, Informa PLC. 2014. http://www.anatomy.tv/anatomytv/html5ui_2018/#/product/har_head_2014/release/2018_04/type/Slides/id/9048010/structureID/-1]. Accessed 8 March 2021.
5. Von Noorden GK, Campos EC. Binocular vision and ocular motility. Theory and management of strabismus. 6th ed. St. Louis: Mosby; 2002.
6. Head and Neck, Superior Oblique Muscle. Primal Picture, Informa PLC. 2014. http://www.anatomy.tv/anatomytv/html5ui_2018/#/product/har_head_2014/release/2018_04/type/Views/id/42948/layer/4/angle/4/structureID/1079308]. Accessed 9 March 2021.
7. Mourits MP, Prummel MF, Wiersinga WM, Koornneef L. Measuring eye movements in Graves' ophthalmopathy. Ophthalmology. 1994;101:1341–6.
8. Garrity JA, Saggau DD, Gorman CA, Bartley GB, Fatourechi V, Hardwig PW, Dyer JA. Torsional diplopia after transantral orbital decompression and extraocular muscle surgery associated with Graves' orbitopathy. Am J Ophthalmol. 1992;113:363–73.
9. Haller T, Furr BA. Fresnel prism use among orthoptists. Am Orthopt J. 2014;64:71–5.
10. Jellema HM, Saeed P, Braaksma-Besselink Y, Schuit A, Kloos R, Mourits MP. Unilateral and bilateral medial rectus recession in Graves' orbitopathy patients. Strabismus. 2014;22:182–7.
11. Esser J, Schittkowski M, Eckstein AK. Graves' orbitopathy: inferior rectus tendon elongation for large vertical squint angles that cannot be corrected with simple muscle recession. Klin Monatsbl Augenheilkd. 2011;228:880–6.
12. Oeverhaus M, Fischer M, Hirche H, Schlütter A, Esser J, Eckstein AK. Tendon elongation with bovine pericardium in patients with severe esotropia after decompression in Graves' orbitopathy: efficacy and long-term stability. Strabismus. 2018;26:62–70.
13. Jellema HM, Saeed P, Mombaerts I, Dolman PJ, Garrity JA, Kazim M, Dhrami-Gavazi E, Lyons C, Nieuwkerk P, Mourits MP. Objective and subjective outcomes of strabismus surgery in Graves' orbitopathy: a prospective multicenter study. Acta Ophthalmol. 2017;95:386–91.
14. Jellema HM, Saeed P, Everhard-Halm Y, Prick L, Mourits MP. Bilateral inferior rectus muscle recession in patients with Graves' orbitopathy: is it effective? Ophthal Plast Reconstr Surg. 2012;28:268–72.
15. Jansen J, Dubois L, Maal TJJ, Mourits MP, Jellema HM, Neomagus P, De Lange J, Hartman LJC, Gooris PJJ, Becking AG. A non-surgical approach with repeated orthoptic evaluation is justified for most blow-out fractures. J Craniomaxillofac Surg. 2020;48:560–8.
16. Walonker AF. Diagnostic evaluation of traumatic cranial nerve palsies. Am Orthopt J. 2004;54:57–61.
17. Silbert DI, Matta NS, Singman EL. Diplopia secondary to orbital surgery. Am Orthopt J. 2012;62:22–8.
18. Gosse EM, Ferguson AW, Gilmour C, MacEwen CJ. Blow-out fractures: patterns of ocular motility and effect of surgical repair. Brit J Oral Maxillofac Surg. 2010;48:40–3.

Open Access This chapter is licensed under the terms of the Creative Commons Attribution 4.0 International License (http://creativecommons.org/licenses/by/4.0/), which permits use, sharing, adaptation, distribution and reproduction in any medium or format, as long as you give appropriate credit to the original author(s) and the source, provide a link to the Creative Commons license and indicate if changes were made.

The images or other third party material in this chapter are included in the chapter's Creative Commons license, unless indicated otherwise in a credit line to the material. If material is not included in the chapter's Creative Commons license and your intended use is not permitted by statutory regulation or exceeds the permitted use, you will need to obtain permission directly from the copyright holder.

Part IV

Positional Aspects of the Globe

Maarten P. Mourits

Learning Objectives
- Exophthalmos and enophthalmos are cardinal features of orbital pathology.
- The most common cause of both bilateral and unilateral exophthalmos is Graves' orbitopathy, whereas an orbital fracture is the most common cause of unilateral enophthalmos.
- Exophthalmos and enophthalmos are clinically assessed with an exophthalmometer.

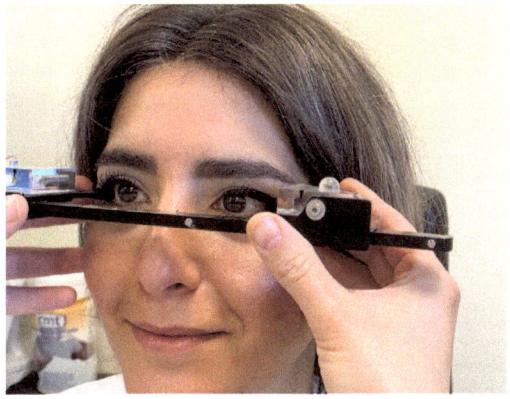

Fig. 7.1 Measuring AGP with a Hertel exophthalmometer

Introduction: The (Axial) Globe Position

The Axial Globe Position (AGP) is defined as the distance between the apex of the cornea and the lateral rim of the orbit in the sagittal plane and it is measured with an instrument called an exophthalmometer (Fig. 7.1). One of the first exophthalmometers was designed by the German ophthalmologist Ernst Hertel (1870–1943) [1], who became so famous that the term 'Hertel value' has become a synonym for AGP.

AGP differs between ethnic groups and between females and males, but within these groups it shows a normal distribution. In Caucasians, AGP varies between 10 and 17 mm in women and between 10 and 20 mm in men. African people tend to have a slightly higher, Asian people a slightly lower AGP. A difference of up to 2 mm in AGP between both globes is generally considered normal. Thus, a difference between both eyes of >2 mm should raise suspicion for a pathological process. Generally, an AGP less than 10 mm is called enophthalmos; AGP >20 mm is called exophthalmos.

The original version of the chapter has been revised. A correction to this chapter can be found at https://doi.org/10.1007/978-3-031-40697-3_23

M. P. Mourits (✉)
Amsterdam University Medical Centers, Location AMC, Amsterdam, The Netherlands

We measured the AGP in 160 healthy Caucasian men and women and compared the results with those of a group of patients with Graves' orbitopathy (GO), and assessed the accuracy of the measurements [2]. We concluded that:

1. Exophthalmometry using a Hertel exophthalmometer is reliable and reproducible.
2. AGP shows a normal distribution in both healthy individuals and patients with GO and is gender dependent.
3. Regression analysis revealed an upper limit of normalcy of 16 mm in females and 20 mm in males.

AGP distribution has been determined in different ethnic groups [3–6]. AGP can also be assessed—albeit less accurately—using digital photography, CT scans, or MRI scans [7]. Conventional exophthalmometers have a number of disadvantages. To overcome these, we developed a parallax-free exophthalmometer (Fig. 7.2) based on the concepts of Davanger [8, 9].

For patients, in whom the lateral wall is missing, i.e. in orbital fractures or after surgery, Naugle developed a superior and inferior orbital rim-based exophthalmometer [10]. Exophthalmos is also called proptosis or (eye)protrusion. Pseudoproptosis refers to a situation that seems to indicate proptosis, which cannot be confirmed by AGP measurement. This is often seen in patients with unilateral upper eyelid retraction (Fig. 7.3). Therefore, AGP should always be measured.

Besides forward displacement, the globe can be pushed downward (hypoglobus), upward

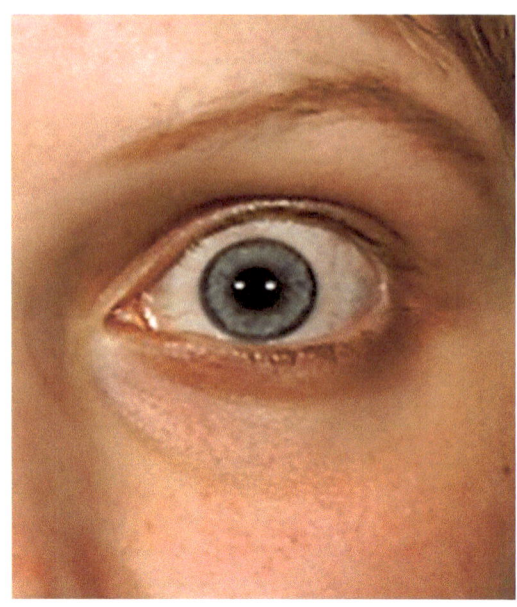

Fig. 7.3 Patient with upper lid retraction, but normal AGP (pseudoproptosis)

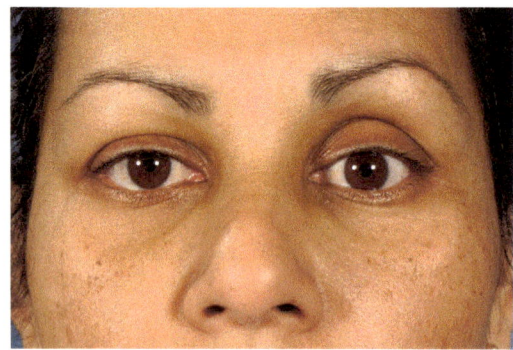

Fig. 7.4 Patient with deep superior sulcus of the left orbit

(hyperglobus), sideward, or any combination. Hypoglobus and enophthalmos are typically seen in orbital floor fractures. A beginning negative discrepancy between the volume of the orbital cavity and the volume of the orbital soft tissues shows by a deep superior sulcus, i.e. the space between the upper lid and superior orbital margin (Fig. 7.4).

The direction of globe displacement gives an indication about the localization of the orbital disease process. For instance, downward displacement suggests an orbital floor fracture or a mucocele of the frontal sinus with a perforated orbital roof, whereas a medial plus downward displacement is

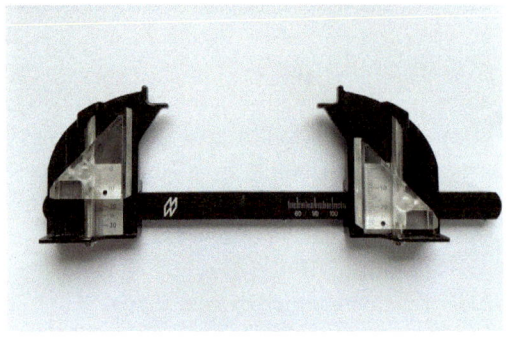

Fig. 7.2 Parallax-free exophthalmometer according to Mourits

indicative for a disease process in the lacrimal gland fossa. Upward displacement with or without proptosis can be seen in diseases of the maxillary sinus, such as shown in Figs. 7.5 and 7.6. Retrobulbar lesions tend to push the globe in a forward direction (Fig.7.7), but often cause other symptoms, such as reduced color vision or a Relative Afferent Pupil Defect (RAPD) in an early stage of the disease due to compression of the optic nerve.

For a correct diagnosis of disorders with exophthalmos and enophthalmos, the following variables are of cardinal importance:

1. Age.
2. Gender.
3. Onset and duration of the complaints.

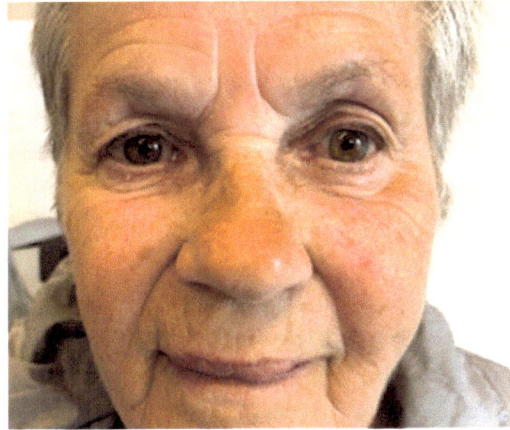

Figs. 7.5 Female patient with a cystic lesion of the left maxillary sinus resulting in orbital floor elevation. (Courtesy of Dr. P. Gooris)

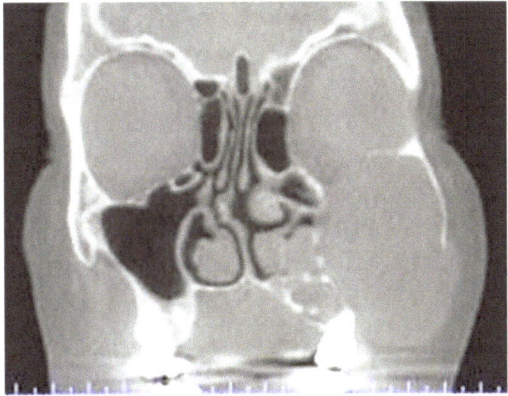

Fig. 7.6 CT scan of the same patient as shown in Fig. 7.5 (Courtesy of Dr. P. Gooris)

4. Eyelid and conjunctiva abnormalities: swelling, redness, ptosis, retraction.
5. Eye motility.
6. Visual acuity, color vision, pupil reaction
7. MRI scans and/or CT scans.
8. Blood tests.

Some phenomena are strongly connected to a particular disease. Thus, an increase of exophthalmos with the Valsalva maneuver suggests an orbital varix and enlarged corkscrew epibulbar vessels are indicative for an arteriovenous fistula.

Diseases with Exophthalmos

The list of diseases and disorders causing exophthalmos is almost infinite. Here, we will shortly discuss the most frequent ones. By far, GO heads the list. It is estimated that >50% of all patients with proptosis suffer from GO [11]. GO will be discussed in Part 6.

Orbital Meningioma

Orbital meningioma is a relatively frequent orbital tumor, with an estimated incidence of approximately sixty cases per year in the Netherlands [12]. The most common presentation is that of an extension into the orbit from an intracranial meningioma located on the outer ridge of the sphenoid bone (Figs. 7.7 and 7.8), hence its name spheno-orbital meningioma (SOM). SOM typically causes hyperostotic changes of the sphenoid wing. These changes result in a reduction of orbital cavity volume and push the globe forward.

Less frequent are meningiomas that arise from the meninges (arachnoid mater) surrounding the optic nerve, the so-called optic nerve sheath meningioma (ONSM, (Fig. 7.9). Because of their localization and direct compression of the optic nerve, they cause visual impairment (reduced (color) vision, reduction of the visual field, RAPD) rather than proptosis in the early stages. Fundoscopy may reveal opticociliary shunt ves-

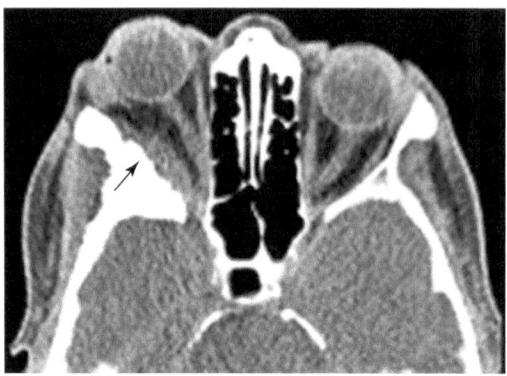

Fig. 7.7 Exophthalmos and hyperostosis (arrow) secondary to intracranial meningioma (which itself is not visible on this CTscan)

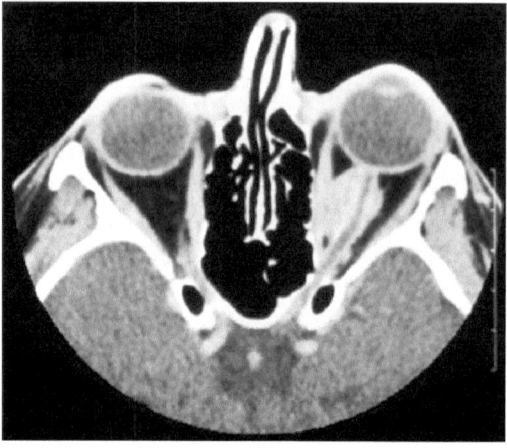

Fig. 7.9 Optic nerve sheath meningioma in the right orbit. Notice the sparing of the optic nerve, a sign that is called the tram-track phenomenon which is present in circa 25% of patients with this kind of tumor

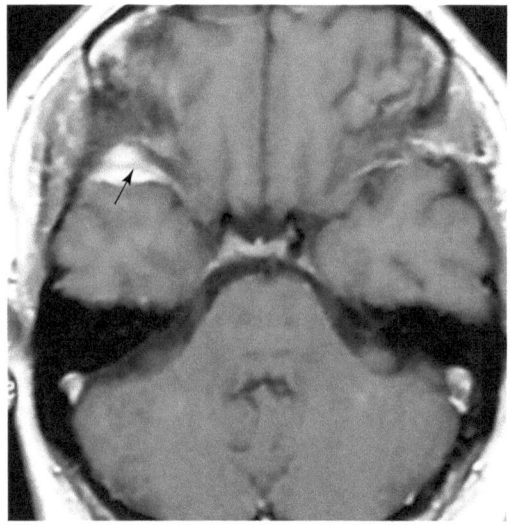

Fig. 7.8 Gadolinium-enhanced MRIscan showing intracranial meningioma (arrow)

sels, together with swelling or pallor (in a later stage) of the optic nerve head.

Orbital meningiomas are usually benign, well-vascularized, slow-growing tumors, that affect—at least in Caucasians—women more often than men. They arise from meningothelial cap cells of the arachnoid villi. They typically become apparent in the fourth to sixth decades. The assumption that hormonal changes play a role in the etiology of meningiomas has never been corroborated. Cranial irradiation, however, is a risk factor for the development of meningiomas.

The diagnosis of SOM is usually straightforward using CT scans and gadolinium-enhanced MRI scans. The diagnosis of ONSM may be challenging, when no clear tram-track phenomenon (Fig. 7.9.) is seen. With a sensitivity of 100% and a specificity of 97% at a threshold uptake ratio of 5.9, somatostatin receptor scintigraphy is a useful additional tool in diagnosing ONSM [13]. The differential diagnosis of ONSM includes: metastasis, lymphoma, sarcoidosis, and other inflammatory diseases.

Meningiomas are managed by careful observation, radiotherapy, or surgery. SOMs with significant morbidity can be reduced in size by a multidisciplinary team consisting of a neurosurgeon and an orbital surgeon via pterional, frontotemporal, or cranio-orbital approaches.

Optic Pathway Glioma (Pilocytic Astrocytoma)

Optic nerve gliomas or optic pathway gliomas (OPG; Fig. 7.10) occur along the entire length of the pre-cortical optic pathways and were usually considered hamartomas, but are now classified as pilocytic astrocytoma's. They arise frequently in association with neurofibromatosis type 1 and then have a better visual prognosis than solitary

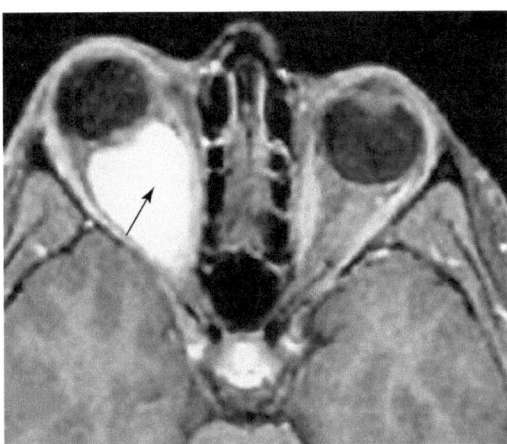

Fig. 7.10 T1-weighted MRI scan showing an optic nerve glioma in a 6-year-old child (arrow)

OPGs [14]. They mostly affect children younger than 10 years of age and 60% of them will eventually have a visual acuity of finger counting or less. Exophthalmos is present in 95% of patients. OPGs will rarely progress to the optic chiasm or the contralateral side. Therefore, observation alone will often be sufficient. Disfiguring proptosis or orbital pain can be a reason for surgical excision. In advanced cases, chemotherapy with vincristine and carboplatin is considered treatment of first choice.

Orbital Cavernous Hemangioma (Orbital Cavernous Venous Malformation)

Orbital cavernous hemangioma (OCH) is a well-encapsulated vascular hamartoma with a spongy structure made up of interconnected vascular channels filled with blood. Cavernous hemangiomas can occur everywhere in the body, but are relatively frequent in the orbit in adults. They sometimes do not cause any complaint and are accidently found in individuals undergoing a screening total body scan. On CT scans, they appear as a well-defined, round to oval homogeneous mass; on T1-weighted MRI scans they produce a homogeneous signal iso-intense to muscle. On T2-weighted scans they are hyperintense to fat and brain.

We described two presentations (Fig. 7.11): [1] the apple-shaped anterior/mid-intraconal, and [2] the pear-shaped apical intraconal type [15]. The former causes a painless exophthalmos and can be easily 'delivered' with some traction of a cryoprobe via a lateral or anterior orbitotomy, as it lies freely in the orbit. The apically located OCH, even if its size is modest, causes reduction of visual acuity and visual field narrowing in the early stages of the disease, because of compression of the optic nerve. A RAPD is usually present. The capsule of this type fuses with adjacent vessels and nerves. Total removal is impeded by poor visualization, lack of maneuvering space, excessive bleeding and risk of damaging of the oculomotor nerves and the ciliary ganglion. Partial removal with or without bipolar cautery shrinkage has been suggested as an alternative procedure, as well as radiotherapy [16, 17].

Orbital Capillary or Infantile Hemangioma

Capillary or infantile hemangiomas are the most common benign vascular orbital neoplasms of childhood, most often located in the superior and anterior orbit or eyelid, and seen more often in girls. On CT scans they appear as a well-defined or infiltrating mass; on T1- and T2-weighted MRI scans they show a heterogeneous signal, hyperintense to muscle and hypo-intense to fat [18]. They typically enlarge during the 6th–12th month and then spontaneously regress until the end of the seventh year of life, when they often have disappeared almost completely. They, therefore, do not need treatment as long as there are no complications. If treatment is required, oral propranolol or topical timolol can be given. Prednisone is also effective, but can cause significant body growth retardation, even if given locally.

Orbital Rhabdomyosarcoma

Rhabdomyosarcoma is the most frequent orbital malignant neoplasm of the orbit in children, with a peak incidence at the age of 7 years and account-

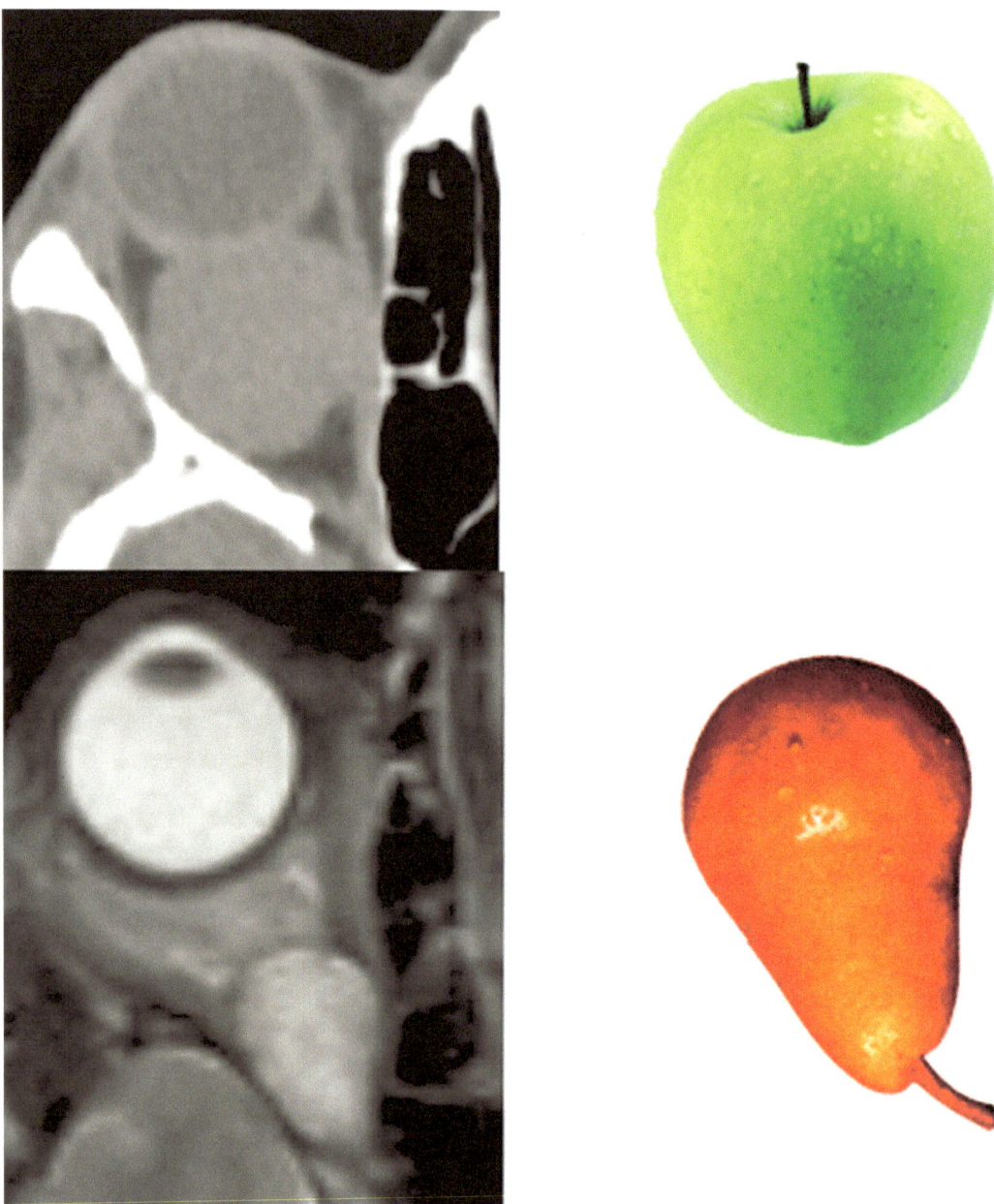

Fig. 7.11 CT scan and T2-weighted MRI scan showing apple- and pear-shaped orbital cavernous hemangiomas

ing for approximately 4–6 new cases per year in the Netherlands [19]. It grows so rapidly that it may mimic an orbital cellulitis or an insect bite (Fig. 7.12). CT scans show a moderately well-defined soft tissue mass. On MRI scans, it is a rather homogeneous irregular mass, more or less iso-intense to muscle tissue. The prognosis has improved over the last sixty years with a survival rate of 30% (after exenteration) around 1960 and a survival rate of 95%, at present, after treatment with vincristine and cyclophosphamide, with or without radiotherapy [20].

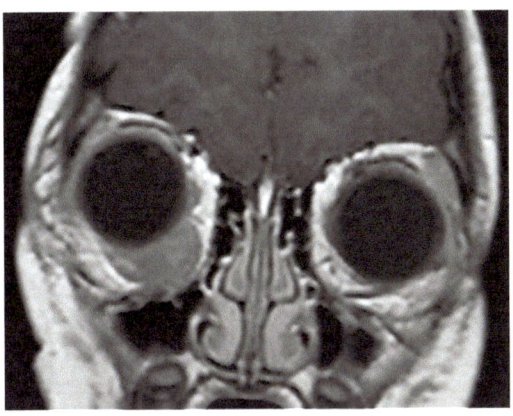

Fig. 7.12 MRI scan of rhabdomyosacoma in the lower medial quadrant of the right orbit in young female

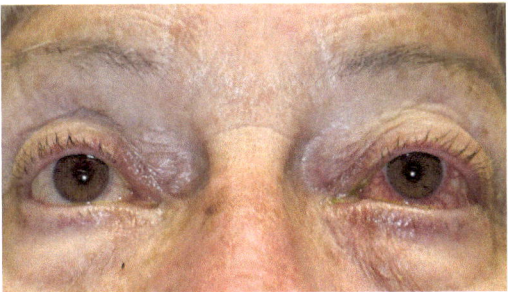

Fig. 7.13 Dilated epibulbar vessels and proptosis in a patient with a traumatic caroticocavernous fistula

Arteriovenous Shunts and Fistulas

A direct connection in the carotid-cavernous area between the arterial and venous systems can cause a pulsating exophthalmos and enlarged epibulbar 'corkscrew' vessels due to a reversal of blood flow in the orbit (Fig. 7.13). Because of an elevated outflow resistance, the intraocular pressure increases and the patient may complain of pain, nausea, and reduced vision. In these 'high-flow' lesions, a pulsating sound (bruit) synchronous to the heart beat may be heard over the orbit or skull. Arteriovenous (AV) fistulas as a result of head trauma arise in a few hours to days.

Slowly evolving proptosis as a result of 'spontaneous' dural AV fistula is seen in older patients with vasculopathy (DM, hypertension). Dural AV fistulas are frequently associated with a dilated superior ophthalmic vein or with supe-rior ophthalmic vein thrombosis [21]. The extraocular muscles are often swollen due to congestion, which may be mistaken for GO. AV shunts are best shown with arteriography. Up to 50% of them disappears spontaneously. As long as they exist, ocular hypertension is treated in the same way as glaucoma. When they do not close or cause significant complications, embolization by an interventional radiologist can be considered.

Orbital Non-Hodgkin Lymphoma

An insidious development of a painless exophthalmos, with a lesion most frequently anteriorly located in the orbit or on the globe, with little or no functional interference, is suggestive for an orbital non-Hodgkin lymphoma NHL [11]. On the globe it appears as a salmon-colored patch (Fig. 7.14). Orbital NHL is mostly seen in patients above 60 years of age, but it is not exceptional in younger patients. Orbital NHLs occur both unilateral and bilateral and compete with sarcoidosis for the second largest cause of bilateral exophthalmos after GO.

Neither CT scans nor MRI scans are conclusive for the diagnosis of NHL. Therefore, a tissue biopsy is always required. The next step is then staging of the disease. Almost 60% of orbital NHLs belong to the extranodal marginal zone B-cell lymphomas of the mucosa-associated lymphoid tissue (MALT) type [22]. These are usually relatively benign in nature. Treatment consists of

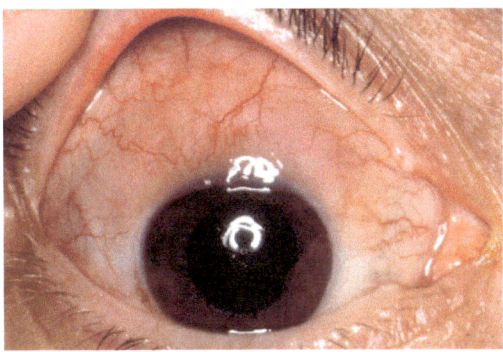

Fig. 7.14 Salmon-colored patch on the globe, suggestive for a non-Hodgkin lymphoma

low-dose radiotherapy, monoclonal antibodies directed to B-lymphocytes (rituximab), or chemotherapy in disseminated forms.

Lacrimal Fossa Lesions

Lesions in the lacrimal fossa tend to push the globe forward, downward, and in the medial direction. In addition, they often cause an S-shaped swelling of the upper eyelid. The most frequent lesions in this area are: dacryoadenitis, lacrimal gland pleiomorphic adenoma, and adenocystic carcinoma, but the lacrimal fossa also may contain NHLs as well as idiopathic orbital inflammatory lesions and many other lesions. The lacrimal gland fossa appears to be a true Pandora's box [23–25].

Dacryoadenitis

Dacryoadenitis shows as an enlarged, hyperdense lacrimal gland with well-demarcated borders on CT scans. The inflammation can be caused by bacteria and viruses, but mostly remains idiopathic. A trial of broad-spectrum antibiotics is recommended. Enlargement of the lacrimal gland is also seen in sarcoidosis.

Pleiomorphic Adenoma

Pleiomorphic adenoma of the lacrimal gland (Figs. 7.15 and 7.16) is a benign lesion, usually of the posterior lobe of the lacrimal gland. It occurs at all ages, but most frequently in adults around the age of 40 years, slightly more in males. On scans, it appears as a well-

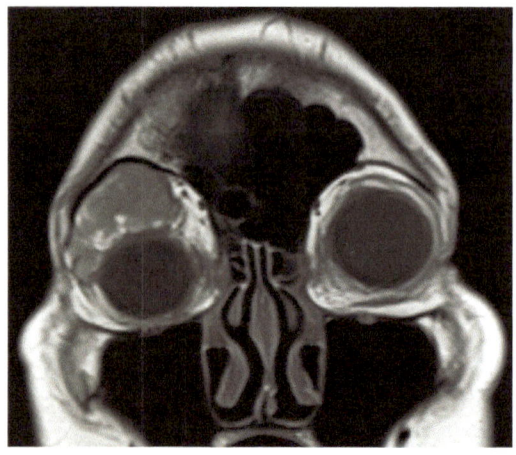

Fig. 7.16 Coronal T1 MRI scan of the same patient showing enlargement of the lacrimal gland

circumscribed round mass in the superolateral quadrant of the orbit. Pressure on the bony orbit may cause indentation, while compression on the globe may cause deformation and choroidal folds. Complete resection including the capsule is required, because recurrence or malignant transformation can otherwise pose significant therapeutical challenges. Consequently, tissue biopsy before removal is contraindicated.

Adenoid Cystic Carcinoma

Adenoid cystic carcinoma is a rarely encountered lesion, despite the fact that it is the most frequent malignant lacrimal gland tumor. It is feared for its bad prognosis. It appears most frequently in the fourth decade. CT scans show bony destruction as a result of its infiltrative growth pattern (Fig. 7.17).

Sarcoidosis

Sarcoidosis is a disease of unknown etiology that may affect multiple organs in the body. In the orbital region, it is most frequently seen in the lacrimal gland. Sarcoidosis is commonly seen in people of African or Scandinavian descent. Similar to GO and orbital NHL, it may present with unilateral or bilateral exophthalmos (Figs. 7.18 and 7.19). Sarcoidosis around the optic nerve causes visual impairment and may be

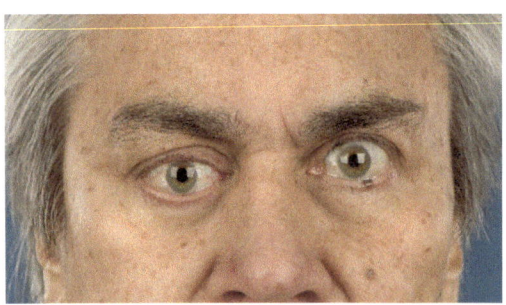

Fig. 7.15 Adult with a pleiomorphic adenoma of the right lacrimal gland

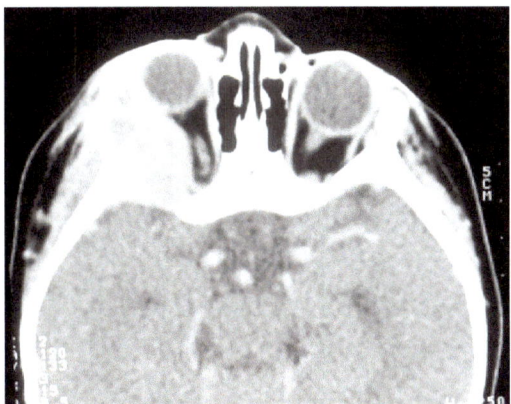

Fig. 7.17 CT scan showing bony destruction in a patient with an adenoid cystic carcinoma of the lacrimal gland

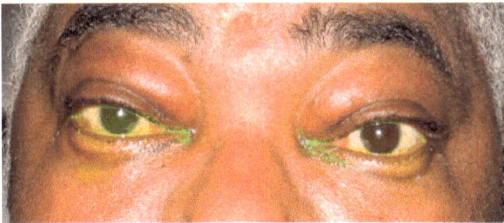

Fig. 7.18 64-year-old patient with bilateral orbital sarcoidosis

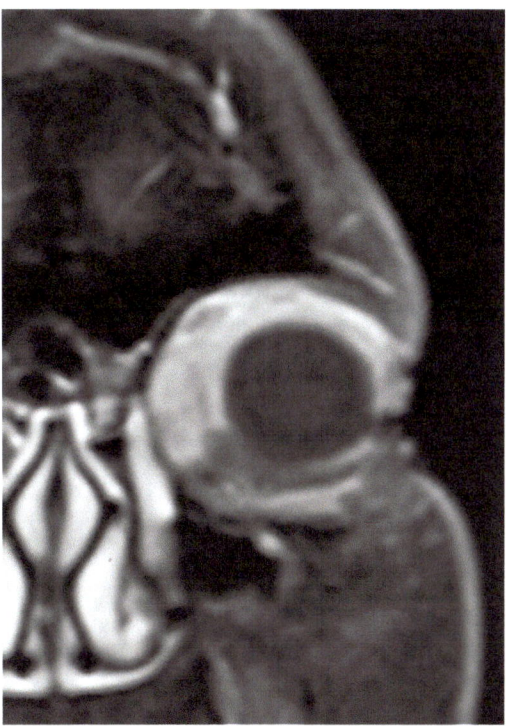

Fig. 7.19 T1 MRIscan of the same patient as shown in Fig. 7.18, showing an indistinct mass around the globe

confused with other forms of optic neuritis. Sarcoidosis can also occur within the globe (uveitis). The diagnosis (increased serum levels of lysozyme and angiotensin-converting enzyme (ACE), gallium scan, biopsy) can be difficult. Treatment consists of prednisone.

Orbital Noninfectious Inflammatory Diseases

To the more common noninfectious inflammatory diseases of the orbit next to sarcoidosis belong granulomatosis with polyangiitis (GPA; formerly called Wegener's granulomatosis), nonspecific idiopathic orbital inflammations (formerly called orbital pseudotumor), and IgG4 lesions.

GPA can be a life-threatening disease, affecting many organs including the upper and lower respiratory tract and the kidneys (glomerulonephritis). GPA is characterized by vasculitis, gran-

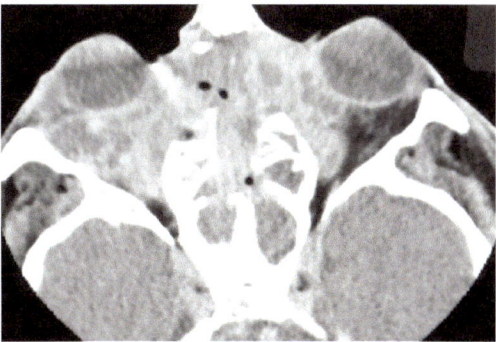

Fig. 7.20 Destructive granulomatosis with poly-angiitis of orbits and nose in a 52-year-old female

ulomatous inflammation, and tissue necrosis. The course is chronic with exacerbations and remissions. Involved orbits show proptosis and visual and motility impairment. CT scans show a destructive mass (Fig. 7.20). High c-ANCA (anti-neutrophil cytoplasmic antibodies) titers are highly specific for GPA. Interestingly, gingival hyperplasia can be one of the early signs of GPA [26]. Classical treatment options are (combinations of) bactrimel, prednisone, azathioprine (Imuran), and cyclophosphamide (Endoxan). More recent treatment options are: mycophenolate mofetil (CellCept), daclizumab, and other monoclonal antibodies.

Idiopathic Orbital Inflammation

Idiopathic orbital inflammations (IOI) occur everywhere in the orbit and are relatively common. By definition, these polymorphous lymphoid infiltrates are restricted to the obit only. Apart from exophthalmos and motility impairment leading to diplopia, they are associated with (severe) pain (Fig. 7.21).

A special form is orbital myositis (Figs. 7.21 and 7.22), in which not only the involved extraocular muscle(s) is/are enlarged, but in more than 50% also the muscular insertion tendons. As such the condition can be distinguished from GO. Orbital myositis can 'jump' from one muscle to another and also to the contralateral orbit. Except for those with excessive fibrosis, IOIs usually respond in a few days to oral prednisone. Response to prednisone, however, does not prove the existence of an IOI. Other entities including orbital NHLs may also improve after steroid treatment. Preferably, a biopsy is taken prior to treatment. Recalcitrant

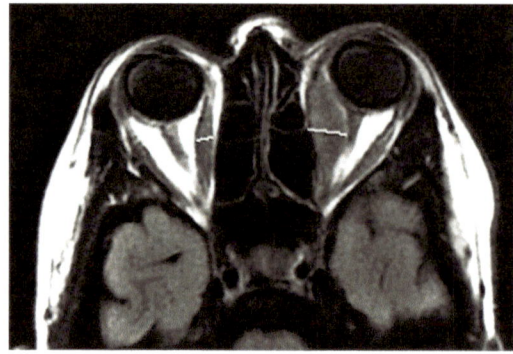

Fig. 7.22 MRI scan of the same patient as shown in Fig. 7.21 with enlarged medial rectus muscle

forms can be treated with monoclonal antibodies such as rituximab.

Fibrous Dysplasia

Fibrous dysplasia is a nonmalignant, nonhereditary developmental disease which leads to the formation of abnormally weak, 'fibrous' bone. It is seen in young adults. The disease manifests itself either as a monostotic form or as a polyostotic form and can be related to endocrine and skin disorders (McCune–Albright syndrome). Involvement of the frontal bone causes facial asymmetry and exophthalmos, and may lead to narrowing of the orbital foramina with compression of vascular and neural structures that run through them. CT scans typically show a 'ground-glass' appearance of the lesion. Surgical treatment requires a multidisciplinary approach.

Neurofibromatosis

Neurofibromatosis type I (NF-1) (formerly called von Recklinghausen disease) is an autosomal hereditary disease that affects 1 in every 3000 births. In 50% it concerns a new mutation. So-called café-au-lait spots on the skin are present from early childhood. A plexiform neurofibroma in the lateral part of an upper eyelid causes a S-shaped swelling. Other manifestations of NF-1 are pulsating exophthalmos due to a defect

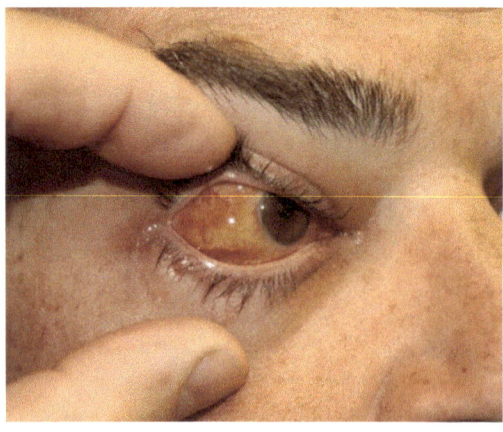

Fig. 7.21 Young adult with orbital myositis, conjunctival swelling and severe pain. All symptoms disappeared in 2 days after treatment with 40 mg prednisone

in de sphenoid bone and neurofibromas of the skin and other parts of the body.

Diseases with Enophthalmos

Orbital Fracture

This is the most common cause of enophthalmos. Orbital fractures are in discussed in Part 5 of this book.

Orbital Varix

Most orbital varices are congenital distensible venous anomalies. In rest they show enophthalmos. Exophthalmos and an unpleasant sensation occur when the vessels of the lesion are filled with blood. This happens, for instance, when the patient bends forward to tie his/her shoelaces or during a Valsalva maneuver. These intermittent swellings of the varix cause atrophy of orbital fat tissue and enophthalmos in rest (Fig. 7.23).

Superficially located varices are seen as dark, dilated vessels of the eyelid or conjunctiva. Combinations of superficial and deeper varices also occur.

Varices are prone to spontaneous bleeding and thrombosis. Increase in size is a typical phenomenon seen in CT scans and MRI scans during the Valsalva maneuver. Management of orbital varices is challenging. Fortunately, most of them do not need treatment.

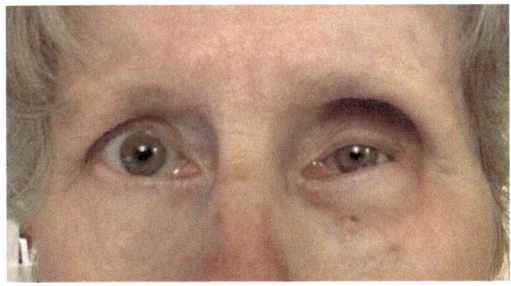

Fig. 7.23 Enophthalmos in a 62-year-old lady with a history of an orbital varix since her early years

Silent Sinus Syndrome

The silent sinus syndrome or imploding antrum syndrome (Figs. 7.24 and 7.25) is a condition in which a chronic negative pressure in the maxillary sinus (due to osteomeatal obstruction) and subsequent reduction in aeration of the antrum causes a downward translation of the orbital floor, together with inward bowing of the medial

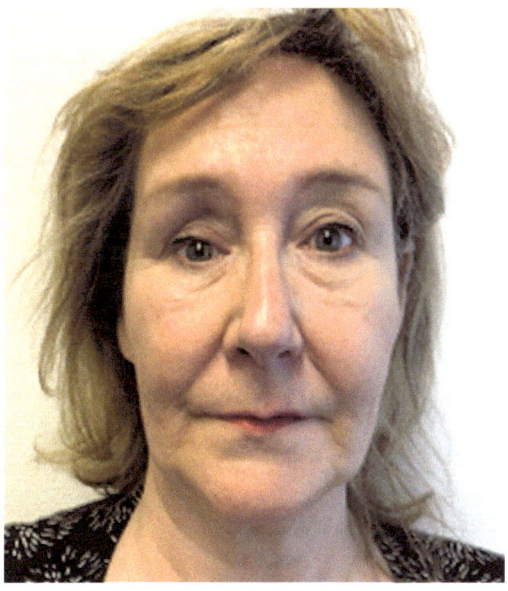

Fig. 7.24 Silent sinus syndrome of the right maxillary sinus. (Courtesy of Dr. P. Gooris)

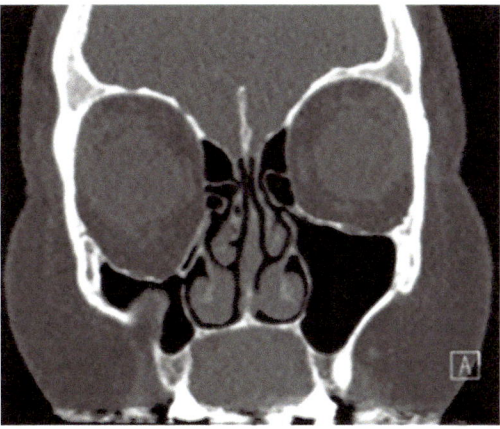

Fig. 7.25 CT scan of the same patient as shown in Fig. 7.24. (Courtesy of Dr. P. Gooris)

and posterolateral walls of the maxillary sinus [27]. These, in turn, cause enophthalmos and hypoglobus. Treatment includes restoration of the normal sinus drainage and reconstruction of the orbital floor.

Metastasis of Mammary Carcinoma

One would think that a metastasis in the orbit would always create proptosis due to the space-occupying nature of the tumor, and in general this is true. However, a scirrhous metastasis of a tubular malignancy of the breast is the exception, as it causes retraction to the globe and impaired motility and, hence, results in enophthalmos (Figs. 7.26). We have seen such presentations as the first sign of mammary carcinoma in women who had a negative mammography [28]. One should realize that the sensitivity of mammography is not higher than 85%. Ophthalmic symptoms are seen in up to 5% of women with otherwise asymptomatic breast cancer.

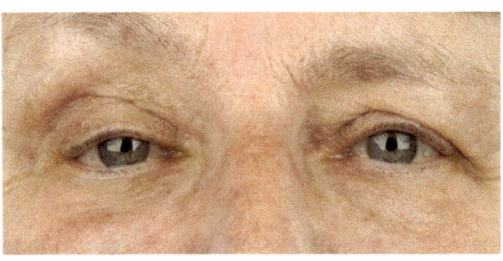

Fig. 7.26 Female with right-sided enophthalmos due to diffuse intraconal metastasis of breast carcinoma

References

1. Wagenmann A. Ernst *Hertel. Graefes*. Arch Ophthalmol. 1943;145:511–3.
2. Mourits MP, Lombardo SH, Van Der Sluijs FA, Fenton S. Reliability of exophthalmos measurement and the exophthalmometry value distribution in a healthy Dutch population and in Graves' patients. An exploratory study. Orbit. 2004;23:161–8.
3. Wu D, Liu X, Wu D, Di X, Guan H, Shan Z, Teng W. Normal values of Hertel exophthalmometry in a Chinese Han population from Shenyang, Northeast China. Sci Rep. 2015;5:8526.
4. Kashkouli MB, Beigi B, Noorani MM, Nojoomi M. Hertel exophthalmometry: reliability and interobserver variation. Orbit. 2003;22:239–45.
5. Gil B, de Montes F, Pérez Resinas FM, Rodríguez García M, González OM. Exophthalmometry in Mexican adults. Rev Investig Clin. 1999;51:341–3.
6. Beden U, Ozarslan Y, Oztürk HE, Sönmez B, Erkan D, Oge I. Exophthalmometry values of Turkish adult population and the effect of age, sex, refractive status, and Hertel base values on Hertel readings. Eur J Ophthalmol. 2008;18:165–71.
7. Pereira TS, Kuniyoshi CH, Leite CA, Gebrim EMMS, Monteiro MLR, Pieroni Gonçalves AC. A comparative study of clinical vs. digital exophthalmometry measurement methods. J Ophthalmol. 2020;2020:article 1397410.
8. Genders SW, Mourits DL, Jasem M, Kloos RJ, Saeed P, Mourits MP. Parallax-free exophthalmometry: a comprehensive review of the literature on clinical exophthalmometry and the introduction of the first parallax-free exophthalmometer. Orbit. 2015;34:23–9.
9. Davanger M. Principles and sources of error in exophthalmometry. A new exophthalmometer. Acta Ophthalmol. 1970;48:625–33.
10. Naugle TC Jr, Couvillion JT. A superior and inferior orbital rim-based exophthalmometer (orbitometer). Ophthalmic Surg. 1992;23(12):836–7.
11. Rootman J. Pathophysiologic patterns of orbital disease. In: Rootman J, editor. Diseases of the orbit: a

multidisciplinary approach. 2nd ed. Philadelphia, PA: Lippincott Williams & Wilkins; 2003. p. 43–52.

12. Claus AB, Bondy ML, Schildkraut JM, Wiemels JL, Wrensch M, Black PM. Epidemiology of intracranial meningioma. Neurosurgery. 2005;57(6):1088–95.

13. Saeed P, Tanck MWR, Freling N, Baldeschi L, Mourits MP, Bennink RJ. Somatostatin receptor scintigraphy for optic nerve sheath meningiomas. Ophthalmology. 2009;116:1581–6.

14. Huang M, Patel J, Patel BC. Optic nerve glioma. In: StatPearls [Internet]. Treasure Island, FL: StatPearls Publishing; 2020.

15. Kloos R, Mourits D, Saeed P, Mourits MP. Orbital apex cavernous hemangioma: beware of the pear. Acta Ophthalmol. 2013;91:328–9.

16. McNab AA, Wright JE. Cavernous haemangiomas of the orbit. Aust New Zeal J Ophthalmol. 1989;17:337–45.

17. Khan AA, Niranjan A, Kano H, Rondziolka D, Flickinger JC, Lunsford LD. Stereotactic radiosurgery for cavernous sinus or orbital hemangiomas. Neurosurgery. 2009;65:914–8.

18. Dutton JJ. Radiology of the orbit and visual pathways. Philadelphia, PA: Saunders Elsevier; 2010.

19. Mourits MP, Koornneef L, Voûte PA. Treatment or orbital rhabdomyosarcoma. Ned Tijdschr Geneeskd. 1985;129:948–51.

20. Jurdy L, Merks JH, Pieters BR, Mourits MP, Kloos RJ, Strackee SD, Saeed P. Orbital rhabdomyosarcomas: A review. Saudi J Ophthalmol. 2013;27:167–75.

21. Van Der Poel NA, De Witt KD, Van Den Berg R, De Win MM, Mourits MP. Impact of superior ophthalmic vein thrombosis: a case series and literature review. Orbit. 2019;38:226–32.

22. Plaisier MB, Sie-Go DMDS, Berendschot TTJM, Petersen EJ, Mourits MP. Ocular adnexal lymphoma classified using the WHO classification: not only histology and stage, but also gender is a predictor of outcome. Orbit. 2007;26:83–8.

23. Fenton S, Slootweg PJ, Dunnebier EA, Mourits MP. Odontogenic myxoma in a 17-month-old child: case report. J Oral Maxillofac Surg. 2003;61:734–6.

24. Hartman LJ, Mourits MP, Canninga-Van Dijk MR. An unusual tumor of the lacrimal gland. Br J Ophthalmol. 2003;87:363.

25. Fenton S, Sie-Go DMDS, Mourits MP. Pleiomorphic adenoma of the lacrimal gland in a teenager, a case report. Eye. 2004;18:77–9.

26. Hanisch M, Fröhlich LF, Kleinheinz J. Gingival hyperplasia as first sign of recurrence of granulomatosis with polyangiitis (Wegener's granulomatosis): case report and review of the literature. BMC Oral Health. 2016;17:33.

27. Hobbs CGL, Saunders MW, Potts MJ. "Imploding antrum" or silent sinus syndrome following nasotracheal intubation. Br J Ophthalmol. 2004;88:974–5.

28. Mourits MP, Saeed P, Kloos JHM. Enophthalmos as first sign of breast cancer. Ned Tijdschr Geneesk. 2015;159:A9114.

Open Access   This chapter is licensed under the terms of the Creative Commons Attribution 4.0 International License (http://creativecommons.org/licenses/by/4.0/), which permits use, sharing, adaptation, distribution and reproduction in any medium or format, as long as you give appropriate credit to the original author(s) and the source, provide a link to the Creative Commons license and indicate if changes were made.

The images or other third party material in this chapter are included in the chapter's Creative Commons license, unless indicated otherwise in a credit line to the material. If material is not included in the chapter's Creative Commons license and your intended use is not permitted by statutory regulation or exceeds the permitted use, you will need to obtain permission directly from the copyright holder.

Ex- and Enophthalmos: Case Reports

8

Peter J. J. Gooris, Gertjan Mensink, and J. Eelco Bergsma

Learning Objectives
- Effect of sinus pathology on change of orbital floor position.
- Absence of direct relationship between slowly progressive change of bony orbital volume and anterior–posterior position of the globe.
- A normal globe position, a degree of enophthalmos or exophthalmos does no inform you in a reliable way of underlying changes of bony orbital volume.

P. J. J. Gooris (✉)
Department of Oral and Maxillofacial Surgery, Amphia Hospital Breda, Breda, The Netherlands

Amsterdam University Medical Centers, Amsterdam, The Netherlands

Department of Oral and Maxillofacial Surgery, University of Washington, Seattle, WA, USA

G. Mensink
Department of Oral and Maxillofacial Surgery, Amphia Hospital Breda, Breda, The Netherlands

University Medical Center Leiden, Leiden, The Netherlands

J. E. Bergsma
Department of Oral and Maxillofacial Surgery, Amphia Hospital Breda, Breda, The Netherlands

Amsterdam University Medical Centers, Amsterdam, The Netherlands

Acta dental school, Amsterdam, The Netherlands

Case Report 1. Enophthalmos

An otherwise healthy 60-year female patient is known with a longstanding persisting sinus pathology. Thereby, she is dissatisfied about the position of her right eye, which is in her perception located downward and backward compared to the opposite site (Figs. 8.1, 8.2, 8.3, 8.4, and 8.5a, b).

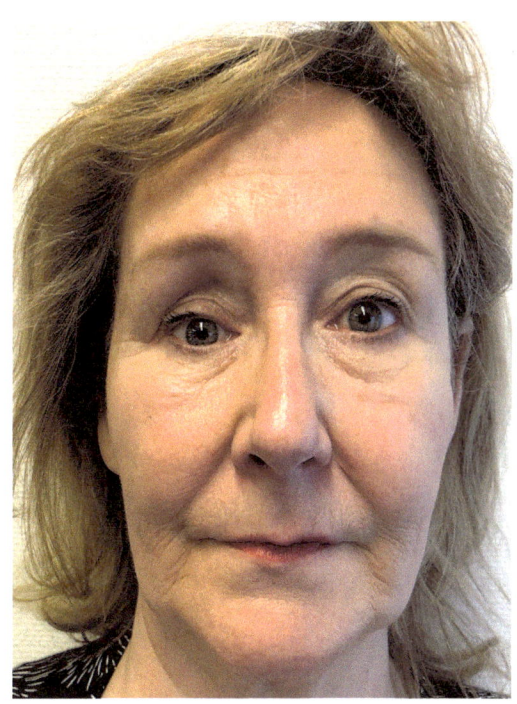

Fig. 8.1 Patient with silent sinus syndrome right antrum: primary gaze

© The Author(s) 2023
P. J. J. Gooris et al. (eds.), *Surgery in and around the Orbit*,
https://doi.org/10.1007/978-3-031-40697-3_8

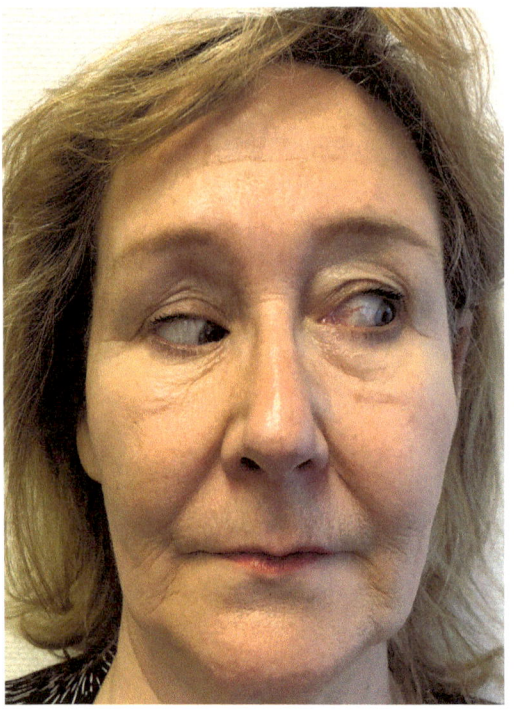

Fig. 8.2 Patient with silent sinus syndrome right antrum: adduction

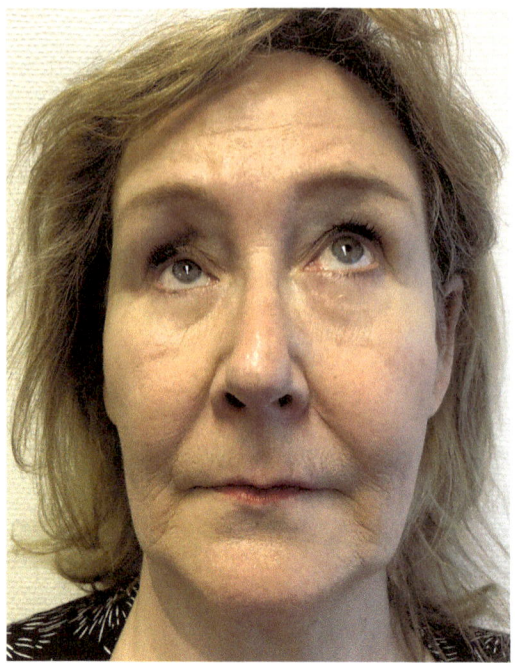

Fig. 8.4 Patient with silent sinus syndrome right antrum: limited upward gaze

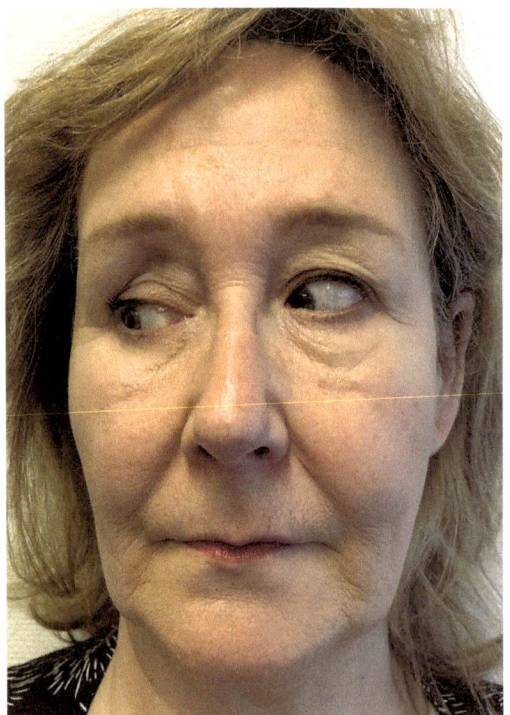

Fig. 8.3 Patient with silent sinus syndrome right antrum: abduction

Because of the extensive and longstanding maxillary sinus pathology, the patient is well known at the ENT Department. Based on the CT's complete obliteration of the infundibulum in combination with a complete sloughing of the remaining maxillary sinus the ENT surgeon suggested a negative sinus pressure which had resulted in a collapse of the maxillary sinus, obvious enlargement of the bony orbital volume as a result of downward displacement of the orbital floor with subsequent hypoglobus and enophthalmos (Figs. 8.6 and 8.7). Based on these findings, a diagnosis of "silent sinus syndrome" was made. This maxillary sinus disease resulted in obvious facial asymmetry.

Clinical examination demonstrated a Hertel of 15/18 mm; also, hypoglobus was confirmed. Because of the deviated position of the left eye in a vertical, horizontal, and A-P direction, subsequent diplopia had developed, most profound in extreme upward gaze. Treatment proposal consisted of a combined ENT & Maxillofacial Surgery approach to correct the negative pressure by a wide infundibulotomy and aeration of the sinus. After the ENT procedure was carried out, aeration of the sinus was established; however, no repositioning of the orbital floor occurred, most likely

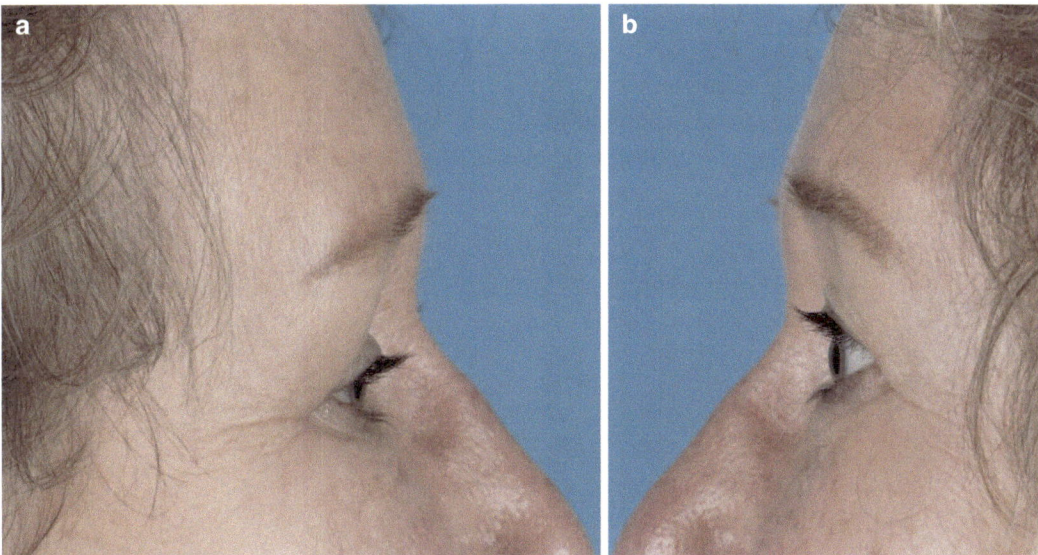

Fig. 8.5 (**a**) Patient with silent sinus syndrome right antrum: lateral view OD. (**b**) Patient with silent sinus syndrome right antrum: lateral view OS

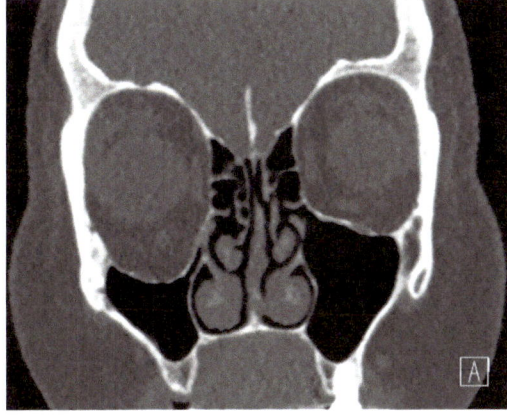

Fig. 8.6 Patient with silent sinus syndrome right antrum: enlarged orbital volume OD, obliteration infundibulum: CT scan: coronal view

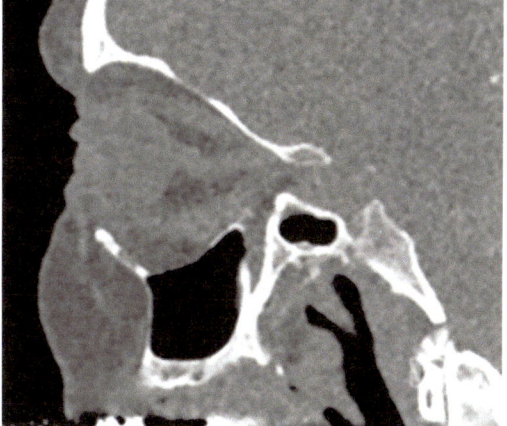

Fig. 8.7 Patient with silent sinus syndrome right antrum: CT scan: sagittal view

due to the already longstanding sinus pathology present. In fact, there was no spontaneous visible effect on the abnormal position of the eye.

Contemplation

As the silent sinus syndrome was present for a long period of time with a significant dislocation of the globe, fibrosis of orbital soft tissue may have occurred. In our opinion, caution is advised not to put too much (acute) tension on the apical orbital soft tissue, i.e., the optic nerve when repositioning the globe as this may cause unfortunate side effects on the neural or neurovascular tissue. We have had a few cases with temporary or persisting loss of vision when repositioning the globe to its former position. In this case, the patient preferred to wait and postpone the surgical repositioning of the globe because of the risks involved of worsening of double vision and/or potential blindness.

Case Report 2. Mild Hyperglobus

This 70-year-old female patient presented with signs and symptoms of recurrent sinusitis and a slowly progressive swelling of the left side of the face.

Clinical examination (Figs. 8.8, 8.9, 8.10, and 8.11) showed a diffuse swelling of the left side of the face including the adjacent lower eyelid; an intact function of NV and NVII was confirmed. There was no diplopia. Hertel measured

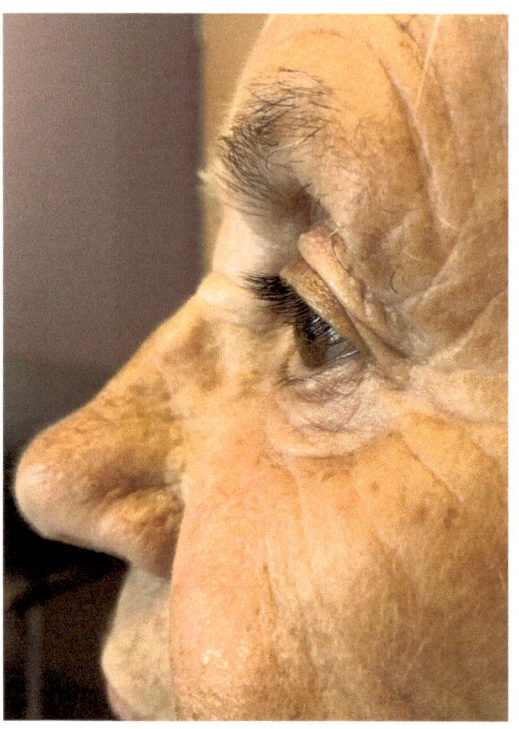

Fig. 8.10 Patient with keratocyst in left antrum: lateral view OS, no exophthalmos present

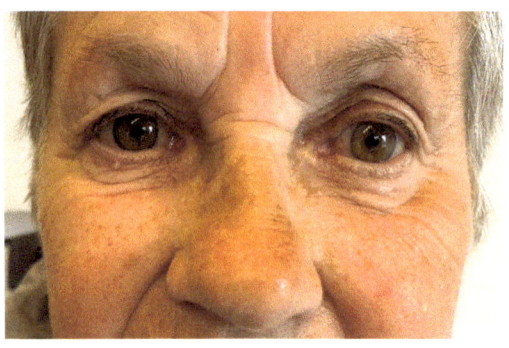

Fig. 8.8 Patient with keratocyst in left antrum: primary gaze, mild hyperglobus OS

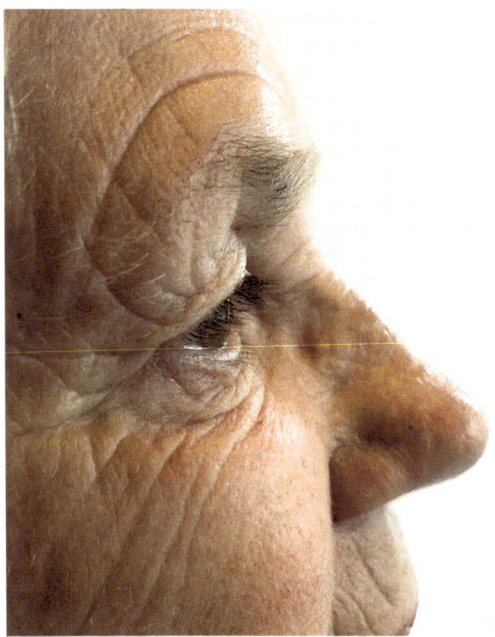

Fig. 8.9 Patient with keratocyst in left antrum: lateral view OD

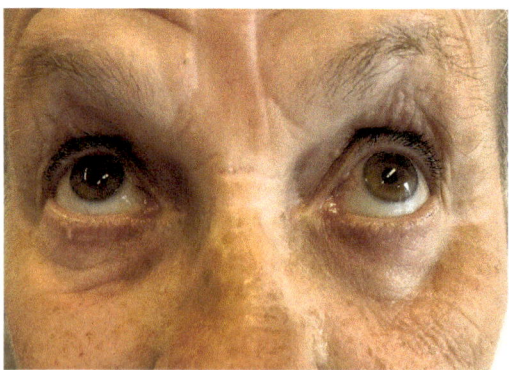

Fig. 8.11 Patient with keratocyst in left antrum: upward gaze, no limitation

15/15 mm. A mild hyperglobus OS was present. Vision was not abnormal.

Intraoral examination showed an elapsed, swollen maxillary vestibule including fluctuation on palpation.

The radiological exam exhibited an extensive cyst-type lesion encompassing the whole left maxillary sinus including orbital floor elevation

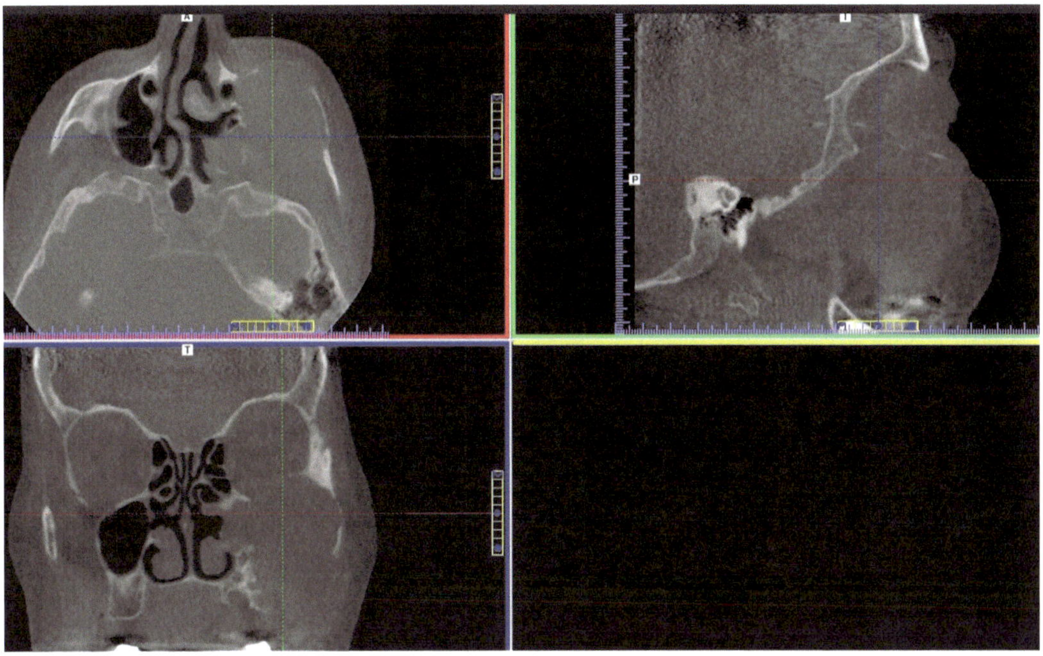

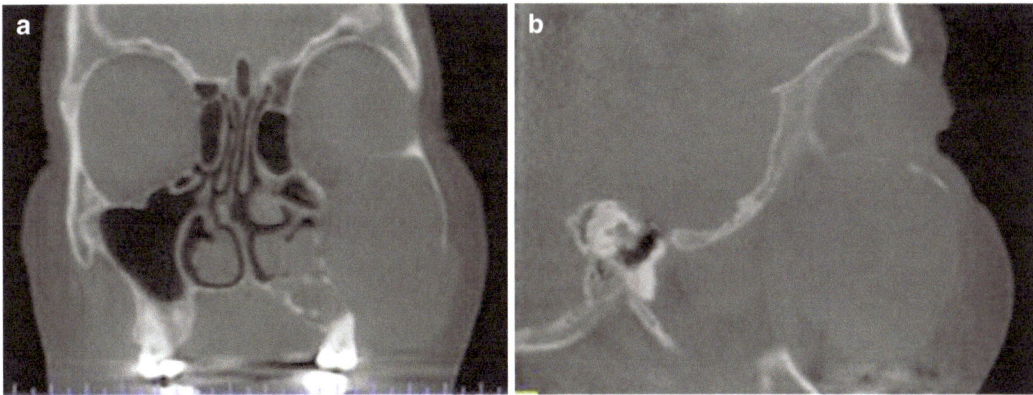

Fig. 8.12 Patient with keratocyst in left antrum: preoperative CT scan: large cystic process with destruction anterior and lateral wall maxilla; elevation orbital floor: corresponding axial, coronal and sagittal view

Fig. 8.13 (**a**) Patient with keratocyst in left antrum: preoperative CT scan: large cystic lesion maxilla, destruction lateral wall maxilla, orbital floor elevation; coronal view. (**b**) Patient with keratocyst in left antrum: preoperative CT scan: Large expanding cystic lesion maxilla with destruction anterior boundary, orbital floor elevation: sagittal view

which resulted in a decrease of the bony orbital volume (Figs. 8.12 and 8.13a, b). On the CT scan, sinus dehiscence due to extensive bony erosion of the anterior and lateral sinus wall as well as partial obliteration of the nasal pathway is diagnosed (Figs. 8.12 and 8.13a). A biopsy was taken: the

diagnosis expanding keratocyst in maxillary sinus with orbital floor elevation as a result of the expanding character of the growing cyst.

Treatment consisted of wide exploration via an intraoral route leaving drains in situ to allow for long-term drainage and carry out thorough

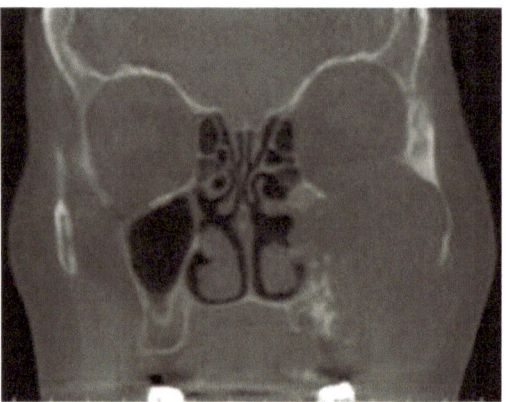

Fig. 8.14 Patient with keratocyst in left antrum: postoperative CT scan: persistent cloudy sinus aspect, persistent orbital floor elevation; normalization nasal pathway: coronal view

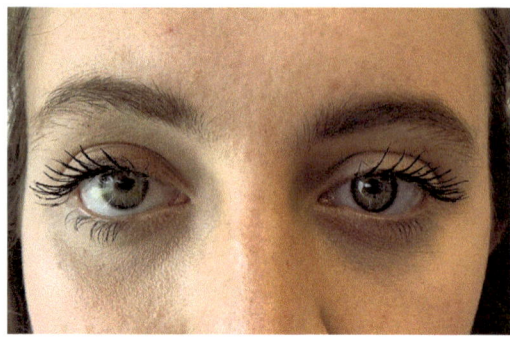

Fig. 8.15 Patient with follicular cyst in right antrum: primary gaze; exophthalmos OD

rinsing of the maxillary cavity. Although the keratocyst treatment was successful so far, cyst excochleation did not result in normalization of the sinus configuration nor of the orbital floor position (Fig. 8.14).

Contemplation

It was interesting to note that, despite the extensive orbital floor elevation and subsequent reduction of orbital bony volume, no proptosis/exophthalmos developed. Apparently, there is sufficient intrinsic capacity of adaptation of orbital soft tissue to cope with slowly progressive bony orbital volume decrease so to minimize exophthalmos to occur. Only a slight hyperglobus was diagnosed. Ultimately, globe displacement was less than one would expect on the basis of orbital bony volume change, i.e., reduction.

Case Report 3. Exophthalmos

A 22-year-old ASA I female patient presented to the Oral and Maxillofacial Clinic with complaints of a slowly progressive swelling of the upper jaw

vestibule on the right side. There was no history of tooth complaints. No recent dental treatment. She also mentioned associated complaints of progressive impairment of air passage through the right side of the nose.

There was no double vision.

Clinical examination showed a very mild buccal swelling of the right side of the face. Asymmetry in globe position, Hertel 19/17 (Fig. 8.15). No diplopia present. Nasal examination showed a deviation of the right lateral nasal wall medially, obliterating the nasal pathway.

Intraoral examination showed a well-maintained "complete" dentition. There was a swollen upper right vestibule, tendency to fluctuation as if the lateral maxillary bony boundary was vanished.

Radiological exam: a CT-scan was obtained: findings: large cyst-type lesion almost completely obliterating the maxillary sinus on the right side including an impacted third maxillary molar which was located high in the cyst in proximity to the orbital floor. There is associated orbital floor elevation (Figs. 8.16 and 8.17). Also, the cyst is expanding medially into the nasal cavity, hereby reducing the nasal passage on the ipsilateral side.

Diagnosis "expanding large follicular cyst related to right-sided impacted third maxillary molar." Associated findings were apparent orbital floor elevation and reduction of nasal passage due to the expanding character of the cyst. Treatment consisted of cyst excochleation and third molar removal via an intraoral approach. The surgical intervention and early follow-up was uneventful.

Two years follow-up showed that there was a good recovery of right-sided airway passage. There was no change in the slight exophthalmos on the affected right side although the orbital floor returned nearly completely to its original position as shown on the postoperative CT scan (Fig. 8.18).

Despite the persistent slight disfiguring eye asymmetry, the patient was satisfied with the end result. There were no complaints of double vision; diplopia was not diagnosed. Minimal asymmetry in Hertel was diagnosed, 19/17 as was diagnosed prior to the cyst development and according to the patient had been present over many years already. As such, there was no desire on the part of the patient to have the existing asymmetry corrected. More so, since orbital floor reconstruction may result in diplopia, refractory to correct.

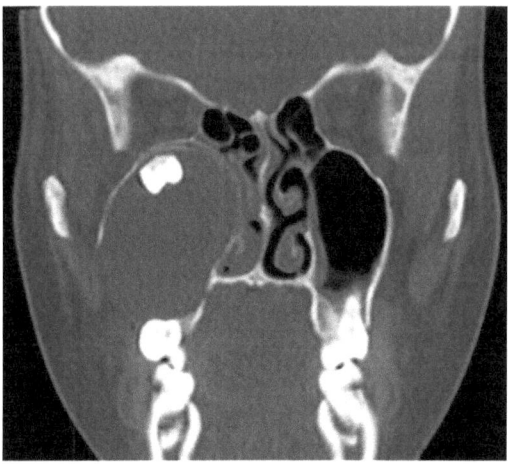

Fig. 8.16 Patient with follicular cyst in right antrum: preoperative CT scan: High positioned impacted third maxillary molar: coronal view

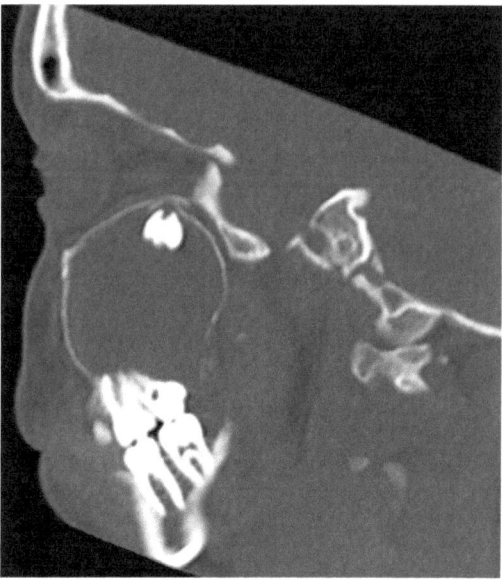

Fig. 8.17 Patient with follicular cyst in right antrum: preoperative CT scan: High impacted maxillary third molar, close to orbital floor: sagittal view

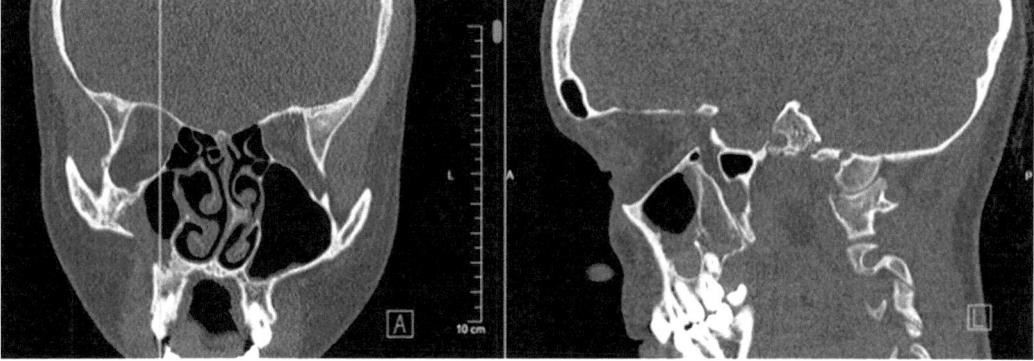

Fig. 8.18 Patient with follicular cyst in right antrum: postoperative CT scan: return of orbital floor to its original position: coronal and associated sagittal view

Contemplation

In this patient, despite the reduction of the bony orbital volume, especially in the apical area, no proptosis/exophthalmos developed; double vision was not reported. This is another example that apparently, orbital soft tissue is capable of adaptation to slowly progressive changes in its surrounding bony volume.

After successful surgical treatment, a nearly complete normalization of the orbital floor to its original position occurred.

Case Report 4. Exophthalmos

An otherwise healthy 10-year-old boy presents to our clinic with mild complaints of a slowly progressing swelling of the left side of the face. There is no associated pain, no tooth-ache in the history. No recent dental treatment. No history of recent trauma. The patient does not complain about eye-related problems.

Extraoral physical examination shows a mild-moderately firm buccal swelling of the left side of the face, no fluctuation, there is no tenderness. The function NV and NVII is intact. Normal vision, no diplopia; however, a slight exoforia OS. Hertel 18/18, no exophthalmos present (Figs. 8.19, 8.20, 8.21, and 8.22).

Intraoral examination showed a well-maintained mixed dentition; firm maxillary vestibular swelling on the left side (Fig. 8.23). No fistula, no fluctuation. In the radiological examination, the orthopantomogram revealed a downward oblique displacement of both bicuspids in the second quadrant as well as an associated cystic type lesion of the left side maxilla (Fig. 8.24).

The additional obtained CT-scan showed an "impacted" second bicuspid in the left side of the maxilla; also a large well-delineated cystic lesion on the left side of the maxilla with almost complete obliteration of the adjacent maxillary sinus and impressive orbital floor elevation OS. This in combination with obliteration of the ipsilateral infundibulum nasi (Figs. 8.25, 8.26, and 8.27).

Diagnosis: "infectious cystic lesion related to impacted bicuspid."

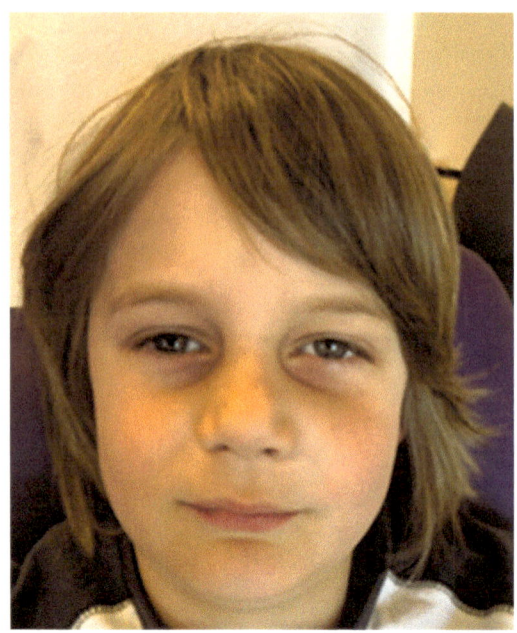

Fig. 8.19 Patient with infectious follicular cyst left antrum: primary gaze

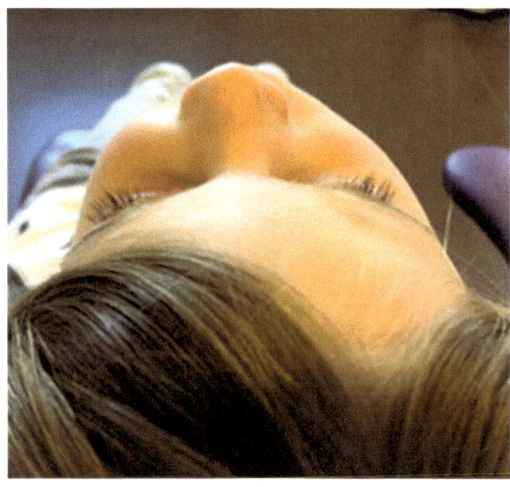

Fig. 8.20 Patient with infectious follicular cyst left antrum: cranial view

Treatment consisted of cyst excochleation and removal of the suspected bicuspid via an intraoral vestibular approach. A limited infundibulotomy was carried out by our ENT colleagues. The follow-up showed an uneventful recovery. Postoperative radiological examination showed a more upright position of the first maxillary left

Fig. 8.21 Patient with infectious follicular cyst left antrum: left lateral view

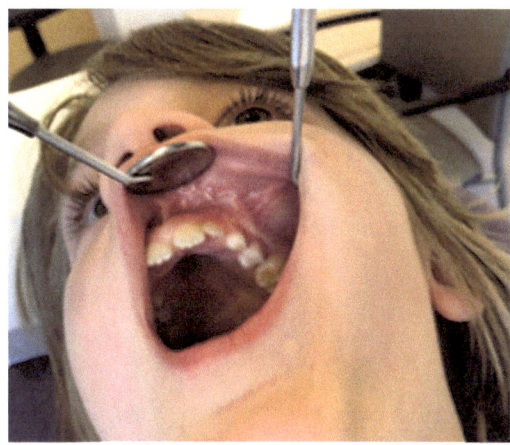

Fig. 8.23 Patient with infectious follicular cyst left antrum: intraoral vestibulum, left upper side

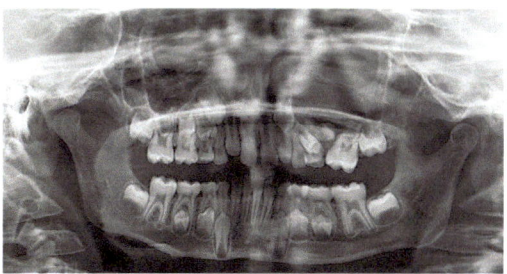

Fig. 8.24 Patient with infectious follicular cyst left antrum: horizontally impacted left maxillary bicuspid: preoperative orthopantomogram

Fig. 8.22 Patient with infectious follicular cyst left antrum: right lateral view

bicuspid and normalization of the orbital floor position and recovery of the maxillary sinus volume; in addition, reactive bone apposition is seen in the antral floor and caudal in the medial and lateral wall of the maxillary sinus (Figs. 8.28, 8.29, and 8.30).

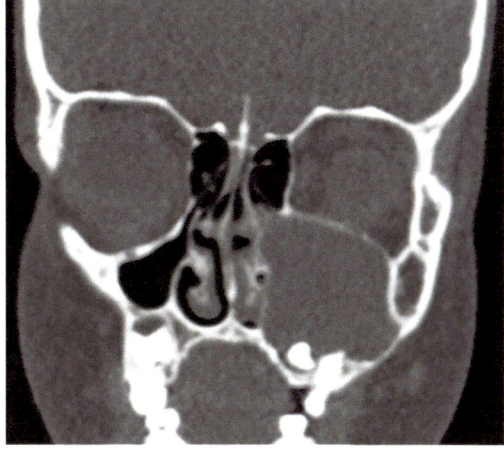

Fig. 8.25 Patient with infectious follicular cyst left antrum: preoperative CT scan: large cyst left maxillary sinus including orbital floor elevation OS: coronal view

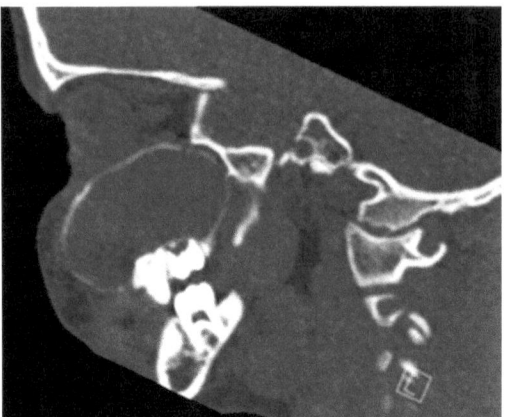

Fig. 8.26 Patient with infectious follicular cyst left antrum: preoperative CT scan: large cyst left maxillary sinus including orbital floor elevation: sagittal view

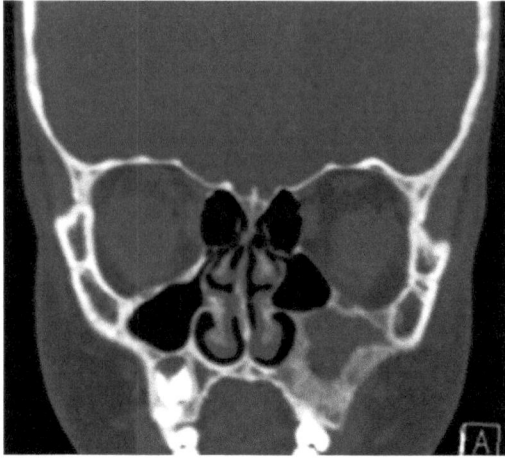

Fig. 8.28 Patient with infectious follicular cyst left antrum: normalization upright position first left maxillary bicuspid: postoperative orthopantomogram

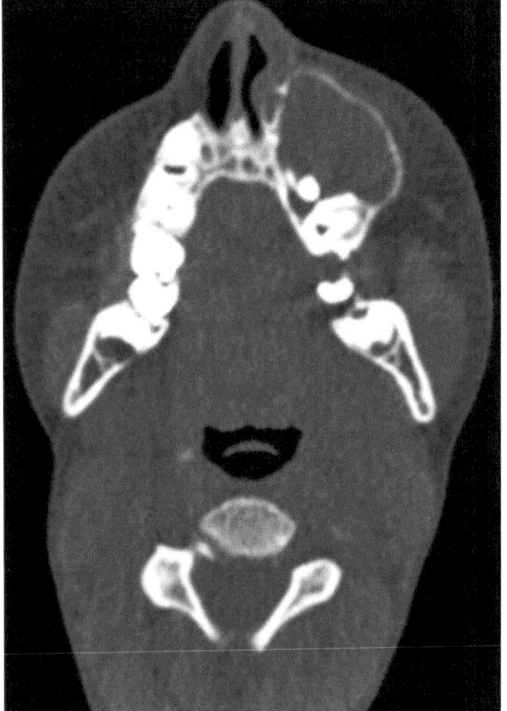

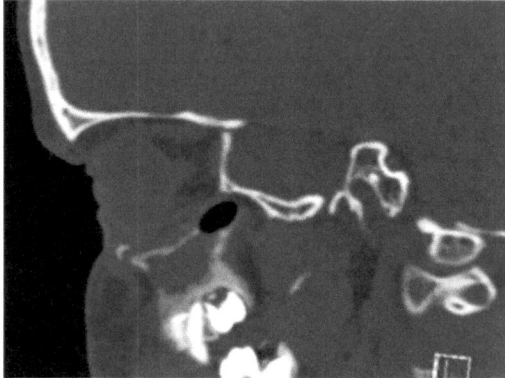

Fig. 8.29 Patient with infectious follicular cyst left antrum: postoperative CT scan: left orbital floor returned to its original position; bone apposition left maxillary sinus walls and floor: coronal view

Fig. 8.27 Patient with infectious follicular cyst left antrum: preoperative CT scan: large left maxillary cyst expanding anteriorly: axial view

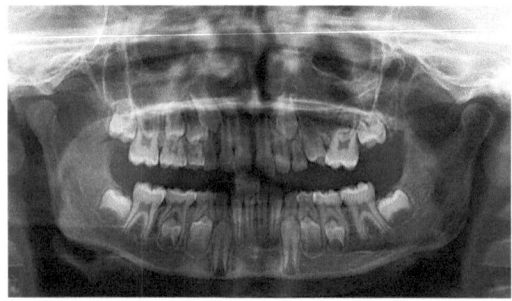

Fig. 8.30 Patient with infectious follicular cyst left antrum: postoperative CT scan: left orbital floor returned to its original position: sagittal view

Contemplation

A large cystic lesion within the maxillary sinus resulted in an impressive elevation of the adjacent orbital floor (Figs. 8.24 and 8.25). However, this narrowing of the orbital bony volume did not result in an outward displacement, proptosis/exophthalmos of the globe. We assume that, because of the slowly progressing pathologic process, the soft tissue of the orbit adapts to its new changing volume, a physiological response to a pathological change.

Also, the orbital floor, which was displaced cranially by the expanding cystic lesion, returned to its pre-pathologic, symmetric position after excochleation of the maxillary cyst with the corresponding adaptive orbital soft tissue changes: no positional changes of the globe were diagnosed during the entire pathologic process, treatment, and post treatment period (Figs. 8.28 and 8.29). We assume, especially in younger children, adaptive changes are capable to adequately cope with volume changes: patient did not report double vision, no exophthalmos did occur.

Annotation

It is interesting to note that, when comparing the radiographic changes in orbital floor position before and after the treatment of the cystic lesion in the maxillary sinus that in the elderly patients the abnormal, displaced position of the orbital floor did not return to its original position while in the younger patient, there is an almost complete return to symmetrical orbital floor position. As can be expected, the younger patients have a higher degree of intrinsic recovery capacity.

Open Access This chapter is licensed under the terms of the Creative Commons Attribution 4.0 International License (http://creativecommons.org/licenses/by/4.0/), which permits use, sharing, adaptation, distribution and reproduction in any medium or format, as long as you give appropriate credit to the original author(s) and the source, provide a link to the Creative Commons license and indicate if changes were made.

The images or other third party material in this chapter are included in the chapter's Creative Commons license, unless indicated otherwise in a credit line to the material. If material is not included in the chapter's Creative Commons license and your intended use is not permitted by statutory regulation or exceeds the permitted use, you will need to obtain permission directly from the copyright holder.

Diagnosis and Clinical Presentation, Workup and Decision-Making of Orbital Fractures

Jesper Jansen, Thomas J. J. Maal, Juliana F. Sabelis, Ruud Schreurs, and Leander Dubois

Learning Objectives

- Knowledge about the anatomy, etiology, and pathophysiology of orbital fractures.
- Recognition of a patient with an orbital fracture: the clinical presentation and symptoms.
- The importance of the orthoptic evaluation and etiology of diplopia, ocular motility disturbance, and binocular single vision (BSV).
- The role of CT imaging and the possibilities of advanced diagnostics in orbital fractures.
- The controversies in clinical decision-making and management of orbital fractures.
- The important aspects of a clinical protocol for orbital fractures.

Introduction

Orbital fractures can be challenging in many aspects. The comminution of the orbital walls can result in the herniation of both adipose and muscle tissue into the adjacent sinuses. This could also affect the periorbita and its function as a suspension system. Initially, the actual damage is difficult to assess due to swelling, contusion of muscles, and the possible presence of a hematoma. This soft tissue trauma may mask the underlying skeletal damage. The indication for orbital reconstruction can be assessed if the right diagnostics are deployed at the right time.

The treatment of orbital fractures varies considerably between specialities and countries. The scientific fundament for surgical indications is weak, as prospective studies based on a clinical

J. Jansen · L. Dubois
Department of Oral and Maxillofacial Surgery, Amsterdam University Medical Centers, Amsterdam, The Netherlands
e-mail: j.jansen@amsterdamumc.nl; l.dubois@amsterdamumc.nl

T. J. J. Maal
3D Lab Department of Oral and Maxillofacial Surgery, Radboud University Medical Center, Nijmegen, The Netherlands
e-mail: thomas.maal@radboudumc.nl

J. F. Sabelis · R. Schreurs (✉)
3D Lab Department of Oral and Maxillofacial Surgery, Amsterdam University Medical Centers, Amsterdam, The Netherlands
e-mail: j.f.sabelis@amsterdamumc.nl; r.schreurs@amsterdamumc.nl

© The Author(s) 2023
P. J. J. Gooris et al. (eds.), *Surgery in and around the Orbit*, https://doi.org/10.1007/978-3-031-40697-3_9

protocol are scarcely available [1–3]. Diplopia is often an indication for orbital reconstruction without clearly defining the nature of the diplopia or motility disorder. Fracture size, another indication for surgery, is difficult to quantify and does not have a high correlation with persistent sequelae [4–7].

In many hospitals, the incidence of solitary orbital floor or medial wall fractures is relatively low, and experience of the individual surgeon is limited. The indication for surgery is partially based on expert opinion as clear guidelines are lacking. With the help of a good and clear clinical protocol, there will be more uniformity concerning the treatment. Since various anatomical and functional systems are affected after a fracture of the orbit, several disciplines (oral and maxillofacial (OMF) surgeon, ophthalmologist, orthoptist, and radiologist) are involved [8]. Each specialist is essential for their field of expertise, and this makes the management of orbital fractures a multidisciplinary matter.

Anatomy

The orbital cavity is composed of seven bones (frontal, lacrimal, ethmoid, zygomatic, maxillary, palatine, and sphenoid) that form a conical shape. The orbital floor and medial wall are relatively thin and are frequently the first to fracture. The orbital soft tissue is composed of the globe and adipose, connective, and muscle tissue. The globe receives support and protection from the adipose and connective tissue. The inferior, superior, medial, and lateral rectus muscles, together with the superior and inferior oblique muscles, move the globe [9, 10].

The connective tissue is a complex framework and can be seen as a suspension system that is important for functioning of the globe. The framework consists of septa, fascia sheets, and ligaments that contain smooth muscle cells. The ligaments support the orbit and hold the soft tissue in position [11]. An example of such a ligament is Lockwood's ligament. This ligament can stabilise the globe in a vertical position and pro-

vide support to the inferior rectus and oblique muscles. Other structures such as the lacrimal system are also important but are usually not affected by an orbital fracture.

Etiology and Pathophysiology

The etiology of orbital fractures is determined by geographic and socioeconomic aspects [12]. The leading causes are interpersonal violence, traffic accidents, and sports. In the younger population, the fractures are mainly caused by interpersonal violence and sports. In the older age groups, they are primarily caused by traffic accidents and falls.

There are two biomechanical models that explain the occurrence of an orbital fracture: the hydraulic and the buckling theory [13, 14]. In the hydraulic theory, the force is directed to the soft tissue of the orbital cavity, increasing the infraorbital pressure. The increased pressure causes the weakest part of the orbital cavity to fracture. The soft tissue is actually pushed through the orbital floor or medial wall. In the buckling theory, the force is directed towards the bone. The strong infraorbital rim does not necessarily fracture, but it transfers the energy to the posterior thin orbital floor, resulting in a fracture. The cause of the fracture may be a combination of both biomechanical principles in most cases.

A distinction must be made between blow-out and trapdoor fractures.

Trapdoor fractures are relatively rare. A part of the orbital content is trapped between the fractured orbital wall in these fractures (Fig. 9.1). This can both be muscle (seldom) or orbital fat with its many septa. Trapdoor fractures occur when the orbital wall is resilient, as is usually the case in children. The pressure wave fractures the bone and the orbital soft tissue bulges into the adjacent sinus. After the pressure drops, the elastic bone bounces back trapping the soft tissue [15]. If an ocular muscle is trapped, there is an indication for immediate surgical intervention, and, in most cases, it is sufficient to free the entrapped tissue. In children, fibrosis and necro-

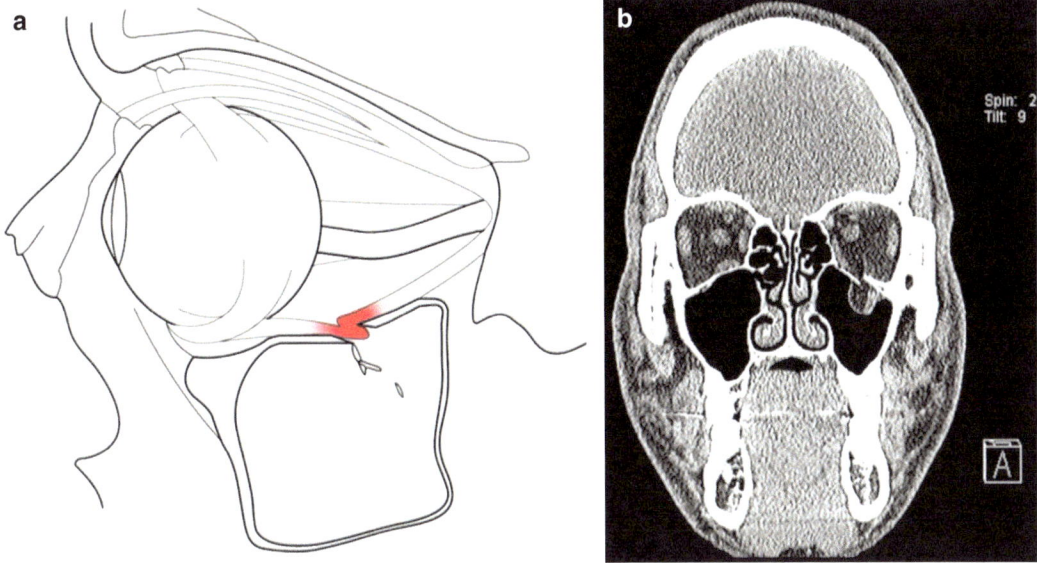

Fig. 9.1 (**a**) Trapdoor fracture; (**b**) CT scan coronal view left orbital floor trapdoor fracture; (c) clinical example limited elevation right globe

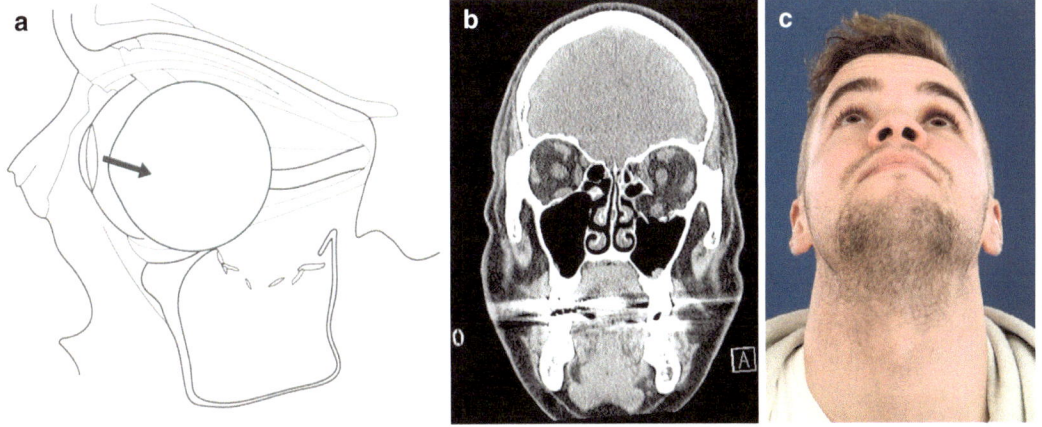

Fig. 9.2 (**a**) Blow-out fracture; (**b**) CT scan coronal view left orbital blow-out fracture; (**c**) clinical example enophthalmos left globe

sis of the rectus muscles can occur quickly. Entrapment of periorbital tissue can also trigger an oculocardiac reflex in addition to pain.

The majority of the orbital fractures are *blow-out fractures* (Fig. 9.2). The increased orbital pressure results in the shattering of one or more orbital walls. The loss of support from the vulnerable floor or medial wall increases orbital volume and can result in protrusion of orbital content into the adjacent sinuses. This can cause displacement of the globe (vertical = hypoglobus; horizon-tal = enophthalmos) and could result in a disturbed ocular motility. Traditionally, a distinction between pure and impure blow-out fractures has been made in the literature. An impure blow-out fracture is caused by concomitant injury of the surrounding bony structures, e.g., the zygomatic complex (ZMC), naso-orbital-ethmoid (NOE) complex or maxillary (Le Fort II/III) fractures. In these combined injuries, the incidence of enophthalmos and hypoglobus is higher through the involvement of an additional orbital wall (lateral

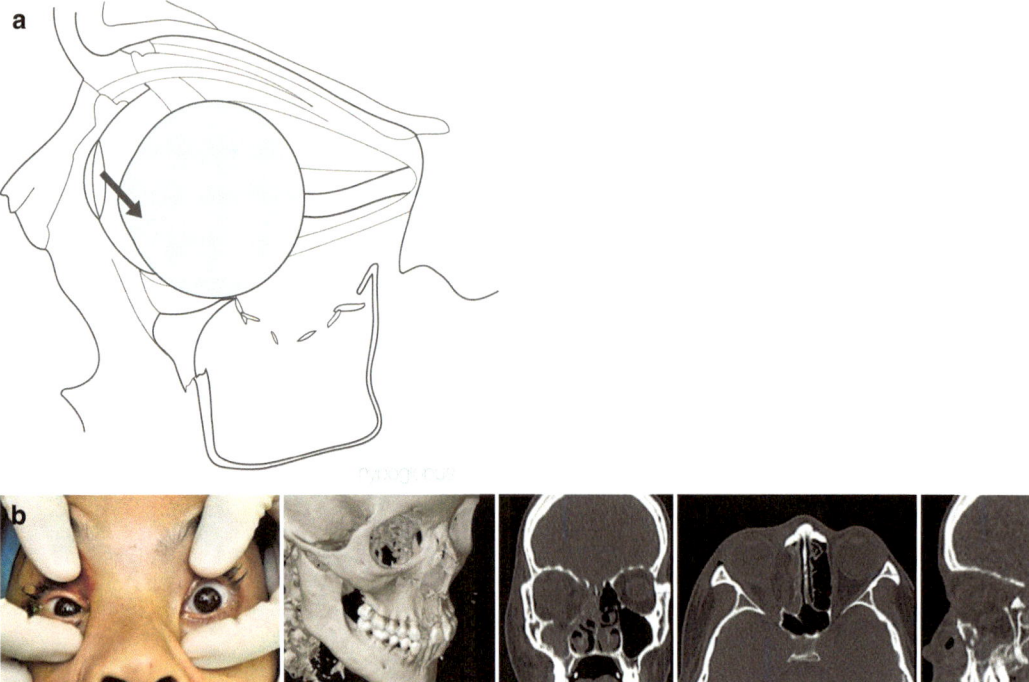

Fig. 9.3 (**a**) Hypoglobus; (**b**) clinical example and CT scan

wall in the case of ZMC fractures) or loss of anterior globe support due to dislocation of the infraorbital rim (Fig. 9.3). A sudden change in the vertical or horizontal position of the globe will result in diplopia in most cases. Although this is a logical consequence, the oculomotor function and degree of diplopia cannot always be directly correlated to the severity of the injury. This adds an extra dimension to the complexity in clinical decision-making, as the direct damage to the periorbital soft tissue caused by the trauma is difficult to assess directly after the trauma [16].

Clinical Presentation

Orbital fractures can be the result of relatively low-impact blunt trauma or a high-energy trauma (HET). Presentation and workup can be different in both categories of trauma patients. The first category will present itself at the general practitioner or the emergency department and the second category in the shock room. Patients brought in after a HET may be unconscious and intubated, making clinical examination more difficult. However, these patients routinely receive a whole-body CT scan, which helps to establish the diagnosis. In these severely injured patients, the risk of vision-threatening emergencies rises. These compromising injuries, e.g., a retrobulbar hematoma, globe perforations or oculocardiac reflex, will be discussed in Chap. 13.

Diagnosing an orbital fracture without imaging can be challenging, as the clinical presentation varies considerably. However, a numb upper lip is highly suspicious for a trauma of the ipsilateral inferior orbital nerve as a consequence of an orbital floor fracture. In addition to the diplopia, motility disturbance, enophthalmos, and hypoglobus, as mentioned earlier, other clinical symptoms are disturbed vision, pain, hyposphagma, periorbital swelling, hematoma, ecchymosis, and infraorbital nerve paresthesia. During inspection and palpation, an asymmetrical osseous projection may be caused by a zygomatic or Le Fort II/III fracture. The pupil reflex should be tested with a penlight exam together with a global assessment of vision, eye motility, and diplopia.

Fig. 9.4 Hertel exophthalmometer in use

Attention should also be paid to signs of retrobulbar hematoma such as proptosis, tense tissue, and severe pain [17]. Enophthalmos can be objectified by clinical examination in bird or submental view or with the help of Hertel exophthalmometer (Fig. 9.4) [18]. Swelling and potential concomitant facial fractures complicate the quantification of enophthalmos. The globe is less visible with swelling of the eyelids and the surrounding tissue. This creates an optical illusion that gives the appearance of an enophthalmos. A Naugle exophthalmometer is an alternative for the Hertel exophthalmometer in case of concomitant injuries.

A multidisciplinary approach is vital in the good clinical management of orbital wall fractures. In general, the OMF surgeon is involved in all patients with facial injuries. During office hours, referral to an orthoptist is recommended to quantify the amount of diplopia and motility disturbance. If required, an ophthalmologist should be consulted for a more extensive assessment of the globe (vision, globe pressure, etc.). Outside office hours, it is advisable to have the ophthalmologist assess the trauma and send the patient to the orthoptist the next working day.

Orthoptic Examination

The orthoptic examination is perhaps the most essential guide in clinical decision-making in orbital fractures. Diplopia and motility disturbance are objectively measured and analysed. Using a motility perimeter, the ductions and the field of binocular single vision (BSV) can be measured (Fig. 9.5) [19]. Average maximum ductions are 40° elevation, 60° depression, and 50° abduction and adduction. With a Hess screen test, it is also possible to measure the deviation and the amount of overaction and underaction of the extraocular muscles. The measurements performed at the first presentation serve as a baseline and possible improvement may be assessed over time. If the values improve significantly in the first 2 weeks after the trauma, the restrictions were probably due to swelling or contusion. If the restrictions persist, an orbital reconstruction may be necessary. A poor BSV at first presentation correlates with persistent sequelae and incomplete recovery. This will be discussed in more detail later in the clinical protocol The orthoptic examination is perhaps the most essential guide in clinical decision-making in orbital fractures. Diplopia and motility disturbance are objectively measured and analysed. Using a motility perimeter, the ductions and the field of binocular single vision (BSV) can be measured (Fig. 9.5) [19]. Average maximum ductions are 40° elevation, 60° depression, and 50° abduction and adduction. With a Hess screen test, it is also possible to measure the deviation and the amount of overaction and underaction of the extraocular muscles. The measurements performed at the first presentation serve as a baseline and possible improvement may be assessed over time. If the values improve significantly in the first 2 weeks after the trauma, the restrictions were probably due to swelling or contusion. If the restrictions persist, an orbital reconstruction may be necessary. A poor BSV at first presentation correlates with persistent sequelae and incomplete recovery. This will be discussed in more detail later in the clinical protocol subsection. The orthoptic examination requires a certain degree of mobility and cooperation that is not always present in trauma patients. Detailed information about the orthoptic examination can be found in Chap. 6.

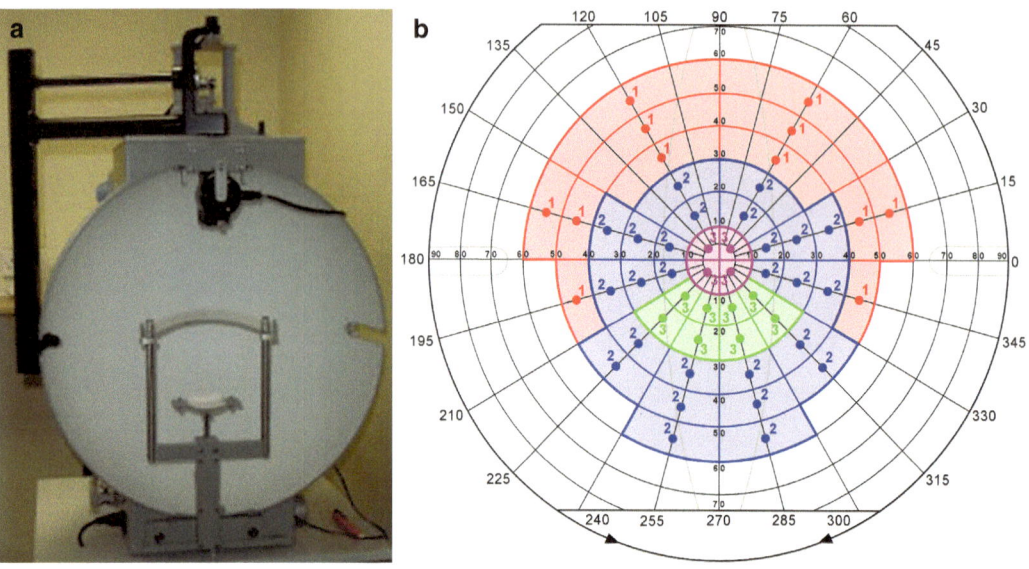

Fig. 9.5 (**a**) Motility perimeter; (**b**) binocular single vision (BSV) score chart

Imaging

In case of an orbital fracture, it is advisable to perform a CT scan (orbital series or complete skull if more facial injuries are suspected 1-millimetre (mm) slices). The size and location of the orbital floor fracture can be best examined in coronal and sagittal view and the medial wall in axial view. Possible entrapment or impingement of soft tissue, a retrobulbar hematoma, emphysema, or stretching of the optic nerve can be observed in soft tissue setting (Fig. 9.6). Based on a CT scan, advanced diagnostics can be performed, a surgical plan can be created, and the scan, can be used during computer-assisted surgery. For these reasons, it is crucial to think ahead about using the proper scan protocol. A Cone Beam CT (CBCT) scan is less appropriate when assessing an orbital fracture. Nevertheless, an intraoperative CBCT scan is ideal for evaluating the orbital implant location in real-time. The clinical decision-making for orbital wall fractures is conducted based on clinical characteristics, both subjective and objective measurements, and CT imaging. In addition to conventional diagnostics, it is also possible to use advanced diagnostics based on a CT scan.

Advanced Diagnostics

Advanced diagnostics can be used to support clinical decision-making. The aim of advanced diagnostics is to extract as much information as possible from the DICOM data and add information if necessary. The CT scan may be viewed as a virtual representation of the patient, subdivided into voxels (three-dimensional pixels). Each voxel has a particular grayscale value corresponding to the X-ray absorption within the voxel's volume. Image viewers allow assessment of the image data in the multiplanar view (two-dimensional axial, coronal, and sagittal slices). Visual comparison between affected and unaffected orbit is feasible, but more sophisticated analysis options only become available if specialised software that allows image manipulation is used (e.g., iPlan Cranial (Brainlab AG, Munich, Germany)).

The workhorse for advanced diagnostics is the segmentation: partitioning the image data in voxels that belong to the same anatomical structure [20]. The easiest segmentation method is thresholding. The user chooses a cut-off HU value; voxels above the threshold are included in the selection, and voxels below are excluded. The selection of bony

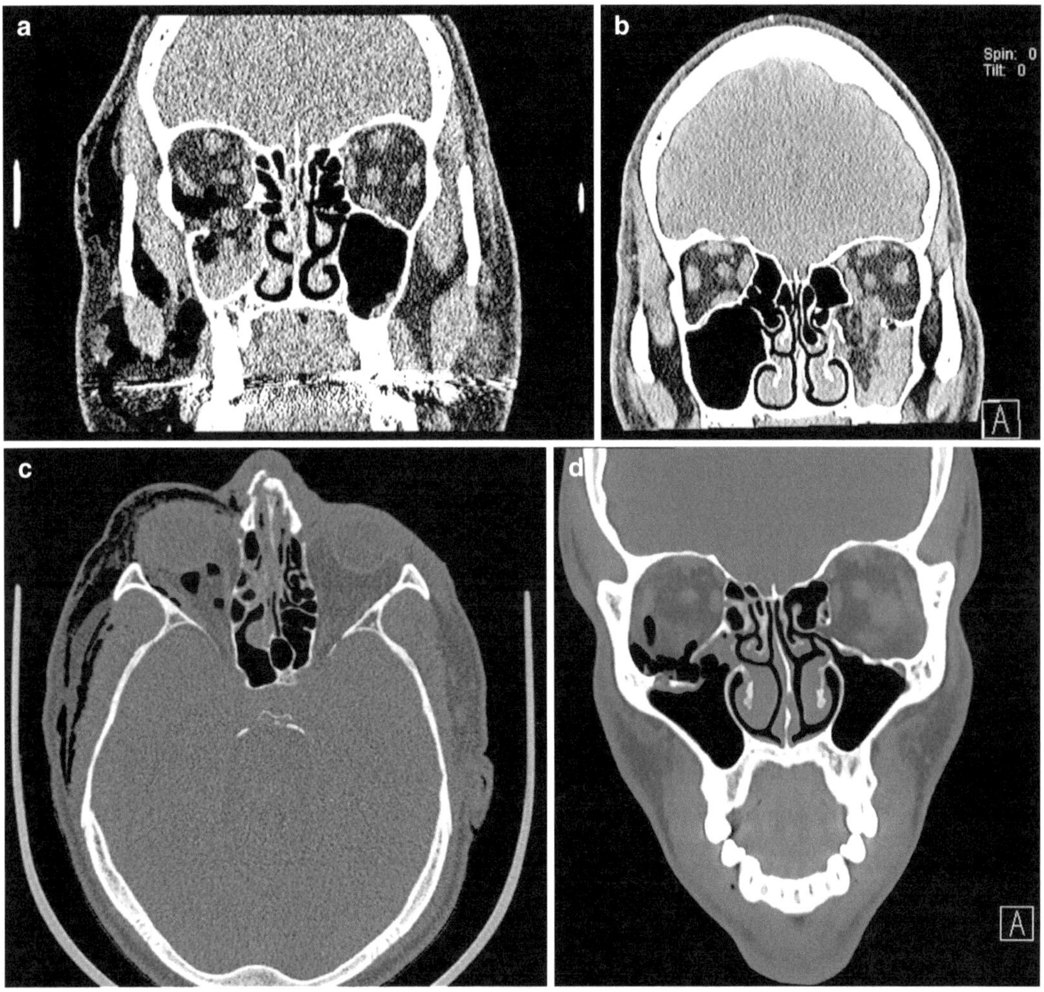

Fig. 9.6 Soft tissue setting of CT scan with (**a**) swelling; (**b**) hematoma formation of the extra ocular muscles; (**c**) emphysema axial view; (**d**) emphysema coronal view

structures within the image data is an example of thresholding. The orbital walls are more difficult to annotate in the image data: the thin bony structures are frequently not completely included in thresholding operations. More sophisticated segmentation approaches are required to fill the gap. Atlas-based segmentation is an example of a segmentation method that can provide an accurate segmentation of the orbit and orbital contents [21–23]. Since segmentation occupies a central position in advanced diagnostics (and subsequent virtual surgical planning), continuous research on even more effective segmentation strategies with minimal user input is performed. Artificial intelligence (AI) is a promising technique that meets these desires and may become available soon.

The segmentation itself is a clarification of the image data, but it does not add new information. The information in the patient model can be expanded by creating a 3D model from the voxel selection; these models can subsequently be manipulated. In unilateral orbital fractures, the unaffected orbit's segmented model may be mirrored (Fig. 9.7) and overlaid on the affected orbit in the multiplanar views. The mirrored model now provides a blueprint for the pretraumatised

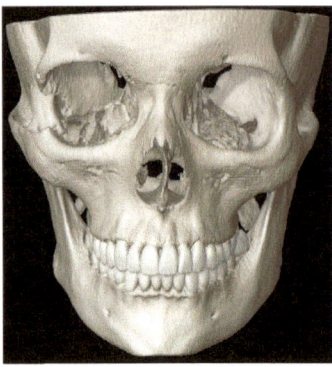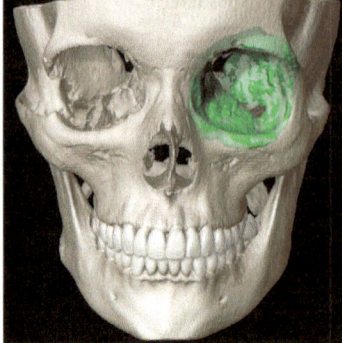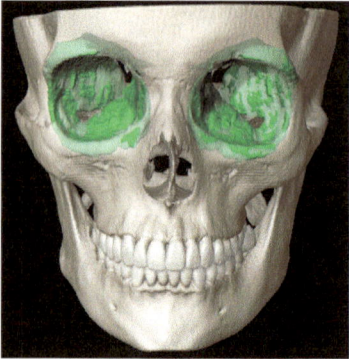

Fig. 9.7 Segmentation and mirroring of the orbit

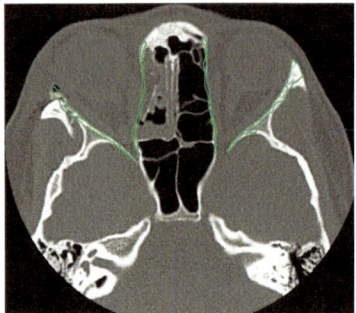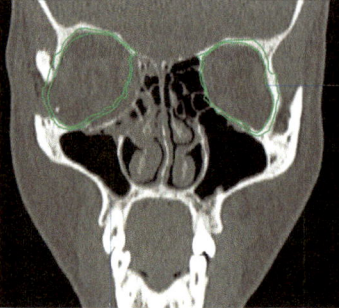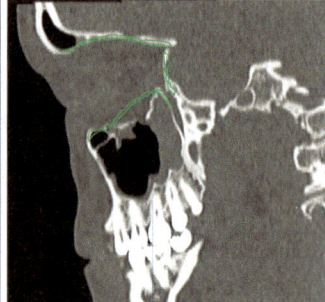

Fig. 9.8 Mirrored orbit visualised in multiplanar views

anatomy of the affected side [21, 24–26]. Comparison between affected and unaffected orbit is much more straightforward than side-to-side visualisation and not affected by angulation of the slice (Fig. 9.8). A more thorough insight into the fracture's location and extent is obtained. A separate segmentation of the orbital contents could quantify the increase of intraorbital volume due to the trauma. Attaching segmentations of surrounding structures to the orbital segmentation model is another manipulation possibility (Fig. 9.9). In the example of a zygomatic complex fracture, the combined model of the orbit, and zygoma provides additional information on displacement of the zygoma (Fig. 9.10). This additional information could affect surgical indication and treatment of the orbit.

The most considerable benefits of advanced diagnostics are improved visualisation of the problem and enhanced analysis methods for objective quantification. With advanced diagnostics, the surgeon can be sure that the treatment decision is supported by all the information present in the image data. If surgery is indicated, advanced diagnostics are the precursors of virtual surgical planning in the computer-assisted surgery workflow. Much information gathered in the advanced diagnostics can be reused in the virtual surgical planning process. The mirrored orbit provides the target anatomy, which is helpful in positioning a virtual implant. Surface models of the orbit, orbital contents, and bony anatomy surrounding the orbit serve as the input for the design of a patient-specific implant. In contrast, the mirrored model of the zygomatic bone already acts as the planning for repositioning of the zygomatic bone. A detailed description of the virtual surgical planning process is provided in Chap. 10.

Clinical Decision-Making

Several specialities are involved in the treatment of orbital fractures, including OMF surgery, ophthal-

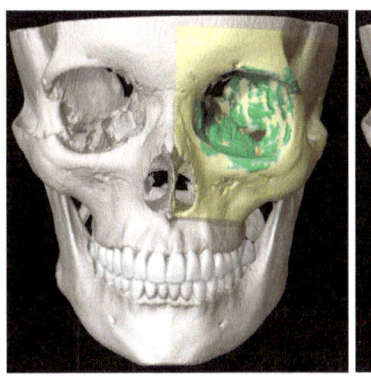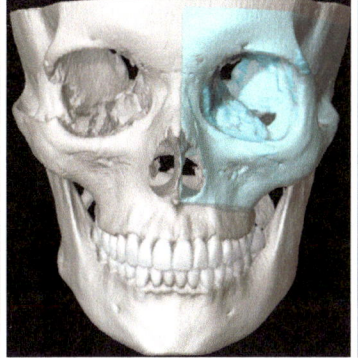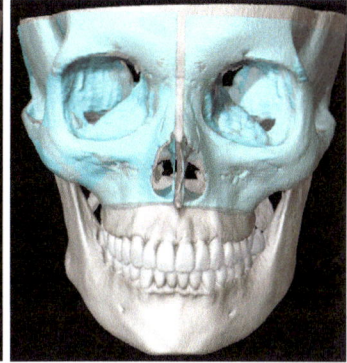

Fig. 9.9 Combining the segmented models of the orbit and surrounding bony structures before mirroring

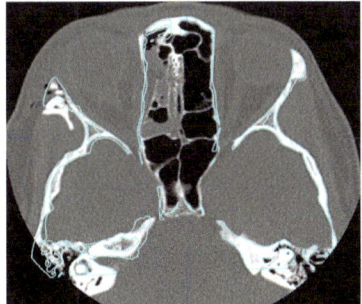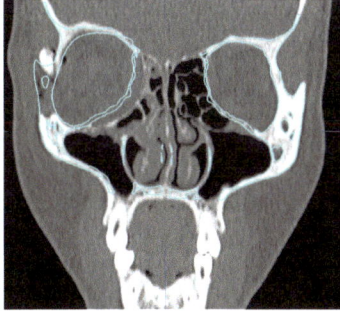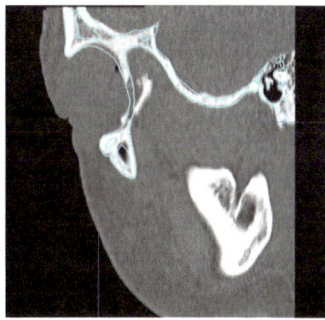

Fig. 9.10 Mirrored orbit + bone model in multiplanar views

mology and, less so, plastic surgery and ear, nose, and throat surgery. In recent decades, there has been a clear trend in the management of orbital fractures, alternating between conservative and surgical treatment preferences. The development of new treatment modalities and types of implants has led to an increase in surgical treatment in the past [27, 28]. Nevertheless, the body's regenerative capacity seems to be significant in orbital fractures, so that surgery is often not required [6].

One of the main topics in the ongoing debate on the management of orbital fractures is the indication for surgery. Most surgeons are apt to repair and reconstruct traumatic injuries of the bony orbit. Evidently, there are some clear and immediate indications for surgery, such as vision-threatening emergencies, significant globe displacement (globe in maxillary sinus), pediatric trapdoor fractures, trapdoor fractures with clear muscle entrapment, or a persistent oculocardiac reflex. Permanent damage to the orbital soft tissue will probably occur without early intervention in these cases. Other indications for orbital

reconstruction are all relative [29]. For most clinicians, it is clear that small asymptomatic orbital wall fractures do not need surgery and larger fractures with early enophthalmos do require an orbital reconstruction. In either case, the lack or presence of clinical symptoms (diplopia or enophthalmos) determine the indication for surgery. True controversy arises in large orbital wall defects without early enophthalmos. Some clinicians will perform surgery to anticipate on expected enophthalmos, although it is uncertain if this will occur. It is the other way around with regard to diplopia, which can resolve spontaneously after the swelling disappears. As there are no clear predictors for enophthalmos or cut-off points for the recovery of diplopia, this discussion will continue. For most surgeons, there is an indication for surgery if the defect size is large (>2 cm^2 or >50% of surface measured on a CT scan) or if diplopia is severe or persistent [3, 30]. Interestingly, the size of the fracture does not necessarily correlate to late enophthalmos and the persistence of diplopia (Fig. 9.11). The litera-

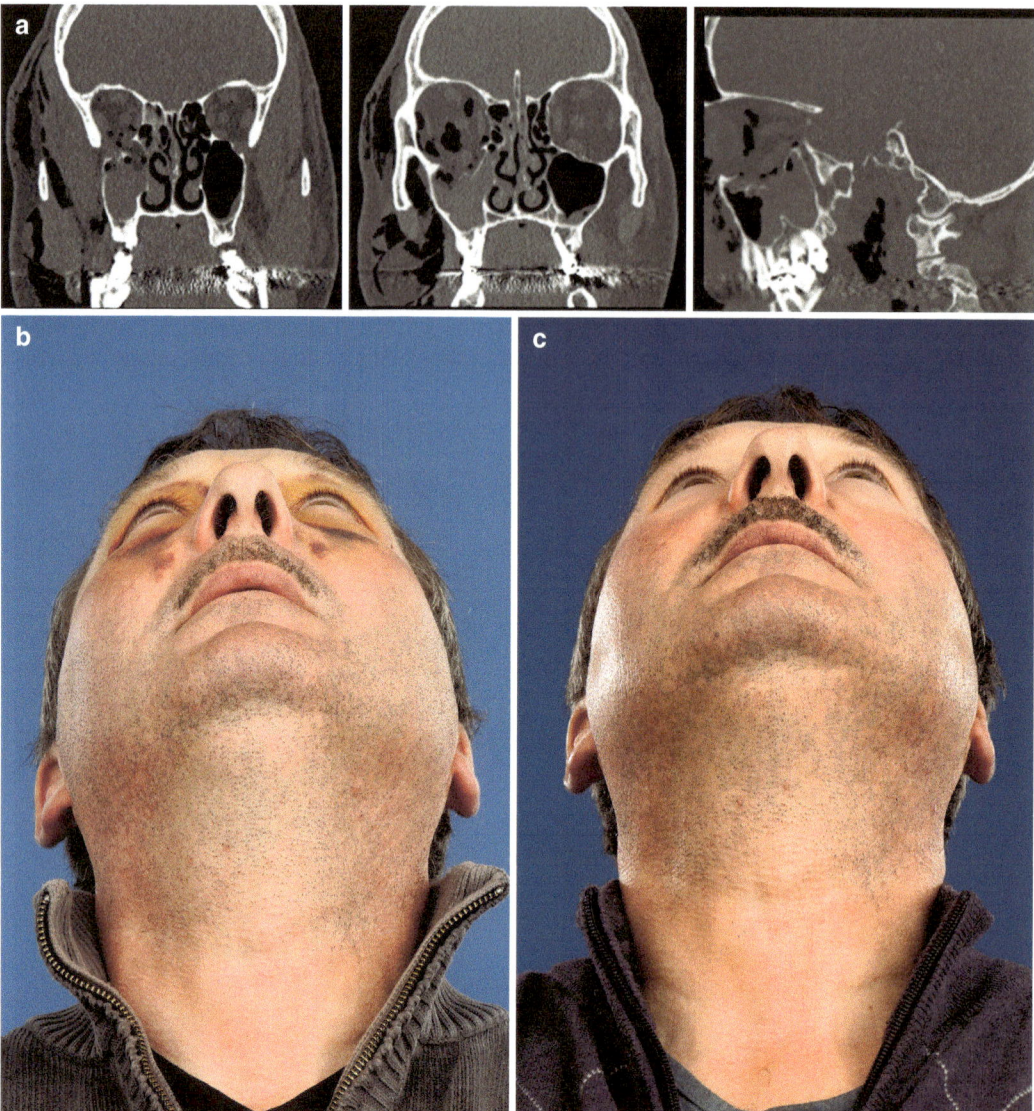

Fig. 9.11 (**a**) Large comminuted fracture without (**b**) early enophthalmos and (**c**) no late enophthalmos or persistent diplopia 6 months after nonsurgical treatment

ture also demonstrates that it is difficult to measure or estimate the defect size, so it seems illogical to base the indication for surgery on the CT scan. Treatment choices can be made more effectively based on clinical symptoms that are present at the time, not what is expected based on the size of the fracture.

The aforementioned considerations are even more complex with diplopia, as there is a great variability in the extent and cause of diplopia and motility restrictions. These are not only caused by entrapment of the inferior rectus muscle or surrounding soft tissue, but also by muscle edema, hemorrhage, and motor nerve palsy. These are conditions that cannot be treated with an orbital reconstruction. In time, some of these deficits may resolve spontaneously. Edema will resolve within several days to weeks, but hemorrhage, fibrosis, and motor nerve palsy can take up to 12 months to recover. The dilemma in clinical decision-making occurs when enophthalmos is minimal or absent, motility has improved, but significant diplopia persists. Diplopia is too complex and multifactorial to rely on subjective observations shortly after trauma. For that reason, multiple studies stress the importance of quantitative evaluation of ocular motility and diplopia [6, 7]. Quantification by a Hess screen test or interpretation of ductions and the field assessment of binocular single vision (BSV) can be extremely helpful in decision-making. Using these measurements, the improvement of ocular motility and diplopia can be objectified over weeks or even months after trauma. In severe diplopia (BSV <60), orbital reconstruction has a significant effect on the outcome [7].

The effect of surgery on the amount of diplopia appears to be limited in mild diplopia (BSV 60–80). In limited diplopia (BSV >80), orbital reconstruction may even worsen the clinical outcome. This stresses the relevance of standard orthoptic evaluation in the workup and decision-making process.

Another critical factor in the decision-making process is the potential adverse effect of delay if the watchful waiting strategy is used. On the one hand, there will be no overtreatment as the diplopia can recover spontaneously over time. On the other hand, there is an assumption that early surgery (<2 weeks) has a better outcome and causes less iatrogenic damage. Nonetheless, there is no solid evidence that delayed surgery has a worse outcome than early intervention [2]. Taking the time for evaluation is, for that reason, recommended in all relative indications.

Clinical Protocol

The clinical protocol presented and suggested here was developed at the Amsterdam UMC [6]. The protocol was part of the Advanced Concept on Orbital Reconstruction (ACOR) research program and had an extensive scientific fundament [27]. The protocol gives the clinician tools for clinical decision-making and contains a flow chart in which nonsurgical and surgical patients are monitored. At each visit, the decision can be

Table 9.1 Indications for orbital reconstruction

Immediate surgery (<24 h)
• Significant globe dislocation
• Pediatric trapdoor fracture
• Persistent oculocardiac reflex
Early surgery (<2 weeks)
• Early enophthalmos (>2 mm) *or*
• Absolute elevation restriction (<15°) *or*
• Absolute abduction restriction (<25°)
Delayed surgery (>2 weeks)
• Late enophthalmos (>2 mm) *or*
• Ductions <8° improvement of most limited duction 2 weeks after trauma *or*
• Persistent limiting diplopia (BSV < 80) *or*
• Other debilitating symptoms (e.g., chronic pain or other controversial indications)

made to opt for surgical intervention based on various indications (Table 9.1).

The emphasis of this protocol is on nonsurgical treatment. The decision for surgery is primarily based on clinical observations and orthoptic measurements and not solely on the size or location of the fracture. The initial clinical presentation can vary widely. Patients with a large fracture may have relatively few symptoms, while a small fracture may cause severe diplopia. As mentioned before, this often does not predict the extent to which the diplopia and restricted ocular motility will recover spontaneously. Frequently, considerable improvement can be observed within 2 weeks of trauma.

First Presentation

The clinical examination should focus on the various symptoms that may indicate an orbital fracture. Objective and standardised measurements are needed to assess the patient properly. If there is a high suspicion of an orbital fracture, a CT scan should be ordered to confirm the diagnosis and determine the severity.

As mentioned before, immediate intervention is necessary in vision-threatening situations. In a trapdoor fracture with oculocardiac reflex or significant globe dislocation (Fig. 9.12), urgent intervention (<24 h) is also required to prevent permanent damage. An orthoptic examination is highly recommended if there is no acute indication for intervention and diplopia or limited motility is

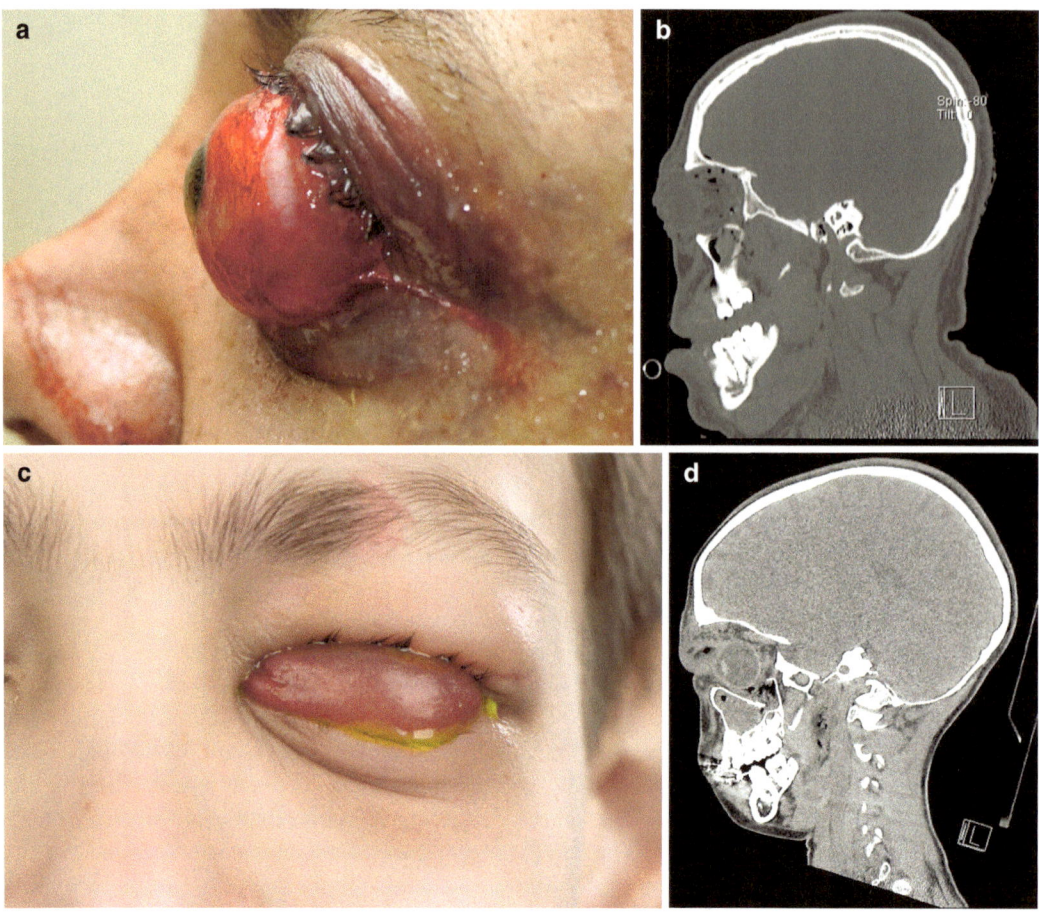

Fig. 9.12 Significant globe dislocations (**a**) clinical example; (**b**) CT scan sagittal view of clinical example; (**c**) clinical example; (**d**) CT scan sagittal view of clinical example

present. If the clinical and orthoptic examination show that there is no early enophthalmos (>2 mm) or an absolute elevation (<15°) or abduction limitation (<25°), nonsurgical treatment with a checkup after 10–14 days can be considered. If early enophthalmos or severe ocular motility restrictions are present, there is an indication for orbital reconstruction within 2 weeks. The expectation is that there is entrapment of the muscles and the connective tissue around the muscles, which will not recover without release.

Photographs taken at the time of initial presentation and subsequent visits can provide the practitioner and the patient with an insight into the recovery and the possible occurrence of cosmetic sequelae/complications, such as scarring or entropion after orbital reconstruction.

Immediately after the trauma, patients are advised to improve the ocular motility and train the ocular muscles by alternately closing one eye several times a day and looking as far as possible in all directions. In this way, the muscles do not slack and work on recovery, the pumping function of the muscles can reduce edema, and possibly fibrosis and adhesions are reduced.

Nonsurgical Follow-up After 2 Weeks

The second visit is meant to evaluate the improvement of motility disturbance and diplopia. If there is insufficient improvement (<8°) of the most limited affected duction or if severe diplopia (BSV < 60) is persistent, there is a strong indication for surgery. In these patients, an intervention is scheduled within 1 week. If it was not possible to perform an orthoptic examination in the first week after trauma, e.g., due to swelling or a noncooperative patient, the same values are used as described earlier (<15° elevation and <25° abduction).

After 2 weeks, it may be expected that most of the swelling has subsided. Conservative treatment can be continued if there is no clear enophthalmos (>2 mm). In these fractures, the periorbita has probably enough resistance to support the soft tissue content of the orbit sufficiently.

Nonsurgical Late Follow-up

If there is no clear indication for surgery, the next visit is scheduled 3 months after trauma and, if indicated, up to 12 months. Incidentally, late enophthalmos may occur, but this is usually within 3 months. If this is considered troublesome by the patient, it is an indication for a secondary reconstruction after the primary nonsurgical treatment failed. If there is persistent mild diplopia (BSV 60–80), surgical intervention can be considered. Diplopia, especially in downward gaze, can be highly disabling. The chance of full recovery is smaller if diplopia is severe (BSV < 60) at first presentation or has been present for a long time.

Follow-up After Orbital Reconstruction

After orbital reconstruction, additional visits are scheduled 1 and 3 weeks after surgery to assess the clinical recovery and, if necessary, manage potential complications after surgery. Due to swelling, ocular motility can still be restricted the first weeks after surgery. An orthoptic examination is scheduled 6 weeks after surgery for that reason. Further visits are similar to the nonsurgical case.

Tables 9.2 and 9.3 indicate which measurements are required for each visit. If the ductions have fully recovered at the 3-month check-up, further orthoptic examinations is performed only on indication. Likewise, the CT scan after 6 months is also only on indication. Often, patients without complaints will not come for a check-up after

Table 9.2 Nonsurgical treatment follow-up

Visit number	1	2	3	4 (optional)	5 (optional)
	First presentation	10–14 days	3 months	6 months	12 months
Clinical examination	X	X	X	X	X
Photography	X	X	X	X	X
Imaging (CT scan)	X			(X)	
Ophthalmic examination	X	X	X	X	X
Orthoptic examination	X	X	X	X	X

Table 9.3 Surgical treatment follow-up

Visit number	1	2	3	4	5	6 (optional)	7 (optional)
	First presentation	Surgery + post-op	10–14 days post-op	6 weeks post-op (± 7 days)	3 months post-op (± 7 days)	6 months post-op (± 7 days)	12 months post-op (± 7 days)
Clinical examination	X	X	X	X	X	X	X
Photography	X		X	X	X	X	X
Imaging (CT scan)	X	X				(X)	
Ophthalmic examination	X			X	X	X	X
Orthoptic examination	X			X	X	X	X
Surgery		X					

3 months, but it is instructive to follow-up patients with complaints for a long time.

To monitor the results of this protocol, we performed a two-center multidisciplinary prospective cohort study [6]. Fifty-eight patients completed the 3, 6, and/or 12 month follow-up. We assessed a full recovery without diplopia or enophthalmos (e.g., >2 mm) in 45 out of these 58 (78%) patients. The other 13 patients had limited diplopia, mainly in extreme upward gaze. Five of those 13 patients did not experience impairment of diplopia in daily life. No patients developed late enophthalmos.

We concluded that a high percentage of patients with orbital floor and/or medial wall fracture recovered spontaneously without lasting diplopia or disfiguring enophthalmos.

Conclusion

Despite the fact that nowadays, we treat orbital fractures in a multidisciplinary setting and know a lot about the anatomy and pathophysiology, it remains difficult to predict the outcome in part of the orbital fractures. By using the appropriate diagnostics at the first presentation, a proper assessment can be made whether there is an indication for surgery. Orthoptic measurements play a crucial role in this. After vision-threatening emergencies have been excluded, the advantages and disadvantages of surgery should be considered. In addition to enophthalmos, persistent diplopia is a sequala of orbital fractures. By performing subjective measurements and comparing them over time, an assessment can be made of the likelihood of full recovery or the necessity for surgery. The described protocol is a tool for future research and further implementation of clinical decision-making in the general clinic.

References

1. Dubois L, Steenen SA, Gooris PJ, Mourits MP, Becking AG. Controversies in orbital reconstruction—I. Defect-driven orbital reconstruction: a systematic review. Int J Oral Maxillofac Surg. 2015;44(3):308–15.

2. Dubois L, Steenen SA, Gooris PJ, Mourits MP, Becking AG. Controversies in orbital reconstruction—II. Timing of post-traumatic orbital reconstruction: a systematic review. Int J Oral Maxillofac Surg. 2015;44(4):433–40.

3. Burnstine MA. Clinical recommendations for repair of isolated orbital floor fractures: an evidence-based analysis. Ophthalmology. 2002;109(7):1207–10. discussion 10-1; quiz 12-3

4. Dubois L, Jansen J, Schreurs R, Habets PE, Reinartz SM, Gooris PJ, et al. How reliable is the visual appraisal of a surgeon for diagnosing orbital fractures? J Craniomaxillofac Surg. 2016;44(8):1015–24.

5. Ploder O, Klug C, Backfrieder W, Voracek M, Czerny C, Tschabitscher M. 2D- and 3D-based measurements of orbital floor fractures from CT scans. J Craniomaxillofac Surg. 2002;30(3):153–9.

6. Jansen J, Dubois L, Maal TJJ, Mourits MP, Jellema HM, Neomagus P, et al. A nonsurgical approach with repeated orthoptic evaluation is justified for most blow-out fractures. J Craniomaxillofac Surg. 2020;48(6):560–8.

7. Alhamdani F, Durham J, Greenwood M, Corbett I. Diplopia and ocular motility in orbital blow-out fractures: 10-year retrospective study. J Craniomaxillofac Surg. 2015;43(7):1010–6.

8. Aldekhayel S, Aljaaly H, Fouda-Neel O, Shararah AW, Zaid WS, Gilardino M. Evolving trends in the management of orbital floor fractures. J Craniofac Surg. 2014;25(1):258–61.

9. Turvey TA, Golden BA. Orbital anatomy for the surgeon. Oral Maxillofac Surg Clin North Am. 2012;24(4):525–36.

10. Gospe SM 3rd, Bhatti MT. Orbital anatomy. Int Ophthalmol Clin. 2018;58(2):5–23.

11. Koornneef L. New insights in the human orbital connective tissue. Result of a new anatomical approach. Arch Ophthalmol. 1977;95(7):1269–73.

12. Bartoli D, Fadda MT, Battisti A, Cassoni A, Pagnoni M, Riccardi E, et al. Retrospective analysis of 301 patients with orbital floor fracture. J Craniomaxillofac Surg. 2015;43(2):244–7.

13. Bullock JD, Warwar RE, Ballal DR, Ballal RD. Mechanisms of orbital floor fractures: a clinical, experimental, and theoretical study. Trans Am Ophthalmol Soc. 1999;97:87–110; discussion−3

14. Ahmad Nasir S, Ramli R, Abd Jabar N. Predictors of enophthalmos among adult patients with pure orbital blowout fractures. PLoS One. 2018;13(10):e0204946.

15. Papadiochos I, Petsinis V, Tasoulas J, Goutzanis L. Pure orbital trapdoor fractures in adults: tight entrapment of perimuscular tissue mimicking true muscle incarceration with successful results from early intervention. Craniomaxillofac Trauma Reconstr. 2019;12(1):54–61.

16. Dubois L, Dillon J, Jansen J, Becking AG. Ongoing debate in clinical decision making in orbital fractures: indications, timing, and biomaterials. Atlas Oral Maxillofac Surg Clin North Am. 2021;29(1):29–39.

17. Girotto JA, Gamble WB, Robertson B, Redett R, Muehlberger T, Mayer M, et al. Blindness after reduction of facial fractures. Plast Reconstr Surg. 1998;102(6):1821–34.

18. Mourits MP, Lombardo SH, van der Sluijs FA, Fenton S. Reliability of exophthalmos measurement and the exophthalmometry value distribution in a healthy Dutch population and in Graves' patients. An exploratory study. Orbit. 2004;23(3):161–8.

19. Sullivan TJ, Kraft SP, Burack C, O'Reilly C. A functional scoring method for the field of binocular single vision. Ophthalmology. 1992;99(4):575–81.

20. Schreurs R, Klop C, Maal TJJ. Advanced diagnostics and three-dimensional virtual surgical planning in orbital reconstruction. Atlas Oral Maxillofac Surg Clin North Am. 2021;29(1):79–96.

21. Mahoney NR, Peng MY, Merbs SL, Grant MP. Virtual fitting, selection, and cutting of preformed anatomic orbital implants. Ophthalmic Plast Reconstr Surg. 2017;33(3):196–201.

22. Jansen J, Schreurs R, Dubois L, Maal TJJ, Gooris PJJ, Becking AG. Orbital volume analysis: validation of a semi-automatic software segmentation method. Int J Comput Assist Radiol Surg. 2016;11(1):11–8.

23. Wagner MEH, Lichtenstein JT, Winkelmann M, Shin HO, Gellrich N-C, Essig H. Development and first clinical application of automated virtual reconstruction of unilateral midface defects. J Craniomaxillofac Surg. 2015;43(8):1340–7.

24. Rana M, Cui CHK, Wagner M, Zimmerer R, Rana M, Gellrich N-C. Increasing the accuracy of orbital reconstruction with selective laser-melted patient-specific implants combined with intraoperative navigation. J Oral Maxillofac Surg. 2015;73(6):1113–8.

25. Karkkainen M, Wilkman T, Mesimäki K, Snäll J. Primary reconstruction of orbital fractures using patient-specific titanium milled implants: the Helsinki protocol. Br J Oral Maxillofac Surg. 2018;56(9):791–6.

26. Scolozzi P. Applications of 3D orbital computer-assisted surgery (CAS). J Stomatol Oral Maxillofac Surg. 2017;118(4):217–23.

27. Schreurs R, Becking AG, Jansen J, Dubois L. Advanced concepts of orbital reconstruction: a unique attempt to scientifically evaluate individual techniques in reconstruction of large orbital defects. Atlas Oral Maxillofac Surg Clin North Am. 2021;29(1):151–62.

28. Gellrich NC, Dittmann J, Spalthoff S, Jehn P, Tavassol F, Zimmerer R. Current strategies in post-traumatic orbital reconstruction. J Maxillofac Oral Surg. 2019;18(4):483–9.

29. Holmes S. Primary orbital fracture repair. Atlas Oral Maxillofac Surg Clin North Am. 2021;29(1):51–77.

30. Christensen BJ, Zaid W. Inaugural survey on practice patterns of orbital floor fractures for American oral and maxillofacial surgeons. J Oral Maxillofac Surg. 2016;74(1):105–22.

Open Access This chapter is licensed under the terms of the Creative Commons Attribution 4.0 International License (http://creativecommons.org/licenses/by/4.0/), which permits use, sharing, adaptation, distribution and reproduction in any medium or format, as long as you give appropriate credit to the original author(s) and the source, provide a link to the Creative Commons license and indicate if changes were made.

The images or other third party material in this chapter are included in the chapter's Creative Commons license, unless indicated otherwise in a credit line to the material. If material is not included in the chapter's Creative Commons license and your intended use is not permitted by statutory regulation or exceeds the permitted use, you will need to obtain permission directly from the copyright holder.

Surgical Treatment of Solitary Orbital Wall Fractures

10

Leander Dubois, Juliana F. Sabelis, Jesper Jansen, Thomas J. J. Maal, and Ruud Schreurs

Learning Objectives

- Support the surgical decision-making in the treatment of orbital fractures with a focus on timing, biomaterials, and approach.
- Master the surgical skills for predictable dissection and reconstruction of the orbit by using the anatomical landmarks.

- Improve basic knowledge about the advantages of virtual treatment planning, navigation, and intraoperative imaging.
- Know when to reconstruct the orbit with patient-specific implants.

Introduction

Fractures of the orbit are relatively common in maxillofacial trauma. They can occur in isolation or involve concomitant skeletal structures as well. There is variation in fracture complexity due to the amount of energy transfer, and clinical presentation can be heterogenous due to soft-tissue involvement. Proper physiological assessment, a full radiological description of the fracture configuration, and orthoptic evaluation form the foundation for decision-making and thereby predictable clinical outcomes [1–6].

The surgeon must combine this information with evidence-based indications for surgery, appropriate timing, knowledge of the different approaches, biomaterials, and advanced technologies. This can be a challenging task, and there is ongoing debate about the optimal management of orbital fractures in facial trauma care. The diversity in training and surgical skills adds a confounding component, since many surgical

L. Dubois (✉) · J. Jansen
Department of Oral and Maxillofacial Surgery,
Amsterdam University Medical Centers, Amsterdam,
The Netherlands
e-mail: l.dubois@amsterdamumc.nl;
j.jansen@amsterdamumc.nl

J. F. Sabelis · R. Schreurs
3D Lab Department of Oral and Maxillofacial
Surgery, Amsterdam University Medical Centers,
Amsterdam, The Netherlands
e-mail: j.f.sabelis@amsterdamumc.nl;
r.schreurs@amsterdamumc.nl

T. J. J. Maal
3D Lab Department of Oral and Maxillofacial
Surgery, Radboud University Medical Center,
Nijmegen, The Netherlands
e-mail: thomas.maal@radboudumc.nl

© The Author(s) 2023
P. J. J. Gooris et al. (eds.), *Surgery in and around the Orbit*,
https://doi.org/10.1007/978-3-031-40697-3_10

specialities manage orbital fractures [6, 7]. In addition, there is a lack of high-quality studies defining the standards of care for the optimal treatment of orbital fractures. The controversies on surgical indication are discussed in Chap. 9; this chapter describes the process after the indication for surgery has been established.

Surgical Decision-Making and Operative Procedure

The goals of surgical treatment of an orbital fracture are clear: the globe needs to be repositioned and the orbital volume restored, in order to recover the ocular function (Fig. 10.1). Although these goals are related to the orbital contents, the restoration of the orbital contour is the first and probably the most predictable step in orbital reconstruction [1].

The shape of the bony orbit and the intricate architecture of the soft tissue pose surgical challenges. Orbital reconstruction is performed in a confined space close to vital and delicate structures, with a limited overview. This presents a risk of iatrogenic damage and surgical complications [1]. Detailed planning and adequate expo-

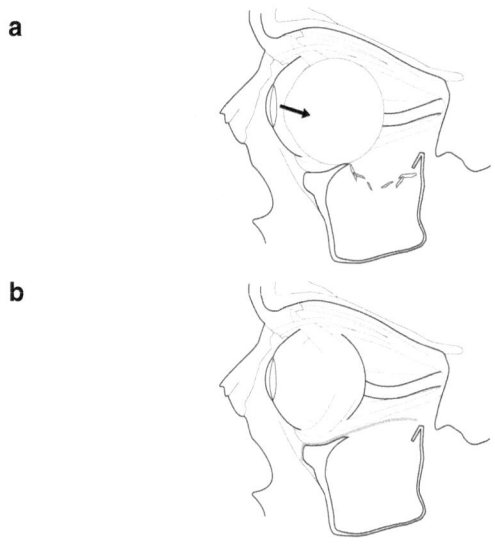

Fig. 10.1 Orbital defect (**a**) and reconstruction of the affected orbital walls by an orbital implant (**b**)

sure of the orbital floor and the medial wall are necessary to avoid complications. Considerations on timing [4–6, 8], biomaterials [4, 6], virtual surgical planning [9, 10], preformed or custom plates (patient-specific implants, PSIs) [10], approach [6, 8], navigation [9, 10], and intraoperative imaging [9] will be discussed in the following paragraphs.

Timing

The literature evidence on ideal timing of the surgical intervention is contradictory and difficult to interpret. Timing of surgery can be differentiated between immediate (<24 h), early (<2 weeks), and delayed treatment (2–6 weeks). Reconstruction after 6 weeks is not considered primary treatment any more, as the soft tissue and bone behave as a revision case [6]. Except for threatening visual emergencies and (pediatric) trapdoor fractures (Chaps. 12 and 13), all indications for surgery are relative [4, 6].

Several surveys on surgical decision-making on orbital fractures indicate that surgeons generally prefer to operate within 2 weeks after the injury [7]. The rationale behind early surgery is that the fracture is more easily accessible [2, 6]. There is less iatrogenic damage, because of the absence of fibrosis, and fewer adhesions in the orbital soft tissue are present. However, there is currently insufficient proof that postponing surgery harms the outcome. As clinical signs and symptoms may change in the days or weeks after trauma, a more delayed approach might influence the type and choice of treatment. The scientific rationale for these timing recommendations is vague considered that some indications for surgery will not always occur (enophthalmos) or will resolve spontaneously over time (diplopia) [5].

From a practical perspective, the surgical approach is easiest performed after ecchymosis and swelling has resolved, which is generally after 4–14 days. The additional time could be used for (virtual) treatment planning and, if indicated, fabrication of a PSI [2, 6, 10].

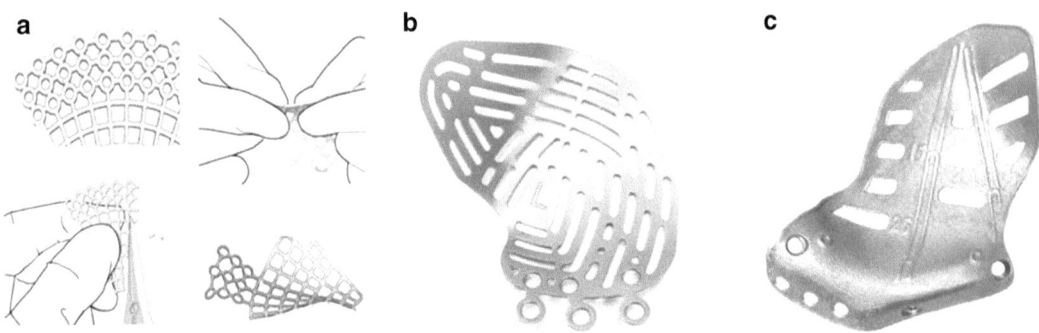

Fig. 10.2 Orbital implant options: flat (**a**), preformed (**b**), patient specific implant (**c**)

Biomaterials

A wide variety of implant materials have been used to reconstruct orbital wall fractures. Implant materials must have specific characteristics to achieve adequate reconstruction of the pretraumatised anatomy and correction of enophthalmos or diplopia. The ideal reconstruction material has perfect architecture or contouring abilities to restore volume and shape, is biocompatible, facilitates drainage of fluids, has no donor site morbidity, is radiopaque, is stable and allows fixation, and is readily available at reasonable cost [3, 11]. Based on their excellent biocompatibility, autologous bone grafts used to be the gold standard. The main disadvantages of autologous bone grafts are donor site morbidity, unpredictable resorption rate (up to 86%), and difficulties in shaping the graft. These drawbacks inspired the development of alloplastic materials that are currently considered the gold standard for reconstruction. Titanium in specific adheres to most of the abovementioned demands and is widely used, either as preformed orbital reconstruction plates, patient-specific implants, or flat titanium meshes for intraoperative moulding (Fig. 10.2).

Virtual 3D Planning and Implant Choice

Since orbital reconstruction is performed in a confined space with a limited overview, a computer-assisted surgery (CAS) workflow can

be of great added value. The first steps in this CAS workflow are generation of the virtual patient model and advanced diagnostics to support diagnosis and indication (Chap. 9). The virtual planning is performed after surgery is indicated, but it utilizes the information already obtained in the advanced diagnostics process. The unaffected mirrored orbit that had been positioned over the affected orbit for obtaining detailed insight into the extent and displacement of the fracture is used as the target anatomy for reconstruction (Fig. 10.3).

The next step is to select an implant that can mimic this target anatomy as closely as possible [12–15]. A virtual model of an implant with a predefined shape, such as a preformed implant, can be imported in dedicated virtual planning software. The 3D model of the implant can be positioned in the virtual patient: the contours of the implant are shown in the 3D model and the multiplanar views. The patient model provides information on existing anatomical structures; visualization of the mirrored orbit's overlay adds the desired reconstruction information to the multiplanar views (Fig. 10.4).

Several positions of the implant may be tested to find the optimal position of the implant. Different aspects are considered to evaluate the position of the implant in the virtual surgical planning. Apart from reconstructing the orbital contour as accurately as possible, it is important that the implant covers the orbital defect, is sufficiently supported by existing bony structures (e.g., support on the dorsal ledge, bony support at

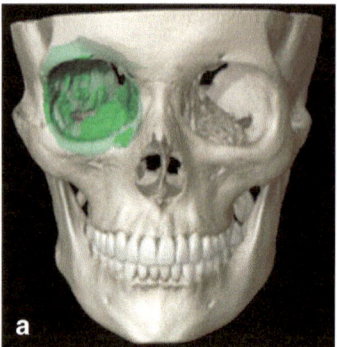 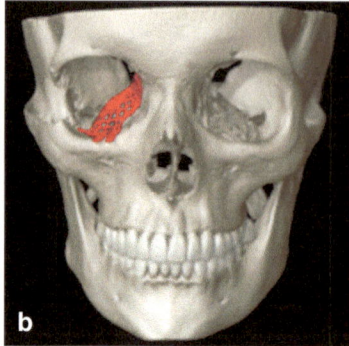 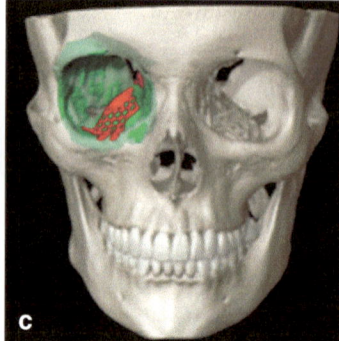

Fig. 10.3 Virtual surgical planning. The mirrored orbital contour is considered the target anatomy for reconstruction (**a**). A virtual model of the implant is imported and positioned in the orbit of the patient (**b**). The mirrored model can be visualized in the positioning process and serves as a blueprint for the target anatomy the implant should reconstruct (**c**)

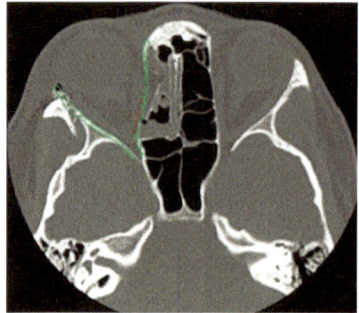 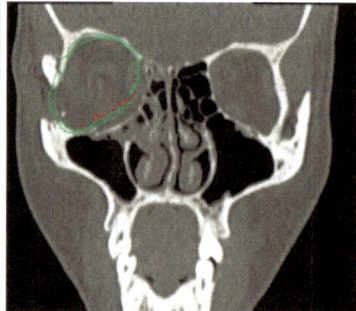 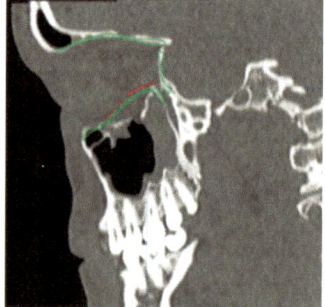

Fig. 10.4 Visualization of the implant model and mirrored orbit in the multiplanar views. The implant follows the mirrored contour nicely, has support on the posterior ledge, and can be fixated on the infraorbital rim

the medial tip of the implant), and allows fixation at the infraorbital rim [16–18]. The planning process may be repeated for preformed implants of different sizes or from different manufacturers, and the implants may be virtually trimmed at predefined locations to evaluate if a reduced size of the implant still meets the positioning requirements. These options ensure that the possibilities of preformed implant reconstruction can be thoroughly evaluated. The optimal implant positioned in the ideal position may be considered the end point of the virtual surgical planning in case of a preformed implant [12, 16, 19].

Preformed or PSI

Preformed titanium implants can provide an adequate reconstruction in the majority of orbital reconstructions. Their predefined shape, which allows virtual surgical planning, is frequently based on the shape of the average orbit and thus provides a satisfactory resemblance to the majority of orbits. The aforementioned trimming at predefined locations increases their fitting potential. In limited cases, an optimal fit cannot be achieved using a preformed implant (Fig. 10.5). An inaccurate virtual fit of preformed implants

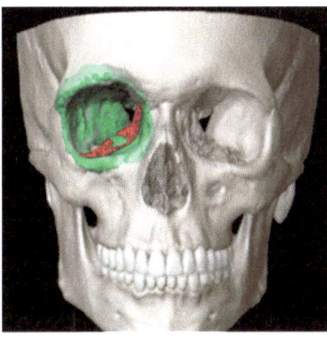
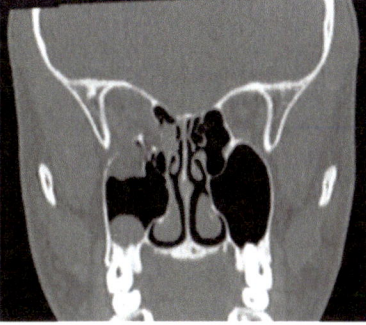
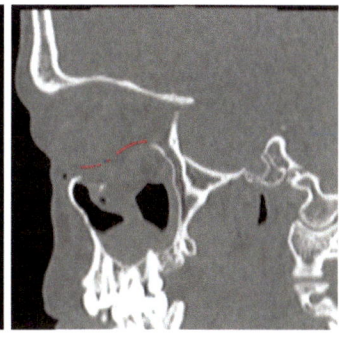

Fig. 10.5 Example of an incorrect fit of a preformed implant. Although the largest implant size was chosen in this example, the implant does not reach the OPPB. The lack of posterior support led to the decision to use a patient-specific implant

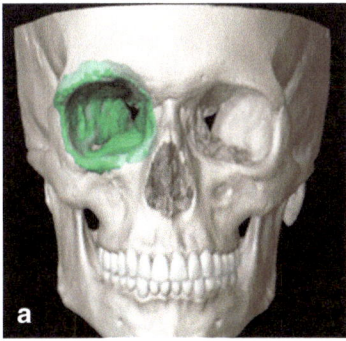
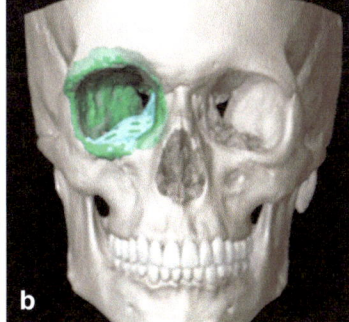
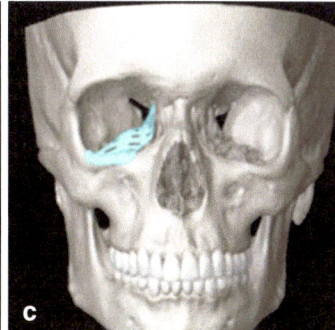

Fig. 10.6 Design of a patient-specific implant. The advanced diagnostics information (**a**) is the basis for designing a patient-specific implant (**b**). Several virtual prototypes may be designed before the virtual planning is completed (**c**) and the final design can be produced

may lead to the decision to use a patient-specific implant (PSI), but a PSI may also be indicated because of clinical considerations such as a need for overcorrecting the volume of the orbit.

The advanced diagnostics information that is utilized in PSI cases is identical to the information used in virtual surgical planning of a preformed implant. The virtual surgical planning process and result differ in a PSI reconstruction; information about anatomical boundaries of the fracture and optimal reconstruction shape is exported from the virtual planning software. Design software is used to model an implant that meets the requirements on defect coverage and reconstruction of the pre-traumatized anatomy

(Fig. 10.6). Prototypes of the implant can be imported into the virtual surgical planning to verify their fit and evaluate the necessity of alterations to the design. If the virtual planner and surgeon are satisfied with the implant design, the virtual model of the PSI may be sent to a manufacturer for fabricating the physical implant in titanium using laser sintering.

A PSI offers enhanced possibilities to tailor the implant's shape to the patient's anatomy (Fig. 10.7). Support on existing bony ledges and an extension over the orbital rim provide feedback on implant positioning to assure that the planned position is reached. The implant's shape can be adjusted to meet the clinical needs of the

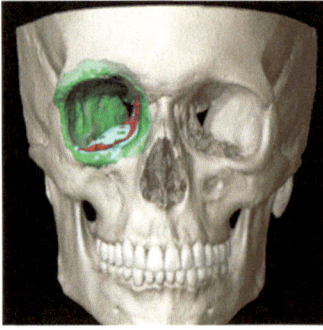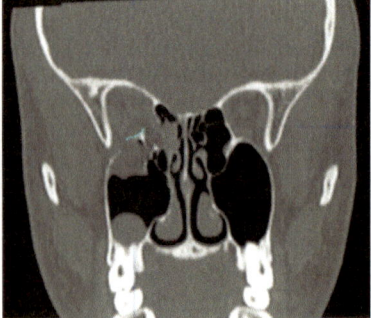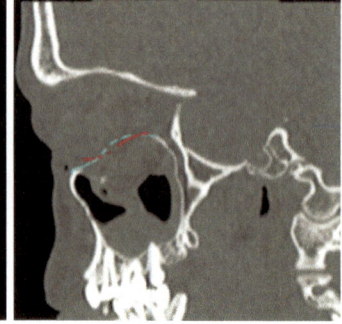

Fig. 10.7 Tailoring the implant to the patient's anatomy and fracture specifics. Compared to the preformed implant, the patient-specific implant exhibits an improved fit on the infraorbital rim. The patient-specific implant is supported by the OPPB, and an overcorrection is embedded in the implant's design

patient as well: an overcorrection in volume or rim height may be embedded to resolve enophthalmos or hypoglobus [14]. Although patient specific implants offer improved positioning and clinical outcome compared to preformed implants, the high cost and logistical demands prohibit their widespread use. Considering these restrictions, the use of PSIs is currently the gold standard only in complex or secondary reconstruction of the orbit [20].

Surgical Technique

Approach

The surgical access to the orbit has undergone a true evolution over the last few decades. In the early nineties, the coronal approach, which was used in neurosurgery, became the working horse in orbital surgery; it was often combined with a skin incision in the lower eyelid. Drawbacks like scar visibility and surgical invasiveness of this extensive procedure have driven the development of more cosmetic incisions around the orbit that deliver at least equivalent but often better surgical access. Widely used approaches to restore the orbital boundaries after an orbital wall fracture are a transconjunctival, subtarsal, or subciliary approach. In the last decade, the transconjunctival incision has gained popularity and has become the standard approach for orbital reconstruction.

The transconjunctival approach has no visible scarring, as it is covered by the lower eyelid. A good exposure can be established and the risk of complications, such as entropion, is relatively low. In theory, two modifications can be used: the preseptal and the retroseptal route. The preseptal route has the advantage that the fat prolapse is limited compared to the retroseptal route. The drawback is that the preseptal route theoretically has a higher chance of postoperative entropion. A modification of the retroseptal route, which partly alleviates the fat prolapse, is explained below.

The approach starts by placing two non-resorbable sutures, one around the inferior rectus muscle and the other on the inferior tarsal plate of the lower eyelid. The first suture is used to raise the globe, so that the upper eyelid can provide protection during surgery. The other one allows the lower eyelid to be everted more easily using the Desmarres retractor. The conjunctiva is stretched and incised with a diathermy or scalpel. The incision should be started approximately four millimetres from the fornix and run from the caruncle to the lateral ligament. The inferior orbital rim should be palpated as a target for the approach. Next, the diathermy can be used to explore the retroseptal loose connective tissue down to the periosteum while keeping tension on the eyelid. The Desmarres retractor is reversed for retaining the lower eyelid, and the periosteum is incised to gain access to the orbital cavity.

A major advantage of the transconjunctival approach is that the medial wall can be exposed with a medial transcaruncular extension and exposure of the floor and rim can be improved

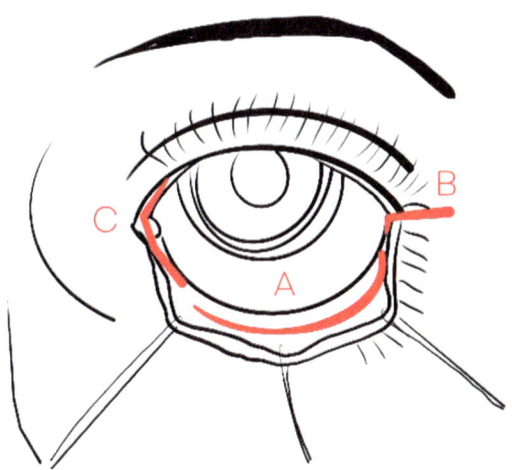

Fig. 10.8 Transconjunctival approach (**a**), lateral extensions(canthotomy/cantholysis) (**b**); medial extension (transcaruncular)

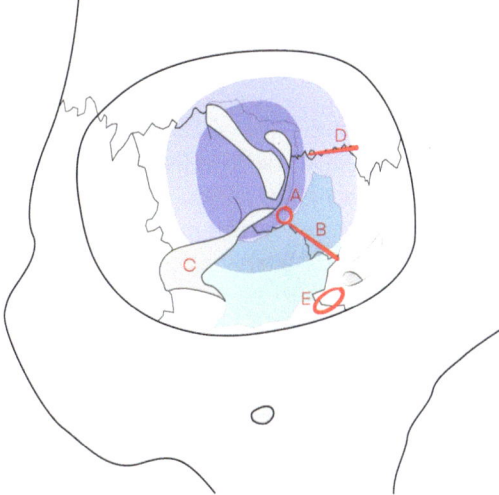

Fig. 10.9 Landmarks in the orbit for orbital reconstruction: medial strut (**a**), orbital process of palatine bone/keyzone (**b**), inferior orbital fissure (**c**), superior strut (**d**), attachment of inferior oblique muscle (**e**)

with a lateral canthotomy or, if necessary, a cantholysis (Fig. 10.8). For the lateral canthotomy, the existing incision can be expanded by stripping the lower part of the lateral canthal tendon, followed by an incision of approximately 1 cm through a natural skin crease. By extending the incision medially through the caruncle, the medial wall up to the roof of the orbit can be reached. The anatomy in this region is complex, with the Horner muscle, medial rectus muscle, and lacrimal drainage system. The incision towards the medial wall must therefore be made through the fibrous part of the caruncle.

Dissection of the Orbit

After subperiosteal access is gained at the inferior orbital rim, the dissection starts laterally. The relatively solid bone of the sphenozygomatic region is intact in most patients, which allows the surgeon to develop the surgical plane relatively easy. The next step is identifying the inferior orbital fissure (Fig. 10.4). The contents may be cauterized by bipolar diathermy. This fissure can be followed to the posterior orbit. On the medial side of the posterior part of the inferior orbital fissure, a ledge may be identified, which is formed by the orbital process of the palatine bone (OPPB). This ledge, or OPPB, consists of relatively solid bone and remains intact in most cases. For that reason, the OPPB is one of the

main pillars of the orbital reconstruction [21]. Beat Hammer identified this region as key zone and Jaquiéry also incorporated this structure in his classification [22]. The location is visualized in Fig. 10.9.

The next steps are medial dissection and finding a way around the fracture. The bony structures in the anterior third of the orbit may be used for this, provided these are unaffected by the fracture. Medial traction is usually enough to mobilize the tissues bulged through the defect. A meticulous dissection with a periosteal elevator is required to find the medial border of the defect. This border can be followed to the OPPB. The medial wall can be extremely thin and can easily be damaged by the exploration of the orbit. Nevertheless, the medial strut, which forms the boundary between the orbital floor and medial wall, is relatively rigid. This region can be identified as a white line when it is intact (Fig. 10.4). In two-wall defects, the medial strut may also be dislocated; a more cranial dissection is advised, and a transcaruncular extension of the transconjunctival incision can be helpful. The orbital contents should be handled very gently. If the orbital contents are released from the fracture gap, it is best to remove remaining sharp edges.

Polydioxanone (PDS) or neuro patties can be used to bundle the contents and keep the soft tissue away during insertion of the reconstruction material.

The aforementioned principles can also be used in case of a trapdoor fracture, but gentle pressure is enough to release the entrapped tissues in most cases. Besides removal of the sharp edges, no further action is necessary.

Relevant Surgical Landmarks Related to the Reconstruction

The globe and the ocular muscles receive ligamental support from the periorbita, which contains a network of connective tissue septa, surrounded by fat. The ligaments are attached to the orbital walls, as described by Koornneef [23]. Unfortunately, the structural integrity of the orbital septae may be compromised as a result of trauma and will certainly be (further) disrupted due to the approach and surgical dissection. Although orbital surgery requires meticulous dissection skills, it remains almost impossible to predictably redress all the soft tissue in the correct position and original dimension. Even after the release of the disrupted or entrapped orbital contents, a certain amount of disturbed soft tissue anatomy, scarring, and fibrosis can definitely be expected.

A properly positioned implant can restore the comminuted bony structures and volume of the orbit. It creates a solid fundament to facilitate free movement of the globe. The orbital implant must be gently inserted between the bone and the soft tissues. The implants require support at the anterior, posterior, lateral, or medial boundaries to obtain a stable position. Several strategic landmarks are fundamental for restoring the appropriate anatomical relations. The orbital process forms the important posterior ledge to dock the implant on. In large two-wall defects, the rotational freedom of the orbital implant is increasing and additional support on the superior strut (border of medial wall and orbital roof) can be of added value to prevent rotational outliers in implant position [24] (Fig. 10.10).

Although the focus lies on the posterior landmarks, the position of the anterior rim is of equal

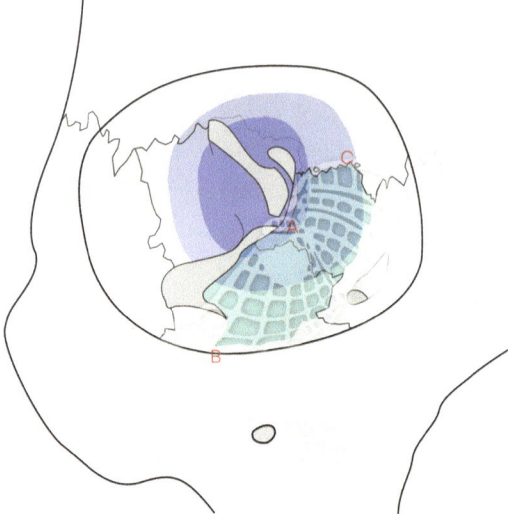

Fig. 10.10 Implant support: ledge (**a**), anterior rim (**b**), medial wall/superior strut (**c**)

importance. In impure blow-out fractures with concomitant facial fractures (ZMC, NOE of Le Fort II/III), the vertical position of the globe is mostly determined by the position of the anterior rim. Suboptimal repositioning of the ZMC automatically leads to a poor position of the orbital implant, especially since the rim is required for fixating the orbital implant anteriorly. This is one of the reasons that the facial pillars must be restored anatomically before the orbit is reconstructed. A more caudally positioned rim will lead to a suboptimal implant position, an increase of orbital volume, and potential enophthalmos or hypoglobus.

Intraoperative Navigation and Imaging

The ideal implant position has been determined during the virtual surgical planning: the goal for the surgeon is to position the implant as close as possible to the planned position during surgery. Because of the confined space and protruding soft tissue, a visual assessment of the implant position is infeasible and realization of the planned position is difficult. The possibility to consult the planning aids the surgeon during reconstruction by creating artificial anatomical landmarks for evaluating the implant position. In the computer-assisted surgery workflow, additional technology may be introduced that can transfer the preoperative planning to the intraoperative setting and provide more sophisticated feedback on implant positioning. Surgical navigation and intraoperative imaging are effective intraoperative technologies in orbital reconstruction; these techniques will be explained in more detail in the following paragraphs.

Navigation

Many people will associate the term navigation with the route guidance in a car. The navigation system tracks the location of the car using GPS, knows the destination, and provides feedback to the driver on the position of the car in relation to the destination. Surgical navigation is a comparable technique, but the working principle is different, as is explained below. Surgical navigation can provide visual and quantitative feedback on the realized implant position with respect to the planned position. Although surgical navigation is an expensive and logistically complex technique, it has shown to improve predictability of orbital reconstruction significantly [25–27].

The navigation system is constituted of the following components: an infrared camera, a patient reference frame, a surgical navigation pointer, and a computer system plus screen (Fig. 10.11). The patient reference frame has reflective marker spheres attached, which allow the camera to detect the position of the frame. The patient reference frame is rigidly attached to the patient's cranium to track the position of the patient within the operation room. Similarly, the navigation instrument has markers attached to track its position. It is essential that the line-of-sight between the camera and the reflective spheres is not blocked, since this would hamper detection of the frame or instrument. The ultimate goal is to visualize the position of the instrument within the virtual planning, but this requires a registration procedure.

The registration procedure establishes relationship between the physical position of the patient and the virtual surgical planning. Registration relies on identifying similarities between the physical patient and the surgical planning. Reference points are generally utilized in facial traumatology. These points are identified in the virtual planning and indicated on the patient. Easy and reproducible identification of these points in the virtual planning and on the

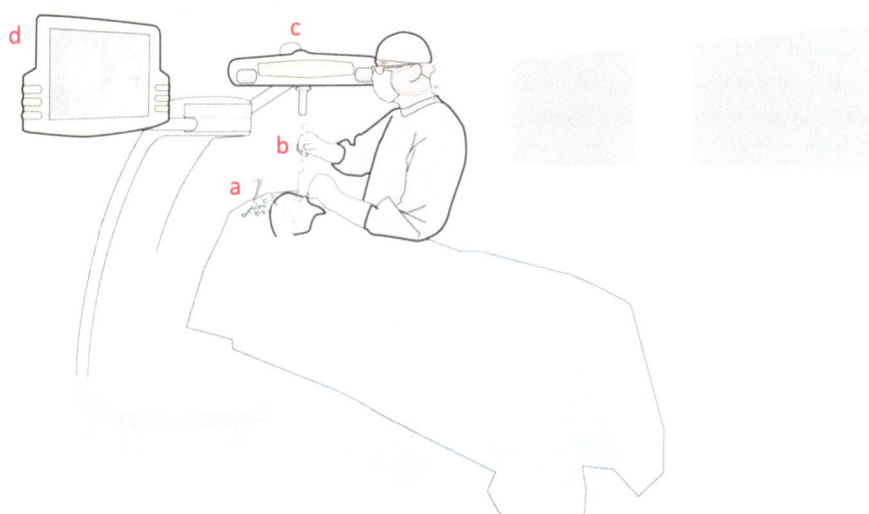

Fig. 10.11 Surgical navigation system with its different components. (**a**) patient reference frame, (**b**) surgical instrument, (**c**) camera, (**d**) screen

patient is essential for an accurate match, which heavily affects the accuracy of the subsequent navigation feedback. Since anatomical landmarks don't meet these indication requirements, the reference points should be artificially created. Bone-anchored titanium screws may be inserted intra-orally or in pre-existent wounds [28]. The screws need to be dispersed as much as possible to ensure a good registration result.

A less invasive method is the use of a dental occlusal splint that embeds the markers [29], but this may require additional imaging (radiation exposure) to the patient since the splint may not be present in the initial computed tomography (CT). A virtual dental registration splint can be obtained by fusing an intraoral scan with the CT; the splint can be 3D printed and used to indicate the registration points on the patient during surgery [30, 31]. This method decreases the radiation exposure, but it does require specific hardware and software. The navigation error yielded by all registration approaches discussed is below 2 mm in the orbital region [28–30]. The decision for a registration method can be made on an individual basis and depends on patient and hospital characteristics.

After the registration procedure, the surgeon can use the navigation to compare the realized implant position to the planned position with the navigation pointer. If the navigation pointer is positioned in the patient, the screen will show the location of the tip of the pointer in the multiplanar view (sagittal, coronal, and axial slices) and in the 3D model of the preoperative CT scan. When the pointer is moved, the view is changed to the slices of the new position. Visualization of the planned implant contour in the multiplanar views helps the surgeon to assess if the planned implant position has been obtained. The reconstructed contour can be evaluated by moving the pointer along the implant, but it is more effective to use predefined marker points on the implant [32]. These marker points can be indicated in the virtual surgical planning, allowing the system to compute the difference in planned and realized marker positions This approach provides quantitative feedback and gives the surgeon an indication on the direction of repositioning.

Even though navigation provides the surgeon with relatively precise information on the implant position, there is some residual error. The most important source of error is the registration procedure. The surgical outcome of orbital surgery should always be validated with a radiological control. This radiological control used to be performed postoperatively. Technological advancements have made cone-beam computed tomography (CBCT) scanners available in different (mobile) setups to facilitate intraoperative imaging [17]. Since the image quality of the mobile CBCT scanners is sufficient, the need for postoperative imaging is eliminated.

Implant positioning can be evaluated on the intraoperative scan for defect coverage and bony support. The intraoperative imaging can also be fused with the virtual surgical planning to provide more thorough feedback on the obtained implant position. The literature has shown that if the implant is positioned without navigation, the surgeon will alter the position of the implant in half of the cases based on intraoperative imaging. Intraoperative imaging significantly improves the implant positioning [33], reduces the need for secondary interventions, and saves operation time [17].

Computer-Assisted Evaluation

Evaluation is an important part of the computer-assisted surgery workflow. The intraoperative or postoperative imaging can be fused with the preoperative 3D planning, which allows objective assessment of the surgical result. The deviation of the final implant position can be quantified and expressed as rotations and translation from the planned position. The volume difference between the unaffected and reconstructed bony orbit may also be assessed. Evaluation and quantification are very instructive and insightful for the inexperienced surgeon to identify errors in planning or surgery. This enhanced knowledge will aid in the planning and surgery of future trauma cases.

Recovery and Follow-up

Patients are hospitalized for one or two nights after orbital reconstruction, depending on postoperative pain and how self-sufficient the patient is. In the first hours after surgery, a retrobulbar hematoma may develop. This can cause compression of the optic nerve and, potentially, loss of vision. For this reason, it is essential to perform pupil function and vision assessment at least hourly in the first 4 h after surgery.

Postoperative swelling and pain, formation of a hematoma, contusion of the ocular muscles, and changes in the position of the globe and muscles generally lead to an increase or at least persistence of symptoms (diplopia and limited eye motility) in the first weeks after surgery. Intraocular swelling can lead to temporary exophthalmos or an elevated position of the globe. The patient should be prepared for this during the preoperative consultation. The patient is advised to start monocular orthoptic exercises three times a day within a few days after the orbital reconstruction to improve motility. These exercises might additionally resolve swelling or avoid early adhesions.

As discussed in the previous chapter, patients are scheduled for follow-up 2 weeks, 6 weeks, and 3 months after surgery. In this period, a significant improvement in ocular motility is often noticed as well as a decrease in diplopia. Both diplopia and motility can be confirmed with objective orthoptic measurements. Further recovery and adaptation will occur up to 1 year after surgery, with possible further subjective improvement of the symptoms.

Despite an anatomical reconstruction of the bony orbit and repositioning of the soft tissue, permanent residual diplopia and limited eye motility may occur. These could be the result of adhesions, local entrapment, or a disruption of the periorbita, the suspension system of the intraocular soft tissue. In general, severe motility restriction and diplopia in central gaze lead to residual symptoms. Enophthalmos can persist due to insufficient restoration of the orbital contours or decrease of soft-tissue volume due to atrophy. Entropion, ectropion, and increased scleral show are complications directly related to the surgical approach. Other possible surgical complications are infraorbital nerve dysesthesia and epiphora.

A well-prepared and performed orbital reconstruction will often lead to a significant improvement of the initial complaints. The globe position will be restored in most cases if the reconstructed orbital wall contours resemble the pretraumatized contours. These contours will form the fundament for redressing of the soft-tissues in order to facilitate ocular motility and diminish the diplopia. A clinical example of clinical improvement after surgical reconstruction is presented in the clinical example.

Clinical Example of Surgical Treatment of a Blow-out Fracture

A 21-year-old female was referred to the department of oral and maxillofacial surgery with a dislocated orbital floor and medial wall fracture (class III), combined with a lateral wall fracture on the right side (Figs. 10.12a–c and 10.13a–c).

Her main clinical problem was a limited elevation with diplopia at elevation and depression, and an evident step at the frontozygomatic suture. There was no significant enophthalmos (Hertel 18/19). Orthoptic evaluation objectified the limited ocular motility at elevation (31° OD / 40° OS) and depression (40° OD / 51° OS), with a binocular single vision (BSV) score of 38/100 points (severe diplopia). The patient was scheduled for early orbital reconstruction with VSP and surgical navigation because of the severe diplopia and dislocated lateral wall.

In the multiplanar reconstructions of the CT scan, the amount of dislocation of the orbital walls can be easily assessed (Figs. 10.14 and 10.15). The DICOM data was imported into the Brainlab software. The advanced diagnostics, with segmentation of the unaffected orbit and mirroring to the affected orbit, are visualized in Figs. 10.15 and 10.16. Several preformed implants were assessed in the virtual surgical planning; the implant of choice at the optimal position is shown in Fig. 10.17a, b. The addi-

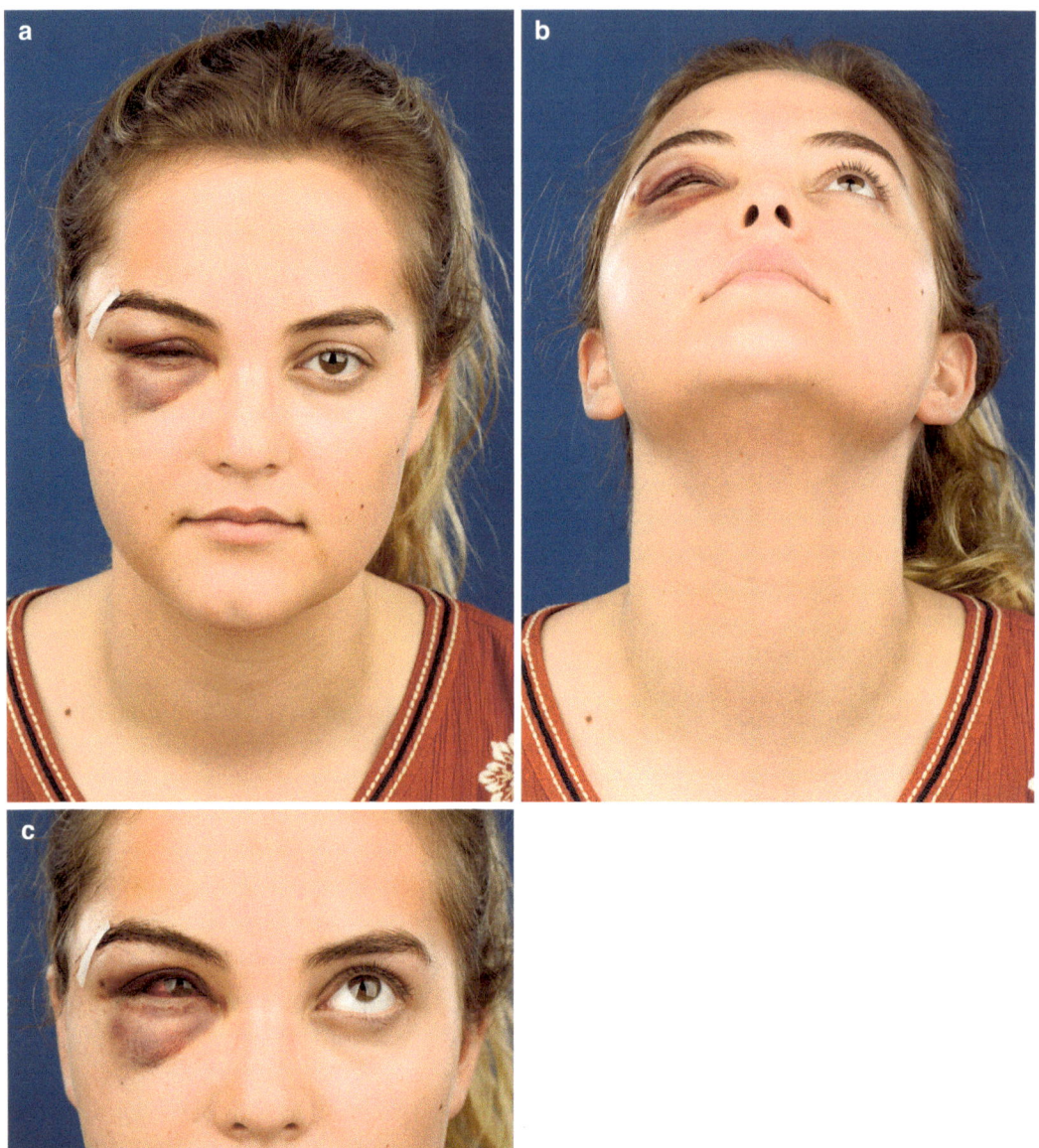

Fig. 10.12 (**a–c**) Clinical appearance at first presentation, (**a**) en face, (**b**) submental, (**c**) elevation

tional screw holes or extensions can be cut beforehand to prevent unnecessary intraoperative implant adjustments.

The orbital defect is reconstructed with a preformed orbital implant. The implant position was controlled with the help of surgical navigation and verified using intraoperative imaging (Fig. 10.18a–c). Superimposition of the intraoperative CBCT scan on the virtual surgical planning enabled direct comparison between planned position and actual result (Fig. 10.19).

The patient was discharged from the hospital 1 day after surgery. She continued her studies 2 weeks after the reconstruction. Normal ocular motility was restored directly after surgery. The

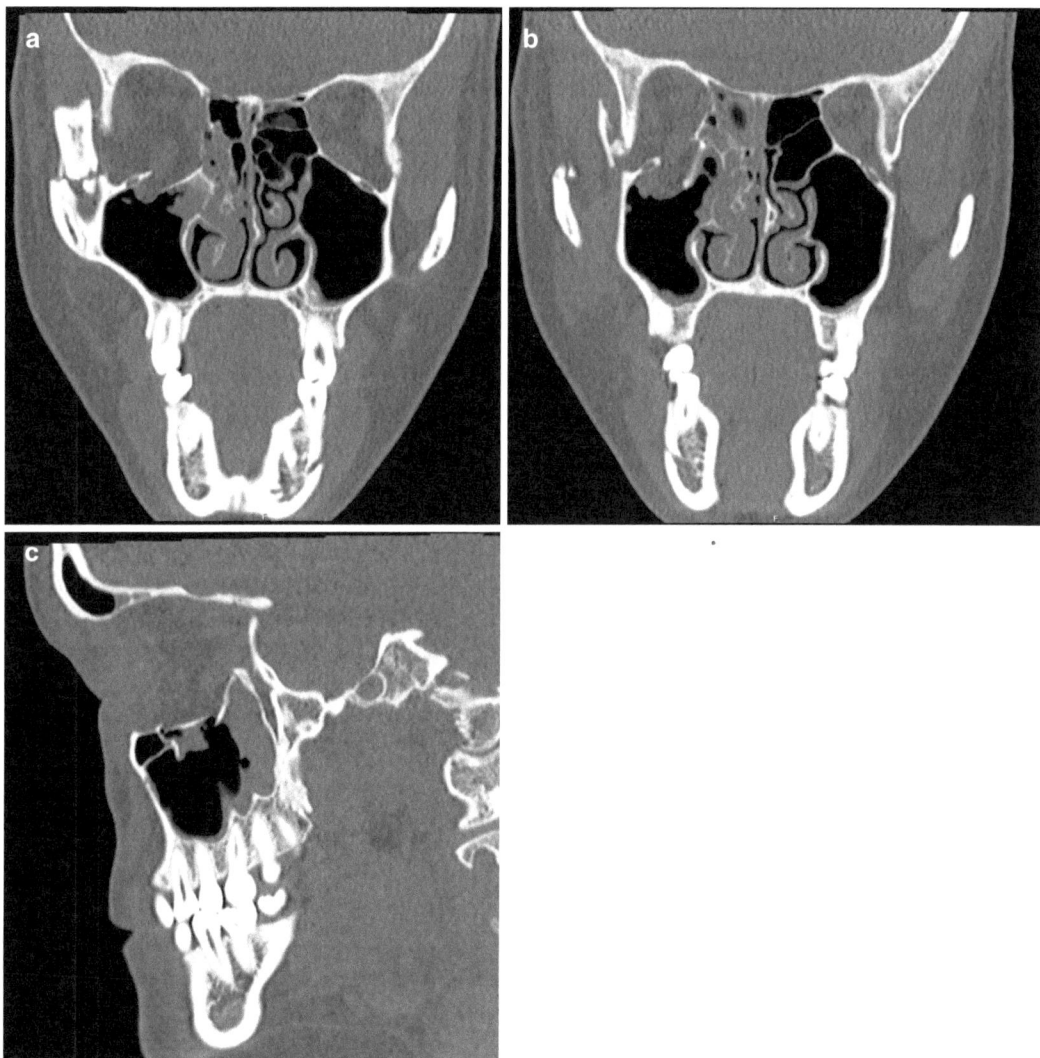

Fig. 10.13 Coronal (**a**, **b**) and sagittal (**c**) views of the CT scan at first presentation

diplopia dissolved in 3 months. During the follow-up of 12 months, no enophthalmos occurred (Fig. 10.20a–c). Figure 10.21a–d shows the improvement of the BSV over time. The patient was satisfied with the recovery of her ocular function and the excellent clinical result.

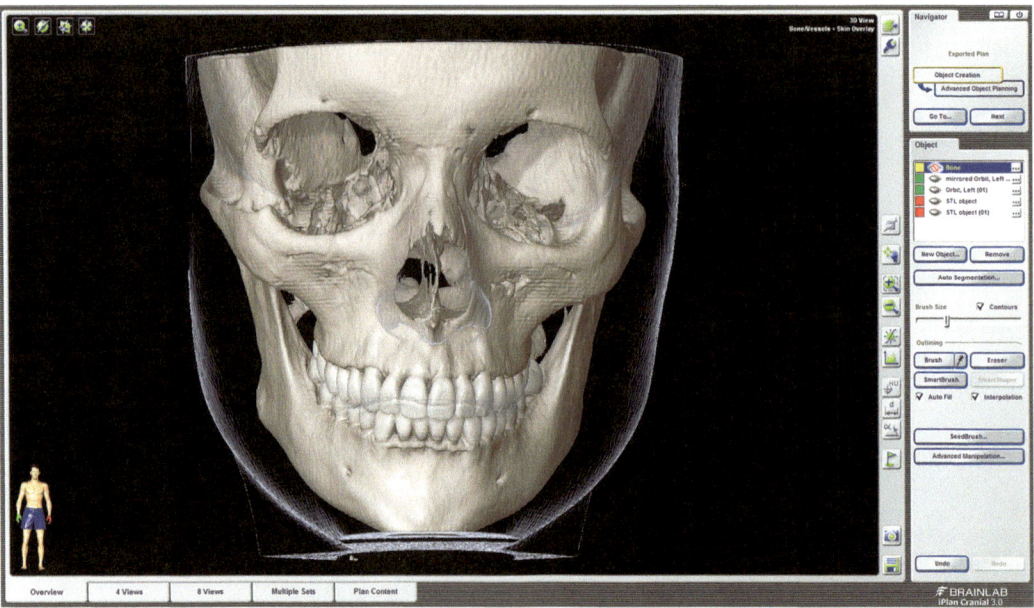

Fig. 10.14 The amount of dislocation of the orbital walls can be easily assessed during the preoperative planning

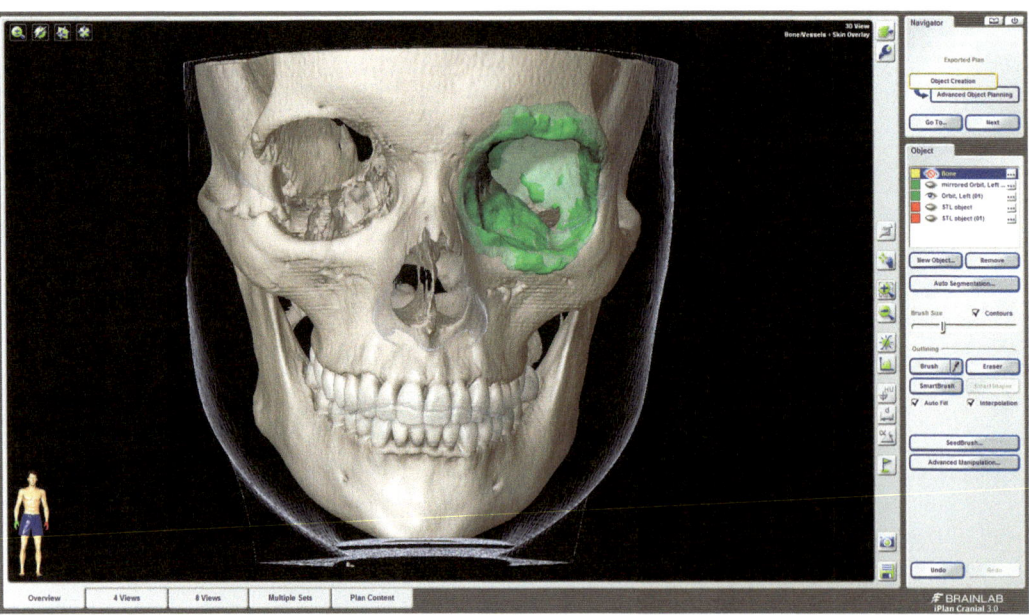

Fig. 10.15 Segmentation of the unaffected side

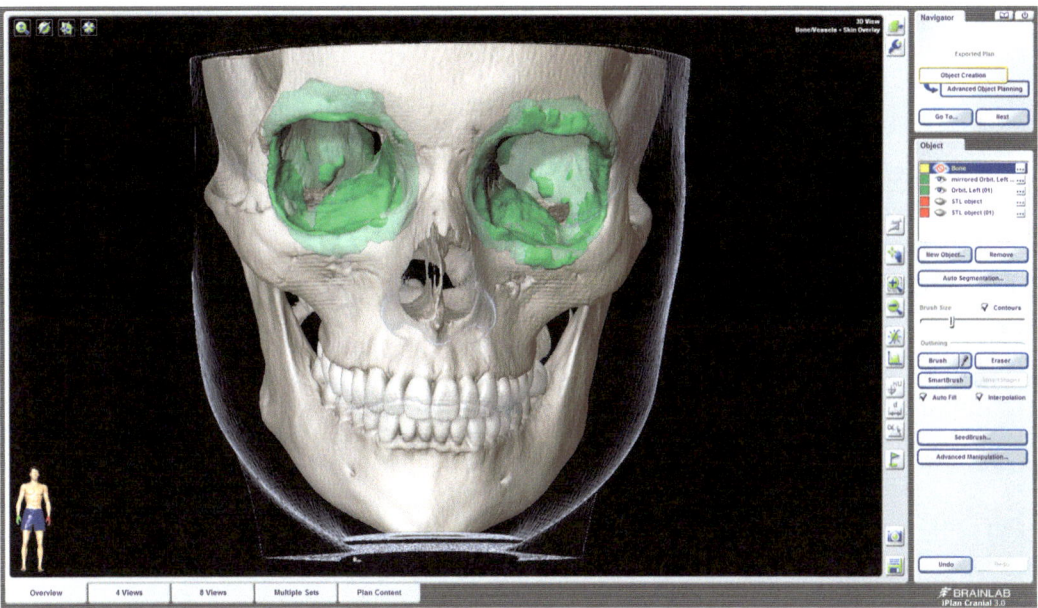

Fig. 10.16 Mirrored to the contralateral side to mimic the pre-traumatized anatomy

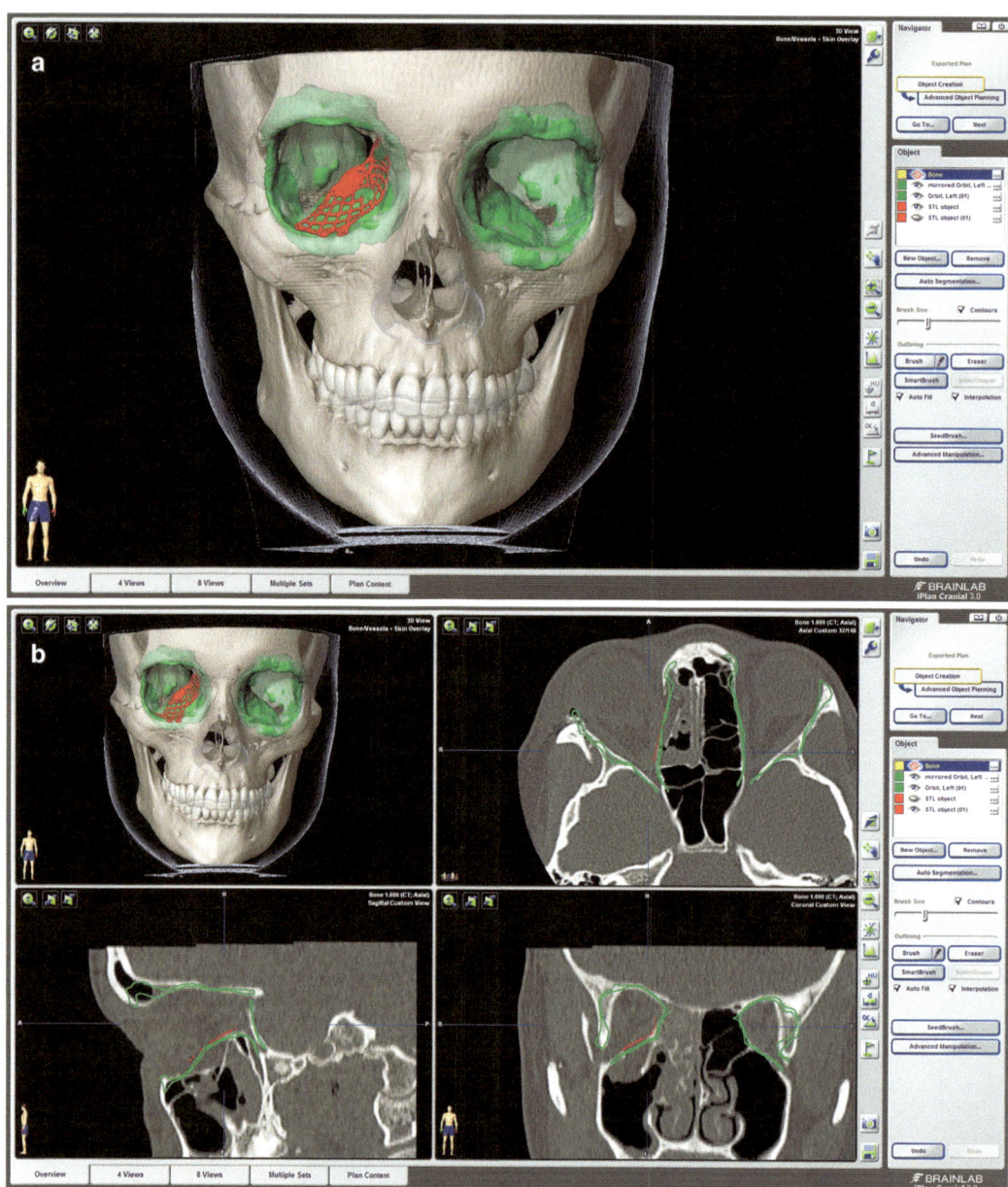

Fig. 10.17 (**a**, **b**) The STL file of the best fitting implant imported into the software

Fig. 10.18 (a–c)
Intraoperative imaging
for quality control

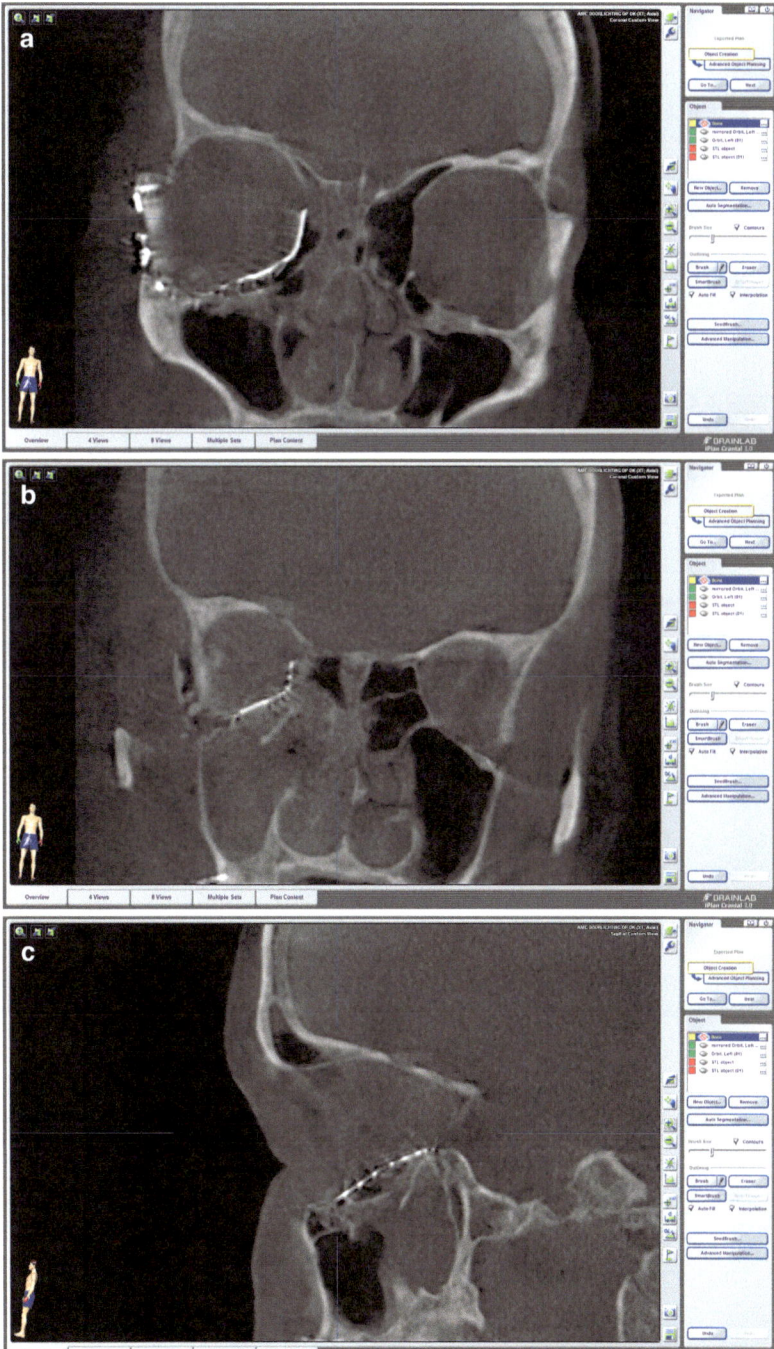

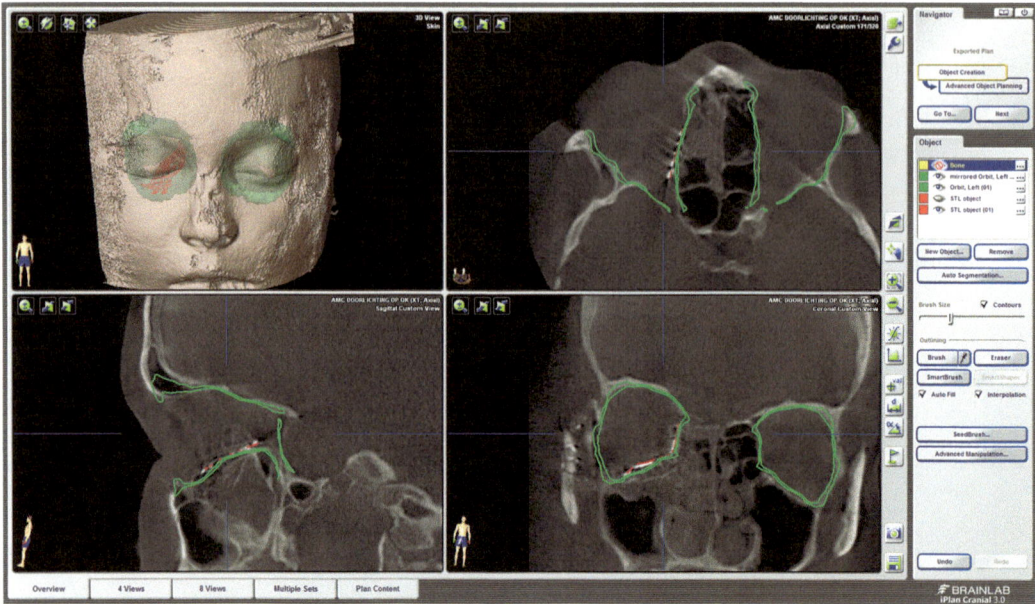

Fig. 10.19 Superimposition of the intraoperative imaging on the virtual surgical planning, which allows one-to-one comparison of the planned and obtained implant position

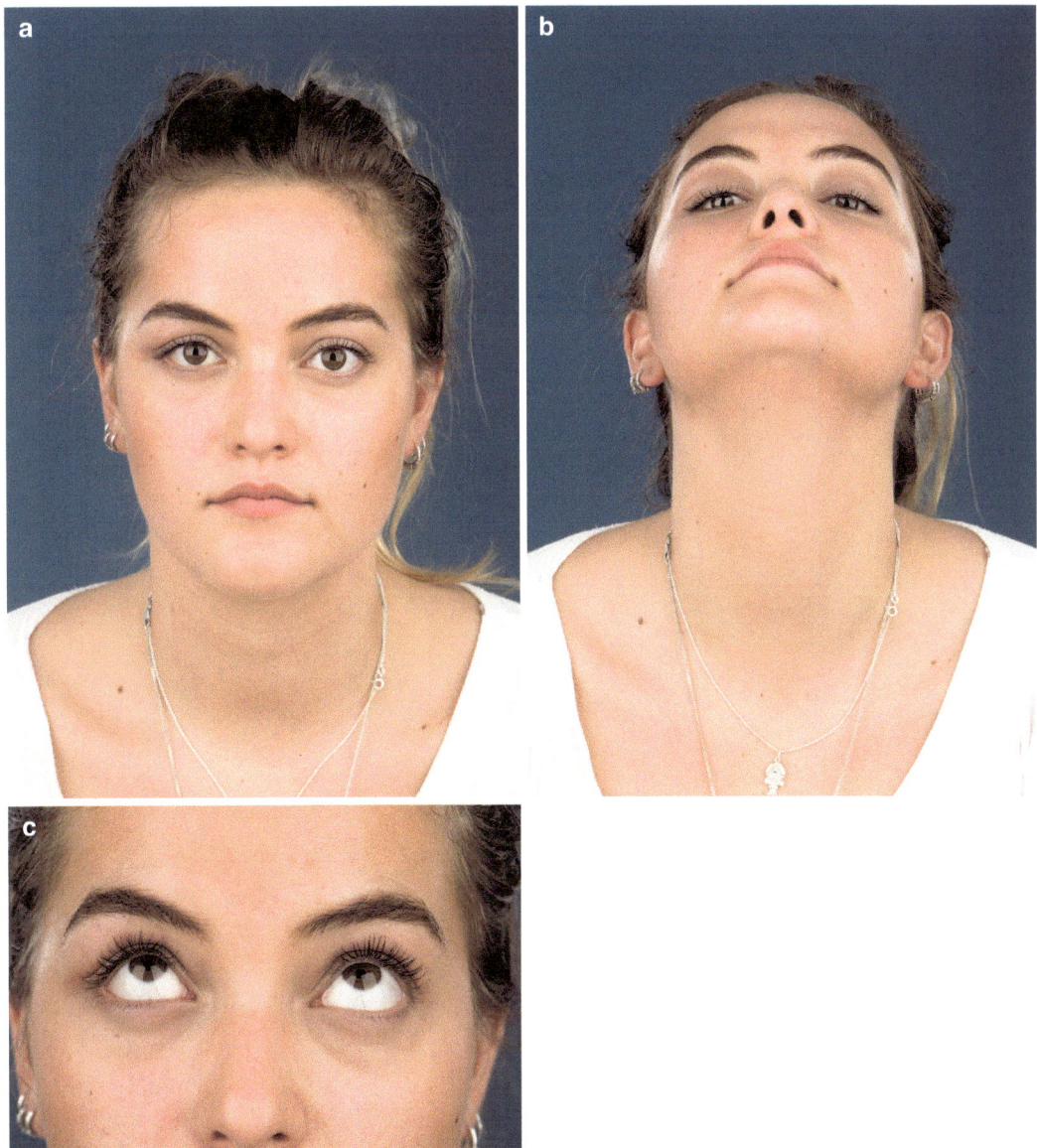

Fig. 10.20 (**a–c**) Clinical result 1 year after surgery, (**a**) en face, (**b**) submental, (**c**) elevation

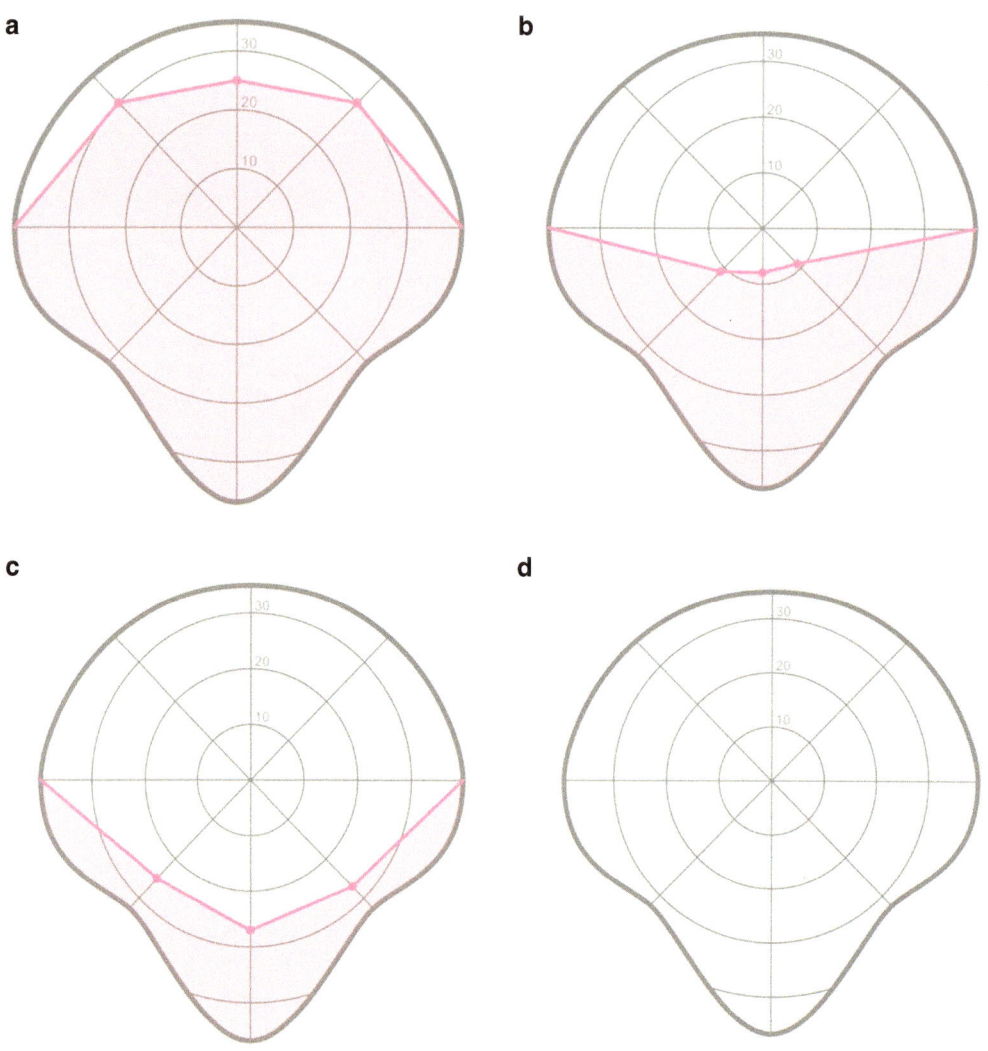

Fig. 10.21 (**a–d**) Binocular single vision (BSV) (**a**) preoperative, (**b**) 2 weeks after surgery, (**c**) 6 weeks after surgery, (**d**) 1 year after surgery

Conclusion

Orbital wall reconstruction can be a complex procedure with variable results. The limited overview, keyhole access and, in complex cases, loss of anatomical references are all challenges that need to be overcome to reconstruct the orbit adequately. Proper clinical decision-making forms the fundament of treatment. The orbital implant positioning should be as accurate and reliable as possible and add to bulb position, ocular movements, soft-tissue volumes, and esthetic outcome. Although meticulous dissection skills of an experienced surgeon are mandatory, a wide range of medical technology can help improve the quality and predictability of treatment further.

References

1. Dubois L, Steenen SA, Gooris PJ, Mourits MP, Becking AG. Controversies in orbital reconstruction-I. Defect-driven orbital reconstruction: a systematic review. Int J Oral Maxillofac Surg. 2015;44(3):308–15.
2. Dubois L, Steenen SA, Gooris PJ, Mourits MP, Becking AG. Controversies in orbital reconstruction-II. Timing of post-traumatic orbital reconstruction: a systematic review. Int J Oral Maxillofac Surg. 2015;44(4):433–40.
3. Dubois L, Steenen SA, Gooris PJ, Bos RR, Becking AG. Controversies in orbital reconstruction-III. Biomaterials for orbital reconstruction: a review with clinical recommendations. Int J Oral Maxillofac Surg. 2016;45(1):41–50.
4. Dubois L, Dillon J, Jansen J, Becking AG. Ongoing debate in clinical decision making in orbital fractures: indications, timing, and biomaterials. Atlas Oral Maxillofac Surg Clin North Am. 2021;29(1):29–39.
5. Jansen J, Dubois L, Maal TJJ, Mourits MP, Jellema HM, Neomagus P, et al. A nonsurgical approach with repeated orthoptic evaluation is justified for most blow-out fractures. J Craniomaxillofac Surg. 2020;48(6):560–8.
6. Holmes S. Primary orbital fracture repair. Atlas Oral Maxillofac Surg Clin North Am. 2021;29(1):51–77.
7. Christensen BJ, Zaid W. Inaugural survey on practice patterns of orbital floor fractures for American oral and maxillofacial surgeons. J Oral Maxillofac Surg. 2016;74(1):105–22.
8. Gooris PJJ, Jansen J, Bergsma JE, Dubois L. Evidence-based decision making in orbital fractures: implementation of a clinical protocol. Atlas Oral Maxillofac Surg Clin North Am. 2021;29(1):109–27.
9. Schreurs R, Becking AG, Jansen J, Dubois L. Advanced concepts of orbital reconstruction: a unique attempt to scientifically evaluate individual techniques in reconstruction of large orbital defects. Atlas Oral Maxillofac Surg Clin North Am. 2021;29(1):151–62.
10. Gander T, Essig H, Metzler P, Lindhorst D, Dubois L, Rücker M, et al. Patient specific implants (PSI) in reconstruction of orbital floor and wall fractures. J Craniomaxillofac Surg. 2015;43(1):126–30.
11. Jansen J. Advanced concepts in orbital wall fractures—virtual surgical planning, intraoperative imaging and clinical management. Amsterdam: University of Amsterdam; 2020.
12. Mahoney NR, Peng MY, Merbs SL, Grant MP. Virtual fitting, selection, and cutting of preformed anatomic orbital implants. Ophthalmic Plast Reconstr Surg. 2017;33(3):196–201.
13. Rana M, Chui CH, Wagner M, Zimmerer R, Rana M, Gellrich NC. Increasing the accuracy of orbital reconstruction with selective laser-melted patient-specific implants combined with intraoperative navigation. J Oral Maxillofac Surg. 2015;73(6):1113–8.
14. Kärkkäinen M, Wilkman T, Mesimäki K, Snäll J. Primary reconstruction of orbital fractures using patient-specific titanium milled implants: the Helsinki protocol. Br J Oral Maxillofac Surg. 2018;56(9):791–6.
15. Scolozzi P. Applications of 3D orbital computer-assisted surgery (CAS). J Stomatol Oral Maxillofac Surg. 2017;118(4):217–23.
16. Bittermann G, Metzger MC, Schlager S, Lagrèze WA, Gross N, Cornelius CP, et al. Orbital reconstruction: prefabricated implants, data transfer, and revision surgery. Facial Plast Surg. 2014;30(5):554–60.
17. Wilde F, Schramm A. Intraoperative imaging in orbital and midface reconstruction. Facial Plast Surg. 2014;30(5):545–53.
18. Rana M, Holtmann H, Rana M, Kanatas AN, Singh DD, Sproll CK, et al. Primary orbital reconstruction with selective laser melted core patient-specific implants: overview of 100 patients. Br J Oral Maxillofac Surg. 2019;57(8):782–7.
19. Jansen J, Schreurs R, Dubois L, Maal TJJ, Gooris PJJ, Becking AG. The advantages of advanced computer-assisted diagnostics and three-dimensional preoperative planning on implant position in orbital reconstruction. J Craniomaxillofac Surg. 2018;46(4):715–21.
20. Schlittler F, Vig N, Burkhard JP, Lieger O, Michel C, Holmes S. What are the limitations of the non-patient-specific implant in titanium reconstruction of the orbit? Br J Oral Maxillofac Surg. 2020;58(9):e80–e5.
21. Gooris PJJ, Muller BS, Dubois L, Bergsma JE, Mensink G, van den Ham MFE, et al. Finding the ledge: sagittal analysis of bony landmarks of the orbit. J Oral Maxillofac Surg. 2017;75(12):2613–27.
22. Jaquiéry C, Aeppli C, Cornelius P, Palmowsky A, Kunz C, Hammer B. Reconstruction of orbital wall defects: critical review of 72 patients. Int J Oral Maxillofac Surg. 2007;36(3):193–9.
23. Koornneef L. New insights in the human orbital connective tissue. Result of a new anatomical approach. Arch Ophthalmol. 1977;95(7):1269–73.
24. Sabelis JF, Youssef SALY, Hoefnagels FWA, Becking AG, Schreurs R, Dubois L. A technical note on multi-wall orbital reconstructions with patient-specific implants. J Craniofac Surg. 2021;33:991.
25. Dubois L, Schreurs R, Jansen J, Maal TJ, Essig H, Gooris PJ, et al. Predictability in orbital reconstruction: a human cadaver study. Part II: navigation-assisted orbital reconstruction. J Craniomaxillofac Surg. 2015;43(10):2042–9.
26. Zavattero E, Ramieri G, Roccia F, Gerbino G. Comparison of the outcomes of complex orbital fracture repair with and without a surgical navigation system: a prospective cohort study with historical controls. Plast Reconstr Surg. 2017;139(4):957–65.
27. Cai EZ, Koh YP, Hing EC, Low JR, Shen JY, Wong HC, et al. Computer-assisted navigational surgery improves outcomes in orbital reconstructive surgery. J Craniofac Surg. 2012;23(5):1567–73.

28. Luebbers HT, Messmer P, Obwegeser JA, Zwahlen RA, Kikinis R, Graetz KW, et al. Comparison of different registration methods for surgical navigation in cranio-maxillofacial surgery. J Craniomaxillofac Surg. 2008;36(2):109–16.

29. Venosta D, Sun Y, Matthews F, Kruse AL, Lanzer M, Gander T, et al. Evaluation of two dental registration-splint techniques for surgical navigation in cranio-maxillofacial surgery. J Craniomaxillofac Surg. 2014;42(5):448–53.

30. Schreurs R, Baan F, Klop C, Dubois L, Beenen LFM, Habets P, et al. Virtual splint registration for electromagnetic and optical navigation in orbital and craniofacial surgery. Sci Rep. 2021;11(1):10406.

31. Zeller AN, Zimmerer RM, Springhetti S, Tavassol F, Rahlf B, Neuhaus MT, et al. CAD/CAM-based referencing aids to reduce preoperative radiation exposure for intraoperative navigation. Int J Med Robot. 2021;17(3):e2241.

32. Dubois L, Essig H, Schreurs R, Jansen J, Maal TJ, Gooris PJ, et al. Predictability in orbital reconstruction. A human cadaver study, part III: implant-oriented navigation for optimized reconstruction. J Craniomaxillofac Surg. 2015;43(10):2050–6.

33. Jansen J, Schreurs R, Dubois L, Maal TJJ, Gooris PJJ, Becking AG. Intraoperative imaging in orbital reconstruction: how does it affect the position of the implant? Br J Oral Maxillofac Surg. 2020;58(7):801–6.

Open Access This chapter is licensed under the terms of the Creative Commons Attribution 4.0 International License (http://creativecommons.org/licenses/by/4.0/), which permits use, sharing, adaptation, distribution and reproduction in any medium or format, as long as you give appropriate credit to the original author(s) and the source, provide a link to the Creative Commons license and indicate if changes were made.

The images or other third party material in this chapter are included in the chapter's Creative Commons license, unless indicated otherwise in a credit line to the material. If material is not included in the chapter's Creative Commons license and your intended use is not permitted by statutory regulation or exceeds the permitted use, you will need to obtain permission directly from the copyright holder.

Orbital Roof Fractures

11

Bram van der Pol, Geert-Jan Rutten,
Peter J. J. Gooris, and J. Eelco Bergsma

Learning Objectives
- Adequate diagnosis of the extent of an orbital roof fracture.
- Intracranial involvement: indication for surgical intervention.

Introduction

Orbital roof fractures are infrequent and are commonly associated with high impact facial trauma. An estimated 1–14.7% of orbital fractures are fractures of the orbital roof [1–4]. These fractures are relatively rare in comparison to fractures of the infero-lateral wall of the orbit and there are limited indications for surgical intervention. In this short overview, we will discuss the indications of surgical reconstruction of orbital roof fractures.

The first line of treatment in orbital roof fractures is often conservative [5]. Management varies based on individual clinical and radiological findings [5–7].

Indication for surgery depends on the symptoms at presentation; skeletal and (brain)soft tissue injury findings on the CT scan, the degree of displacement of bone fragments, the degree of intracranial involvement and the involvement of the superior orbital rim. The majority of these surgical procedures will be a collaboration between surgeons of different background in neurosurgery, maxillofacial surgery, plastic surgery and ophthalmology.

Background

Isolated orbital roof fractures are uncommon since the roof is well protected by the sturdy supraorbital rim, the zygoma, and more posteriorly the sphenoid bone. Most superior orbital wall fractures are caused by high-energy impact trauma to the head and are more frequent in the paediatric population; in the younger paediatric

B. van der Pol · G.-J. Rutten
Department of Neurosurgery, St. Elisabeth-Tweesteden Hospital Tilburg, Tilburg, The Netherlands
e-mail: bram.vanderpol@etz.nl; g.rutten@etz.nl

P. J. J. Gooris (✉)
Department of Oral and Maxillofacial Surgery, Amphia Hospital Breda, Breda, The Netherlands

Amsterdam University Medical Centers, Amsterdam, The Netherlands

Department of Oral and Maxillofacial Surgery, University of Washington, Seattle, WA, USA

J. E. Bergsma
Department of Oral and Maxillofacial Surgery, Amphia Hospital Breda, Breda, The Netherlands

Amsterdam University Medical Centers, Amsterdam, The Netherlands

Acta dental school, Amsterdam, The Netherlands
e-mail: EBergsma@amphia.nl

© The Author(s) 2023

P. J. J. Gooris et al. (eds.), *Surgery in and around the Orbit*,
https://doi.org/10.1007/978-3-031-40697-3_11

population, fractures of the orbital roof are the most common type of fracture in this region, due to the neurocranium – face ratio and the associated characteristics of the orbit at the younger age [8, 9] (Chap. 12). As the frontal sinus pneumatizes during age, the transmission of force from the superior orbital rim to the anterior cranial base diminishes: concordantly, orbital roof fractures become less frequent during adulthood. In adults, an orbital roof fracture is often associated with more extensive cranial-facial and brain injury than in children. In children, these orbital fractures are typically isolated (linear skull fracture) and treated conservatively accept when there is a risk of a growing skull fracture, which can result as a "late" complication of a dural laceration. In such paediatric head trauma cases, after several months, herniation of (cerebral) brain tissue through the skull-dura defect may occur. Typically, a delayed onset develops of neurological deficits [1, 5]. The majority (>75%) of adult orbital roof fractures are associated with traumatic brain injury and infringe on the cranial integrity potentially leading to CSF fistula, pneumocephalus and secondary meningitis [2, 5, 6].

Presentation

Despite the infrequent presentation of orbital roof fractures, if not well recognized ophthalmological, neurological, functional and aesthetic complications and associated morbidity may potentially result. The characteristic symptoms of orbital roof fractures are orbito-frontal skin lacerations, peri-orbital hematoma, oedema and orbital emphysema. Furthermore, displacement of the globe in any direction may occur (exophthalmos, enophthalmos, hyperglobus or hypoglobus). In addition, limitation of eye movement or gaze restriction, diplopia or visual impairment, i.e. loss of vision from a raised intraorbital pressure and altered sensation in the supraorbital and supratrochlear skin region, facial asymmetry are described (see patient III). The most common intracranial injury is the presence of a CSF leak. Intracranial haemorrhage may be present as well as prolapse of brain tissue. A (late) developing pulsatile exophthalmos in case of an orbital meningoencephalocele may occur.

Several classifications of orbital roof fractures have been described: the blow-in fracture, with inferior displacement of the orbital roof, the blow-up fracture, with superior displacement of the orbital roof-rim into the anterior cranial fossa, supraorbital rim fracture and frontal sinus fracture with involvement of the adjacent orbital roof [5–7].

Management

When an orbital roof fracture is suspected, assessment of the (associated) injuries by a multidisciplinary team must be done rapidly to avoid secondary damage to the brain tissue, orbit and globe [1–3]. Treatment is individually tailored. Generally, a nonsurgical approach is warranted in the absence of intracranial, sinus or orbital symptoms. The choice of surgical versus conservative treatment starts with an accurate assessment of the injury. An adequate work-up must include neurological and ophthalmological assessment. When neurological injury is suspected or proven, prompt consultation by a neurosurgical team is indicated. The standard Glasgow Coma Scale workup is essential. The radiological imaging of choice is a thin (1 mm) high-resolution CT scan with 3D-reconstructions [7]. (Chap. 4) In orbital roof fractures, MR imaging is of lesser importance and contra-indicated when magnetic foreign bodies are suspected [2–4]. Immediate surgical intervention (decompression) is needed in a minority of patients with a retrobulbar hematoma or threatening orbital compartment syndrome and consists of urgent lateral canthotomy or cantholysis. In the short-term post traumatic period, patients should be advised not to blow their nose as this may result in emphysema or pneumocephalus.

Skeletal Injury: Isolated Orbital Roof Fracture

Isolated fractures of the orbital roof (Figs. 11.1 and 11.2) are rare and can usually be treated conservatively in the absence of diplopia, rectus muscle impingement or persistent cerebrospinal fluid (CSF) leak. In the case of CSF fistula with oculorrhea or rhinorrhea, a period of watchful waiting for at least 48 hours is advisable because

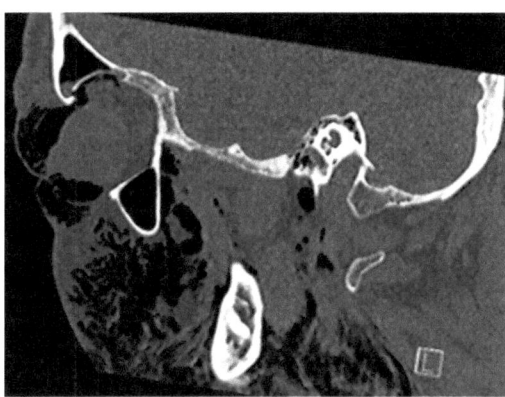

Fig. 11.1 CT scan: sagittal slide showing isolated displaced orbital roof fracture, wide spread emphysema present

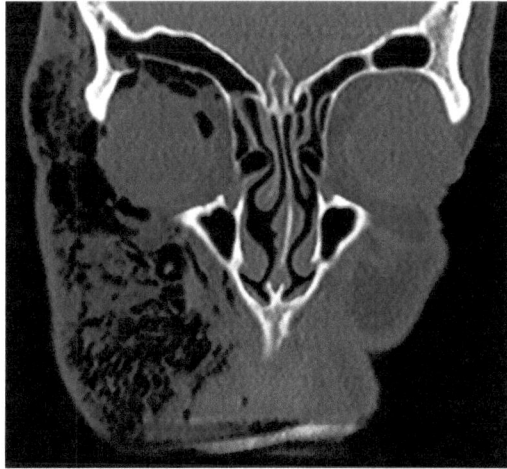

Fig. 11.2 CT scan: coronal slide showing isolated orbital roof fracture

the majority of cases resolves spontaneously [2, 10]. The efficacy of lumbar drainage and prophylactic antibiotics in these cases with CSF leaks is still subject of debate [2, 4].

Skeletal Injury: Concomitant Skull Fractures

When there is an associated fracture of the adjacent supra-orbital rim or frontal sinus wall, surgical intervention is often indicated, more so when displacement of bony fragments is present (see patient III). When the posterior table of the frontal sinus wall is involved in the fracture, craniali-

sation should be considered. Other, frequent bone injuries consist of a fracture of the zygoma and ethmoid.

Concomitant Ophthalmologic and Neurologic Injuries

In patients presenting with major ophthalmologic signs and/or clear ocular injury, urgent surgery is warranted to prevent further deterioration especially in cases with a substantial displacement of bone fragment intra-orbitally or intracranially [11]. The timing of surgery also depends on the need for urgent treatment of concomitant (brain) injury. Surgery of solitary orbital roof fractures with displacement of the orbital rim usually takes place in the sub-acute stage after the initial swelling has subsided.

Oculorrhea and/or Rhinorrhea

As stated above, in the case of CSF fistula with oculorrhea or rhinorrhea, a period of watchful waiting is advisable for at least 48 h because the majority resolve spontaneously [10]. In case of persistent leakage, surgery is indicated. Surgical nuances are variable and case-specific, but all consist of a subcranial approach to the orbital roof through a bitemporal (coronal) or superior-orbital incision. After resection of bone fragments and reconstruction of dura, orbital roof reconstruction with titanium screws with plates and/or mesh systems yield an effective result but is not a necessity when the orbital rim is intact [6].

Care is taken to avoid perioperative entrapment of the superior rectus and superior oblique muscle. Postoperative CSF leakage is rare but should be watched for; therefore, MRI and fluid analysis may be indicated.

To summarize, the indications for surgical reconstruction of superior orbital wall fractures are as follows:

1. Concomitant intracranial injury or penetrating orbital roof fragments intracranially.
2. Visual impairment or globe displacement.
3. Persistent CSF leakage warranting reconstruction of the frontal sinus.

4. Dislocation or fragmentation of the superior orbital rim.
5. Growing fracture in the paediatric population.

Illustrative Cases

Patient I

A first illustrative case is a healthy young man working on a construction site who was hit by a large metal rod and presented to our ER with an orbito-frontal skin laceration but without neurological or visual deficits (Fig. 11.3). A CT-scan

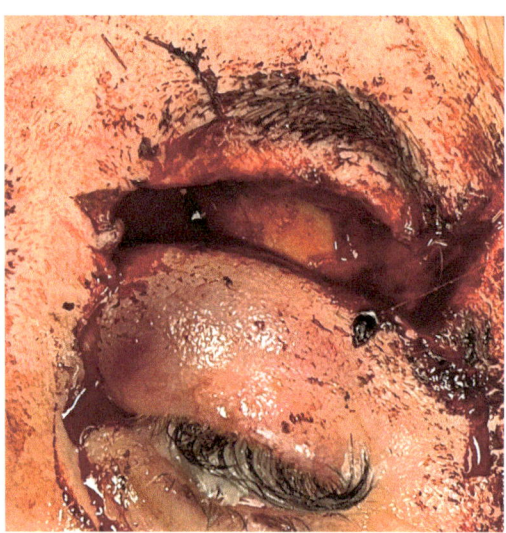

Fig. 11.3 Clinical image frontal laceration

including 3D reconstruction revealed a comminuted fracture of the left orbital rim, frontal sinus and superior orbital roof (Figs. 11.4 and 11.5). We opted for surgery because of the comminution and extent of the orbital roof fracture in combination with the orbital rim and frontal sinus. Intervention was carried out in the acute phase because of a large skin laceration necessitating primary closure and cosmetic reasons. He was treated in a joint venture of the neuro- and maxillofacial surgeons with primary open fracture reduction, plate fixation and dural repair through the existing laceration (Figs. 11.6 and 11.7). No prophylactic antibiotics were administered. He made an uneventful recovery and was discharged after 5 days.

Patient II

A second case is an elderly cyclist who fell off his e-bike and presented with headache, vomiting and peri-orbital swelling but without ophthalmologic symptoms. On radiologic examination, a right orbito-frontal fracture with involvement of the orbit and frontal sinus was evident as well as a pneumocephalus indicating traumatic fisteling of the frontal sinus and the intracranial space (Figs. 11.8 and 11.9). The patient was operated by the neuro- and maxillofacial surgeons on the second day after trauma when the initial swelling had partially subsided. Repositioning of the bone fragments as well as reconstruction and craniali-

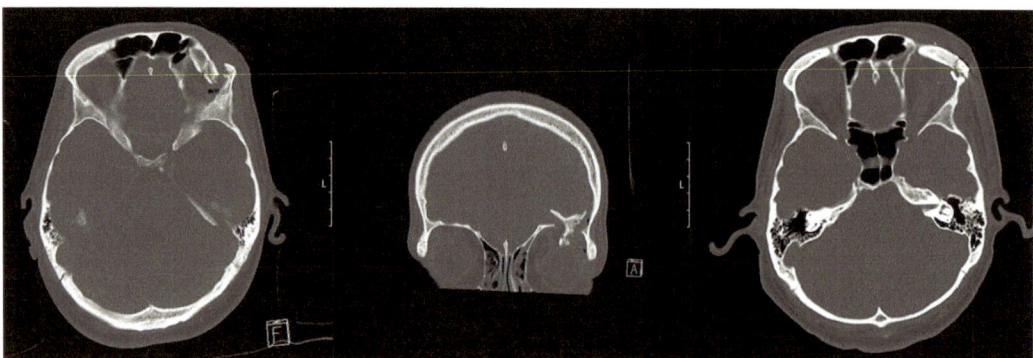

Fig. 11.4 CT scan revealing an orbito-frontal fracture with displacement on the left side with involvement of the orbital roof and frontal sinus: axial and coronal view (patient Fig. 11.3)

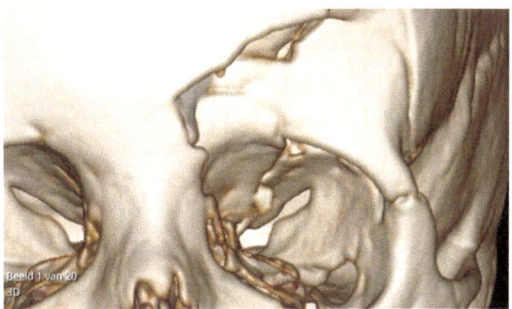

Fig. 11.5 CT scan-3D reconstruction revealing an orbito-frontal fracture with involvement of the orbital roof and frontal sinus including (patient Fig. 1.3)

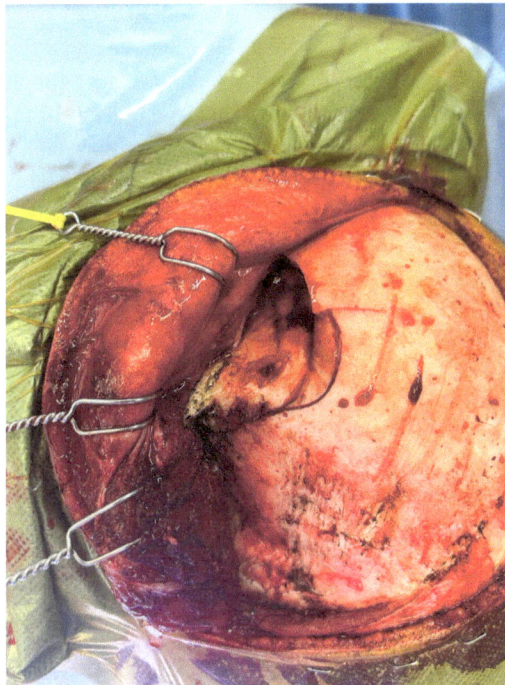

Fig. 11.6 Preoperative view comminuted superior orbital rim, anterior table frontal sinus and associated orbital roof fracture (patient Fig. 11.3)

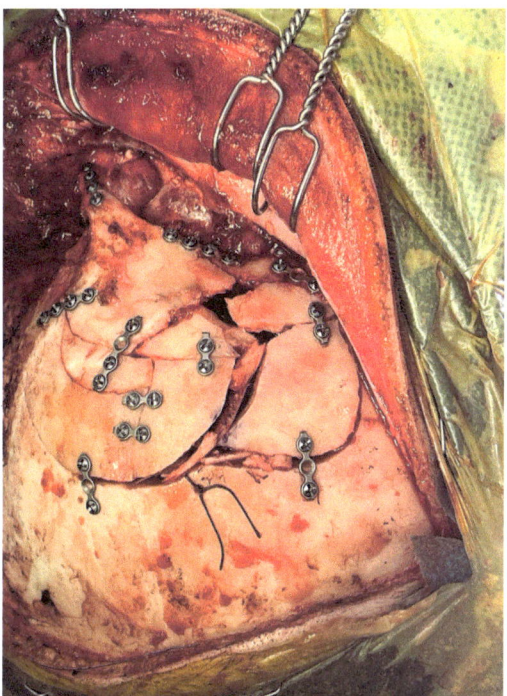

Fig. 11.7 Postoperative open reduction and internal fixation of comminuted superior orbital rim and frontal sinus anterior table fracture (patient Fig. 11.3)

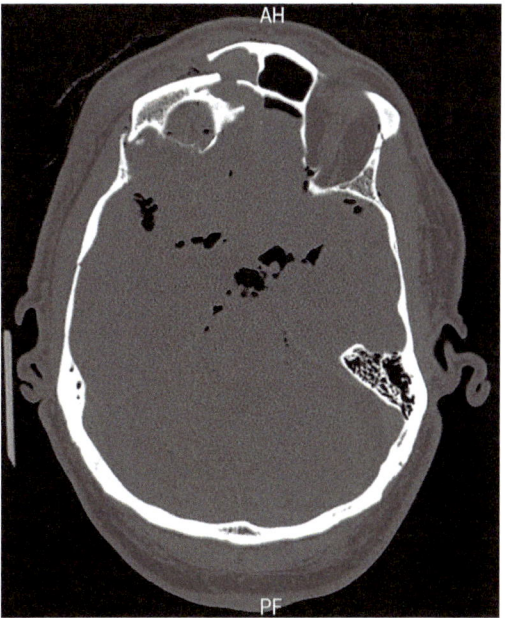

Fig. 11.8 CT scan revealing an orbito-frontal fracture, displacement present, with involvement of the orbital roof and frontal sinus

sation of the frontal sinus was preformed through a coronal approach. The surgery was effective in treating the cosmetic defect but afterwards, he developed nasal CSF leakage which was treated with 3 days of external lumbar CSF drainage and prophylactic antibiotics. On discharge, he was diagnosed with epilepsy and mild cognitive dysfunctions which were not present on initial pre-

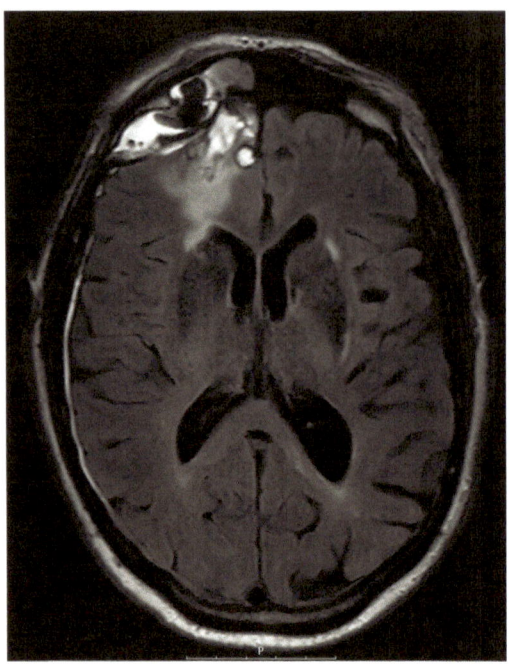

Fig. 11.9 CT scan revealing an orbito-frontal fracture with involvement of the orbital roof and frontal sinus including a pneumocephalus

sentation. This case illustrates that clinical outcome after orbital roof fractures is mostly dependent on the extent of the traumatic injury to the brain.

Patient III

In a third case, a 34-year-old male patient presented to the ER after being involved in a bike accident and fell with the tempo-frontal side of his head onto the pavement. The patient complaint about severe headache, no vomiting. There was no double vision. Facial examination revealed some flattening of the involved forehead, more evident on the right side. Also, evident enophthalmos OD was measured, which was, according to the patient not present prior to the accident. No diplopia was reported, BSV was present and all ductions were normal. A CT scan showed an extensive comminuted fracture of the supraorbital rim as well as a blow-out fracture of the orbital roof (Fig. 11.10). Because of the extent of the

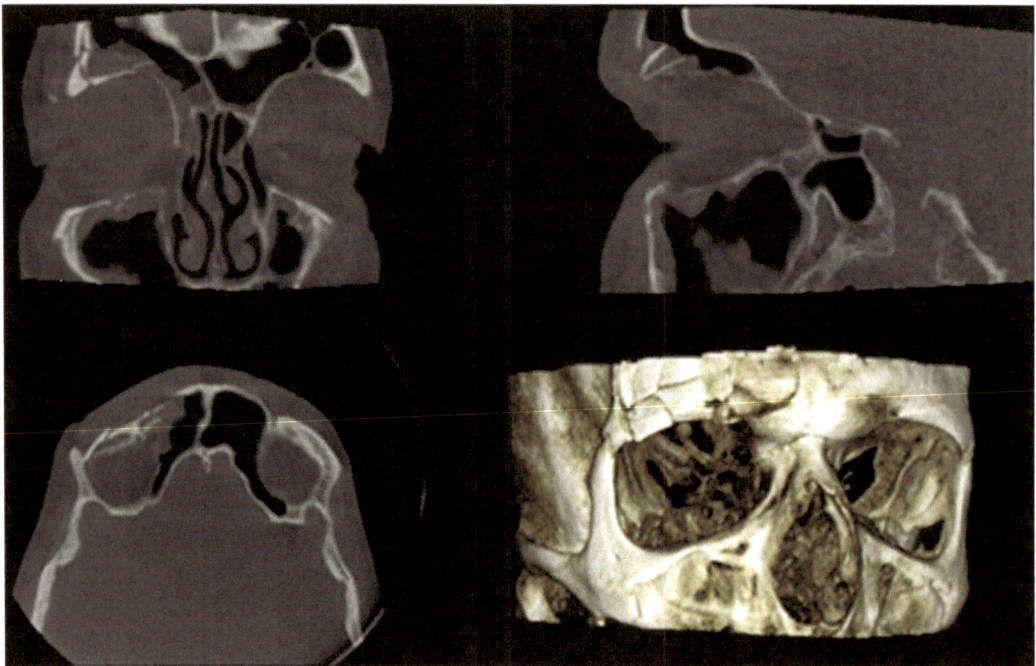

Fig. 11.10 CT scan, coronal, sagittal and axial view including 3-D reconstruction showing an extensive supraorbital rim fracture in combination with a cranially displaced orbital roof fracture (patient Fig. 11.10)

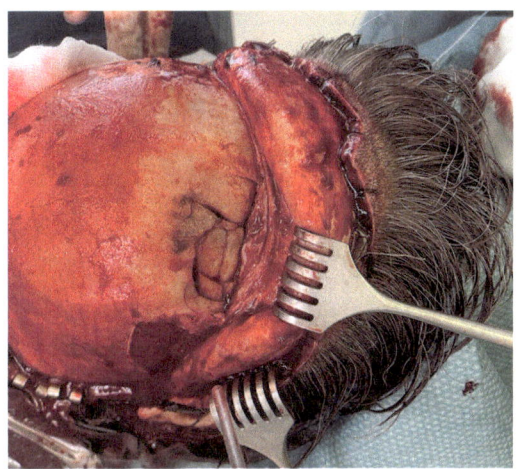

Fig. 11.11 Bicoronal incision providing wide access to the fracture of the supraorbital rim and associated orbital roof OD (patient Fig. 11.10)

comminution of the rim fracture, the displaced orbital roof fracture and the resultant enophthalmos, open reduction and internal fixation was carried out using a coronal approach which allowed for wide exposure of the fracture site (Fig. 11.11). Miniplate fixation was applied (Fig. 11.12a, b). Postoperatively, facial symmetry was accomplished with proper globe position, no double vision or neurological symptoms occurred.

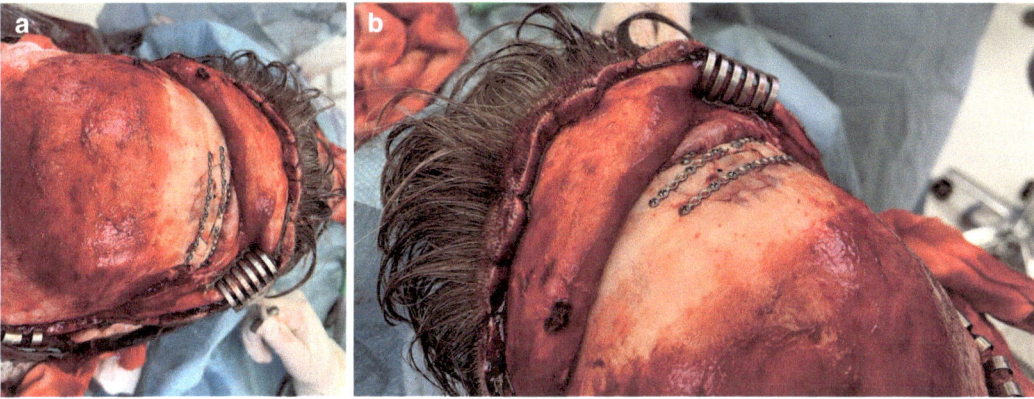

Fig. 11.12 (**a**) Open reduction and internal fixation with miniplates of comminuted supraorbital fracture (patient Fig. 11.10). (**b**) Open reduction and internal fixation with miniplates of comminuted supraorbital fracture (patient Fig. 11.10)

Conclusion

The evaluation and treatment of complex orbital fractures is best done in a medical centre with experience in craniofacial trauma and by a multi-disciplinary team. The indication of surgical treatment of orbital roof fractures is limited to fractures with substantial dislocation of bone fragments that result in penetration of the orbit and/or cranial fault, visual disturbance or cosmetic defects. The surgical goals are reduction of penetrating bone fragments, dural repair and (cosmetic) reconstruction of orbit and frontal sinus. Although scientific evidence to guide treatment choices is limited to case series and expert opinions, both surgical and conservative treatment are effective in selected patients. The clinical outcome of these patients depends mostly on collateral damage due to traumatic brain injury and visual disturbances.

References

1. Righi S, Boffano P, Guglielmi V, Rossi P, Martorina M. Diagnosis and imaging of orbital roof fractures: a review of the current literature. Oral Maxillofac Surg. 2015;19:1–4.
2. Cossman J, Morrison S, Taylor H, Salter A, Klinge P, Sullivan S. Traumatic orbital roof fractures: interdisciplinary evaluation and management. Plast Reconstr Surg. 2015;133:335e–43e.
3. Lozada K, Cleveland P, Smith J. Orbital Trauma Semin Plast Surg. 2019;33:106–13.
4. Lucas J, Allen M, Nguyen B, Svider P, Folbe A, Carron M. Orbital roof fractures: an evidence-based approach. Fac Plast Surg Aesth Med. 2020;22:471–80.
5. Connon F, Austin S, Nastri A. Orbital roof fractures: a clinically based classification and treatment algorithm. Craniomaxillofac Trauma Reconstruction. 2015;8:198–204.
6. Manolidis S, Weeks BH, Kirby M, Scarlett M, Hollier L. Classification and surgical management of orbital fractures: experience with 111 orbital reconstructions. J Craniofac Surg. 2002;13(6):726–37; discussion 738
7. Righi S, Boffano P, Guglielmi V, Rossi P, Martorina M. Diagnosis and imaging of orbital roof fractures: a review of the current literature. J Oral Maxillofac Surg. 2015 Mar;19(1):1–4.
8. Stotland MA, Do NK. Pediatric orbital fractures. J Craniofac Surg. 2011 Jul;22(4):1230–5.
9. Oppenheimer AJ, Monson LA, Buchman SR. Pediatric orbital fractures. Craniomaxillofac Trauma Reconstruction. 2013 Jan;6:9–20.
10. Oakley GM, Orlandi RR, Woodworth BA, Batra PS, Alt JA. Management of cerebrospinal fluid rhinorrhea: an evidence -based review with recommendations review. Int Forum Allergy Rhinol. 2016 Jan;6(1):17–24.
11. Santamaria J, Mehta A, Reed D, Blegen H, Bishop B, Davies B. Orbital roof fractures as an indicator for concomitant ocular injury. Graefe's Arch Clin Exper Opthal. 2019;257:2541–5.

Open Access This chapter is licensed under the terms of the Creative Commons Attribution 4.0 International License (http://creativecommons.org/licenses/by/4.0/), which permits use, sharing, adaptation, distribution and reproduction in any medium or format, as long as you give appropriate credit to the original author(s) and the source, provide a link to the Creative Commons license and indicate if changes were made.

The images or other third party material in this chapter are included in the chapter's Creative Commons license, unless indicated otherwise in a credit line to the material. If material is not included in the chapter's Creative Commons license and your intended use is not permitted by statutory regulation or exceeds the permitted use, you will need to obtain permission directly from the copyright holder.

Peter J. J. Gooris, Maarten P. Mourits,
Gertjan Mensink, and J. Eelco Bergsma

Learning Objectives

- Orbital fractures in children exhibit different features in children as compared to the adult population: children are no small adults.
- A white eye after blunt trauma in a child can be deceiving. A trapdoor fracture may betray itself only by a limitation of elevation. An attempt to elevate the eyes may result in a cardiac arrest.
- Such a trapdoor trauma requires instant intervention.
- Nausea caused by trapdoor fractures may be confused with concurrent head injury i.e., concussion.
- A growing skull fracture can be a late complication in case of orbital roof fracture involvement.

P. J. J. Gooris (✉)
Department of Oral and Maxillofacial Surgery,
Amphia Hospital Breda, Breda, The Netherlands

Amsterdam University Medical Centers, Amsterdam,
The Netherlands

Department of Oral and Maxillofacial Surgery,
University of Washington, Seattle, WA, USA

M. P. Mourits
Amsterdam University Medical Centers, Location
AMC, Amsterdam, The Netherlands

G. Mensink
Department of Oral and Maxillofacial Surgery,
Amphia Hospital Breda, Breda, The Netherlands

University Medical Center Leiden,
Leiden, The Netherlands

J. E. Bergsma
Department of Oral and Maxillofacial Surgery,
Amphia Hospital Breda, Breda, The Netherlands

Amsterdam University Medical Centers, Amsterdam,
The Netherlands

Acta dental school, Amsterdam, The Netherlands

Introduction

Epidemiology

In children, a traumatic injury to the craniofacial skeleton can result in facial fractures but may also affect and disturb facial growth [1, 2]. Fortunately, facial fractures in children are relatively rare and are less common than in the adult population [2, 3]. The incidence of facial fractures varies with age. Only approximately 1% of all facial fractures occur in children under the age of 1 year. The majority of fractures are observed in children within the age group of 13–18 years of age; boys are twice as much involved in facial fractures as girls are [1–6]. In children, under the age of 16, the overall incidence of orbital wall fractures is 5–25% of all facial fractures, which is lower than the incidence in the adult population

© The Author(s) 2023
P. J. J. Gooris et al. (eds.), *Surgery in and around the Orbit*,
https://doi.org/10.1007/978-3-031-40697-3_12

(around 50%) [1–5]. As explained further on, a skull fracture may occur in the very young and can include a fracture of the anterior skull base resulting in intracranial injury [1, 2].

A contiguous orbital roof fracture is more common in very young children [7–9]. However, the incidence of orbital roof fracture is likely to be underreported [8, 9]. Although fractures of the orbital floor and medial wall are diagnosed at any age, there is an increase of isolated fractures of the orbital floor with age and development [3, 5]. In case of an orbital floor fracture in children, more than 50% entrapment of orbital soft tissue is encountered [1–6, 10]. Not surprisingly, the mechanism of the craniofacial impact, differs with age (Table 12.1):

Growth and Development and Subsequent Different Kind of Patterns of Injury

In the pediatric population, there is a relatively high proportion of cancellous, richly vascular bone and growing sutures of cartilaginous structure which is responsible for the characteristic of elasticity of the young growing bone. In adults, the bone becomes more compact, dense and rigid. The craniofacial skeleton undergoing mineralization, most profoundly at the age 2–3 years, changes from an elastic to a rigid structure with age. The elastic structure still has the intrinsic capacity to deform or buckle instead of fracture when force is applied resulting in less fractures.

Depending on age, the size and shape, the anatomy-proportion of the skull changes. The orbital floor itself deepens and becomes less steep from lateral to medial with age. The lowest point along the orbital floor shifts posteriorly [3, 4]. Neurocranial growth is continuous

and stimulated by the enlarging brain. There is a preponderance of this growth mainly in the first 2 years, after which it gradually decreases over the following years. Facial skeletal growth is discontinuous, is multifactorial and varies in location and direction [2, 3]. Orbital depth reaches 90% of adult dimensions at age 6 and 95% at age 12, which is analogous to the cranial growth [3, 11, 12]. The fastest growth of the orbit is within 12–24 months; after age 6, the rate of expansion declines (Table 12.2) [11, 12].

Pneumatization of the paranasal sinuses develops during childhood. While in utero, the sinuses and nasal cavity are in fact mucosal tissue within cancellous bone and form one single structure. During development, the ethmoid, frontal and maxillary sinus subdivide in a predictable sequence [2] (Table 12.3).

The *frontal sinus* pneumatization evolves at age 7, completes before adulthood at around age 16 [1, 2, 7, 9]. Radiographically, this pneumatization becomes "visible" at age 8. The lack of pneumatization in young children allows for more direct transmission of force to the supraorbital rim which extends directly posterior to the anterior cranial base and orbital roof with subsequent an increased risk of an orbital roof fracture as a result [2, 3, 6, 9]. Once frontal sinus pneumatization has been completed, less force is directly transmitted to the anterior cranial base and impact forces are dissipated.

Table 12.2 Average orbital volume (mL) [12]

At birth	9–15
6 years	20
Mature age	25–28

Table 12.3 Growth and development [1, 2]

	Years of age of development	Adult size reached at (year)
Maxillary sinus	0–3 and 7–12	16
Ethmoid sinus	1–12	12
Frontal sinus	7–16	16

Table 12.1 Type of impact related to age

Young child	Impact as a result of daily activities, like a fall. More skull fractures and orbital roof fractures
Older child	Impact as a result of sport, traffic or violence and sport related injuries, results in orbital floor and medial wall fractures

Ethmoid air cells already present at birth gradually grow to an adult size at age 12 (Fig. 12.1). During the continuous pneumatization and subsequent expansion, the medial orbital wall becomes progressively thinner as the lamina papyracea and thus more susceptible for orbital wall fracture in adulthood [2, 3].

Maxillary sinus development is biphasic. Its growth peaks at age 0–3 and at age 7–12. The maxillary sinus is initially located medial to the orbit, by age 4, it develops more infero-laterally, expands at age 12 and reaches its adult size at age 16. Eventually, both the changes in bone morphology of the craniofacial skeleton and the sinus development during growth will affect how the force of impact will be transmitted and how this

will result in a variable fracture pattern (Table 12.4).

In the group till the age of 7, due to a higher cranial to face ratio (Table 12.5) which results in a proportionately larger neurocranium i.e., more exposure of the frontal bone, head trauma will more often result in a skull and orbital roof fracture rather than into a fracture of the facial complex [2, 5, 7–9] (Figs. 12.2 and 12.3).

With exceeding age (older than 7), isolated fractures of the lateral wall are declining in frequency because of its increase in thickness and non-sinus boundary.

The probability of fractures of the orbital floor does not exceed that of the orbital roof until age 7 [1, 2, 4, 7, 10]. It is said that the orbital floor fracture only "starts" at age 3–4 because of the pneumatization of the maxillary sinus [2, 9, 10]. Unerupted maxillary dentition in the undeveloped maxillary sinus also resists orbital floor fractures in young children, especially under the age of 7 [2] (Figs. 12.4, 12.5 and 12.6).

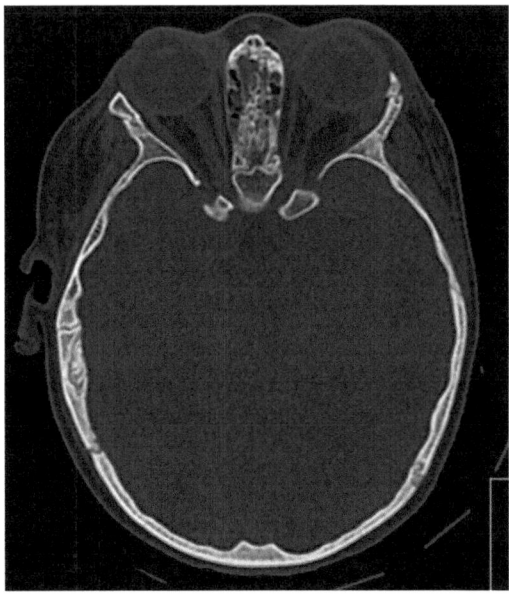

Fig. 12.1 The early presence of ethmoid air cells, axial view

Table 12.4 Relation between age and facial fracture pattern

Birth—till age 7	Orbital roof > orbital floor	Incomplete pneumatization frontal sinus High cranium-face ratio presence unerupted maxillary dentition
Age 7—adulthood	Orbital floor > orbital roof Increase medial wall	Completion maxillary sinus pneumatization Eruption maxillary dentition completion ethmoid sinus pneumatization

Table 12.5 Anatomical changes during maturation "size ratio" in growth and adult size

	Ratio neurocranium:face	% of Adult size neurocranium	% of Adult size face
Birth	8:1	35%	25%
2 years		75%	70%
5 years	4:1		80%
10 years		95%	
Mature	2:1	100%	100%

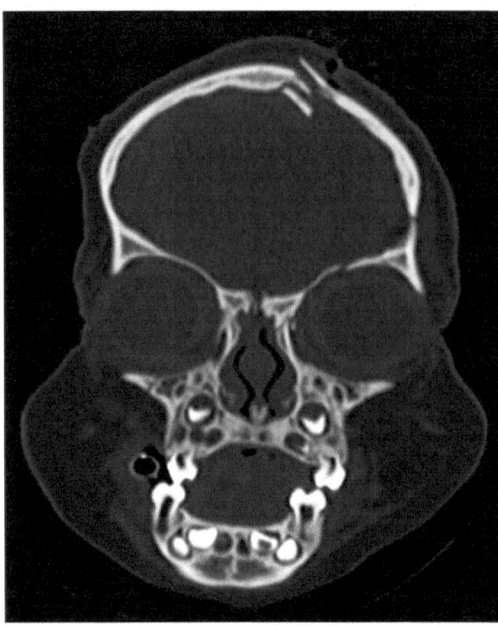

Fig. 12.2 Fracture of the anterior skull and orbital roof in the very young child after fall, coronal view

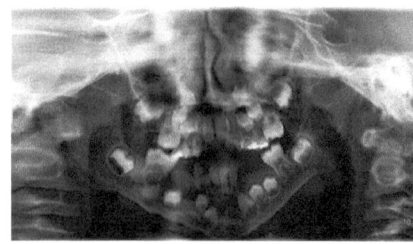

Fig. 12.4 The high maxillary cuspid location during mixed dentition: eye–teeth. Panoramic view

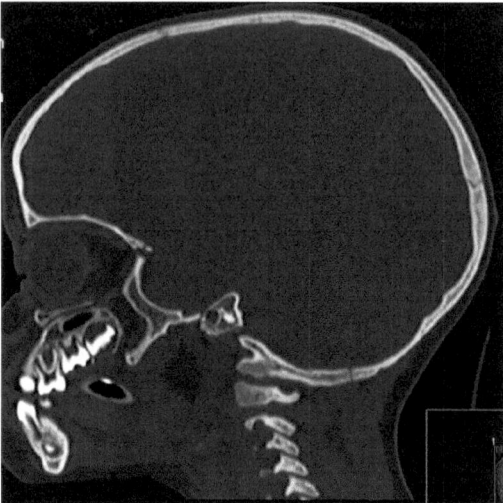

Fig. 12.5 The small maxillary sinus at early age and the high position of the mixed dentition just below the orbital floor, sagittal view

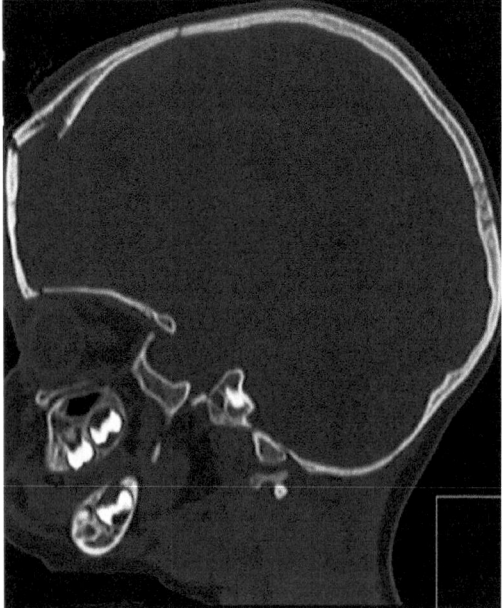

Fig. 12.3 Fracture of the anterior skull and orbital roof in the very young child after fall, sagittal view

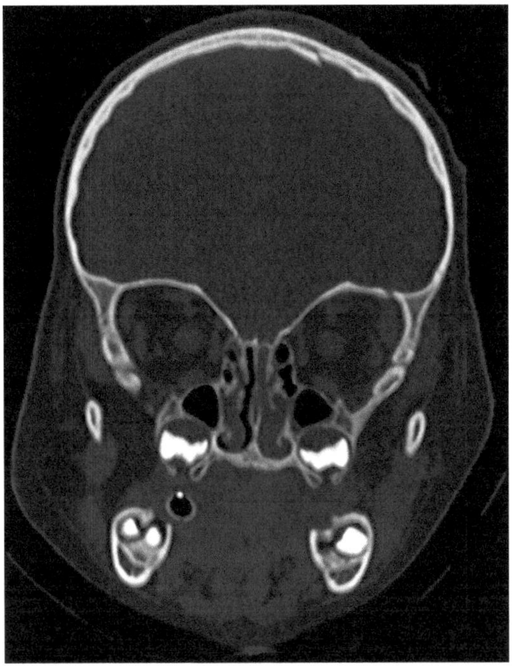

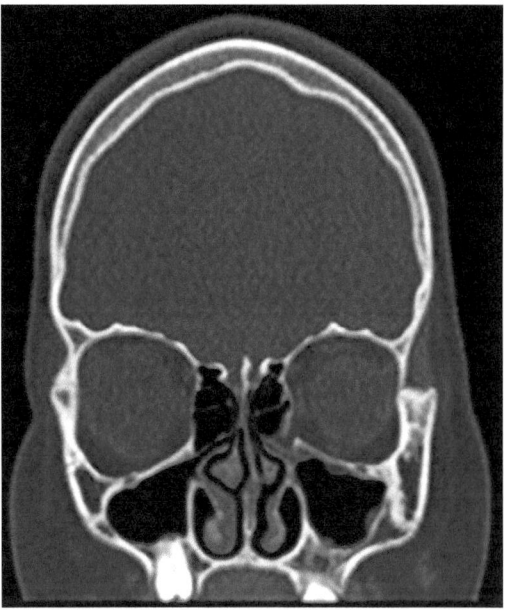

Fig. 12.7 Coronal view trapdoor fracture orbital floor OS; subtle tear-drop sign present; small accompanying low medial wall fracture

Fig. 12.6 View of early, developing stage ethmoid air cells and maxillary sinus, high position mixed dentition, coronal view

The Orbital Floor Fracture in Children: "An Evolving Pattern"

Fracture patterns and susceptibility of orbital fractures change with age. As stated above, this is the combined result of a change in anatomy on the one hand, and a physiological change due to growth and development during maturation on the other. Besides these anatomical changes, physiology during growth will affect the mechanical properties of the craniofacial skeleton. Cancellous immature elastic bone develops into rigid mature bone. Elastic bone will absorb energy differently compared to rigid bone. The elastic, flexible bone that comprise the immature orbital floor is able to deform more than adult compact bone when traumatic force is applied. Because of the flexibility, the orbital floor may bend rather than fracture and if a fracture does occur, the intrinsic elastic property allows for the tendency to recoil. The thick and elastic periosteum may also contribute to the trapdoor mecha-

nism of the orbital wall involved [13]. But before the fracture recoils, soft tissue may herniate and remain entrapped after the hinge-fracture returns to its original position [1, 3, 4, 6]. Mature bone in these cases is much more prone to fracture without subsequent recoil. In the adult case, we are mostly dealing with an open floor fracture with downward displacement of the orbital content, clearly visible on the coronal image of the CT scan. However, in the younger still growing pediatric population, when the child presents with clinical symptoms of an orbital floor fracture, often hardly any findings of displacement of the orbital floor are diagnosed on the CT scan (Fig. 12.7). The thin not fully mineralized bone may be hard to recognize on the scan images or only a tear-drop sign may be present, suggesting orbital soft tissue to prolapse (Figs.12.7 and 12.8). This *trapdoor phenomenon* causes an acute mechanical failure in vertical gaze. Apart from periosteal lining and orbital fat, the inferior rectus- and inferior oblique muscle may become entrapped, the muscle component causing more pronounced inability of vertical globe motility [1 4, 6, 7, 13 19].

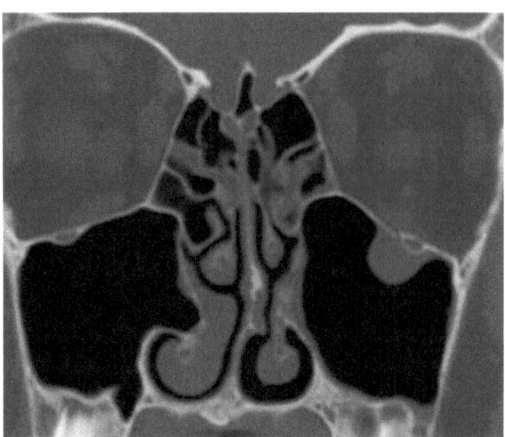

Fig. 12.8 Teardrop sign as a result of an orbital floor trapdoor fracture OS, coronal view

Trapdoor Fracture of the Orbital Floor: Findings

Early Findings

Clinical symptoms:

- Few/absence of peri-orbital signs of facial trauma (ecchymosis), subconjunctival hemorrhage (white eyed orbital fracture)
- Pain/decreased sensation infraorbital nerve supply region
- Lack of enophthalmos
- Marked impaired ocular mobility: limited vertical gaze (Fig. 12.9)
- Head posture (torticollis) to counteract double vision (Fig. 12.10)
- More rarely: oculo-cardiac/oculo-vagal reflex as a result stimulation of the ophthalmic division NV—afferent reticular formation—visceral motor nuclei N Vagus: efferent limb to cardiac system [7, 16–19]. There also may be traction on baroreceptors potentially present in the orbital soft tissue
- Nausea-vomiting-vertigo/bradycardia-hypotension/syncope (watch for potential arrhythmias)

As explained above, due to the elasticity of the bone in children and the ability to recoil, impact to the orbit results in a pure and linear fracture of the orbital floor and peri-orbital content may become entrapped resulting in an acute restric-

tion of eyeball elevation: the patient experiences double vision.

Once the peri-orbital lining is disrupted, extra-conal fat and the highly organized connective tissue septa, an accessory locomotor system can herniate in pathologic circumstances like blow-out fractures and can account for the motility disturbances in these cases [20]. As a result, upward gaze is severely restricted and can luxate the oculocardiac reflex. Posturing of the head will reduce the diplopia. Traction on the orbital soft tissue, the extra ocular muscles (EOM) or peri-orbital fat lining stimulates the afferent ophthalmic division of NV resulting in nausea, vomiting and a vaso-vagal (including bradycardia-syncope) as a response [3, 6, 16–19]. *The nausea may be confused with concurrent head injury* i.e., *concussion*. Spontaneous resolution is highly unlikely, surgical intervention should be employed preferably within 12–24 h [1, 3, 19, 21–24]. As stated above, a typical "finding" is the just subtle or absence of skeletal radiological CT findings which may lead to misdiagnosis (Fig. 12.7). Despite multislice (1.0 mm thickness, 1.0 mm increment) computed tomography, CT images are restricted in revealing orbital soft tissue entrapment [18, 24]. In the examination of the patient, the clinical presentation often with marked limitation of globe elevation should outweigh the radiographical (non)findings (Figs. 12.7 and 12.9a, b). When in doubt, an additional MRI can depict more precisely the extent and differentiation of the injured orbital soft tissue.

Late Findings

When no proper surgical intervention is carried out within time, as a result of persisting ischemia, necrosis of herniated, incarcerated orbital soft tissue may develop [4, 18, 21, 23–25]. This is especially true for connective tissue septae, orbital fat and fascial muscle sheet. Fibrosis and finally scarring result in persisting or potentially permanent vertical motility restriction of the globe [21, 23, 24, 26, 27]. Ischemia of developing orbital

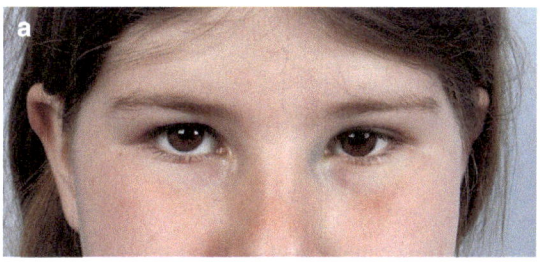
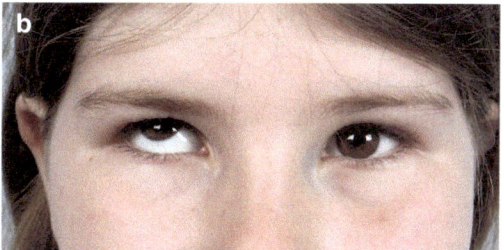

Fig. 12.9 (**a**) Preoperatively clinical view: primary gaze. (**b**) Preoperatively clinical view: limited upward gaze OS

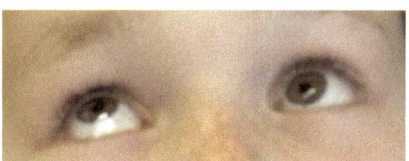

Fig. 12.10 Young boy with torticollis, compensatory head tilt to limit double vision

soft tissue and EOM tissue is more likely to result in incapacitating ocular motility: one can assume that the growing potential to an adult structure is irreversibly interrupted by the longer standing incarceration or strangulation of especially inferior oblique-inferior rectus muscle resulting in a complication very difficult to correct. It is reported that, in the younger patient group, diplopia takes more time to resolve and that they have more persistent problems [6, 28]. So, if left untreated, permanent restriction in ocular motility may result.

In some cases, patients are referred after several weeks; meanwhile, the patient may develop a torticollis to compensate for the double vision (Fig.12.10).

Variations in Orbital Wall Fractures

Blow-Up Fracture, Blow-In Fracture

A blow-up fracture involves the superior displacement of the orbital roof into the anterior cranial fossa [4]. A blow-in fracture describes an inferior displacement of the orbital roof [4]. In case of an orbital roof fracture, watch for dis-

placement and possible accompanying dural tearing resulting in a possible leptomeningeal cyst (encephalocele) or (progressive) pulsatile exophthalmos; proptosis, vertical dystopia may develop. Rarely, a progressive, growing orbital roof skull fracture is seen, which may still develop months after head injury has occurred [2, 8, 9, 29, 30].

Pure Versus Impure Orbital Floor Fracture

A distinction can be made between a pure or indirect (solely orbital floor fracture) and impure or direct (orbital floor in conjunction with other fractures) orbital floor fracture [2].

Open Door Fracture Versus Trapdoor

Opposite of the trapdoor fracture is the "open door" fracture, a floor fracture without entrapment which is more common in de adult population [2]. When the children grow older, chances of an open door, blowout fracture increase and enophthalmos may result. The term "blow-out" fracture had already been introduced by Smith and Regan in 1951 [14].

Medial Wall Fracture

The medial wall fracture in older children is similar to adults and is described in Chap. 10.

Complex (Multi) Fracture

Orbital fractures can of course also be part of a complex Naso-Orbito-Ethmoid (NOE) fracture, a midface Lefort fracture (<5%) or a relatively simple zygomatic complex fracture (16%), fortunately rare in younger children [2, 7].

Tests and Treatment Principles

The management of growing individuals who present with an orbital fracture requires a customized approach, adjusted according to the growing individual. In the evaluation of the patient, apart from a change in vision, critical aspects consist of globe motility disturbances and enophthalmos or hypoglobus.

Thorough examination by the OMF surgeon, the ophthalmologist and orthoptist should be carried out (Fig. 12.11a). However, a complete examination can be difficult to obtain in the young, sometimes obstinate uncooperative patient. A "white-eye" orbital fracture may even lead to denial diagnosis and doctor's delay in adequate treatment [25, 26]. A CT scan should be obtained in case of suspicion of an orbital wall fracture. Displacement of bone structures provides a simple diagnosis but often the findings in children are limited to a tear-drop sign (Fig. 12.8). Moreover, CT images may also incorrectly deny the existence of a fracture (Fig. 12.7). A 3D reconstruction can support a more accurate diagnosis. Nonetheless, the clinical findings are in the lead when it comes to a treatment plan.

When no acute enophthalmos, hypoglobus, diplopia or entrapment is present, these fractures can be treated conservatively, closely monitored during surveillance. When diplopia is present, orthoptic evaluation is mandatory prior to surgery to classify the extent of the motility disorder and compare these initial findings with future recovery development (Fig.12.11a–c). When on initial presentation, there is double vision in *many or all* directions, this is most likely to be caused by swelling i.e., oedema instead of entrapment of orbital soft tissue. Allow some time to recover and follow closely. When double vision is seen in just a *few or one* direction, entrapment is the most obvious diagnosis which warrants immediate intervention [24, 29]. Once entrapment is diagnosed, surgical intervention of especially the incarcerated tissue should be done within 12–24 h [1–4, 7, 14, 15, 19, 21, 24, 25, 27]. The primary goal is to release the entrapped orbital soft tissue (Figs. 12.12 and 12.13). Because of the recoil of the linear trapdoor fracture, there is hardly any need to restore the orbital floor in such cases (Fig. 12.13). If intervention is carried out instantly, a complete recovery within days is very likely (Figs. 12.11c and 12.14a, b). Another indication for immediate or early intervention is the presence of acute enophthalmos >2 mm and hypoglobus. The Hertel exophthalmometer is used to measure the extent of the enophthalmos.

When during follow-up the initial (peri)orbital swelling has subsided and clinical signs and symptoms of limited recovery of ocular motility or enophthalmos >2 mm and hypoglobus become apparent, *early intervention* (2–14 days) may be indicated. Again, the definition, measurement and reduction of double vision i.e., diplopia should always be objectivated by careful orthoptic evaluation [31].

Delayed intervention (2–3 weeks) may be performed when there is insufficient recovery of double vision. However, a delayed or late intervention will generally result in a poorer outcome, especially in the younger generation. Moreover, the high bone and soft tissue turnover in the younger patient group challenges the intervention at a later stage.

Late enophthalmos can be an indication for *late intervention* (>3 weeks).

Fig. 12.11 (**a**) Orthoptic evaluation: Hess preoperatively: severe limitation of upward gaze OS, confined limitation of depression OS, near (30 cm distance) little exophoria, small right—hyperphoria (no double vision), far (5 m distance) little right—hypertropia (double vision), overshoot upward gaze OD. (**b**) Orthoptic evaluation: Hess 2 weeks postoperatively, restore eye motility, however not yet complete recovery upward gaze OS, recovery limitation of depression OS, still some overshoot OD. (**c**) Orthoptic evaluation: Hess 3 months postoperatively, full recovery of limitation of elevation OS

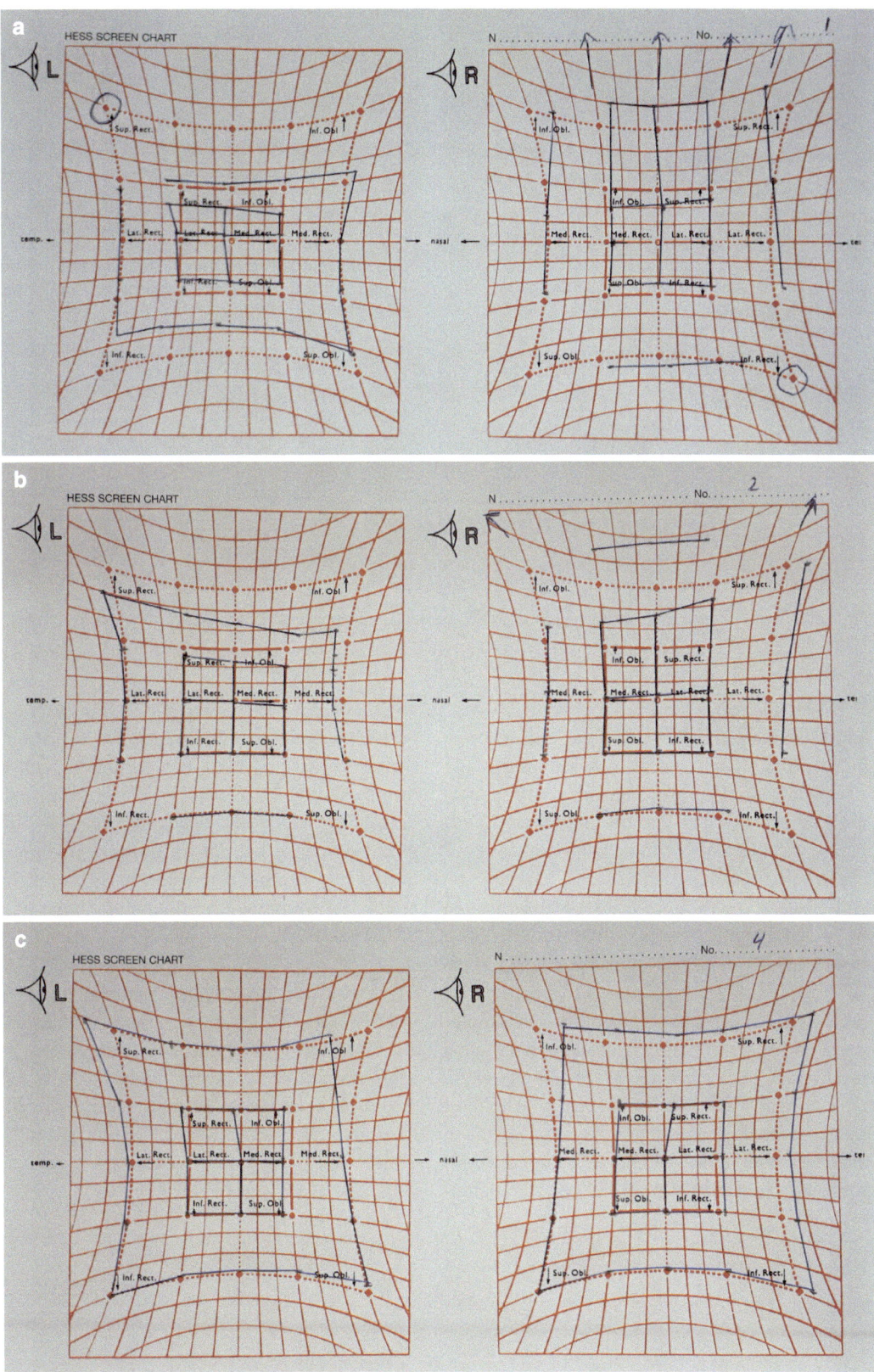

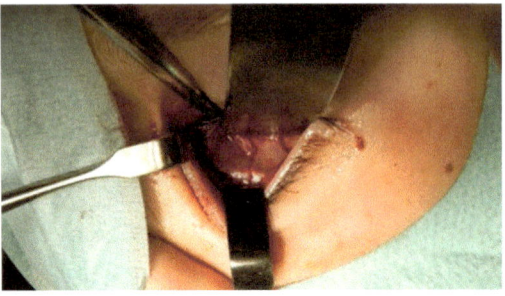

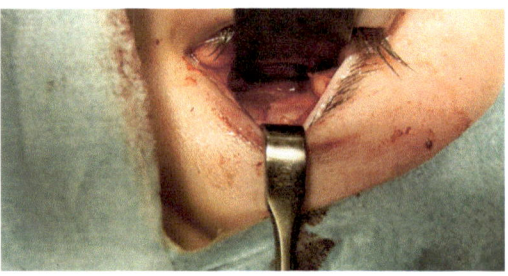

Fig. 12.12 Intraoperative view orbital floor OS: herniated, incarcerated orbital soft tissue in linear trapdoor orbital floor fracture (patient Fig. 12.9)

Fig. 12.13 Intraoperative view orbital view OS: retrieved orbital soft tissue from linear trapdoor orbital floor fracture (patient Fig. 12.9)

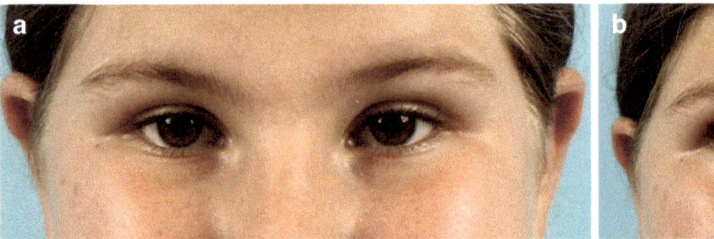

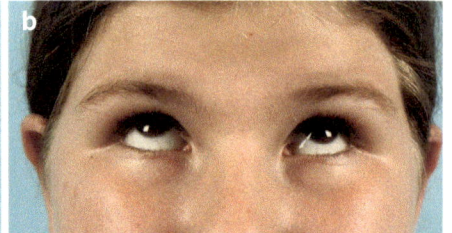

Fig. 12.14 (**a**) Postoperatively 2 weeks: primary gaze: binocular single vision. (**b**) Postoperatively 2 weeks: recovery limitation elevation OS

When treatment is indicated, even in children, a transconjunctival approach gives ample access to the orbital floor. Nevertheless, when surgically intervening in the growing orbit, we should always keep in mind that the orbital wall morphology is different in children and further development is still to come [1, 32].

Releasing the entrapped soft tissue will often be enough treatment; orbital floor reconstruction is seldom necessary. If a larger floor defect needs coverage to prevent recurrence of herniation, we preferably use an autologous graft. The autologous grafts are instantly available, have ideal mechanical properties, revascularization potential and adaptation to orbital tissue with a minimal immune response. In the growing individual, we are hesitant to use foreign or non-degradable materials. We tend not to use alloplastic materi-

als, there is an increased rate of infection and possible migration of the reconstruction material. Screw fixation of Med Por or titanium plates (mesh) should be avoided in the growing individual [4]. Resorbable materials are more suitable but may cause an inflammatory response during the resorption process which negatively affects the surrounding orbital soft tissue [33]. Intra-operatively, a presurgical and postsurgical forced duction test is conducted (Fig.12.15). Postoperatively, neurologic, ophthalmologic and orthoptic follow-up are necessary. Instructions are given not to blow the nose and patients are cautioned to avoid sneezing: both can cause subcutaneous or intra-orbital emphysema. Generally, if surgical intervention is carried out in time, a good final outcome can be expected (Fig. 12.16a–g).

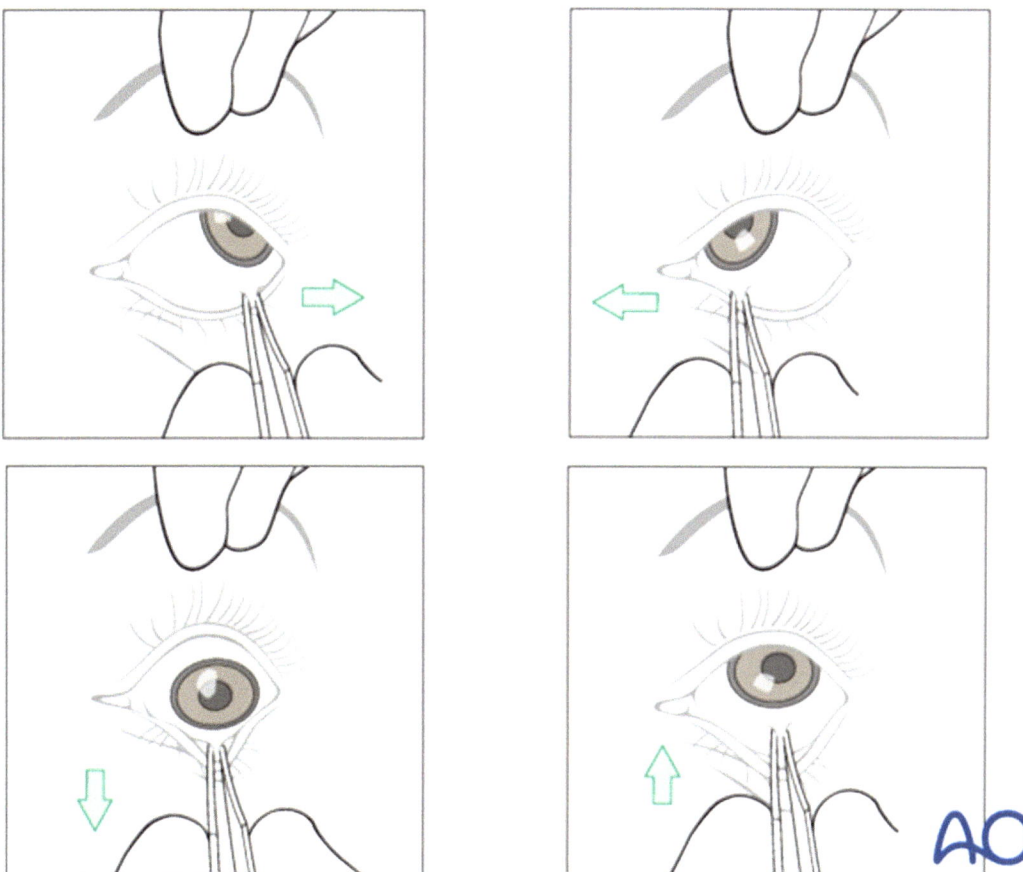

Fig. 12.15 AO illustration of forced duction test (with permission from AO Foundation)

Post-operative Warning Signs

Increasing pain or reduced visual acuity or inadequate pupil reaction require instant re-examination and exploration. A compartment syndrome of the orbital apex or retrobulbar hemorrhage is a serious threat for vision. Orbital roof fractures can be associated with a fracture of the anterior skull base, the dura lining may be torn resulting in CSF leakage. Thus, in the long term, the development of a growing orbital roof fracture is exceptional [29].

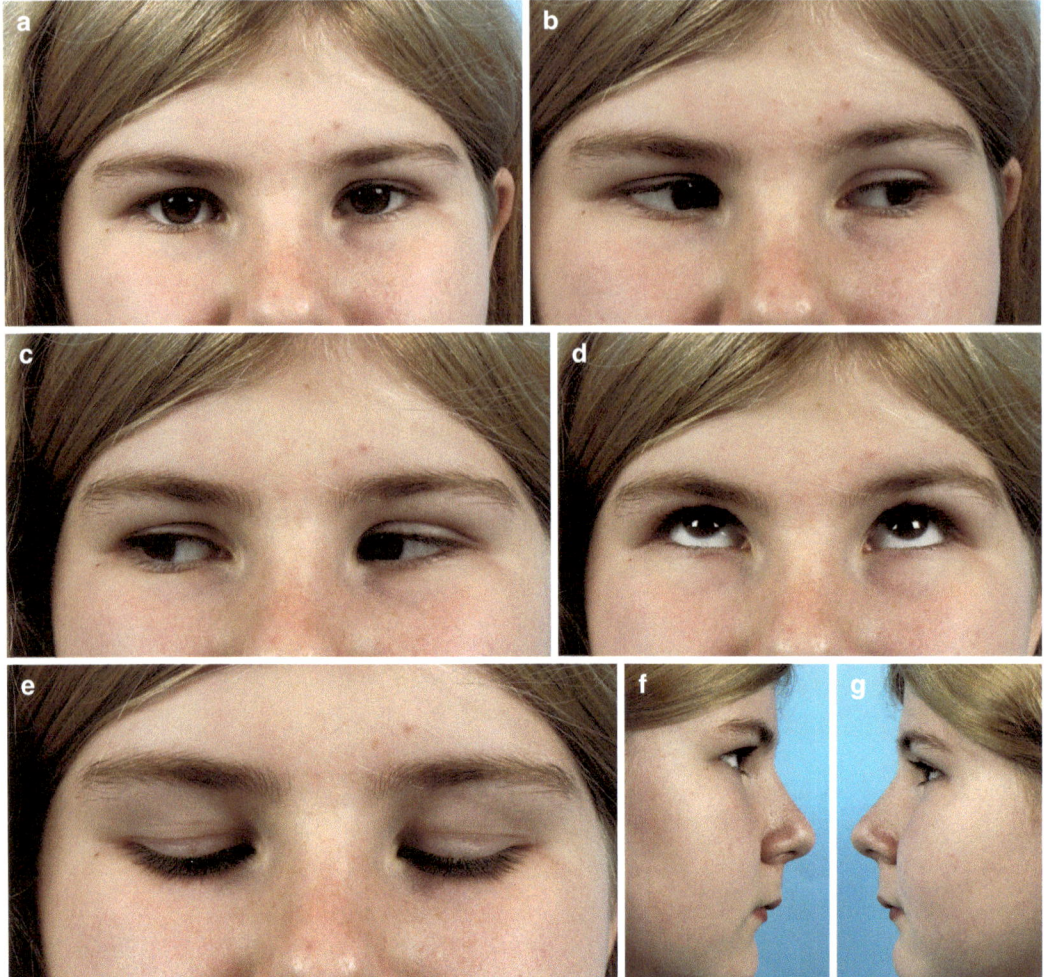

Fig. 12.16 (**a**) Postoperatively 2 years: primary gaze: binocular single vision. (**b**) Postoperatively 2 years: abduction, single vision. (**c**) Postoperatively 2 years: adduction, single vision. (**d**) Postoperatively 2 years: elevation, unlimited. (**e**) Postoperatively 2 years: depression, unlimited. (**f**) Postoperatively 2 years: A-P globe position 17 mm Hertel OD. (**g**) Postoperatively 2 years: A-P globe position 17 mm Hertel OD

References

1. Stotland MA, Do NK. Pediatric orbital fractures. J Craniofac Surg. 2011;22:1230–5.
2. Oppenheimer AJ, Monson LA, Buchman SR. Pediatric orbital fractures. Craniomaxillofac Trauma Reconstr. 2013;6:9–20.
3. Gerber B, Kiwanuka P, Dhariwal D. Orbital fractures in children: a review of outcomes. Br J Oral Maxillofac Surg. 2013;51:789–93.
4. Joshi S, Kassira W, Thaller SR. Overview of pediatric orbital fractures. J Craniofac Surg. 2011;22:1330–2.
5. Sirichai P, Anderson PJ. Orbital fractures in children: 10 years experience from a tertiary centre. Br Oral Maxillofac Surg. 2015;53(10):938–42.
6. Heggie AA. Isolated orbital floor fractures in the paediatric patient: case series and review of management. Int J Oral Maxillofac Surg. 2015;44:1250–4.
7. Cobb ARM, Jeelani NO, Ayliffe PR. Orbital fractures in children. Br Oral Maxillofac Surg. 2013;51:41–6.
8. Greenwald MJ, Boston D, Pensler JM, Radkowski MA. Orbital roof fractures in childhood. Ophthalmology. 1989;96(4):491–6.
9. Messinger A, Radkowski MA, Greenwald MJ, Pensler JM. Orbital roof fractures in the pediatric population. Plast Reconstr Surg. 1989;84:213–6.
10. Koltai PJ, Amjad I, Meyer D, Feustel PJ. Orbital fractures in children. Arch Otolaryngol Head Neck Surg. 1995;121:1375–9.
11. Chang JT, Morrison CS, Styczynski R, Mehan W, Sullivan SR, Taylor HO. Pediatric orbital depth and

growth: a radiographic analysis. J Craniofac Surg. 2015;26(6):1988–91.

12. Escaravage GK, Dutton J. Age-related changes in the pediatric human orbit on CT. Ophthalmic Plast Reconstr Surg. 2013;29:150–6.

13. Guyot L, Lari N, Benso-Layoun C, Denis D, Chossegros C, Thiery G. Orbital fractures in children. J Fr Ophthalmol. 2011;34(4):265–74.

14. Smith B, Regan WF Jr. Blow-out fracture of the orbit; mechanism and correction of internal orbital fracture. Am Ophthalmol. 1957;44:733–9.

15. Losee JE, Afifi A, Jiang S, Smith D, Chao MT, Vecchione L, Hertle R, Davis J, Naran S, Hughes J, Paviglianiti J, Deleyiannis FW-B. Pediatric orbital fractures: classification, management, and early follow up. Plast Reconstr Surg. 2008;122:886–97.

16. Sires BS, Stanley RB, Levine LM. Oculocardiac reflex caused by orbital floor trapdoor fractures: an indication for urgent repair. Arch Ophthalmol. 1998;116:955–6.

17. Cohen S, Garret C. Paediatric orbital floor fractures: nausea/vomiting as signs of entrapment. Otolaryngol Head Neck Surg. 2003;129:43–7.

18. Bansagi Z, Meyer D. Internal orbital fractures in the paediatric age group: characterization and management. Ophthalmology. 2000;107:829–36.

19. Egbert JE, May K, Kersten RC, Kulwin DW. Paediatric orbital floor fracture. Ophthalmology. 2000;107:1875–9.

20. Koornneef L. Orbital septae: anatomy and function. Ophthalmology. 1979;86(5):876–80.

21. Smith B, Lisman R, Simonton J. Volkmann's contracture of the extraocular muscles following blow out fractures. Plast Reconstr Surg. 1984;74:200–5.

22. de Man K, Wijngaarde R, Hes J, de Jong PT. Influence of age on the management of blow-out fractures. Int Oral Maxillofac Surg. 1991;20:330–6.

23. Wei LA, Durairaj VD. Pediatric orbital floor fractures. J AAPOS. 2011;15:173–80.

24. Parbhu KC, Galler KE, Li C, Mawn LA. Underestimation of soft tissue entrapment by computed tomography in orbital floor fractures in the pediatric population. Ophthalmology. 2008;115:1620–5.

25. Jordan DR, Allen LH, White J, Harvey J, Pashby R, Esmaeli B. Intervention within days for some orbital floor fractures: the white-eyed blowout. Ophthalmic Plast Reconstr Surg. 1998;14:379–90.

26. Lane K, Penn RB, Bilyk JR. Evaluation and management of pediatric orbital fractures in a primary care setting. Orbit. 2009;26(3):183–91.

27. Karthik R, Cynthia S, Vivek N, Saravanan C, Prashanthi G. Intraoperative findings of extraocular muscle necrosis in linear orbital trapdoor fractures. J Oral Maxillofac Surg. 2019;77(6):1229.e1–8. https://doi.org/10.1016/j.joms.2019.02.033.

28. Cope MR, Moos KF, Speculand B. Does diplopia persist after blow out fractures of the orbital floor in children. Br J Oral Maxillofac Surg. 1999;37:46–51.

29. Suri A, Mahapatra AK. Growing fractures of the orbital roof. A report of two cases and a review. Pediatr Neurosurg. 2002;36(2):96–100.

30. Coon D, Yan N, Ones D, Howell LK, Grant MP, Redett RJ. Defining pediatric orbital roof fractures: patterns, sequelae, and indications for operation. Plast Reconstr Surg. 2014;134(3):442e–8e.

31. Jansen J, Dubois L, Maal T, Mourits MP, Ellema HM, Neomagus P, Lange de J, Hartman LJC, Gooris P, Becking AG. A nonsurgical approach with repeated orthoptic evaluation is justified for most blow-out fractures. J Craniomaxillofac Surg. 2020;48:560–8.

32. Nagasao T, Hikosaka M, Morotomi T, Nagasao M, Ogawa K, Nakajima T. Analysis of the orbital floor morphology. J Craniomaxillofac Surg. 2007;35:112–9.

33. Bergsma JE. Poly(L-lactide) implants in repair of the orbital floor. A 5 years animal study. Cells Mater. 1994;4:31–6.

Open Access This chapter is licensed under the terms of the Creative Commons Attribution 4.0 International License (http://creativecommons.org/licenses/by/4.0/), which permits use, sharing, adaptation, distribution and reproduction in any medium or format, as long as you give appropriate credit to the original author(s) and the source, provide a link to the Creative Commons license and indicate if changes were made.

The images or other third party material in this chapter are included in the chapter's Creative Commons license, unless indicated otherwise in a credit line to the material. If material is not included in the chapter's Creative Commons license and your intended use is not permitted by statutory regulation or exceeds the permitted use, you will need to obtain permission directly from the copyright holder.

Maarten P. Mourits, Peter J. J. Gooris,
and J. Eelco Bergsma

Learning Objectives
- At present, there is no treatment for traumatic optic neuropathy. However, spontaneous improvement is often seen.
- Treatment of a tight orbit due to raised intraorbital pressure as a result of fast increase of the intraorbital volume by, for instance, retrobulbar hemorrhage consists of immediate lateral canthotomy and cantholysis.

M. P. Mourits (✉)
Amsterdam University Medical Centers, Location AMC, Amsterdam, The Netherlands

P. J. J. Gooris
Department of Oral and Maxillofacial Surgery, Amphia Hospital Breda, Breda, The Netherlands

Amsterdam University Medical Centers, Amsterdam, The Netherlands

Department of Oral and Maxillofacial Surgery, University of Washington, Seattle, WA, USA

J. E. Bergsma
Department of Oral and Maxillofacial Surgery, Amphia Hospital Breda, Breda, The Netherlands

Amsterdam University Medical Centers, Amsterdam, The Netherlands

Acta dental school, Amsterdam, The Netherlands

Introduction

Loss of vision is the most dramatic outcome of disorders of the eye and of orbital trauma and orbital diseases. One of the most common causes of loss of vision is trauma or disease of the eyeball itself, such as perforation of the cornea or globe, inflammation of the globe (endophthalmitis) or massive intraocular hemorrhage, but these conditions are beyond the scope of this book. Brain trauma or diseases may also result in blindness. Here, we will focus on situations that may cause blindness, which are located outside the globe itself and are due to changes of the optic nerve. The optic nerve is sensitive for shock and compression. Blunt trauma of the orbit can cause blindness due to contusion of the optic nerve. A number of intraorbital changes, that give an increase in intraocular tension may cause optic neuropathy, eventually leading to loss of visual functions.

Traumatic Optic Neuropathy

Even rather mild contusion of the orbit can cause optic neuropathy, which is thought to be the result of a transmitted shock to the optic canal resulting in edema of the intracanalicular part of the optic nerve, compression, ischemia and loss of axons. This condition is called indirect traumatic optic neuropathy (TON) and can stand alone or present

© The Author(s) 2023
P. J. J. Gooris et al. (eds.), *Surgery in and around the Orbit*,
https://doi.org/10.1007/978-3-031-40697-3_13

in combination with orbital fractures. The term direct TON is reserved for situations in which after trauma, orbital bony fragments have injured the optic nerve. The majority of patients with TON are young males [1].

Patients with TON complain of blurred vision, and/or decreased color vision, and/or decreased visual field. On inspection, the pupil on the affected side is wide and not responding to light or a RAPD (Chap. 7) is present. After some weeks to months, the optic nerve head can become pale as a sign of atrophy.

In order to decease optic nerve edema, patients with TON have been treated with high doses of prednisone or with optic canal decompression, but neither appeared to result in a better outcome than no treatment at all (e.g., observation) [2, 3]. Moreover, the Corticosteroid Randomization for Acute Head Trauma (CRASH) trial found an increased rate of death among patients with acute head trauma treated with high-dose corticosteroids compared to placebo-treated patients [1]. Hence, no treatment exists at this time for patients with TON. However, some spontaneous improvement is to be expected in up to 57% [2].

Retrobulbar Hemorrhage

A so-called retrobulbar hemorrhage (RH) in fact is a bleeding that can occur at any place in the orbit and not just posterior to the globe. Because the orbit is surrounded by bony walls at the medial, lateral, superior and inferior side and by the orbital septum in the front, a compartment syndrome with increasing intraorbital pressure develops if the intraorbital volume increases. Typical causes of increasing intraorbital volume are enlargement of the extraocular muscles and fat increase as seen in patients with Graves' orbitopathy (Chap. 14), subperiosteal empyema and orbital abscess in retroseptal orbital cellulitis (Chap. 17) and hemorrhages due to orbital varices or venous malformations and to orbital trauma. RH after retrobulbar nerve blocks for intraocular surgery have been reported in 55 out of 12,500 patients, e.g., in 0.44% [4]. RH causing blindness has been described after elective sur-

gery, such as, orbital decompression, blepharoplasty and even after fine-needle aspiration [5–8].

Patients with RH complain about extreme pain and blurred vision. On inspection, there is swelling of the eyelids, increasing exophthalmos, impaired motility, a RAPD that turns into a wide pupil not responding to light and finally a tight orbit with no light perception.

The best option for treatment is a lateral canthotomy and inferior cantholysis of the lateral ligament. The upper and lower lid are cut in a skin fold with sharp scissors in the lateral angle, where the eyelids fuse (Fig. 13.1).

One blade of scissors is then placed under the lower arm of the lateral canthal ligament and cut. Immediately, the lower lid gives way and the intraorbital tension decreases followed by a narrowing of the dilated pupil. Local anesthesia is not necessarily used. Infiltration with anesthetic fluid takes time and would increase the pressure even more. In addition, the pain caused by the cut is quickly compensated as the pain caused by the high pressure decreases [9, 10]. This wound usually heals beautifully even without stitches. Cantholysis for RH should be done as soon as possible if a tight orbit is found. RH can occur together with other manifestations of orbital trauma such as orbital fractures. Further clinical and radiological evaluation of the patient, however, should be postponed until a cantholysis has been performed.

Cantholysis is generally advised to do within 24 hours after the onset of the proptosis, but we have seen patients with a complete recovery of visual functions, who had been operated after

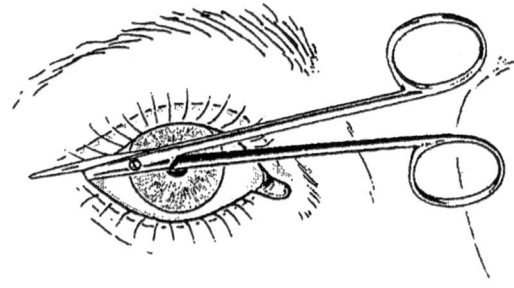

Fig. 13.1 Lateral cantholysis

more than 24 hours. So, it is worth the effort to do a cantholysis in any patient with a tight orbit [11]. The significance of a cantholysis in a tight orbit is comparable to a tracheotomy in obstructed airways. Tracheotomy saves lives; cantholysis saves eyes.

An alternative for a cantholysis is a horizontal incision through the eyelid halfway the lid margin and the eyebrow. The cut should have a length of 1–2 cm and be deep enough to pass through the orbital septum (Chap. 22), which will be apparent from a prolapse of pre-aponeurotic fat through the wound. Steroids or tension lowering eyedrops are of little use. If no or insufficient response is seen after cantholysis or incision of the orbital septum, other causes of optic injury have to be excluded and orbital decompression can be considered.

Conclusion

It must be clear that the clinician should be alert on signs and symptoms of trauma or compression of the optic nerve which often warrants immediate intervention. Depending on the cause of the optic nerve damage, decompression and or steroid therapy can be considered.

References

1. Edwards P, Arango M, Balica L, Cottingham R, El-Sayed H, Farrell B, Fernandes J, Gogichaisvili T, Golden N, Hartzenberg B, Husain M, Ulloa MI, Jerbi Z, Khamis H, Komolafe E, Laloë V, Lomas G, Ludwig S, Mazairac G, de Muñoz Sanchéz ML, Nasi L, Olldashi F, Plunkett P, Roberts I, Sandercock P, Shakur H, Soler C, Stocker R, Svoboda P, Trenkler S, Venkataramana NK, Wasserberg J, Yates D, Yutthakasemsunt S, CRASH Trial Collaborators. Final results of MRC CRASH, a randomised placebo-controlled trial of intravenous corticosteroid in adults with head injury-outcomes at 6 months. Lancet. 2005;365(9475):1957–9.
2. Levine LA, Beck RW, Joseph MP, Seiff S, Kraker R. The treatment of traumatic optic neuropathy: the international optic nerve trauma study. Ophthalmology. 1999;106(7):1268–77.
3. Wilhelm H. Traumatic optic neuropathy—the present state. Klin Monatsbl Augenheilkd. 2004;221(8):702–5.
4. Edge KR, Nicoll JM. Retrobulbar hemorrhages after 12,500 retrobulbar blocs. Anesth Analg. 1993;76(5):1019–22.
5. Hartley JH Jr, Lester JC, Schatten WE. Acute retrobulbar hemorrhage during elective blepharoplasty. Its pathophysiology and management. Plast Reconstr Surg. 1973;52:8–15.
6. Teng CC, Reddy S, Wong JJ, Lisman RD. Retrobulbar hemorrhage nine days after cosmetic blepharoplasty resulting in permanent visual loss. Ophthalmic Plast Reconstr Surg. 2006;22(5):388–9.
7. Long JC, Ellis PP. Total unilateral visual loss following orbital surgery. Am J Ophthalmol. 1971;71(Suppl):218–20.
8. Liu D. Complications of fine needle aspiration biopsy of the orbit. Ophthalmology. 1985;92:1768–71.
9. Goodall KL, Brahma A, Bates A, Leatherbarrow B. Lateral canthotomy and inferior cantholysis: an effective method of urgent orbital decompression for sight threatening acute retrobulbar haemorrhage. Injury. 1999;30(7):485–90.
10. Mourits MP, de Groot JA. Retrobulbar haemorrhage. Ned Tijdschr Geneeskd. 1995;25(139):361–3.
11. Dixon JL, Beams OK, Levine BJ, Papas MA, Passarello BA. Visual outcomes after traumatic retrobulbar hemorrhage are not related to time or intraocular pressure. Am J Emerg Med. 2020;38(11):2308–12.

Open Access This chapter is licensed under the terms of the Creative Commons Attribution 4.0 International License (http://creativecommons.org/licenses/by/4.0/), which permits use, sharing, adaptation, distribution and reproduction in any medium or format, as long as you give appropriate credit to the original author(s) and the source, provide a link to the Creative Commons license and indicate if changes were made.

The images or other third party material in this chapter are included in the chapter's Creative Commons license, unless indicated otherwise in a credit line to the material. If material is not included in the chapter's Creative Commons license and your intended use is not permitted by statutory regulation or exceeds the permitted use, you will need to obtain permission directly from the copyright holder.

Graves' Disease: Introduction

Maarten P. Mourits

Learning Objectives
- Graves' orbitopathy is a common auto-immune disease that affects women 5–6 times more than men.
- The natural course of Graves' orbitopathy is characterized by a phase of increasing severity, a plateau phase and a phase of decreasing severity.
- Graves' orbitopathy starts with an increase of extraocular muscle volume, which is usually accompanied by signs of inflammation.
- Smoking, old age, high antibody levels and male gender are the most important risk factors to develop severe Graves' orbitopathy. Thus, women more often have GO than men, but when men have GO, they mostly have a more severe form.

Incidence, Nomenclature, History

With an estimated incidence of 45 (36.8 females, 8.3 males) per 100,000 persons per year, Graves' disease (GD) is the most common autoimmune disease [1]. GD (or: Basedow's disease) serves as an umbrella term for Graves' thyroid disease, Graves' eye/orbit disease, pretibial myxedema, and acropachy. These subentities can occur together or separately.

Graves' thyroid disease (GTD) is the most frequent of them. Although the cause is still unknown, the body produces TSH-receptor-binding immunoglobulins (TBII) which either stimulate, suppress or do not influence thyroid function. As a result, circa 85% of patients with GTD are hyperthyroid, 5–10% are hypothyroid, and the remainder are euthyroid.

Depending upon the criteria used, approximately 50% of GD patients have Graves' eye/orbit disease, called Graves' Orbitopathy or Ophthalmopathy in most European countries (GO), Thyroid Associated Orbitopathy (TAO) in the UK, and GO or Thyroid Eye Disease (TED) in the United States of America.

Pretibial myxedema, that is reddish, thickened areas on the lower legs, often associated with pruritis, and acropachy, i.e., soft-tissue swelling of the hands and clubbing of the fingers due to periostitis of the metacarpal bones, is rare. When present together with GO, the symptoms are usually more severe [2].

The oldest descriptions of what we now call GD, among which a combination of goiter, tachycardia and exophthalmos, the so-called Merseburger trias, are from the Welshman Caleb Hillier Parry (Fig. 14.1, 1786, published in 1825), the Dubliner Robert James Graves (Fig. 14.2, 1835) and the German Carl Adolph von Basedow (Fig. 14.3, 1840).

M. P. Mourits (✉)
Amsterdam University Medical Centers, Location AMC, Amsterdam, The Netherlands

© The Author(s) 2023
P. J. J. Gooris et al. (eds.), *Surgery in and around the Orbit*,
https://doi.org/10.1007/978-3-031-40697-3_14

Fig. 14.1 Caleb Parry (taken from Wikipedia—public domain)

Fig. 14.3 Carl von Basedow (taken from Wikipedia—public domain)

Temporal Relationship Go with GTD, Risk Factors

The majority of patients develop GTD *and* GO within a time span of 12 months; in a minority of patients, the onset of the eye complaints precedes the thyroid disease, whereas in some more patients, the eye disease starts after the onset of GTD. Approximately, 15% develop GO more than 5 years after the onset of their GTD [3]. Patients may develop GO in the absence of hyperthyroidism. This is called euthyroid GO.

GO is 5–6 times more frequent in females. Although GO may start at any age (from 0 to 100), it is most frequent in women between 30 and 60 years of age. Men often develop GO at older age and suffer from more severe forms. GO at childhood is extremely rare; soft tissue involvement and proptosis are the more frequent signs in children with GO [4].

Fig. 14.2 Robert Graves (taken from Wikipedia—public domain)

Risk factors to develop GO are as follows

1. Smoking [5]
2. Female gender
3. Radioactive iodine treatment [6]
4. Iatrogenic hypothyroidism [6]

Risk factors to develop severe GO are as follows

1. Smoking
2. Male gender
3. Old age
4. Concomitant diabetes mellitus [7]
5. Concomitant pretibial dermopathy and/or acropachy [2]
6. High antibody titers [8, 9]

Natural Course

As Rundle described in 1957 [10], untreated GO has a biphasic course. It starts with a period of increasing severity, then reaches a plateau phase and subsequently a phase of spontaneous improvement (Fig. 14.4), although a complete disappearance of symptoms and signs is mostly not to be expected.

The initial phase is called the active phase and is usually accompanied by signs of inflammation, such as red, swollen eyelids, pain and worsening (Chap. 15). During the inactive phase, signs of inflammation disappear; exophthalmos, eyelid retraction and motility impairment (might) remain. The duration of the several phases varies

considerably. A total length of 3–5 years is no exception.

Exophthalmos in GO is the result of two pathological processes in the orbit: an increase of orbital contents by neo-adipogenesis and an increase of muscle volume (Fig. 14.5). Recently, we have studied the course of mild GO in terms of volume changes. We, therefore, had adopted and validated a CT-based method (Mimics, Materialise) for the calculation of orbital soft tissue volumes [11]. We found that an increase in muscle volume is an early phenomenon and coincides with inflammatory changes. After about 1 year, muscle volume deceases slightly (with 1 cm^3), but fat volume starts to increase over the next 4 years (with almost 2 cm^3) and then stabilizes. Interestingly, this fat volume increase does not seem to be related to inflammatory changes [12].

Differences in the duration and severity of these two processes might explain the different presentations of GO. Of interest is also that smoking is associated with both muscle volume increase and inflammatory changes [13].

A strong relationship has been found between smoking and the severity of GO (Fig. 14.6) [5].

In 95 newly diagnosed patients with GO, we found that in 25% of them, the muscle and fat volumes were within the limits of normalcy; 61% of the patients only had an increase of muscle volume, 5% only an increase in fat volume and 9% demonstrated an increase in both muscle and fat volume [14].

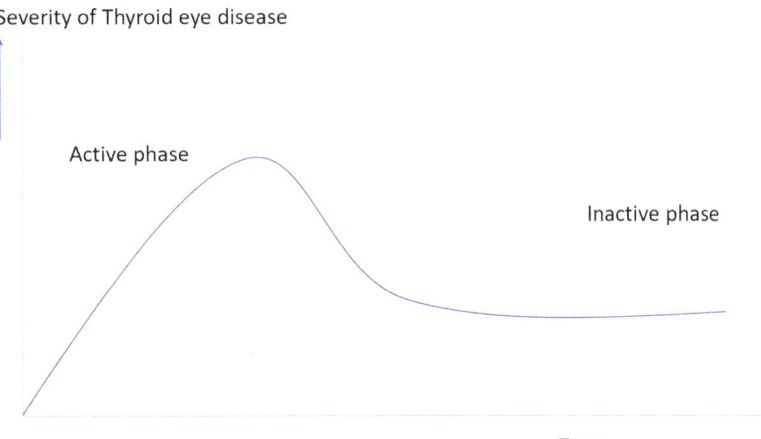

Fig. 14.4 Natural course of GO: severity versus time

Severity of Thyroid eye disease

Active phase

Inactive phase

Time ⟶

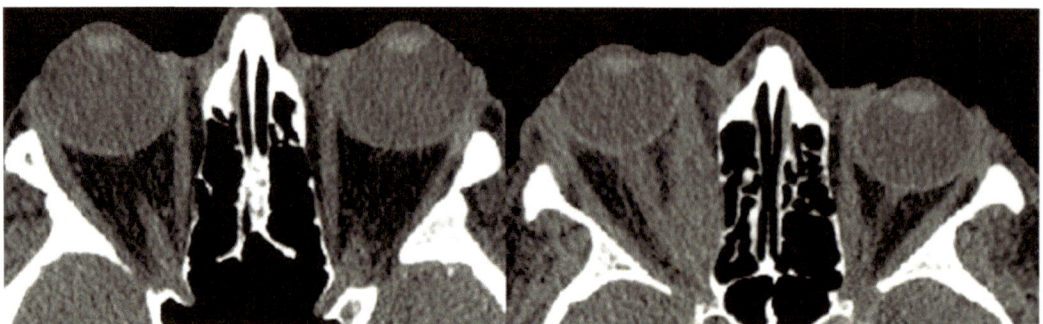

Fig. 14.5 Exophthalmos in GO due to fat (left) and muscle (right) volume

Severity of Graves' orbitopathy

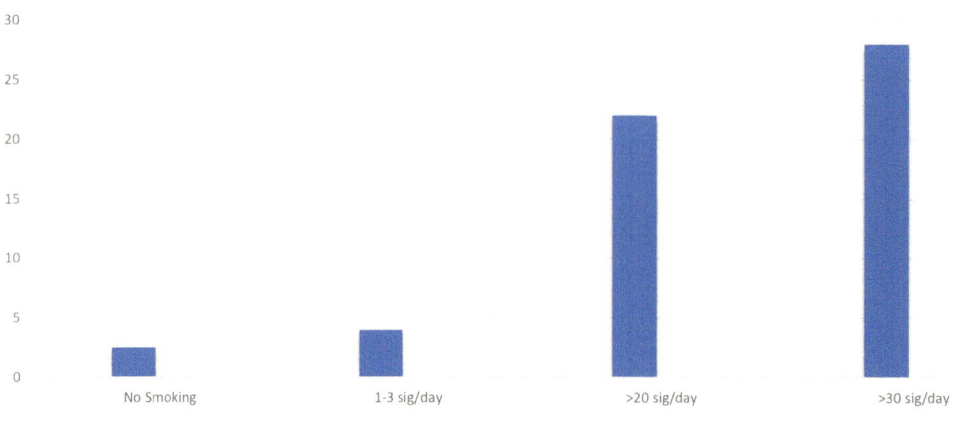

Odds ratio's between smoking habbits

Fig. 14.6 Smoking (*X*-axis) versus disease severity (*Y*-axis) after Prummel et al. (1993) [5]

Auto-antibodies, such as TBII, correlate with the severity and activity of GO [15, 16] and have prognostic significance for the course of GO [16]. Iatrogenic hypothyroidism and treatment with radio-active iodine may worsen the orbitopathy [17, 18].

Once GO has burnt out, it is unlikely to flare up again. However, we have seen recurrences as late as 7 years after the first episode [19].

During pregnancy, TBII serum levels decrease and the orbitopathy declines. However, postpartum antibody titers often increase to levels that exceed pre-pregnancy values and can cause a recurrence of GO and a clinical relapse of disease activity [20].

References

1. Furszyfer J, Kurland LT, McConahey WM, Elveback LR. Graves' disease in Olmsted County, Minnesota, 1935 through 1967. Mayo Clin Proc. 1970;45:636–44.
2. Fatourechi V, Bartley GB, Eghbali-Fatourechi GZ, Powell GC, Ahmed DD, Garrity JA. Graves' dermopathy and acropachy are markers of severe Graves' ophthalmopathy. Thyroid. 2003;13:1141–4.
3. Wiersinga WM, Smit T, Van Der Gaag R, Koornneef L. Temporal relationship between the onset of Graves' ophthalmopathy and onset of thyroidal Graves' disease. J Endocrinol Investig. 1988;11:615–9.
4. Chan W, Wong GH, Fan SD, Cheng AC, Lam DS, Ng IS. Ophthalmopathy in childhood Graves' disease. Br J Ophthalmol. 2002;86:740–2.
5. Prummel MF, Wiersinga WM. Smoking and the risk of Graves' orbitopathy. JAMA. 1993;269:479–82.

6. Li HX, Xiang N, Hu WK, Jiao XL. Relation between therapy options for Graves' disease and the course of Graves' ophthalmopathy: a systematic review and meta-analysis. J Endocrinol Investig. 2016;39:1225–33.

7. Kalmann R, Mourits MP. Diabetes mellitus: a risk factor in patients with Graves' orbitopathy. Br J Ophthalmol. 1999;83:463–5.

8. Eckstein A, Esser J, Mann K, Schott M. Clinical value of TSH receptor antibodies measurement in patients with Graves' orbitopathy. Pediatr Endocrinol Rev. 2010;7(Suppl 2):198–203.

9. Eckstein AK, Plicht M, Lax H, Neuhauser M, Mann K, Lederbogen S, Heckmann C, Esser J, Morgenthaler NG. Thyrotropin receptor autoantibodies are independent risk factors for Graves' ophthalmopathy and help to predict severity and outcome of the disease. J Clin Endocrinol Metab. 2006;91:3464–70.

10. Rundle FF. Management of exophthalmos and related ocular changes in Graves' disease. Metabolism. 1957;6:36–48.

11. Regensburg NI, Kok PH, Zonneveld FW, Baldeschi BL, Saeed P, Wiersinga WM, Mourits MP. A new and validated CT-based method for the calculation of orbital soft tissue volumes. Invest Ophthalmol Vis Sci. 2008;49:1758–62.

12. Potgieser PW, De Win MML, Wiersinga WM, Mourits MP. Natural course of mild Graves' orbitopathy: increase of orbital fat but decrease of muscle volume with increased fatty degeneration during a 4-year follow-up. Ophthalmic Plastic Reconstr Surg. 2019;35:456–60.

13. Regensburg NI, Wiersinga WM, Berendschot TTJM, Saeed P, Mourits MP. Effect of smoking on orbital fat and muscle volume in Graves' orbitopathy. Thyroid. 2011;21:177–81.

14. Regensburg NI, Wiersinga WM, Berendschot TTJM, Potgieser P, Mourits MP. Do subtypes of Graves' orbitopathy exist? Ophthalmology. 2011;118:191–6.

15. Morris JC, Hay ID, Nelson RE, Jiang NS. Clinical utility of thyrotropin receptor antibody assays: comparison of radioreceptor and bioassay methods. Mayo Clin Proc. 1988;63:707–12.

16. Gerding MN, Van Der Meer JWC, Broenink M, Bakker O, Wiersinga WM, Prummel MF. Association of thyrotropin receptor antibodies with the clinical features of Graves' ophthalmopathy. Clin Endocrinol. 2000;52:267–71.

17. Prummel MF, Wiersinga WM, Mourits MP, Koornneef L, Berghout A, Van Der Gaag R. Effect of abnormal thyroid function on the severity of Graves' ophthalmopathy. Arch Intern Med. 1990;50:1098–101.

18. Tallstedt L, Lundell G, Torring O, Wallin G, Ljunggren JG, Blomgren H, Tabe A. Occurrence of ophthalmopathy after treatment for Graves' hyperthyroidism. The Thyroid Study Group. N Engl J Med. 1992;326:1733–8.

19. Kalmann R, Mourits MP. Late recurrence of unilateral Graves' orbitopathy on the contralateral side. Am J Ophthalmol. 2002;133:727–9.

20. Rotondi M, Cappelli C, Pirali B, Pirola I, Magri F, Fonte R, Castellano M, Rosei EA, Chiovato L. The effect of pregnancy on subsequent relapse from Graves' disease after a successful course of antithyroid drug therapy. J Clin Endocrinol Metab. 2008;93:3985–8.

Open Access This chapter is licensed under the terms of the Creative Commons Attribution 4.0 International License (http://creativecommons.org/licenses/by/4.0/), which permits use, sharing, adaptation, distribution and reproduction in any medium or format, as long as you give appropriate credit to the original author(s) and the source, provide a link to the Creative Commons license and indicate if changes were made.

The images or other third party material in this chapter are included in the chapter's Creative Commons license, unless indicated otherwise in a credit line to the material. If material is not included in the chapter's Creative Commons license and your intended use is not permitted by statutory regulation or exceeds the permitted use, you will need to obtain permission directly from the copyright holder.

Maarten P. Mourits

Learning Objectives

- There is an abundance of symptoms and signs in Graves' orbitopathy. Nevertheless (or because of this), the diagnosis can be challenging
- In order to start the right treatment, the patient must be classified in terms of disease severity and activity
- The orbitopathy has a huge impact on the life of patients with Graves' orbitopathy. Generally, the changed appearance is perhaps the most significant of all aspects of Graves' orbitopathy, although not every patient would easily admit that

Symptoms and Signs

Undoubtedly, *exophthalmos* is the best-known sign of patients with Graves' orbitopathy (GO). It occurs in 60% of patients diagnosed with GO. Exophthalmos means an axial globe position above the limit of normal (Chap. 7). However, *upper eyelid retraction* is more common and is seen in at least 90% of the patients [1–3]. The

M. P. Mourits (✉)
Amsterdam University Medical Centers, Location AMC, Amsterdam, The Netherlands

combination of eyelid retraction and exophthalmos is responsible for the change in externality. The resulting *'startling appearance'* is the source of social disregard and reduced self-acceptance. Patients avoid being photographed and feel they do not look what they are or were.

Frank upper lid retraction may be preceded by scleral show in downgaze, called *von Graefe's sign*: when looking downward, the upper lid is not exactly following the eyeball and the white sclera above the corneal limbus (or iris) becomes visible.

Due to the increased surface of the eye exposed to the outside world, the eye starts to *feel gritty*, may start *tearing* spontaneously, while bright light is less tolerated (*photophobia*). The patient tries to compensate for these inconveniences by squeezing with the eyelids, which in turn leads to *frowns on the forehead* and negatively contributes to the changed appearance (Figs. 15.1 and 15.2). Apart from upper lid retraction, there may be *lower lid retraction* and *lagophthalmos* (the eyes cannot be closed completely), which contributes to the evaporation of the tear film, eventually resulting in breakdown of the corneal epithelium. If not treated adequately and in time, a corneal ulcer may develop that can cause permanent blindness.

In the next phase, the eyelids become *swollen*, either due to edema or to an increase of fat tissue. The conjunctiva also becomes swollen (*chemosis*), and epibulbar *blood vessels dilate*,

© The Author(s) 2023
P. J. J. Gooris et al. (eds.), *Surgery in and around the Orbit*,
https://doi.org/10.1007/978-3-031-40697-3_15

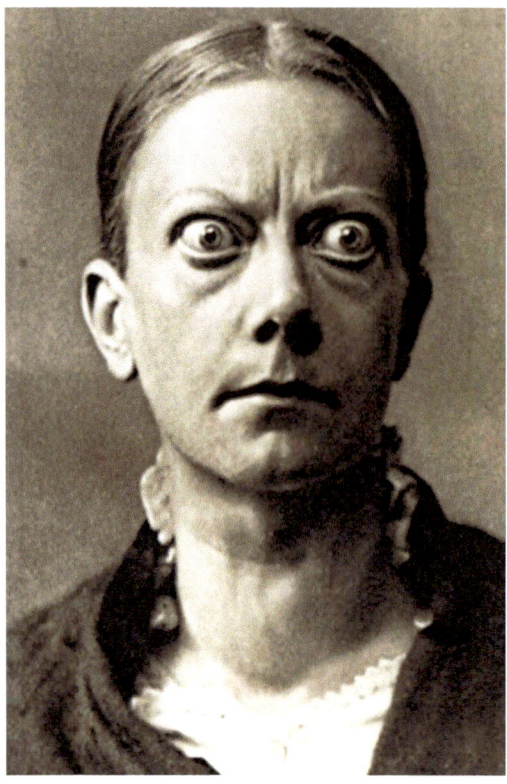

Fig. 15.1 Historic photograph of a woman with upper lid retraction and exophthalmos. Note the frowns on her forehead (taken from Wikipedia—public domain)

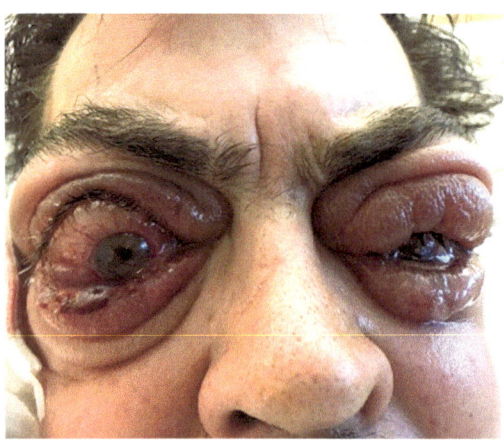

Fig. 15.2 Male patient with severe and vision-threatening GO: extreme proptosis, eyelid retraction and swelling. He reported a markedly decreased quality of life

causing a reddish appearance. Around 40% of GO-patients show *motility impairment*. Because the inferior rectus muscle is most often involved,

an *elevation impairment* is most frequently seen. When the inferior rectus of both eyes is equally involved, the patient will compensate for the inability to elevate her/his eyes by keeping the head back (*torticollis*). This head posture may (wrongly) be perceived as an arrogant attitude. When the inferior recti muscles are unequally affected, *double vision* becomes inevitable as long as both eyes have good vision. Apart from vertical impairment of eye movements, motility reduction can occur in any direction of gaze.

In the most severe presentation of GO, called Dysthyroid Optic Neuropathy (DON) [4], a number of symptoms become evident that are related to compression or stretching of the optic nerve due to increased soft tissue volumes in the orbit. These are *reduced color vision (especially in the blue/yellow axis), reduced visual acuity, reduced visual fields and*—in case of predominantly unilateral GO—*a Relative Afferent Pupil Defect (RAPD*; the pupil at the most affected side becomes larger rather than smaller when illuminating each eye alternately). Ophthalmoscopy shows a *swollen optic nerve head* and *choroidal folds*.

The most distressing sign of GO is *globe luxation*: the globe becomes positioned anterior to the eyelids. This can happen with sagging eyelids. Gentle pressure puts the eye back in place.

Typical of GO is the *diversity of clinical signs and symptoms*. Typical also is the *asymmetry*. GO is the most common cause of *bilateral*, but also of *unilateral,* exophthalmos. But even in bilateral GO, the presentation of the two sides may differ considerably.

Diagnosis, Imaging and Differential Diagnosis

Despite the multitude of symptoms and signs, a diagnosis of GO remains difficult. In 70% of patients referred to a specialized center for Graves' disease, the question was: 'Is this GO?' In such a center, an ophthalmologist together with an endocrinologist were able to rule out GO in about 10%, and to confirm GO in >80%.

Diagnosis of GO is based on the combination of the following items:

1. Upper lid retraction
2. Exophthalmos
3. Eye movement reduction, especially elevation restriction
4. (Family) history of Graves' disease
5. Enlarged extraocular muscle(s) on CT scan or MRI scan
6. Presence of auto-antibodies: TSH-receptor-binding immunoglobulins (TBII), thyroid peroxidase antibodies (anti-TPO)

Eyelid swelling is deliberately not included in this list, because it is a non-specific sign which can also be seen in healthy elderly people and is often seen in smokers. Strict criteria for a definition of eyelid swelling do not exist, and therefore, the prevalence of eyelid swelling is unknown. However, it is important to consider GO in patients with non-understood eyelid swelling, even when other manifestations of GO are lacking. Blepharoplasty to correct eyelid swelling in patients with a beginning GO easily results in eyelid retraction and it will be difficult for the surgeon to explain afterwards that eyelid retraction would have occurred anyway (e.g., without a previous operation).

As described above, the following items are paramount in the diagnosis of GO

1. Upper eyelid retraction can be congenital or the result of a contralateral blepharoptosis. It can also be caused by eyelid fibrosis secondary to trauma or surgery. It is sometimes seen in patients having undergone a (too enthusiastically performed) blepharoplasty.
2. Exophthalmos is seen in many conditions, such as shown in Chap. 7. The combination of upper eyelid retraction and exophthalmos is extremely suggestive for GO. In contrast, the combination of blepharoptosis and exophthalmos is highly suggestive for other diseases than GO.
3. Acquired motility impairment with or without diplopia is also very suggestive for GO. In the early stages, patients typically complain about double vision in the early morning or when tired. Another cause of acquired double vision is a palsy of the fourth cranial nerve (trochlear nerve), seen after trauma, in patients with diabetes mellitus or otherwise healthy, though elderly people. In contrast to patients with GO, in order to compensate for their double vision, they tilt their head down.
4. GO is assumed to originate from genetic mutations that predispose to disease and interact with environmental factors, such as infections, through epigenetic modification [5]. Whether true or not, GO is seen in families with kinswoman that have or have had GD far more frequently than expected by coincidence. Obviously, the presence of GD in relatives, only supports, but does not prove GO.
5. As discussed above, extraocular muscle enlargement is an early phenomenon and occurs in up to 70% of the patients [6]. The inferior and medial recti are most frequently involved. One or more muscles may be enlarged, but it is uncommon for the lateral rectus muscle to be solely enlarged (in contrast to the inferior and medial recti). In contrast to myositis, in which the lateral rectus muscle is often affected as the only extraocular muscle, the tendons of the muscles in GO are not thickened. Enlargement of the tendons together with evident pain is very suggestive for myositis or other forms of idiopathic orbital inflammation (Chap. 5). Swelling of the extraocular muscles can also be seen in caroticocavernous fistulas. The muscles are swollen due to congestion.
6. Both CT and MRI scans are helpful in making a diagnosis of GO. CT scans have the advantage of a clear visualization of the orbital bones, which is useful in preparing for orbital decompression. The disadvantage is the irradiation burden. MRI scans have no radiation burden at all and can help to distinguish between active and burnt-out GO (see next session). Both CT scans and MRI scans show enlargement of the extraocular muscles, apical crowding and stretching of the optic nerve and tenting of the posterior sclera, when pres-

ent. Moreover, these scans also help to rule out other diseases than GO. TBII and/or anti-TPO serum levels are elevated in more than 95% of patients with GO [7].

Severity

In each patient with GO, the disease severity grade and the activity stage have to be assessed. For grading the disease severity, the Werner's NOSPECS classification has been used [8], but the EUGOGO classification is easier and probably more relevant [3]. Three classes of severity are distinguished:

1. Mild GO: There are only mild symptoms and signs that are not or hardly interfering with daily life activities. These could be eyelid retraction, mild proptosis, mild motility impairment without significant diplopia and/or mild corneal involvement (stippling of the corneal epithelium). Approximately, 60% of all patients with GO have this mild form. Treatment consists of lubricants and dark glasses. Smoking must be discouraged. Otherwise, a wait-and-see policy is justified. If there are signs of disease activity, selenium can be considered [9]. When the situation is stable for 6 months, corrective surgery can be started.
2. Moderate GO: There is significant eyelid swelling and/or proptosis. Evident restriction of eye movements causes torticollis and/or diplopia. Daily life activities are seriously affected. Around 35% of patients seen at tertiary referral centers suffer from this form. Especially in this group, assessment of disease activity is of the utmost importance. Patients with active GO are initially treated with immunosuppressive therapy. When the disease has become or is stable, rehabilitating surgical steps can be undertaken.
3. Severe or vision-threatening GO: Also called Dysthyroid Optic Neuropathy (DON) [5]. Due to extreme swelling of the orbital soft tissues in combination with a tight orbital septum and subsequent raised intra-

orbital pressure [10], the optic nerve becomes stretched or compressed, eventually leading to blindness. This is an exceptional form, which only occurs in about 5% of patients. Interestingly, the amount of proptosis is usually not excessive. A tight orbital septum is believed to prevent auto-decompression, thus causing less exophthalmos and DON [11]. Patients with DON always have active GO. Immediately, immunosuppressive treatment should be started. When there is no or an insufficient response, orbital decompression should be performed. Even in patients with longstanding blindness due to DON, orbital decompression may result in the return of vision. A corneal ulcer can also lead to vision loss. In countries with good and accessible health care, it is rarely seen. Finally, globe luxation is also vision-threatening and needs immediate correction.

Activity

Also, the disease activity has to be scored. As discussed above, the initial phase of Rundle's curve is associated with inflammation and worsening of signs and symptoms.

Immunosuppressive treatments, such as glucocorticosteroids or orbital irradiation, are thought to be effective only in the active stage of the disease. On the other hand, certain surgical treatments, such strabismus surgery, should only be undertaken when the disease is quiet and no changes have been observed for at least 3–6 months. Thus, it is important to distinguish patients who have active and patients who have burnt-out GO. Patients with moderate or severe as well as active GO should be treated with immunosuppressive therapy, whereas patients with burnt-out GO can be scheduled for surgical intervention(s).

To assess the activity of GO, the Clinical Activity Score (CAS) has been developed [12]. The CAS (Table 15.1) is based on the classic signs of inflammation, e.g., rubor (redness), dolor (pain), tumor (swelling) and function laesa

Table 15.1 The Clinical Activity Score (CAS) for GO by Mourits et al. (1989) [12]

Pain
1. Painful, oppressive feeling on or behind the globe
2. Pain on attempted up-, side- or downgaze
Redness
3. Redness of eyelids
4. Diffuse redness of the conjunctiva
Swelling
5. Chemosis (edema of the conjunctiva)
6. Swollen caruncle or plica
7. Edema of the eyelids
8. Increase of proptosis of »2 mm in the last 1–3 months
Impaired function
9. Worsening of motility in any direction of »5° in the last 1–3 months
10. Decrease of visual acuity of »1 line on the Snellen chart in the last 1–3 months (using a pinhole)

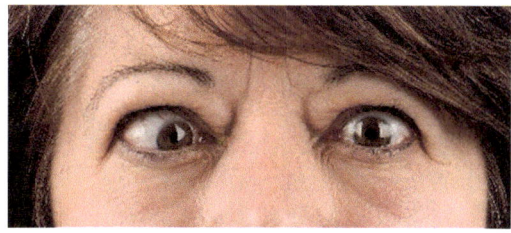

Fig. 15.4 A 44-year-old female with remaining upper lid retraction and swelling, and unchanged convergent strabismus, in burnt-out GO, CAS = 1

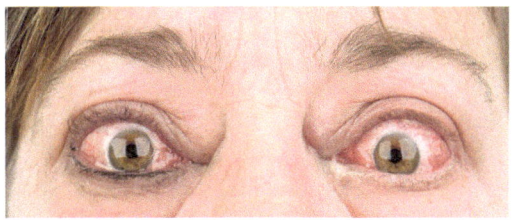

Fig. 15.3 A 40-year-old female with active GO: pain, eyelid swelling, redness conjunctiva and chemosis; CAS = 4

(impaired function). The fifth classic sign of Celsius has not been included, because the CAS is a clinical score, that—for the sake of convenience—should be applied without special instruments such as a heat detector (Figs. 15.3 and 15.4).

For each of the signs present, one point is given. The sum of these points defines the activity score. Items 1–7 can be scored after one visit; items 8–10 require at least two subsequent visits with an interval of at least 1 month.

Our initial study showed that patients with a CAS of three or more out of 10 responded well to immunosuppressive treatment, while those with less than three points were likely not to respond [12]. The CAS has been criticized for several rea-

sons. Firstly, all features are given equal weighting, and it is not clear whether this is appropriate. Secondly, the first seven items are subjective. Thirdly, the scoring is binary. Some improvement is not sufficient to alter the score of each item. Dickinson and Perros, therefore, developed for the EUGOGO a photographic atlas of graded inflammatory signs, which is available at www.eugogo.org, to overcome some of these drawbacks [13]. In spite of its disadvantages, the CAS has become quite popular and a fixed item of randomized clinical trials [14–16]. With some effort, observers reached agreement in 86% of cases [17]. In a subsequent study, we showed that, using a cut-off point of 4 on a scale of 10, the positive predictive value of the CAS was 80%, the negative predictive value 64%, the sensitivity 55% and the specificity 86% [18]. It should be realized that response to prednisone or irradiation as a quality measure of the CAS is a derivative and does not take into account that prednisone is ineffective in some individuals. We found that intravenously administered prednisone stabilizes active GO in 85% and reduces the severity of GO in 38% of the treated patients. It appears that some patients do not respond to prednisone at all, and this becomes clear already after several weeks [19, 20].

The intrinsic value of the CAS has been further established by its relationship with TBII serum levels, ultrasonography and octreotide uptake on octreotide scintigraphy, all of which are considered parameters of disease activity [21–23]. Over the years, there has been a tendency to use the first seven items of the CAS only, which has the advantage that a score can be assessed on the same day during the first visit. The 7-item CAS, however, has never been evaluated for its predicting value.

Impact and Quality of Life

The changed appearance, complaints of painful eyes, double vision and blurred vision have a major impact on the patient with GO. Parties are avoided, driving becomes difficult or impossible; the hot flush caused by the hyperthyroidism can easily lead to arguments. Not infrequently, patients with a severe form of GO often lose both their partner and their job. The long duration of more serious forms of GO (up to 5 years) plays a role in this. Patients with GO score their situation worse than patients with type-I diabetes mellitus do.

Gerding et al. demonstrated that the quality of life in patients with GO is markedly decreased [24]. Forty percent of our patients reports limitations of daily life activities, such as driving a car or leisure activities, 80% mentions reduced self-confidence and 44% complains about some kind of social isolation.

Terwee et al. developed a disease-specific quality-of-life questionnaire for patients with GO, the so-called GO-QOL, which is very helpful in studies to perceive the disease from the patient's point of view [25].

References

1. Bartley GB, Fatourechi V, Kadrmas EF, Jacobsen SJ, Ilstrup DM, Garrity JA, Gorman CA. Clinical features of Graves' ophthalmopathy in an incidence cohort. Am J Ophthalmol. 1996;121:284–90.
2. Bartley GB, Gorman CA. Diagnostic criteria for Graves' ophthalmopathy. Am J Ophthalmol. 1995;119:792–5.
3. European Group on Graves' Orbitopathy (EUGOGO), Wiersinga WM, Perros P, Kahaly GJ, Mourits MP, Baldeschi L, Boboridis K, Boschi A, Dickinson AJ, Kendall-Taylor P, Krassas GE, Lane CM, Lazarus JH, Marcocci C, Marino M, Nardi M, Neoh C, Orgiazzi J, Pinchera A, Pitz S, Prummel MF, Sartini MS, Stahl M, Von Arx G. Clinical assessment of patients with Graves' orbitopathy: the European Group on Graves' Orbitopathy recommendations to generalists, spe-

cialists and clinical researchers. Eur J Endocrinol. 2006;155:387–9.
4. McKeag D, Lane C, Lazarus JH, Baldeschi L, Boboridis K, Dickinson AJ, Hullo AI, Kahaly G, Krassas G, Marcocci C, Marinò M, Mourits MP, Nardi M, Neoh C, Orgiazzi J, Perros P, Pinchera A, Pitz S, Prummel MF, Sartini MS, Wiersinga WM, European Group on Graves' Orbitopathy (EUGOGO). Clinical features of dysthyroid optic neuropathy: a European group on Graves' Orbitopathy (EUGOGO) survey. Br J Ophthalmol. 2007;91:455–8.
5. Tober Y, Huber A. The etiology of autoimmune thyroid disease: a story of genes and environment. J Autoimmun. 2009;32:231–9.
6. Regensburg NI, Wiersinga WM, Berendschot TTJM, Potgieser P, Mourits MP. Do subtypes of Graves' orbitopathy exist? Ophthalmology. 2011;118:191–6.
7. Paunkovic J, Pauncovic N. Does auto-antibody-negative Graves' disease exist? A second evaluation of the clinical diagnosis. Horm Metab Res. 2006;38:53–6.
8. Werner SC. Modification of the classification of the eye changes of Graves' disease. Am J Ophthalmol. 1977;83:725–7.
9. Marcocci C, Kahaly GJ, Krassas GE, Bartalena L, Prummel M, Stahl M, Altea MA, Nardi M, Pitz S, Boboridis K, Sivelli P, Von Arx G, Mourits MP, Baldeschi L, Bencivelli W, Wiersinga WM, European Group on Graves' Orbitopathy. Selenium and the course of mild Graves' orbitopathy. N Engl J Med. 2011;364:1920–31.
10. Otto AJ, Koornneef L, Mourits MP, Deen-Van LL. Retrobulbar pressures measured during surgical decompression of the orbit. Br J Ophthalmol. 1996;80:1042–5.
11. Koornneef L. Eyelid and orbital fascial attachments and their clinical significance. Eye. 1988;2:130–4.
12. Mourits MP, Koornneef L, Wiersinga WM, Prummel MF, Berghout A, Van Der Gaag R. Clinical criteria for the assessment of disease activity in Graves' ophthalmopathy: a novel approach. Br J Ophthalmol. 1989;73:639–44.
13. Dickinson AJ, Perros P. Controversies in the clinical evaluation of active thyroid-associated orbitopathy: use of a detailed protocol with comparative photographs for objective assessment. Clin Endocrinol. 2001;55:283–303.
14. Mourits MP, Van Kempen-Harteveld ML, García MB, Koppeschaar HP, Tick L, Terwee CB. Radiotherapy for Graves' orbitopathy: randomised placebo-controlled study. Lancet. 2000;355:1505–9.
15. Douglas RS, Kahaly GJ, Patel A, Sile S, Thompson EHZ, Perdok R, Fleming JC, Fowler BT, Marcocci C, Marinò M, Antonelli A, Dailey R, Harris GJ, Eckstein A, Schiffman J, Tang R, Nelson C, Salvi M, Wester S, Sherman JW, Vescio T, Holt RJ, Smith TJ. Teprotumumab for the treatment of active thyroid eye disease. N Engl J Med. 2020;382:341–52.

16. Kahaly GJ, Riedl M, König J, Pitz S, Ponto K, Diana T, Kampmann E, Kolbe E, Eckstein A, Moeller LC, Führer D, Salvi M, Curro N, Campi I, Covelli D, Leo M, Marinò M, Menconi F, Marcocci C, Bartalena L, Perros P, Wiersinga WM, European Group on Graves' Orbitopathy (EUGOGO). Mycophenolate plus methylprednisolone versus methylprednisolone alone in active, moderate-to-severe Graves' orbitopathy (MINGO): a randomised, observer-masked, multicentre trial. Lancet Diabetes Endocrinol. 2018;6:287–98.

17. Gerding MN, Prummel MF, Kalmann R, Koornneef L, Wiersinga WM. The use of colour slides in the assessment of changes in soft-tissue involvement in Graves' ophthalmopathy. J Endocrinol Investig. 1998;21:459–62.

18. Mourits MP, Prummel MF, Wiersinga WM, Koornneef L. Clinical activity score as a guide in the management of patients with Graves' ophthalmopathy. Clin Endocrinol. 1997;47:9–14.

19. Yang M, Wiersinga WM, Soeters MR, Mourits MP. What is the aim of immunosuppressive treatment in patients with Graves' orbitopathy? Ophthalmic Plast Reconstr Surg. 2014;30:157–61.

20. Bartalena L, Veronesi G, Krassas GE, Wiersinga WM, Marcocci C, Marinò M, Salvi M, Daumerie C, Bournaud C, Stahl M, Sassi L, Azzolini C, Boboridis KG, Mourits MP, Soeters MR, Baldeschi L, Nardi M, Currò N, Boschi A, Bernard M, Von Arx G, Perros P, Kahaly GJ, European Group on Graves' Orbitopathy (EUGOGO). Does early response to intravenous glucocorticoids predict the final outcome in patients with moderate-to-severe and active Graves' orbitopathy? J Endocrinol Investig. 2017;40:547–53.

21. Gerding MN, Van Der Meer JW, Broenink M, Bakker O, Wiersinga WM, Prummel MF. Association of thyrotrophin receptor antibodies with the clinical features of Graves' ophthalmopathy. Clin Endocrinol. 2000;52:267–71.

22. Gerding MN, Prummel MF, Wiersinga WM. Assessment of disease activity in Graves' ophthalmopathy by orbital ultrasonography and clinical parameters. Clin Endocrinol. 2000;52:641–6.

23. Gerding MN, Van Der Zant FM, Van Royen EA, Koornneef L, Krenning EP, Wiersinga WM, Prummel MF. Octreotide-scintigraphy is a disease-activity parameter in Graves' ophthalmopathy. Clin Endocrinol. 1999;50:373–9.

24. Gerding MN, Terwee CB, Dekker FW, Koornneef L, Prummel MF, Wiersinga WM. Quality of life in patients with Graves' ophthalmopathy is markedly decreased: measurement by the medical outcomes study instrument. Thyroid. 1997;7:885–9.

25. Terwee CB, Gerding MN, Dekker FW, Prummel MF, Wiersinga WM. Development of a disease specific quality of life questionnaire for patients with Graves' ophthalmopathy: the GO-QOL. Br J Ophthalmol. 1998;82:773–9.

Open Access This chapter is licensed under the terms of the Creative Commons Attribution 4.0 International License (http://creativecommons.org/licenses/by/4.0/), which permits use, sharing, adaptation, distribution and reproduction in any medium or format, as long as you give appropriate credit to the original author(s) and the source, provide a link to the Creative Commons license and indicate if changes were made.

The images or other third party material in this chapter are included in the chapter's Creative Commons license, unless indicated otherwise in a credit line to the material. If material is not included in the chapter's Creative Commons license and your intended use is not permitted by statutory regulation or exceeds the permitted use, you will need to obtain permission directly from the copyright holder.

Etiology and Pathogenesis of Graves' Orbitopathy

16

Wilmar M. Wiersinga

Learning Objectives
- Mechanistic explanation of eye changes
- Orbital fibroblasts as target cells
- Excessive glycosaminoglycan production
- Swelling of extraocular muscles and orbital fat
- Autoimmune thyroid disease
- Role of TSH receptor antibodies
- Interaction with IGF-1 receptor
- Role of genetic and environmental factors

Introduction

The etiology of Graves' orbitopathy (GO) is still incompletely understood, but many aspects of the pathogenesis have been unraveled over the last decades. Family members of Graves' patients have an increased risk to develop the same disease, and smoking is a strong risk factor both to provoke GO and to develop severe forms of GO. The mechanistic explanation of the orbital reshaping in GO is now well-understood, whereas significant progress has been made in our understanding of immunological and molecular changes leading to GO.

Mechanistic Explanation of Eye Changes in Graves' Orbitopathy

The hallmark of Graves' orbitopathy (GO) is swelling of extraocular muscles and orbital fat, evident from orbital imaging with CT-scans or MRI. Swollen retrobulbar tissues are due to local inflammation and production of excessive amounts of hydrophilic glysoaminoglycans (particularly hyaluronan) causing edema. When the extraocular muscles are affected, it leads to dysfunction due to a failure of relaxation. It limits movement into the field of the ipsilateral antagonist, which, if asymmetrical, gives rise to double vision [1]. The human orbit is a tight space, completely surrounded by bone except anteriorly. Here, instead of bone, there is a fascial sheet across the orbital opening called the anterior orbital septum. Because of the bony surroundings, swollen retrobulbar tissues have no other outlet than pushing the globe forwards causing exophthalmos [2].

The development of exophthalmos has even been termed "nature's own decompression". A tight anterior orbital septum might preclude proptosis and nature's own decompression, resulting in higher retrobulbar pressures and optic neurop-

W. M. Wiersinga (✉)
Department of Internal Medicine-Endocrinology, Amsterdam University Medical Centers, Location AMC, Amsterdam, The Netherlands
e-mail: w.m.wiersinga@amsterdamumc.nl

© The Author(s) 2023

P. J. J. Gooris et al. (eds.), *Surgery in and around the Orbit*,
https://doi.org/10.1007/978-3-031-40697-3_16

athy. This mechanism is supported by measuring retrobulbar pressure with an intraorbitally positioned micropressure transducer. In GO patients with sight loss due to dysthyroid optic neuropathy (DON), retrobulbar pressure ranged from 17 to 40 mmHg (mean 29 mmHg) falling to 9–12 mmHg upon decompression; in GO patients with marked exophthalmos but without DON, retrobulbar pressure was 9–11 mmHg, with no reduction upon surgical decompression [3]. Pressure on the optic nerve leads to colour impairment, altered pupil responses and loss of vision. In contrast, patients with equivalent intraorbital soft tissue swelling but with a lax anterior orbital septum will "self-decompress" to develop exophthalmos with no or limited rise in retrobulbar pressure. Thus, patients with muscle restriction but without exophthalmos are at risk for DON [1]. Inflammation in the eyelids causes visible edema, erythema and festoons. Upper eyelid retraction is multifactorial, due to a combination of increased sympathetic stimulation of Müller's muscle, contraction of the levator muscle due to its direct involvement, and scarring between the lacrimal gland fascia and levator which specifically gives rise to lateral flare [1]. Corneal signs are secondary phenomena: a wide palpebral aperture leads to increased tear evaporation, which combined with poor blinking causes superficial punctate erosions and symptoms of surface irritation [1].

Recent volumetric studies of Caucasian human orbits have quantified the size of orbital contents. Bony orbital cavity volume (OV) in control males is greater than in control females (28.9 cm^3 vs 24.9 cm^3, $p < 0.001$); likewise, orbital fat volume in males is greater than in females (16.2 cm^3 vs 14.0 cm^3, $p < 0.001$), and the same is true for muscle volume (4.2 cm^3 vs 3.7 cm^3, $p < 0.001$) [4]. The sex difference disappears when applying the ratio of fat volume to orbital volume (FV/OV 0.56 ± 0.11 vs 0.56 ± 0.10) and of muscle volume to orbital volume (MV/OV 0.15 ± 0.02 vs 0.15 ± 0.02). There exists a direct correlation between age and FV/OV and a weaker indirect correlation between MV/OV and age, indicating an increase of fat

volume and a decrease of muscle volume with advancing age [4]. Age-specific reference ranges of FV/OV and MV/OV ratios could thus be established [4]. These data have been applied to a series of 95 consecutive referrals with untreated GO. All patients were Caucasians and had definite GO, but the patient mix was heterogenenous ranging from mild to severe ophthalmopathy [5]. In 25% of patients, neither FV nor MV was exceeding the upper normal limit, 5% had only an increased FV, 61% had only increased MV, and 9% had both increased FV and MV. It should be realized that patients without increased fat or muscle volumes (values lower than the 97.5th percentile of the reference interval) could have volumes which increased recently to values still within the $P_{2.5}$–$P_{97.5}$ reference interval. Comparing patients with and without increased MV (MV/OV 0.24 vs 0.16 $p < 0.00$; FV/OV 0.59 vs 0.60, NS), patients with increased MV were older (52 year vs 45 year), had more proptosis (22 mm vs 20 mm), more impaired ductions (e.g. elevation 40° vs 45°), higher diplopia scores (1 vs 0), and higher serum TBII concentrations (9.7 vs 4.2 U/L). Relative to patients without increased FV, patients with increased FV (FV/OV 0.81 vs 0.57, $p < 0.00$; MV/OV 0.22 vs 0.20, NS) had more proptosis (24 mm vs 21 mm) and less diplopia (score 0 vs 1) [5]. Whereas these associations make sense, the differential involvement of orbital fat and extraocular muscle enlargement in GO remains incompletely understood [6]. It could exist a number of GO variants, each with a different immunopathogenesis: one with predominant muscle enlargement and another with predominant fat enlargement, and mixtures. An alternative hypothesis is that the increase in orbital fat is just a late phenomenon in the natural history of GO [6]. This view is strengthened by the finding that a long duration of GO at diagnosis is associated with a greater orbital fat volume as compared to a shorter duration (FV/OV >1 year 0.65 vs FV/OV <1 year 0.55, $p = 0.004$), whereas muscle volume was similar (MV/OV >1 year 0.22 vs MV/OV <1 year 0.21, NS) [6]. The hypothesis is further supported by a 4-year follow-up of patients with mild GO who did not require specific treatment: fat volume was greater at follow-up (FV/OV at baseline 0.57 vs FV/OV at 4 year 0.65, $p < 0.000$) whereas muscle volume was smaller at follow-up (MV/OV at baseline 0.17 vs MV/OV at 4 year 0.14, $p < 0.000$) [7, 8]. The swelling of orbital fat appears to be a rather late phenomenon, in line with the transition of a subset of orbital fibroblasts into mature adipocytes (adipogenesis) during the immunopathogenesis of GO.

Immunopathogenesis of Graves' Orbitopathy

Orbital fibroblasts (OF) are considered the target cells of the autoimmune attack in GO. For retrobulbar T cells from GO patients recognize autologous OF (but not eye muscle extracts) in a MHC class I restricted manner, and proliferate in response to autologous proteins from OF (but not from orbital myoblasts). Conversely, OF proliferate in response to autologous T cells dependent on MHC class II and CD40-CD40L signaling [9]. OF are capable, when stimulated, to produce excessive amounts of glycosaminoglycans (GAG), notably hyaluronic acid (HA). GAGs are very hydrophilic compounds and thus attract much water, resulting in edematous swelling. GAGs accumulate in the endomysial space between muscle fibers. There is no increase in the number of muscle fibers and no ultrastructural damage to the muscle cells themselves in GO (except in very advanced cases when some damage may be seen). OF investing the extraocular muscle fibers and residing within the orbital connective/fatty tissue are heterogeneous cells, which can be classified based on the presence or absence of the cell surface glycoprotein CD90 also known as thymocyte antigen-1 (Thy-1) [10, 11]. OF expressing this antigen are capable of excessive HA production and are abundant in the extraocular muscles. Conversely, OF within the orbital connective/fatty tissue are Thy-1 negative and characteristically undergo adipogenesis under appropriate conditions.

Immunocompetent cells in the orbital tissues consist of CD4+ and CD8+ T cells, few B cells, monocytes, and abundant macrophages [12]. Many of these cells are activated memory cells, frequently located adjacent to blood vessels. Macrophages, but also B cells, might present the responsible autoantigen—most likely the TSH receptor- to T cells, which are then stimulated to recognize OF. Activated infiltrating T cells produce cytokines and chemokines capable of remodeling orbital tissues. The cytokine profile in the early stages of GO is predominantly derived from T_h1 cells, whereas cytokines are mostly derived from T_h2 cells in patients with a GO duration >2 years [13]. Data suggest GO is primarily a T-cell mediated disease, initiated by the migration of T-helper cells into the orbit. Cytokines induce expression of immunomodulatory proteins on orbital endothelial cells and fibroblasts, such as HLA, hsp-72 and several adhesion molecules, generating T-cell migration. Cytokine-activated OF synthesize chemo attractants, IL6 and RANTES, perpetuating the immune attack.

Macrophages may present antigen to T-cells (CD40L) through provision of costimulatory and proinflammatory cytokines. Activated T-cells bind to CD40+ fibroblasts, inducing proinflammatory compounds (like cytokines, COX2, PGE2) and excessive GAG production. A subset of OF may differentiate into mature adipocytes, associated with increased expression of TSH receptors [14].

Fibrocytes are bone marrow derived cells from the monocyte lineage, expressing CD45, CD34, CXCR4, collagen 1, Tg and functional TSH receptors. Circulating fibrocytes are highly abundant in GO patients, and seem to infiltrate orbital connective tissue where they might transition to CD34+ OF [15]. Teprotumumab, a human monoclonal IGF-1R blocking antibody, attenuates the actions of IGF-1 and TSH in fibrocytes; specifically, it blocks the induction of proinflammatory cytokines by TSH [16]. It raises the question about the nature of the autoantigen in Graves' orbitopathy.

TSH Receptors Are the Major Autoantigen in Graves' Orbitopathy

TSH receptors (TSHR) in human retrobulbar tissues were detected in 1993, in GO patients to a greater degree than in healthy persons [17]. It raised the hypothesis that TSHR autoimmunity (the hallmark of Graves' hyperthyroidism) could also impact the orbit. Full-length functional TSH receptors are expressed on OF [18], in active stages of the disease to a greater extent than in the inactive stages, and directly related to IL-1β levels in the orbit [19]. Graves' immunoglobulins (isolated from the serum of patients with Graves' hyperthyroidism) as well as M22 (a monoclonal TSH receptor stimulating antibody) recognize TSH receptors on OF as evident from increased cAMP and hyaluronan production in cell cultures of differentiated human OF [9]. Another post-TSHR signaling pathway runs not via cAMP but through PI3K (phospho-inositide 3-kinase) resulting in adipogenesis [10, 20].

Clinical studies support the role of TSHR. First, serum TSHR-Ab concentrations are higher in patients with both Graves' hyperthyroidism and GO than in patients with Graves' hyperthyroidism without GO [21]. Second, serum TSHR-Ab are directly related to the activity and severity of GO [22]. Third, the higher serum TSHR-Ab, the higher the risk of an unfavorable course of GO [23]. Fourth, serum TSHR-Ab fall after thyroidectomy or treatment with antithyroid drugs, but increase substantially after radioactive iodine therapy. [131]I therapy (but not thyroidectomy or antithyroid drugs) is associated with development or worsening of GO in about 15% of patients [24, 25]. One could argue that not all GO patients have concomitant Graves' hyperthyroidism: about 10% of all GO patients are euthyroid or even hypothyroid. However, serum TSHR-Ab can be detected in the majority if not all of such patients. But the most convincing argument for the TSH receptor as the primary autoantigen in GO is derived from experimental animal studies. Genetic immunization of mice with the TSH receptor A-subunit plasmid have produced a fair animal model of GO [26, 27]. Despite inherent limitations to the mouse model

(e.g., the lateral wall of the orbit in rodents is not made of bone but consists of a connective tissue septum), the model resembles reasonably well the human condition.

Cross-Talk Between TSH Receptors and IGF-1 Receptors

Another autoantigen in GO might be the insulin like growth factor-1 receptor (IGF-1R). IGF-1R are indeed upregulated on OF of GO patients, but serum IGF-1 and IGF-binding proteins in GO patients are normal. Graves' immunoglobulins are able to induce hyaluronan synthesis in OF; the effect can be blocked by monoclonal IGF1-R blocking antibodies, suggesting a pathway independent of the TSHR [28]. The authors suggested Graves' IgG contain antibodies stimulating the IGF-1R. IGF-1R antibodies were detected in 10% of GO patients and in 11% of controls [29]. IGF-1R antibodies were not related to GO disease activity or severity. Their serum concentrations were rather constant, demonstrating relatively stable expression over time. IGF-1R antibodies failed to stimulate IGF-1R autophosphorylation but instead inhibited IGF-1-induced signaling. Available data do not support the idea that stimulating IGF-1R antibodies are involved in the pathogenesis of GO [30]. Further studies provided proof that TSHR stimulating antibodies do not activate IGF-1 receptors [31]. Convincing evidence that IGF-1 receptors do not act as primary autoantigens in GO is obtained from animal studies applying genetic immunization with IGF-1Rα plasmid [27]. These mice generated high levels of IGF-1Rα antibody, but did not develop apparent pathology. Of interest is that some animals immunized with TSHR A-subunit developed low-titer IGF1-Rα antibodies shortly after immunization. It can be concluded that the TSHR and likely its A-subunit, is the primary autoantigen in GO. TSHR are "the culprit as well as the victim" not only in Graves' hyperthyroidism, but also in Graves' orbitopathy [32].

Functional interactions between TSHR and IGF-1R have been demonstrated in cultured OF obtained from GO patients. Simultaneous activation by TSH and IGF-1 synergistically increases hyaluronan secretion in OF [33]. Blockade of IGF-1R inhibits both hyaluronan synthesis and Akt phosphorylation induced by the monoclonal TSHR stimulating antibody M22 or IGF-1 [34]. Such studies suggest a crosstalk between TSHR and IGF-1R [35]. The TSHR is a heptahelical G-protein coupled receptor, whereas the IGF-1R is a receptor tyrosine kinase. Despite these fundamental structural differences, both receptors are phosphorylated by G-protein receptor kinases, which enables β-arrestin binding [36]. Arrestins mediate receptor internalization and also activate the mitogen-activated protein kinase (MAPK) pathway. TSHR must neighbor IGF-1R for crosstalk in GO fibroblasts to occur, and this depends on arrestin-β-1 acting as a scaffold [37]. TSH activates both G-proteins and β-arrestin, suggesting the different signals in the TSHR are propagated in differentiated intramolecular pathways [36]. A signalosome has been proposed, in which arrestin-β-1 mediates proximity and crosstalk of the TSHR and IGF-1R [38]. Combination therapy targeting simultaneously TSHR and IGF-1R might be more effective than targeting either receptor alone (the TSH receptor by either small molecule TSHR antagonists or monoclonal TSHR blocking antibodies, and the IGF-1R by monoclonal IGF-1R blocking antibodies) [36, 38].

Genes and Environment

A number of susceptibility genes for Graves' disease has been identified, notably TSHR, HLA, CTLA4, PTPN22, CD40 and FRCL3. However, the frequency of particular polymorphisms in these susceptibility genes do not differ substantially between Graves' hyperthyroid patients with and without Graves' orbitopathy [39]. In the absence of clear genetic markers specific for GO, it is difficult to explain why, among Graves' hyperthyroid patients with the same level of TSH receptor antibodies, some will develop GO whereas others will not. TSHR immunoglobulins are heterogeneous, comprising antibodies that stimulate or block the TSH receptor or antibodies that are neutral with respect to their binding to

the TSHR; the various classes of TSHR-Ab might have differential effects on orbital tissues.

Alternatively, environmental factors could play a role, and in this respect, it is noteworthy that smoking greatly increases the risk for GO (odds ratio 7.7, 95% CI 4.3–13.7) [40]. The proportion of GO among all referred patients with Graves' hyperthyroidism decreased from 57% in 1960 to 35% in 1990, and to 29% in 2010 [41, 42]. The secular decline in smoking is likely causally related to a lower prevalence of GO in the last decades [43]. GO patients referred in 2012 had less severe and less active ophthalmopathy than those referred in 2000 [44]. Orbital muscle volume is larger in GO patients who are current smokers compared to GO patients who are never smokers or ex-smokers; orbital fat volume had no relationship with smoking behaviour [45].

A new risk factor for GO might be hypercholesterolaemia [46]. Statin use (for >60 days in the past year vs <60 days or nonuse) was associated with a 40% decreased hazard of GO (adjusted HR 0.60, CI 0.37–0.93) in a population of patients with newly diagnosed Graves' hyperthyroidism [47]. No significant association was found for the use of nonstatin cholesterol-lowering medication or cyclooxygenase 2 inhibitors and the development of GO. Subsequent studies confirmed high serum cholesterol levels as a potential risk, at least in patients with a relatively short duration of Graves' disease [48, 49]. The pleiotropic anti-inflammatory actions of statins, which are not related to their cholesterol-lowering action, could be key in explaining the association with GO [46].

References

1. Dickinson AJ, Hintschich C. Clinical manifestations. In: Wiersinga WM, Kahaly GJ, editors. Graves' orbitopathy. 3rd ed. Basel: Karger Publishers; 2017. p. 1–25.
2. Prummel MF, Koornneef L, Mourits MP, Heufelder AE, Wiersinga WM. Introduction to Graves' ophthalmopathy. In: Prummel MF, editor. Recent developments in Graves' ophthalmopathy. Dordrecht: Kluwer Academic Publishers; 2000. p. 1–14.

3. Otto AJ, Koornneef L, Mourits MP, Deen-van L. Retrobulbar pressures measured during surgical decompression of the orbit. Br J Ophthalmol. 1996;80:1042–5.

4. Regensburg NI, Wiersinga WM, van Velthoven MEJ, Berendschot TTJM, Zonneveld FW, Baldeschi L, et al. Age and gender-specific reference values of orbital fat and muscle volumes in Caucasians. Br J Ophthalmol. 2011;95:1660–3.

5. Regensburg NI, Wiersinga WM, Berendschot TTJM, Potgieser P, Mourits MP. Do subtypes of Graves' orbitopathy exist? Ophthalmology. 2011;118:191–6.

6. Wiersinga WM, Regensburg NI, Mourits MP. Differential involvement of orbital fat and extraocular muscles in Graves' ophthalmopathy. Eur Thyroid J. 2013;2:14–21.

7. Potgieser PW, Wiersinga WM, Regenburg NI, Mourits MP. Some studies on the natural history of Graves' orbitopathy: increase in orbital fat is a rather late phenomenon. Eur J Endocrinol. 2015;173:149–53.

8. Potgieser PW, de Win MMML, Wiersinga WM, Mourits MP. Natural course of mild Graves' orbitopathy: increase of orbital fat but decrease of muscle volume with increased muscle fatty degeneration during a 4-year follow-up. Ophthalmic Plast Reconstr Surg. 2019;35:456–60.

9. Prabhakar BS, Bahn RS, Smith TJ. Current perspective on the pathogenesis of Graves' disease and ophthalmopathy. Endocr Rev. 2003;24:802–35.

10. Bahn RS. Pathogenesis of Graves' orbitopathy. In: Bahn RS, editor. Graves' disease. A comprehensive guide for clinicians. New York: Springer; 2015. p. 179–86.

11. Smith TJ, Koumas L, Gagnon A, Bell A, Sempowski GD, Phipps RP, et al. Orbital fibroblast heterogeneity may determine the clinical presentation of thyroid-associated ophthalmopathy. J Clin Endocrinol Metab. 2002;87:385–92.

12. Hai YP, Lee ACH, Frommer L, Diana T, Kahaly GJ. Immunohistochemical analysis of human tissue in Graves' orbitopathy. J Endocrinol Investig. 2020;43:123–37.

13. Aniszewski JP, Valyasevi RW, Bahn RS. Relationship between disease duration and predominant orbital T cell subset in Graves' ophthalmopathy. J Clin Endocrinol Metab. 2000;85:776–80.

14. Valyasevi RW, Erickson DZ, Harteneck DA, Dutton CM, Heufelder AE, Jyonouchi SC, et al. Differentiation of human orbital preadipocyte fibroblasts induces expression of functional thyrotropin receptor. J Clin Endocrinol Metab. 1999;84:2257–62.

15. Smith TJ. TSH-receptor expressing fibrocytes and thyroid-associated ophthalmopathy. Nat Rev Endocrinol. 2015;11:171–81.

16. Chen H, Mester T, Raychaudhuri N, Kauh CY, Gupta S, Smith TJ, et al. Teprotumumab, an IGF-1R monoclonal blocking antibody inhibits TSH and IGF-1 action in fibrocytes. J Clin Endocrinol Metab. 2014;99:E1635–40.

17. Feliciello A, Porcellini A, Ciullo I, Bonavolonta G, Avvedimento EV, Fenzi G. Expression of the thyrotropin-receptor mRNA in healthy and Graves' disease retrobulbar tissue. Lancet. 1993;342:337–8.

18. Bahn RS, Dutton CM, Natt N, Joba W, Spitzweg C, Heufelder AE. Thyrotropin receptor expression in Graves' orbital adipose/connective tissues: potential autoantigen in Graves' ophthalmopathy. J Clin Endocrinol Metab. 1998;83:998–1002.

19. Wakelkamp IMMJ, Bakker O, Baldeschi L, Wiersinga WM, Prummel MF. TSH-R expression and cytokine profile in orbital tissue of active vs. inactive Graves' ophthalmopathy patients. Clin Endocrinol. 2003;58:280–7.

20. Kumar S, Nadeem S, Stan MN, Coenen M, Bahn RS. A stimulatory TSH receptor antibody enhances adipogenesis via phosphoinositide 3-kinase activation in orbital preadipocytes from patients with Graves' ophthalmopathy. J Mol Endocrinol. 2011;46:155–63.

21. Vos XG, Smit N, Endert E, Tijssen JG, Wiersinga WM. Frequency and characteristics of TBII-seronegative patients in a population with untreated Graves' hyperthyroidism. Clin Endocrinol. 2008;69:311–7.

22. Lytton SD, Ponto KA, Kanitz M, Matheis N, Kohn LD, Kahaly GJ. A novel thyroid stimulating bioassay is a functional indicator of activity and severity of Graves' orbitopathy. J Clin Endocrinol Metab. 2010;95:2123–31.

23. Eckstein AK, Plicht M, Lax H, Neuhäuser M, Mann K, Lederbogen S, et al. Thyrotropin receptor autoantibodies are independent risk factors for Graves' ophthalmopathy and help to predict severity and outcome of the disease. J Clin Endocrinol Metab. 2006;91:3464–70.

24. Laurberg P, Wallin G, Tallstedt L, Abraham-Nordling M, Lundell G, Torring O. TSH-receptor autoimmunity in Graves' disease after therapy with antithyroid drugs, surgey, or radioiodine: a 5-year prospective randomized study. Eur J Endocrinol. 2008;158:69–75.

25. Acharya SH, Avenell A, Philip S, Burr J, Bevan JS, Abraham P. Radioiodine therapy (RAI) for Graves' disease (GD) and the effect on ophthalmopathy: a systematic review. Clin Endocrinol. 2008;69:943–50.

26. Berchner-Pfannschmidt U, Moshkelgosha S, Diaz-Cano S, Edelmann B, Görtz G-E, Horstmann M, et al. Comparative assessment of female mouse model of Graves' orbitopathy under different environments, accompanied by proinflammatory cytokine and T-cell responses to thyrotropin receptor antigen. Endocrinology. 2016;157:1673–82.

27. Moshkelgosha S, So P-W, Deasy N, Diaz-Cano S, Banga JP. Cutting edge: retrobulbar inflammation, adipogenesis, and acute orbital congestion in a preclinical female mouse model of Graves' orbitopathy induced by thyrotropin receptor plasmid—in vivo electroporation. Endocrinology. 2013;154:3008–15.

28. Smith TJ, Hoa N. Immunoglobulins from patients with Graves' disease induce hyaluronan synthesis in their

orbital fibroblasts through the self-antigen, insulin-like growth factor-1 receptor. J Clin Endocrinol Metab. 2004;89:5076–80.

29. Minich WB, Dehina N, Welsink T, Schwiebert C, Morgenthaler NG, Köhrle J, Eckstein A, Schomburg L. Autoantibodies to the IGF-1 receptor in Graves' orbitopathy. J Clin Endocrinol Metab. 2013;98:752–60.

30. Krieger CC, Neumann S, Gershengorn MC. Is there evidence for IGF-1R stimulating antibodies in Graves' orbitopathy pathogenesis? Int J Mol Sci. 2020;21:E6561.

31. Marcus-Samuels B, Krieger CC, Boutin A, Kahaly GJ, Neumann S, Gershengorn MC. Evidence that Graves' ophthalmopathy immunoglobulins do not directly activate IGF-1 receptors. Thyroid. 2018;28:650–5.

32. Chen CR, Pichurin P, Nagayama Y, Latrofa F, Rapoport B, McLachlan SM. The thyrotropin receptor autoantigen in Graves' disease is the culprit as well as the victim. J Clin Invest. 2003;111:1897–904.

33. Krieger CC, Neumann S, Place RF, Marcus-Samuels B, Gershengorn MC. Bidirectional TSH and IGF-1 receptor cross-talk mediates stimulation of hyaluronan secretion by Graves' disease immunoglobulins. J Clin Endocrinol Metab. 2015;100:1071–7.

34. Kumar S, Iyer S, Bauer H, Coenen M, Bahn RS. A stimulatory thyrotropin receptor antibody enhances hyaluronic acid synthesis in Graves' orbital fibroblastst: inhibition by an IGF-1 receptor blocking antibody. J Clin Endocrinol Metab. 2012;97:1681–7.

35. Wiersinga WM. Autoimmunity in Graves' ophthalmopathy: the result of an unfortunate marriage between TSH receptors and IGF-1 receptors? J Clin Endocrinol Metab. 2011;96:2386–94.

36. Krause G, Eckstein A, Schülein R. Modulating TSH receptor signaling for therapeutic benefit. Eur Thyroid J. 2020;9(Suppl.1):66–77.

37. Krieger CC, Boutin A, Jang D, Morgan SJ, Banga JP, Kahaly GJ, et al. Arrestin-β-1 physically scaffolds TSH and IGF-1 receptors to enable crosstalk. Endocrinology. 2019;160:1468–79.

38. Neumann S, Krieger CC, Gershengorn MC. Targeting TSH and IGF-1 receptors to treat thyroid eye disease. Eur Thyroid J. 2020;9(Suppl.1):59–65.

39. Bednarczuk T, Gopinath B, Ploski R, Wall JR. Susceptibility genes in Graves' ophthalmopathy: searching for a needle in a haystack? Clin Endocrinol. 2007;67:3–19.

40. Prummel MF, Wiersinga WM. Smoking and risk of Graves' disease. JAMA. 1993;269:479–82.

41. Perros P, Kendall-Taylor P. Natural history of thyroid eye disease. Thyroid. 1998;8:423–5.

42. Tanda ML, Piantanida E, LiParulo L, Veronesi G, Lai A, Sassi L, et al. Prevalence and natural history of Graves' orbitopathy in a large series of patients with newly diagnosed Graves' hyperthyroidism seen in a single centre. J Clin Endocrinol Metab. 2013;98:1443–9.

43. Wiersinga WM. Smoking and thyroid. Clin Endocrinol. 2013;79:145–51.

44. Perros P, Zarkovic M, Azzolini C, Ayvaz G, Baldeschi L, Bartalena L, et al. PREGO (presentation of Graves' orbitopathy) study: changes in referral patterns to European Group on Graves' Orbitopathy (EUGOGO) centres over the period from 2000 to 2012. Br J Ophthalmol. 2015;99:1531–5.

45. Regensburg NI, Wiersinga WM, Berendschot TTJM, Saeed P, Mourits MP. Effect of smoking on orbital fat and muscle volume in Graves' orbitopathy. Thyroid. 2011;21:177–81.

46. Lanzolla G, Vannucchi G, Ionni I, Campi I, Sileo F, Lazzaroni E, Marino M. Cholesterol serum levels and use of statins in Graves' orbitopathy: a new starting point for the therapy. Front Endocrinol. 2020;10:933. https://doi.org/10.3389/fendo.2019.00933.

47. Stein JD, Childers D, Gupta S, Talwar N, Nan B, Lee BJ, et al. Risk factors for developing thyroid-associated ophthalmopathy among individuals with Graves' disease. JAMA Ophthalmol. 2015;133:290–6.

48. Sabini E, Mazzi B, Profilo MA, Mautone T, Casini G, Rochi R, et al. High serum cholesterol is a novel risk factor for Graves' orbitopathy: results of a cross-sectional study. Thyroid. 2018;28:386–94.

49. Lanzolla G, Sabini E, Profilo MA, Mazzi B, Sframeli A, Rochi R, et al. Relationship between serum cholesterol and Graves' orbitopathy (GO): a confirmatory study. J Endocrinol Investig. 2018;41:1417–23.

Open Access This chapter is licensed under the terms of the Creative Commons Attribution 4.0 International License (http://creativecommons.org/licenses/by/4.0/), which permits use, sharing, adaptation, distribution and reproduction in any medium or format, as long as you give appropriate credit to the original author(s) and the source, provide a link to the Creative Commons license and indicate if changes were made.

The images or other third party material in this chapter are included in the chapter's Creative Commons license, unless indicated otherwise in a credit line to the material. If material is not included in the chapter's Creative Commons license and your intended use is not permitted by statutory regulation or exceeds the permitted use, you will need to obtain permission directly from the copyright holder.

Medical Management of Graves' Orbitopathy

17

Wilmar M. Wiersinga

Learning Objectives

- Relevance of smoking cessation
- Relevance of restoration and maintenance of euthyroid function
- Utility of local nonspecific measures
- Mild GO: wait-and-see policy or selenium
- Moderate-to-severe GO: steroids, if insufficient response, consider retrobulbar irradiation, cyclosporin, rituximab, or teprotumumab
- Very severe GO (DON, dysthyroid optic neuropathy): high-dose steroids, if insufficient response, urgent orbital decompression

Introduction

The overall management of GO is based on three pillars: (1) smoking cessation, (2) restoration and maintenance of euthyroid function, and (3) management of the orbital disease itself. Here, we distinguish between (a) local measures to be applied in any stage of GO, (b) medical interventions to be applied in the active stage of GO, and (c) surgical interventions to be applied in the inactive stage of GO (like orbital decompression, eye muscle surgery, and eyelid surgery) (Fig. 17.1). GO guidelines are available [1]. One of the recommendations is to refer GO patients— except for the mildest cases- to combined thyroid-eye clinics or specialized centers providing endocrinological and ophthalmological expertise. There is some evidence that in doing so, outcomes are improved [2–4]. Patients often need moral and psychological support. Thyroid and/or GO patient associations and contact with fellow patients could be very important in helping patients to cope with their disease [4].

W. M. Wiersinga (✉)
Department of Internal Medicine-Endocrinology,
Amsterdam University Medical Centers, Location
AMC, Amsterdam, The Netherlands
e-mail: w.m.wiersinga@amsterdamumc.nl

© The Author(s) 2023
P. J. J. Gooris et al. (eds.), *Surgery in and around the Orbit*,
https://doi.org/10.1007/978-3-031-40697-3_17

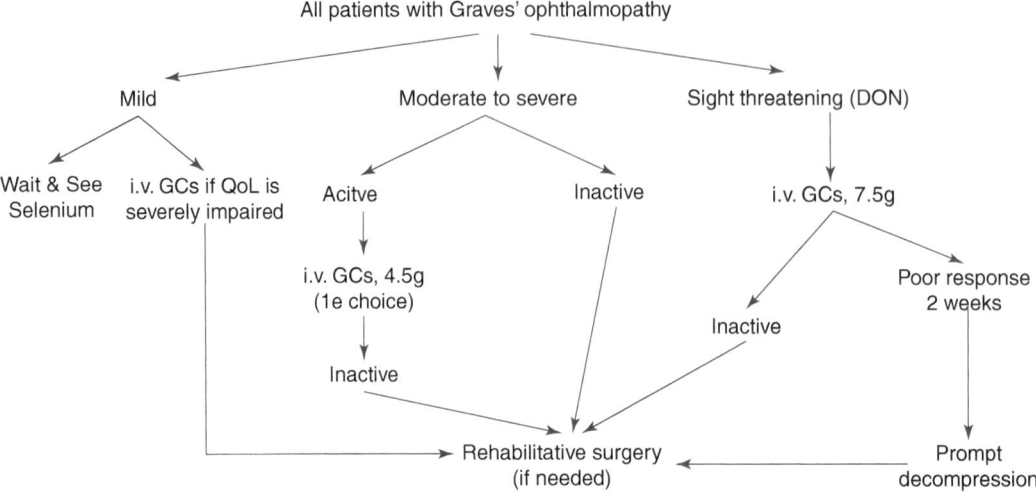

Fig. 17.1 Management of Graves' orbitopathy

Smoking Cessation

To convince people to stop smoking is rarely successful. However, the deleterious effects of smoking on thyroid eye disease in particular may increase the likelihood that GO patients are susceptible to take the advice to stop smoking [5]. Current smoking increases the risk of Graves' hyperthyroidism about twofold and of Graves' ophthalmopathy about threefold. Smokers tend to have more severe GO. Smoking increases the risk of developing or worsening of eye changes about fourfold after ^{131}I therapy of Graves' hyperthyroidism. The outcome of GO treatment with glucocorticoids or retrobulbar irradiation is less favorable in smokers relative to nonsmokers. Refraining from smoking might reduce the risk of developing exophthalmos and diplopia. The regarding guideline reads "We recommend that physicians urge all patients with Graves' hyperthyroidism, irrespective of the presence or absence of GO, to refrain from smoking, if necessary with the help of specialized smoking cessation programs or clinics" [1].

Restoration and Maintenance of Euthyroidism

GO occurs mostly in the presence of Graves' hyperthyroidism, whereas a minority (about 5–10%) presents with euthyroidism or hypothyroidism. Both hyperthyroidism and hypothyroidism have a negative impact on GO, and prompt restoration and maintenance of euthyroidism is indicated [1, 6]. Hypothyroidism should be treated with levothyroxine tablets, but making the appropriate choice how to treat Graves' hyperthyroidism under these circumstances is more complicated. This is caused by worsening of GO after radioactive iodine treatment in about 15% [7]. Risk factors for such worsening are recent onset GO, active GO, severe hyperthyroidism, high TSHR-Ab (TSH-receptor antibodies), and smoking. Worsening can be prevented with oral prednisone starting with a daily dose of 0.3–0.5 mg/kg (or lower doses in low-risk patients) for 6–12 weeks. Consequently, ^{131}I therapy is not preferred in active moderate-to-severe GO and DON. [131]I therapy in active mild GO can be used with steroid prophylaxis. Patients with inactive GO can safely receive radioiodine without steroid cover, as long as post-radioiodine hypothyroidism is avoided and other risk factors (particularly smoking) are absent. Antithyroid drugs can be used irrespective of active or inactive GO. Some centers continue antithyroid drugs as long as medical or surgical treatment of GO is required, and discontinue antithyroid drugs never or only after 2–4 years when GO has arrived in its late inactive stages [8, 9]. If recurrent hyperthyroidism occurs after stopping antithyroid drugs, radioiodine might be given successfully [8]. Total thyroidectomy could be effective treatment of GO at least theoretically by removing all thyroid antigens and thyroid-infiltrating lymphocytes, especially when done at a rather early stage of the disease. Trials investigating early thyroid ablation in GO patients (^{131}I therapy + thyroidectomy vs ^{131}I therapy alone) have not shown convincingly the superiority of thyroidectomy [10].

Local Measures

GO patients may benefit from simple local measures [11]. Sunglasses are helpful against photophobia and reflex tearing in response to wind when outdoors. Dark glasses may be also comfortable for patients who are self-conscious of their changed appearance and prefer to hide their eyes. Ocular surface symptoms are common, related to dry eyes secondary to increased lid aperture, exophthalmos, and reduced blink rate. Artificial tears help to control surface symptoms during the day, while gels protect the corneal surface during the night. It is recommended that all GO patients are treated extensively with non-preserved artificial tears with osmoprotective properties. Raising the head of the bed and diuretics have been advocated as a mean to reduce periorbital oedema, but good evidence for their efficacy in this respect is lacking. Botulinum toxin type A has been used to reduce lid retraction by one or more injections of 5 IU into the levator complex of the upper lid. Botulinum toxin has also been used successfully for chemical denervation of the overactive supercilii muscle to address mid-brow frowning and glabellar lines in GO patients. Botulinum toxin might be applied to reduce diplopia by injecting the drug into the appropriate extraocular muscle; the effect lasts for about 12 weeks but nevertheless is often appreciated by patients to bridge the time until GO has become inactive and corrective eye muscle surgery can be done. Prisms might be considered as an alternative to manage troublesome diplopia in the active stage of GO.

Mild Graves' Orbitopathy

Mild GO is characterized by ophthalmological features which have only a minor impact on daily life, insufficient to justify immunosuppressive or surgical treatment. Usually, there exists one or more of the following features: minor lid retraction (<2 mm), mild soft-tissue involvement, exophthalmos <3 mm above normal for race and gender, no or intermittent diplopia, and corneal exposure responsive to lubricants [1]. In most patients with mild GO, a "wait-and-see" strategy is sufficient. To refrain from smoking, restoration of euthyroid function and appropriate local measures are usually sufficient. It also helps that there exists a tendency to spontaneous improvement in the natural history of GO. Selenium supplementation has been formally recommended in mild GO [1]. In a randomized double-blind placebo-controlled study, patients with mild GO received either 100 μg selenium twice daily or placebo for 6 months [12]. At 6 months, quality-of-life as assessed by the disease-specific GO-QoL and overall ocular involvement had improved more often in the selenium group compared to the placebo group (61% vs 36%, $p < 0.001$); after withdrawal of selenium, the improvement was maintained at 12 months. The progression of GO to more severe forms was significantly lower in the selenium group than in the placebo group

(7% vs 26%). No selenium-related adverse events were observed. Enrolled patients mostly came from marginally selenium-deficient areas in Europe [13]. It is unclear whether selenium supplementation is beneficial and safe in selenium-sufficient areas. In longstanding inactive mild GO, there is no evidence that selenium is effective. Selenium is incorporated as selenocysteine in selenoproteins, which play a major role in the maintenance of the cellular redox state [1]. Increased generation of reactive oxygen species may be involved in the pathogenesis of GO, providing a biologic rationale for the application of selenium [13, 14]. Occasionally, objectively mild GO has such a profound impact on quality-of-life that these cases might be considered not mild but moderate-to-severe: it would qualify for immunosuppressive treatment if active, or rehabilitative surgery if inactive [1].

Moderate-to-Severe Graves' Orbitopathy

Moderate-to-severe GO is characterized by ophthalmological features which have sufficient impact on daily life to justify the risks of immunosuppression (if active) or surgical intervention (if inactive). Usually, there exists two or more of the following features: lid retraction ≥2 mm, moderate or severe soft tissue involvement, exophthalmos ≥3 mm above normal for race and gender, inconstant or constant diplopia. Sight-threatening GO should be absent [1]. Essential for an appropriate choice of treatment is the assessment of the disease activity of GO. This can be done by the clinical activity score (CAS), based on the classical signs of inflammation (dolor, rubor, and tumor): a CAS <3 likely indicates inactive GO and a CAS ≥3 active GO [15]. Ever since the MRC report of Sir Brain in 1955, glucocorticoids have been the mainstay of medical treatment for advanced forms of GO [16]. Glucocorticoids are more effective than placebo (response rates 83% vs 11%) [17], intravenous methylprednisolone is more effective and better tolerated than oral prednisone (response rates 77% vs 51%) [18], and intravenous methyl-

prednisolone + mycophenolate is more effective than iv methylprednisolone alone (response rate 71% vs 53%) [19], as demonstrated in successive RCTs [1]. A commonly used dosing schedule for oral prednisone is 60 mg for 2 weeks, 40 mg for 2 weeks, 30 mg for 4 weeks, 20 mg for 4 weeks, and then tapering down to zero dose in 8 weeks. The optimal dose of intravenous methylprednisolone pulse therapy has been investigated in a RCT [20]. The preferred scheme was 500 mg iv once weekly for 6 weeks followed by 250 mg once weekly for another 6 weeks, a cumulative dose of 4.5 g [20]. A lower cumulative dose of 2.25 g was less effective, but a higher cumulative dose of 7.5 g might be used in the most severe cases. Mycophenolate can be added to the iv methylprednisolone pulse therapy (cumulative dose 4.5 g given in 12 weeks) in a daily dose of 360 mg twice daily for 24 weeks [19]. Steroids are effective especially against soft tissue involvement (pain, redness, and oedema) and double vision, but their effect on exophthalmos reduction is often disappointing. The side effects of steroids are well known, but in the early days of intravenous pulses of methylprednisolone, rare but very severe side effects occurred (like liver failure and cardiovascular events) [21]. Current recommendations therefore are that iv pulses of steroids should not be used in patients with recent viral hepatitis, significant liver dysfunction, severe cardiovascular morbidity, or psychiatric disorders [1, 22]. Cumulative doses exceeding 8 g should be avoided, and daily doses should be restricted to ≤1 g and infused slowly. Diabetes and hypertension should be controlled. Using this strategy, steroids will inactivate GO in most patients, allowing rehabilitative surgery as required in the late inactive fibrotic stages of the disease. Not all patients with active moderate-to-severe GO respond to GO, and a flare-up of the ophthalmopathy after an initial response to steroids is frequently seen. It could be seen that combining iv steroids with orbital irradiation or mycophenolate reduces the relapse rate. In case of a partial or no response to steroids, shared decision-making is recommended to select a second-line therapy [1]. Several options are available if GO is still active (Fig. 17.2). (1) A second

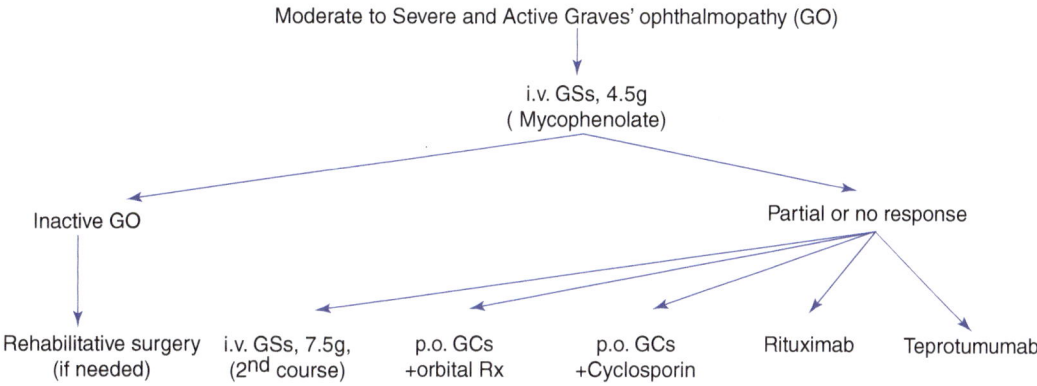

Fig. 17.2 Management of active moderate-to-severe Graves' orbitopathy, and treatment options in case of partial or no response to steroids

course of iv methylprednisolone pulses if the patient tolerates it, and a cumulative dose of >8 g is avoided. (2) Low-dose oral prednisone (e.g. 20 mg daily) + orbital irradiation (administered e.g. in 10 daily doses of 2 Gy over 2 weeks). Retrobulbar irradiation in GO is more effective than sham irradiation according to two RCTs [23, 24]. (3) Low-dose oral prednisone + cyclosporine (e.g. starting dose 5 mg/kg), which has shown to be effective in two RCTs [25, 26]. (4) Rituximab in a dose of 1000 mg given twice over 2 weeks, causes depletion of B-cells (also in the orbit) and modulation of all B-cell functions. The results of two RCTs in active moderate-to-severe GO have been conflicting: one study compared rituximab with placebo and found no difference [27] and the other study compared rituximab with iv methylprednisolone pulses and found rituximab slightly better in improving CAS, ocular motility and QoL [28]. The discrepancy is incompletely understood [29], but on the other hand, the number of positive responses on even single doses of 500 mg or even 100 mg rituximab in open studies cannot be denied. It appears to justify rituximab therapy in case of steroid failure. A serious but fortunately rare adverse event of rituximab is a cytokine storm leading to the subacute development of dysthyroid optic neuropathy (DON) [30]. Monoclonal antibodies against tumor necrosis factor-alpha (anti-TNFα) have been tried in active moderate-

to-severe GO with disappointing results [1]. Monoclonal antibodies against interleukin-6 receptors (anti-IL-6R, tocilizumab) have been given with remarkable success in GO patients not responding to steroids [31]; these results await confirmation by independent groups. A major breakthrough has been the recent development of monoclonal antibodies blocking the IGF-1 receptors on orbital fibroblasts and lymphocytes. Teprotumumab (anti-IGF-1R monoclonal antibody) in a dose of 10 mg/kg for the first infusion and 20 mg/kg for the next seven infusions or placebo was administered intravenously once every 3 weeks for 24 weeks in 170 patients with active moderate-to-severe GO [32, 33]. Responders were 73% in the teprotumumab and 14% in the placebo group. The response involved improvement in CAS, proptosis, double vision, and GO-QoL. Especially, the reduction in exophthalmos was impressive: the mean reduction in proptosis by week 24 ranged from 2.95 to 3.32 mm. That extent of proptosis reduction is not often seen upon steroid treatment. The safety profile of the drug is reassuring. It is not known whether or not teprotumumab will have the same great effect on proptosis reduction if given during the late inactive fibrotic stage of GO. Teprotumumab has been approved for adult GO by the FDA in the USA. Application in Europe awaits head-to-head comparison with iv steroids, as well as cost-effectiveness studies.

Very Severe Graves' Orbitopathy

Very severe GO or sight-threatening GO includes patients with dysthyroid optic neuropathy (DON) and/or corneal breakdown. There has been just one RCT on the treatment of DON, comparing high-dose steroid pulses and immediate orbital decompression surgery [34]. About 83% of the surgically treated patients failed to improve visual acuity within 2 weeks vs 44% in the medically treated patients; all patients improved after switching to the other treatment modality. Consequently, the recommendation is to start immediately with very high doses of methylprednisolone: 1000 mg iv daily for 3 consecutive days during the first week, to be repeated in the second week. If response is absent or poor at day 14, urgent orbital decompression is required. Severe corneal exposure should be treated medically or by means of progressively more invasive surgeries in order to avoid progression to corneal breakdown [1].

Conclusion

The outcome of Graves' orbitopathy has improved greatly over the last few decades. Improvement has been caused not only by earlier diagnosis and treatment of associated Graves' hyperthyroidism, but also by early referral to specialized thyroid/eye clinics and a structured approach in the diagnosis and management of the ophthalmopathy itself. Further improvement is to be expected as Graves' orbitopathy is to a certain extent a preventable disease (stop smoking!), and a real causal therapy of Graves' disease will be available in the next few years directed at blocking signaling via the TSH receptor.

References

1. Bartalena L, Baldeschi L, Boboridis K, Eckstein A, Kahaly GJ, Marcocci C, et al. The 2016 European Thyroid Association/European Group on Graves' Ophthalmopathy guidelines for the management of Graves' orbitopathy. Eur Thyroid J. 2016;5:9–26.

2. Estcourt S, Hickey J, Perros P, Dayan C, Vaidya B. The patient experiences of services for thyroid eye disease in the United Kingdom: results of a nationwide survey. Eur J Endocrinol. 2009;161:483–7.

3. Perros P, Dayan CM, Ezra D, Estcourt S, Kickey J, Lazarus JH, et al. Management of patients with Graves' orbitopathy: initial assessment, management outside specialized centres and referral pathways. Clin Med. 2015;151:173–8.

4. Perros P, Zarkovic M, Azzolini C, Ayvaz G, Baldeschi L, Bartalena L, et al. PREGO (presentation of Graves' orbitopathy) study: changes in referral patterns to European Group on Graves' Orbitopathy (EUGOGO) centres over the period from 2000 to 2012. Br J Ophthalmol. 2015;99:1531–5.

5. Wiersinga WM. Smoking and thyroid. Clin Endocrinol. 2013;79:145–51.

6. Prummel MF, Wiersinga W, Mourits M, Koornneef L, Berghout A, van der Gaag R. Amelioration of eye changes of Graves' ophthalmopathy by achieving euthyroidism. Acta Endocrinol. 1989;121(Suppl. 2):185–9.

7. Traisk E, Tallstedt L, Abraham-Nordling M, Andersson T, Berg G, Calissendorff J, et al. Thyroid-associated ophthalmopathy after treatment for Graves' hyperthyroidism with antithyroid drugs or iodine-131. J Clin Endocrinol Metab. 2009;94:3700–7.

8. Elbers L, Mourits M, Wiersinga W. Outcome of very long-term treatment with antithyroid drugs in Graves' hyperthyroidism associated with Graves' orbitopathy. Thyroid. 2011;21:279–83.

9. Laurberg P, Berman DC, Andersen S, Bulow-Pedersen I. Sustained control of Graves' hyperthyroidism during long-term low-dose antithyroid drug therapy in patients with severe Graves' orbitopathy. Thyroid. 2011;21:951–6.

10. Menconi F, Leo M, Vitti P, Marcocci C, Marino M. Total thyroid ablation in Graves' orbitopathy. J Endocrinol Investig. 2015;38:809–15.

11. Boboridis K, Anagnostis P. Local treatment modalities. In: Wiersinga WM, Kahaly GJ, editors. Graves' orbitopathy. A multidisciplinary approach—questions and answers. 3rd ed. Basel: Karger; 2017. p. 202–6.

12. Marcocci C, Kahaly GJ, Krassas GE, Bartalena L, Prummel M, Stahl M, e t al. Selenium and the course of mild Graves' orbitopathy. N Engl J Med. 2011;364:1920–31.

13. Rayman MP. The importance of selenium in human health. Lancet. 2000;356:233–41.

14. Marcocci C, Bartalena L. Role of oxidative stress and selenium in Graves' hyperthyroidism and orbitopathy. J Endocrinol Investig. 2013;36(Suppl):15–20.

15. Terwee CB, Dekker FW, Mourits MP, Gerding MN, Baldeschi L, Kalmann R, et al. Interpretation and validity of changes in scores on the Graves' ophthalmopathy quality of life questionnaire (GO-QoL) after different treatments. Clin Endocrinol. 2005;54:391–8.

16. Brain SR. Cortisone in exophthalmos: report on therapeutic trial of cortisone and corticotrophin (ACTH) in

exophthalmos and exophthalmic ophthalmoplegia by a panel appointed by the Medical Research Council. Lancet. 1955;268:6–9.

17. Van Geest RJ, Sasim IV, Koppeschaar HP, Kalmann R, Stravers SN, Bijlsma WR, Mourits MP. Methylprednisolone pulse therapy for patients with moderately severe Graves' ophthalmopathy: prospective, randomized, placebo-controlled study. Eur J Endocrinol. 2008;15(8):229–37.

18. Kahaly GJ, Pitz S, Hommel G, Dittmar M. Randomized, single-blind trial of intravenous versus oral steroid monotherapy in Graves' orbitopathy. J Clin Endocrinol Metab. 2005;90:5234–40.

19. Kahaly GJ, Riedl M, Konig J, et al. Mycophenolate plus methyl prednisolone versus methylprednisolone alone in active, moderate-to-severe Graves' orbitopathy (MINGO): a randomised, observer-masked, multicentre trial. Lancet Diab Endocrinol. 2018;6:287–98.

20. Bartalena L, Krassas GE, Wiersinga W, Marcocci C, Salvi M, Daumerie C, et al. Efficacy and safety of three different cumulative doses of intravenous methylprednisolone for moderate to severe and active Graves' orbitopathy. J Clin Endocrinol Metab. 2012;97:4454–63.

21. Marino M, Morabito E, Brunetto MR, Bartalena L, Pinchera A, Marcocci C. Acute and severe liver damage associated with intravenous glucocorticoid pulse therapy in patients with Graves' ophthalmopathy. Thyroid. 2004;14:403–6.

22. Marcocci C, Watt T, Altea MA, Rasmussen AK, Feldt-Rasmussen U, Orgiazzi J, et al. Fatal and nonfatal adverse events of glucocorticoid therapy for Graves' orbitopathy: a questionnaire survey among members of the European Thyroid Association. Eur J Endocrinol. 2012;166:247–53.

23. Mourits MP, van Kempen-Harteveld ML, Garcia MB, Koppeschaar HP, Tick L, Terwee CB. Radiotherapy for Graves' orbitopathy: randomised placebo-controlled study. Lancet. 2000;355:1505–9.

24. Prummel MF, Terwee CB, Gerding MN, Baldeschi L, Mourits MP, Blank L, et al. A randomized controlled trial of orbital radiotherapy versus sham irradiation in patients with mild Graves' ophthalmopathy. J Clin Endocrinol Metab. 2004;89:15–20.

25. Kahaly G, Schrezenmeir J, Krause U, Schweikert B, Meuer S, Muller W, et al. Ciclosporin and prednisone v. prednisone in the treatment of Graves' ophthalmopathy: a controlled, randomized and prospective study. Eur J Clin Investig. 1986;16:415–22.

26. Prummel MF, Mourits MP, Berghout A, Krenning EP, van der Gaag R, Koornneef L, Wiersinga WM. Prednisone and cyclosporine in the treatment of severe Graves' ophthalmopathy. N Engl J Med. 1989;321:1353–9.

27. Stan MN, Garrity JA, Carranza Leon BG, Prabin T, Bradley EA, Bahn RS. Randomized controlled trial of rituximab in patients with Graves' orbitopathy. J Clin Endocrinol Metab. 2015;100:432–41.

28. Salvi M, Vannucchi G, Campi I, Curro N, Dazzi D, Simonetta S, et al. Efficacy of B-cell targeted therapy with rituximab in patients with active moderate to severe Graves' orbitopathy: a randomized controlled study. J Clin Endocrinol Metab. 2015;100:422–31.

29. Stan MN, Salvi M. Rituximab therapy for Graves' orbitopathy—lessons from randomized control trials. Eur J Endocrinol. 2017;176:R101–9.

30. Krassas GE, Stafilidou A, Boboridis KG. Failure of rituximab treatment in a case of severe thyroid ophthalmopathy unresponsive to steroids. Clin Endocrinol. 2010;72:853–5.

31. Perez-Moreiras JV, Gomez-Reino JJ, Maneiro JR, Perez-Pampin E, Romo Lopez A, Rodriguez Alvarez FM, et al. Efficacy of tocilizumab in patients with moderate-to-severe corticosteroid-resistant Graves' orbitopathy: a randomized clinical trial. Am J Ophthalmol. 2018;195:181–90.

32. Smith TJ, Kahaly GJ, Ezra DG, Fleming JC, Dailey RA, Tang RA, et al. Teprotumumab for thyroid-associated ophthalmopathy. N Engl J Med. 2017;376:1748–61.

33. Douglas RS, Kahaly GJ, Patel A, Sile S, Thompson EHZ, Perdok R, et al. Teprotumumab for the treatment of active thyroid eye disease. N Engl J Med. 2020;382:341–52.

34. Wakelkamp IM, Baldeschi L, Saeed P, Mourits MP, Prummel MF, Wiersinga WM. Surgical or medical decompression as a first-line treatment of optic neuropathy in Graves' ophthalmopathy? A randomized controlled trial. Clin Endocrinol. 2005;63:323–8.

Open Access This chapter is licensed under the terms of the Creative Commons Attribution 4.0 International License (http://creativecommons.org/licenses/by/4.0/), which permits use, sharing, adaptation, distribution and reproduction in any medium or format, as long as you give appropriate credit to the original author(s) and the source, provide a link to the Creative Commons license and indicate if changes were made.

The images or other third party material in this chapter are included in the chapter's Creative Commons license, unless indicated otherwise in a credit line to the material. If material is not included in the chapter's Creative Commons license and your intended use is not permitted by statutory regulation or exceeds the permitted use, you will need to obtain permission directly from the copyright holder.

Surgical Treatment of Graves' Orbitopathy

Maarten P. Mourits, Peter J. J. Gooris, and J. Eelco Bergsma

Learning Objectives
- Surgical rehabilitation of the patient with GO follows four steps: orbital decompression, strabismus surgery, eyelid lengthening procedures and blepharoplasty [1].
- Each previous step may influence a subsequent step.
- Surgery is performed during the quiescent stage of GO. There is one exception: dysthyroid optic neuropathy that is not responding to medical treatment.

- Not a single orbital decompression technique is best. Probably the best is to tailor the orbital decompression technique to the patient and to the surgeon [2].

Introduction and History

In the first half of the nineteenth century, Caleb Hillier Parry, Robert James Graves and Carl Adolph von Basedow, independently described signs and symptoms of a new disease, which we now call Graves' disease [3–5]. From the very beginning, a wide variety of treatment modalities have been considered for this disease, of which surgical intervention played an important role. An interesting paper about the pioneers of Graves' disease was published by Alper in 1995 [6].

Management of patients with Graves' orbitopathy (GO), nowadays, consists of medical treatment (Chap. 17), if patients have active and moderately to severe disease, or surgical treatment, if patients have burnt-out (Chap. 17) disease. The most important aim of medical treatment is inactivation of the disease process, whereas surgical treatment aims at functional and cosmetical rehabilitation. Usually, a period of 3–6 months of unchanging symptoms and signs is awaited before surgery is undertaken. The only exception is dysthyroid optic neuropathy (DON),

M. P. Mourits
Amsterdam University Medical Centers, Location AMC, Amsterdam, The Netherlands

P. J. J. Gooris
Department of Oral and Maxillofacial Surgery, Amphia Hospital Breda, Breda, The Netherlands

Amsterdam University Medical Centers, Amsterdam, The Netherlands

Department of Oral and Maxillofacial Surgery, University of Washington, Seattle, WA, USA

J. E. Bergsma (✉)
Department of Oral and Maxillofacial Surgery, Amphia Hospital Breda, Breda, The Netherlands

Amsterdam University Medical Centers, Amsterdam, The Netherlands

Acta dental school, Amsterdam, The Netherlands
e-mail: ebergsma@amphia.nl

© The Author(s) 2023
P. J. J. Gooris et al. (eds.), *Surgery in and around the Orbit*,
https://doi.org/10.1007/978-3-031-40697-3_18

not responding to medical treatment. Immediate orbital decompression is indicated in such cases. The surgical rehabilitation of a patient with GO often requires four steps: orbital decompression, strabismus surgery, eyelid lengthening and blepharoplasty procedures. As each previous step may influence a subsequent step, this sequence of surgeries has to be respected. To get an impression how many treatment modalities an 'average' GO-patient, referred to a tertiary referral institute, undergoes, we evaluated a 1-year's cohort of 95 consecutive new patients. After a follow-up of 5 years, 24 patients had been treated with intravenously administered prednisolone pulse treatment, 50 had undergone orbital decompression, 39 one or more strabismus corrections, and 51 eyelid corrections (at the last visit to the clinic, 78% of patients had no complaints anymore, 16% had diplopia in the extremes of gaze, 5% had diplopia in all directions and one had complaints of facial pain). In conclusion, 50% of them underwent orbital decompression.

In GO, there is a discrepancy between the bony orbital volume and its contents (Figs. 18.1 and 18.2). Nature itself tries to compensate by 'spontaneous' orbital decompression; the walls, especially the medial wall, become concave [7]. This phenomenon, unfortunately, is rare and its effects limited. The solution, therefore, is either making the orbital bony volume bigger by creating an orbital fracture (Fig. 18.3), or decreasing the orbital fat volume; in other words, removing orbital fat.

In 1888, Krönlein excised an orbital tumor via a lateral orbitotomy, thus after having

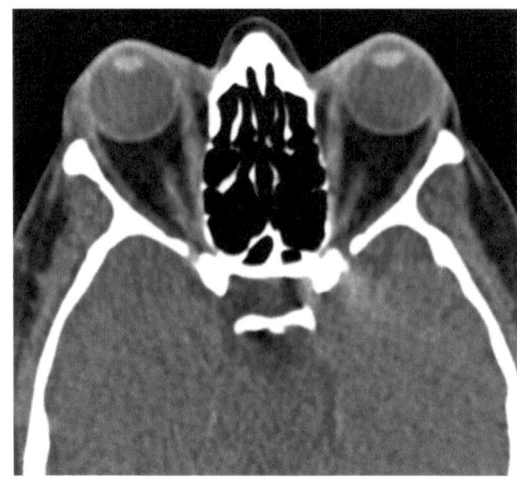

Fig. 18.2 Increase of orbital contents due to fat tissue increase

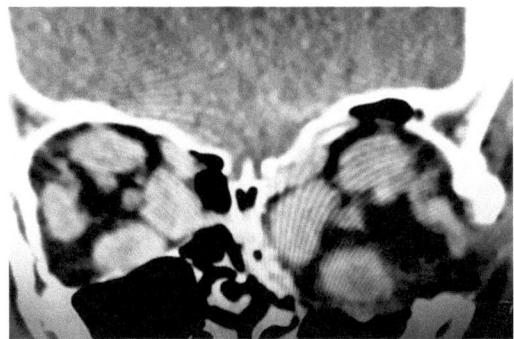

Fig. 18.3 Enlarged left orbit after extensive 3-wall decompression

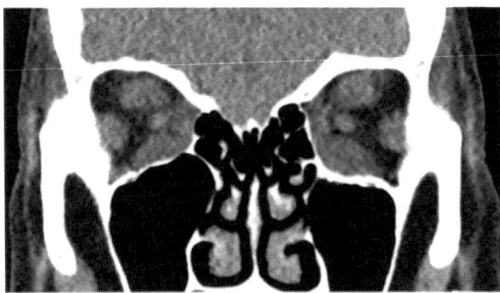

Fig. 18.1 Increase of orbital contents due to enlargement of the extraocular muscles

removed an orbital bony boundary [8]. Dollinger, in 1890, was the first to perform an orbital decompression in a patient with GO [9]. Hirsch, in 1930, described a route via the canine fossa to approach the orbital floor to correct 'eines excessiven Exophthalmus' and, in 1950, to treat 'Malignant Exophthalmos' [10, 11] Naffziger, in 1931, adapted a transcranial technique and removed the orbital roof for progressive exophthalmos following thyroidectomy [12]. Sewall and Kistner, in 1936, approached the orbit via the medial wall [13]. Walsh and Ogura described a transantral approach for orbital decompression [14]. Hence, over time, approaches via all four orbital walls have been explored. The history of more than hundred years of orbital decompression has been recorded by Daniel Rootman [15].

Sole fat removal to correct exophthalmos in Graves' disease has been described by Moore in 1920 [16], but had initially a rather questionable reputation in Western-Europe due to a lack of pertinent data. This changed when better studies were published [17, 18]. It was reported that 1 cc of fat removal corresponded with 1 mm of proptosis reduction [19]. Several studies on fat removal for GO-patients came from Asian countries [20, 21]. The outcomes of these studies cannot be simply applied on patients in Western countries as their orbits differ from Asian orbits and perhaps the presentation of GO in Asians also differs from that in Westerners [22]. Moreover, it is difficult to understand how fat removal alone could reduce the exophthalmos when muscle enlargement is the principle cause of exophthalmos [23]. Here, we enter a major problem in the literature and our knowledge of orbital decompression in GO-patients. Although this literature is abundant (> than 1000 Pubmed publications on orbital decompressive surgery in GO or TED), almost all studies are retrospective and not comparative. Outcome criteria are not well-defined and evaluation of complications differ significantly. These shortcomings have made that there has been little progress in our understanding of orbital decompression despite the huge numbers of publications. One of the very view comparative studies on orbital decompression is a EUGOGO-study [2] with 18 different approaches in 139 patients, that shows that the more orbital walls are removed, the more proptosis reduction can be expected, no matter what technique has been used. In addition, fat removal, adds to the proptosis reduction.

Next to fat removal and bony orbital decompression, there is a third way to camouflage disfiguring proptosis: by enlargement of the orbital rim [24]. This technique has never found many followers as far as we know. In conclusion, the overwhelming majority of orbital decompressions is a bony decompression, with or without fat-removal.

Bony Orbital Decompression: Indications and Approaches

The strongest indication for GO is DON not responding to other treatment modalities [25]. Orbital decompression appears to be very effective in restauration of visual functions [25]. After Leo Koornneef, in Amsterdam, in 1985, had introduced the 3-wall coronal approach (Fig. 18.4), we started to do orbital decompressions for disfiguring proptosis as well [26, 27]. Other indications were persistent pain or a corneal ulcer as a complication of severe proptosis and lagophthalmos or the very exceptional globe (sub-) luxation.

The most common techniques and approaches at that time were the transantral/nasal (Ogura-Walsh) [14], the Lynch (medial), the lateral (with or without the orbital rim), the translid and the transconjunctival approaches. Or combinations of these, such as the 'balanced' decompression (medial and lateral wall, in order to decrease iat-

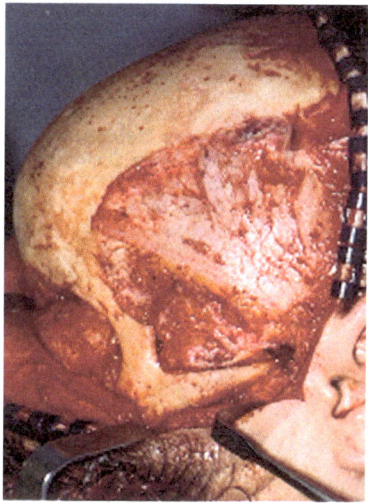

Fig. 18.4 The coronal approach

rogenic diplopia) [28]. Other techniques that quickly gained popularity were the deep lateral wall approach via a superior eyelid crease incision [29, 30] and the endoscopic approach [31]. Basically, with these techniques, one to three orbital walls are (partly) removed, the periosteum is incised and the orbital fat is allowed to herniate into the surrounding cavities (sinuses) or in later years is removed. The techniques differ in the way the walls are approached. Some complications are specifically related to the chosen technique. For instance, postoperative numbness of the cheek is caused by dissection of the bony canal of the infraorbital nerve and thus only seen after orbital floor decompression. A complication that is seen after any kind of orbital decompression is induced or worsened diplopia. Minor modifications have been proposed in order to improve the outcome or to reduce complications. It is assumed that in DON especially the medial wall has to be removed in order to get the best results, because the medial wall is closest to the medial rectus muscle, which is thought to contribute the most to compression of the optic nerve. However, a reduction alone of intra-orbital pressure, obtained by any orbital decompression, appears already to be effective [26, 32].

In line with these modifications are the 'balanced' decompression [28] and the decompression with leaving the periosteum intact [33], which are assumed to cause less diplopia. These assumptions sound plausible, but we do not know if they are true, as comparative studies with well-described inclusion and outcome criteria and fixed protocols for i.e., the assessment of diplopia have never been carried out. To provide such well-described criteria, we have suggested to use the Goldman perimeter and Sullivan's score (see: Chap. 6), with which diplopia can be scored in an easy and reproducible fashion [34].

Clearly, the goals of orbital decompressive surgery are normalization of the visual functions (if they were impaired) and restauration of the premorbid face. It is often thought that the more proptosis reduction the better, but this is not true. Far more important is a symmetrical position of the globes. Symmetrical Hertel values are not always obtained after a single operation and redo-decompressions can result in a satisfactory outcome in the end.

Orbital decompression is one step in the functional and cosmetic rehabilitation of the patient with GO, but not the only one. Strabismus and eyelid correction contribute markedly to the final result. Thus, the final evaluation cannot be made before all steps have been taken [35].

For the reasons aforementioned, we cannot say which decompression is best. We, ourselves, started with the coronal approach but shifted to the 'swinging eyelid' technique, which was described by McCord already in 1981 [36]. Before our change, we compared both techniques in a prospective way. We found that both techniques, in our hands, were equally effective, but the swinging eyelid is much faster, less imposing to the patient than the coronal, with a shorter hospitalization time and less morbidity [37]. For the medial wall we chose an additional retrocaruncular approach.

Swinging Eyelid Technique and Retrocaruncular Approach

To obtain decompression of the globe, an increase of volume of the bony orbit usually combined with a decrease of periorbital fat is the recommended treatment. Hence, an easy access to the orbital walls, but also to the periorbital fat is mandatory. Especially, the transconjunctival approach combined with a swinging eyelid incision gives in our opinion an optimal approach to the orbital floor, the medial and the lateral wall. If the more posterior and superior regions of the medial orbit have to be approached, an additional transcaruncular approach can be done (Fig. 18.5).

During any orbital decompression, the pupil has to be checked repeatedly throughout the entire procedure. We therefore do not recommend a non-transparent cornea-eye shield. At regular intervals, the retractors should be released and the pupil should be checked. If the pupil dilates and anisocoria develops, the procedure must be interrupted. This unwanted effect on the pupil is most likely caused by too much and or too long pressure on the ciliary ganglion located

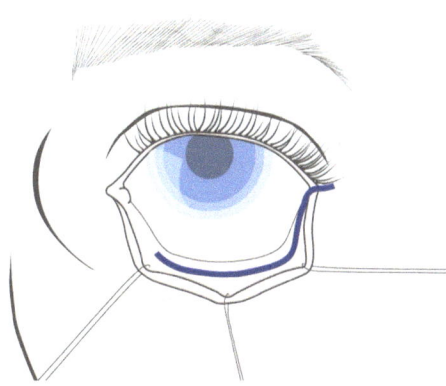

Fig. 18.5 Transconjunctival approach for orbital decompression (with permission from *AO Foundation*)

in the orbital apex. Continued elevated intraorbital pressure may damage the parasympathetic pathways in the ganglion resulting in deficient innervation to the sphincter pupillae muscle and to the ciliary muscle. Pupillary dilatation will occur and should be considered a warning sign to relief apical pressure (Chap. 5). Limited alertness to this sign may result in permanent pupillary dilatation and loss of accommodation. During the entire procedure, the cornea is kept wet, either by eyedrops or by gels or by ointment.

We prefer diathermal cutting with a protective sleeve exposing just about 5 mm of the needle, thus reducing the risk of collateral thermal damage to the eye. The lateral eyelid technique is relatively straight forward. The procedure begins with a cantholysis: an incision is made in the lateral canthal fold of about 8–10 mm, dissecting the dermis, the inferior lateral canthal tendon and the periosteum (Fig. 18.6). After dissection, the lower eyelid gives way, exposing the conjunctiva. Subsequently, a flexible metal spatula puts tension on the conjunctiva on the posterior side of the infra orbital rim. A Desmarres retractor is placed on the anterior part of the orbital rim, thus creating a stretched path for dissection of the conjunctiva and periosteum (Fig. 18.7). The incision can be either pre- or post-septal. Subsequently, the incision is extended on the medial side up to the lacrimal punctum (Fig. 18.8). Leaving the retractors in place, the orbit itself can be accessed. Starting on top or just

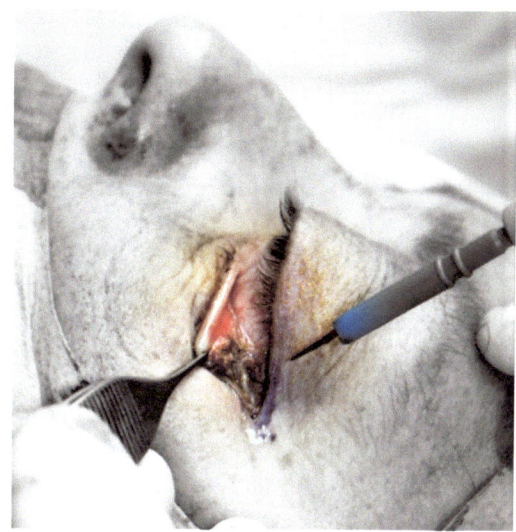

Fig. 18.6 The cantholysis

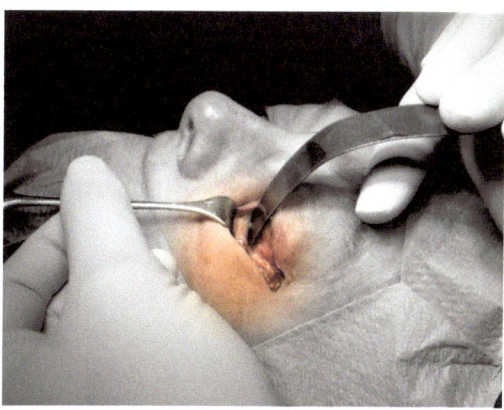

Fig. 18.7 Exposure of the infraorbital rim and incision of the conjunctiva

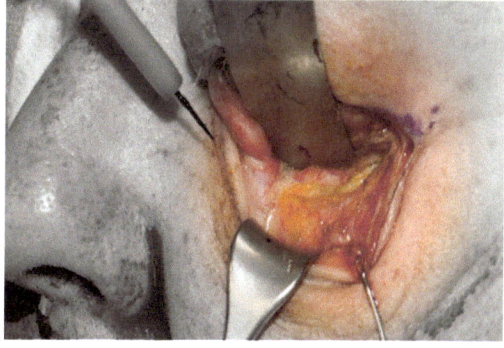

Fig. 18.8 Access to the orbital contents

posteriorly of the orbital rim, the periorbit can be elevated from the lateral wall, the orbital floor and the lower part of the medial wall. The flexible spatula can be replaced to gently lift or shift the bulbus to reveal and give access to the different orbital walls (Fig. 18.9). Care is taken not to detach the inferior oblique muscle located medially of the lacrimal punctum. Now that the different orbital walls are accessible, osteotomies can be made perpendicular to the orbital rim and parallel to the infraorbital nerve using a piezotome (Fig. 18.10), the instrument which allows us to precisely make the necessary bone cuts and helps control to leave the bone-boundary of the infraorbital canal intact. Fragments of the thin orbital floor can be removed easily. In our opinion, it is paramount to preserve a small bony bridge covering the infraorbital nerve both to protect the nerve

itself (Fig. 18.11), as well as to prevent postoperative hypoglobus.

The lateral wall can be approached in a similar way using the piezotome to set out the boundaries for the osteotomy. The bone of the lateral wall exists of two bony plates (Chap. 2) and is much more rigid and sturdy compared to the orbital floor. Therefore, an osteotome has to be used as well to mobilize the different parts of the wall (Fig. 18.12). Care should be taken not to damage the temporal muscle; this will lead to profuse bleeding. Tamponade with Surgicel or Spongostan will normally resolve this bleeding. The bone

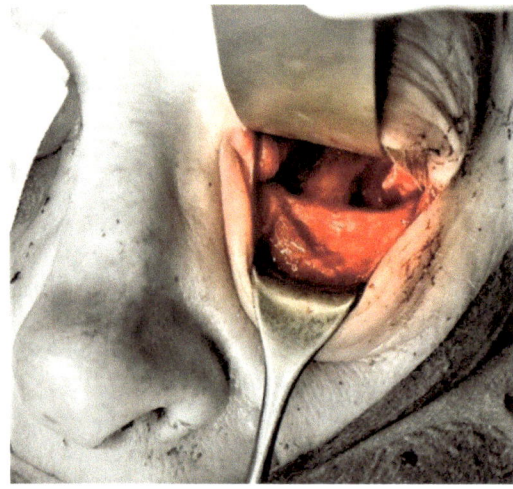

Fig. 18.11 Bony bridge covering the infraorbital nerve is kept intact

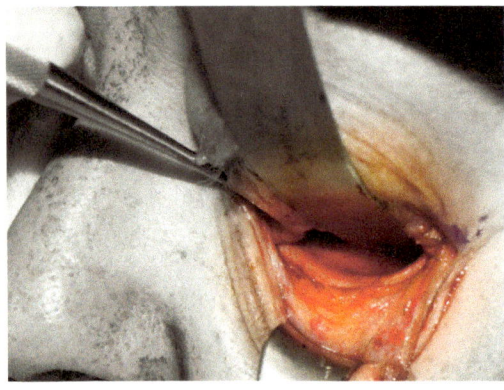

Fig. 18.9 Access to the orbital walls

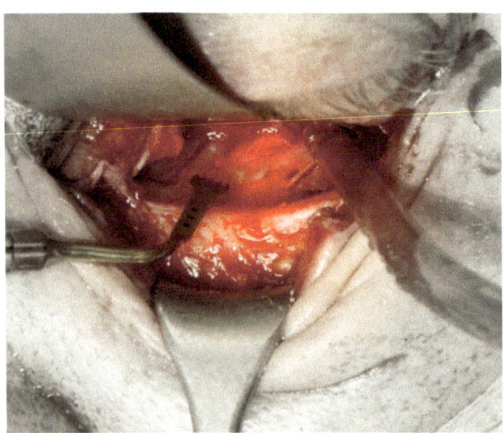

Fig. 18.10 Use of piezotome

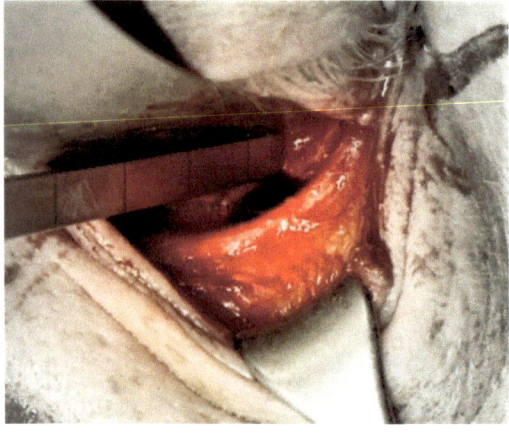

Fig. 18.12 Use of osteotome to remove lateral wall from inside out

removal of the orbital floor can be continuous with the lateral wall, no strut has to be preserved. On the medial side, however, a key area of bone has to be preserved for the proper support and position of the globe and to prevent iatrogenic obliteration of the infundibulum.

The swinging eyelid technique will provide ample approach of the anterior and inferior part of the medial wall. If further decompression is needed, an additional retrocaruncular approach can be used. To facilitate this approach, a suture is placed in both the lower and upper eyelid that are subsequently used as retractor (Fig. 18.13). An additional retractor is placed medially to expose the caruncle. A vertical incision is made just lateral/posterior of the caruncle using diathermy. Subsequently, blunt dissection is performed posteriorly till the medial wall is exposed. The exposure will be posterior to the lacrimal sac. Using this technique, a good exposure of the more posterior and superior parts of the medial wall is accomplished. Care should be taken to locate and preserve the anterior and posterior ethmoidal arteries or obtain good cauterization. As a rule of thumb, 2–3 mm of the most medial part of the orbital floor and the 2–3 mm of the most inferior part of the medial wall should be kept in place. After the bony decompression, further decompression is obtained by periorbital fat removal. The first step is careful dissection of the periorbita in a perpendicular plane to the orbital rim. Subsequently using Metzenbaum scissors,

the periorbita is further opened allowing the periorbital fat to bulge into the newly created space of the bony orbit. The amount of fat to be harvested is individually tapered (Fig. 18.14). Special care is taken not to damage the orbital muscles or the peri-muscular fat as this will enhance the risk of contusion and adhesions of these delicate orbital soft tissues potentially resulting in an increase of postoperative diplopia.

After hemostasis and conformation of the proper symmetrical position of the globes, the wounds can be closed. Using a resorbable 5 × 0 suture, the closure starts with aligning the lateral canthus. The suture runs through the periosteum, the inferior tarsal tendon and the superior tarsal tendon. Care is taken to position the upper eyelid just anterior of the lower eyelid (Fig. 18.15). After correct alignment of the lateral canthus, the dermis is closed with a non-resorbable 6 × 0 suture. There is a lot of controversy regarding the

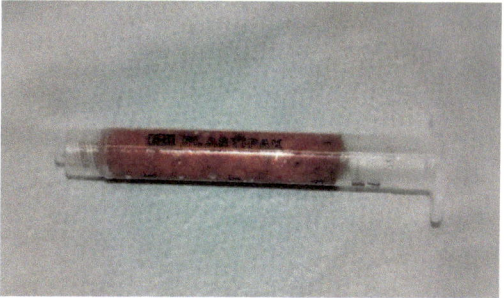

Fig. 18.14 Assessment of quantity of removed fat

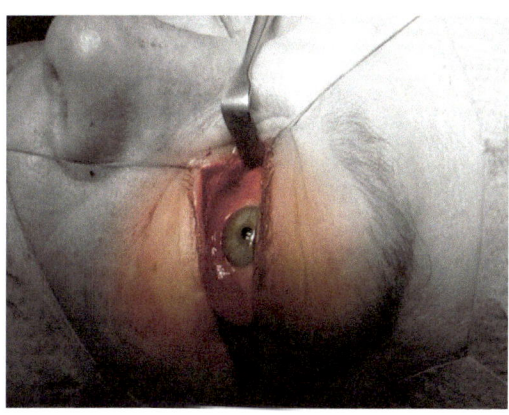

Fig. 18.13 Access to the medial wall

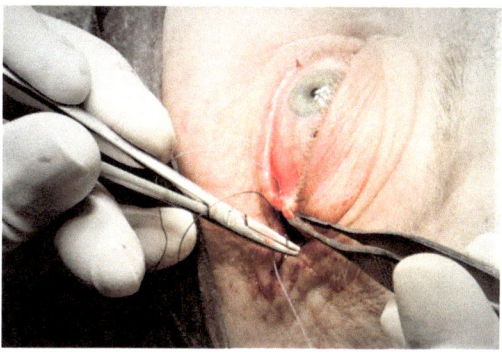

Fig. 18.15 Repositioning of the lateral canthus

closure of the conjunctiva, we prefer not to close the conjunctiva to lessen the chance of creating an entropion and no adverse reaction have been noticed. Figure 18.16 shows the results of a 3-wall orbital decompression (plus eyelid corrections) as described above.

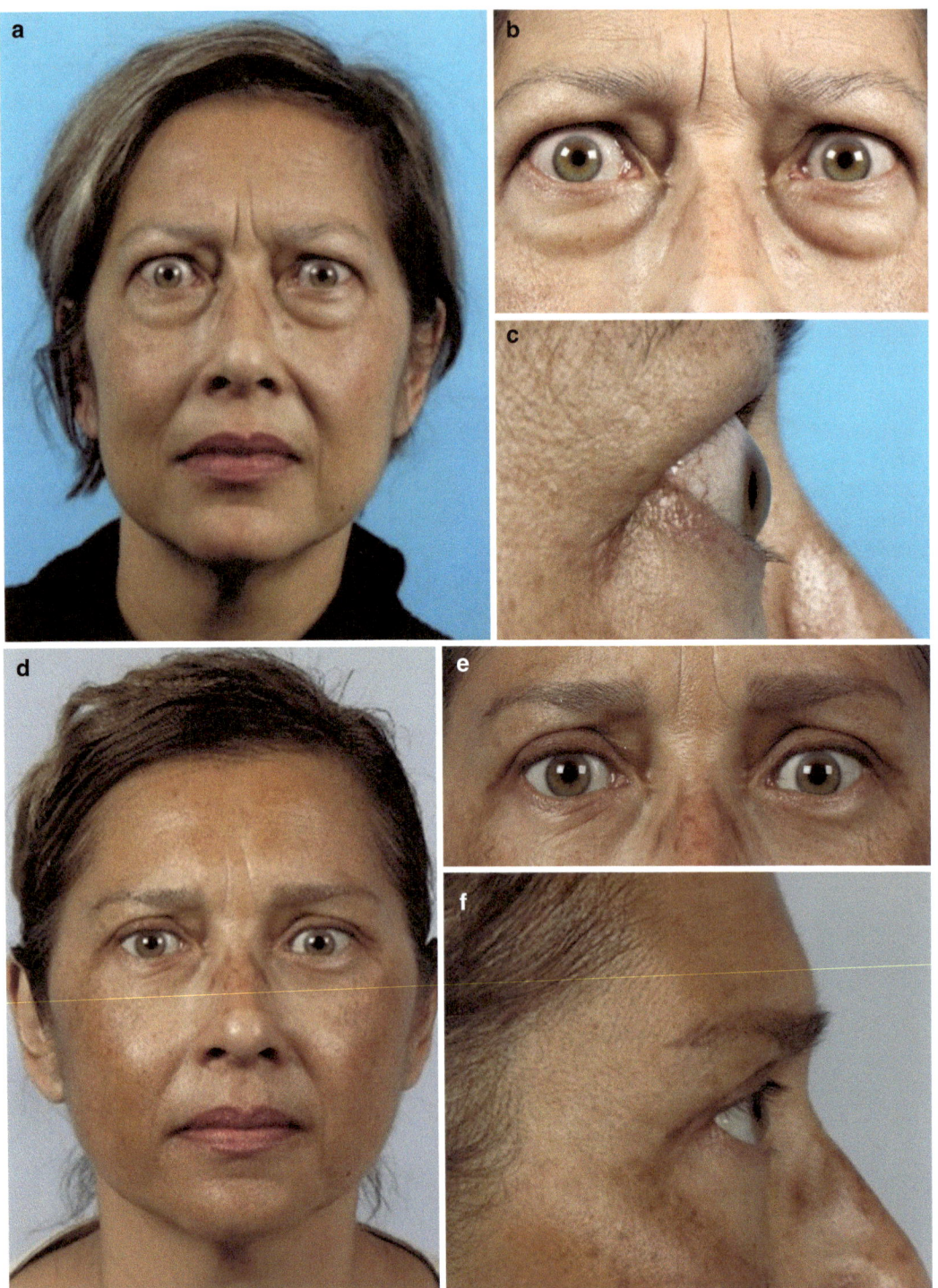

Fig. 18.16 (**a–f**) 52 year-old lady, who underwent bilateral 3-wall orbital decompression and eyelid corrections. Hertel values dropped from 22 to 18 mm and from 23 to 19 mm. Diplopia disappeared completely

Complications

The most feared complication of orbital decompression, no doubt, is a retrobulbar hemorrhage causing blindness. We have seen retrobulbar bleedings at the time the patients awoke from their general anesthesia, started to cough and raised their intraorbital pressure. Proptosis quickly recurred and the orbit felt very stiff. Immediate evacuation of the blood reversed the threatening situation. Therefore, operated patients have to be watched closely at the moment of extubation, but also during the next 24 h postoperative. We have never seen hemorrhages after that period of time. The most frequent complication is induction or worsening of diplopia [38]. Here, a lot of misunderstanding exists, because everyone defines these conditions differently. We, therefore, suggested fixed models to assess diplopia and changes of diplopia [34], but till present, these have reached few followers. Numbness of the region innervated by the infraorbital nerve is rather common after floor decompressions, but mostly resolves with time. Obstruction of the ostiomeatal complex after medial wall and/or floor decompression can cause drainage problems (Fig. 18.17) and recurrence of exophthalmos and retrobulbar pain. Preservation of the medial orbital strut prevents this complication. Re-opening of the passage of the maxillary sinus to the nose and aeration of the sinus cures the problem. Hypoglossus is another complication of orbital floor decompression, however since the sagittal bone strut i.e., the roof of the infraorbital canal is left intact, no postoperative hypoglobus has been observed.

CSF-leakage and anosmia are rare complications after decompressions including the medial wall. Remarkably, the incidence of postoperative infections of the orbits are extremely low. There is no need to prescribe peri-operative antibiotics, prophylactically.

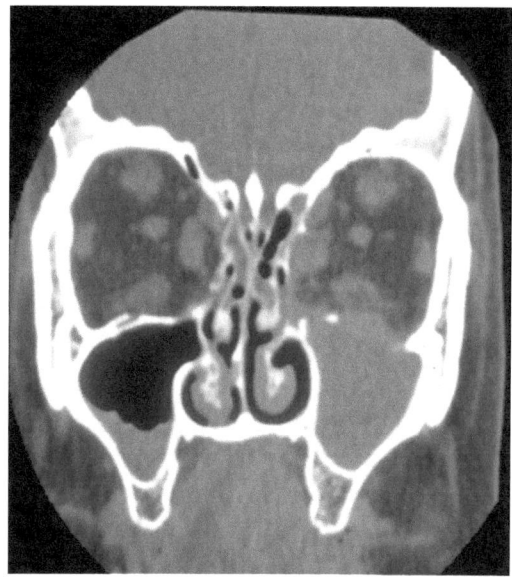

Fig. 18.17 Obstruction of the left antrum after medial wall and floor orbital decompression

Conclusion

Since the fifties of the last century, all over the world, numerous bony orbital decompressions have been performed. What can be said is that all bony decompressions are more or less effective in the restauration of the visual functions and in reduction of proptosis and -at the same time- are safe [2]. The chance of persistent blindness is around 1% [39]. There are many complications described, but most of them are infrequent and -moreover-can be taken care off.

References

1. Shorr N, Seiff SR. The four stages of surgical rehabilitation of the patient with dysthyroid ophthalmopathy. Ophthalmology. 1986;93(4):476–83.
2. European Group on Graves' Orbitopathy (EUGOGO), Mourits MP, Bijl H, Altea MA, Baldeschi L, Boboridis K, Currò N, Dickinson AJ, Eckstein A, Freidel M,

Guastella C, Kahaly GJ, Kalmann R, Krassas GE, Lane CM, Lareida J, Marcocci C, Marino M, Nardi M, Mohr C, Neoh C, Pinchera A, Orgiazzi J, Pitz S, Saeed P, Salvi M, Sellari-Franceschini S, Stahl M, von Arx G, Wiersinga WM. Outcome of orbital decompression for disfiguring proptosis in patients with Graves' orbitopathy using various surgical procedures. Br J Ophthalmol. 2009;93(11):1518–23.

3. Parry CH. Collections from the unpublished medical writings of the late Caleb Hillier Parry, vol. 2. London: Underwoods; 1825. p. 11–124.

4. Graves RJ. New observed affection of the thyroid gland in females. (Clinical lectures). Lond Med Surg J. 1835;7:516–7.

5. Von Basedow KA. Exophthalmus durch Hypertrophie des Zellgewebes in der Augenhöhle. Wochenschr Helik. 1840;6:197–204.

6. Alper MG. Pioneers in the history of orbital decompression for Graves' ophthalmopathy. Doc Ophthalmol. 1995;89:163–71.

7. Vasanthapuram VH, Naik M. Spontaneous orbital wall decompression in thyroid eye disease: a rare feature. Ophthalmic Plast Reconstr Surg. 2021;37(1):e7.

8. Krönlein R. Zur Pathologic und Behandlung der Dermoidcysten der Orbita. Beitrage zur Klein Chir. 1888;4:149.

9. Dollinger J. Die Druckentlastung der Augenhöhle durch Entfernung der äußeren Orbitalwand bei hochgradigem Exophthalmus (Morbus basedowii) und konsekutiver Hornhauterkrankung. Deutsche Medizin Wochenschr. 1911;41:1888.

10. Hirsch O, Urbanek AG. Behandlung eines exzessiven Exophthalmus (Basedow) durch Entfernung von Orbitalfett von der Kieferhöhle. Aus Monatsschrift f Ohren-heilkunde und Laryngo-Rhinologie. 1930;64:212–3.

11. Hirsch O. Surgical decompression of malignant exophthalmos. Arch Otolaryngol. 1950;51:325–34.

12. Naffziger HC. Progressive exophthalmos following thyroidectomy: its pathology and treatment. Ann Surg. 1931;94:582–4.

13. Kistner FB. Decompression for exophthalmos: report of three cases. JAMA. 1939;112:37–8.

14. Walsh TE, Ogura JH. Transantral orbital decompression for malignant exophthalmos. Laryngoscope. 1957;67:544–68.

15. Rootman DB. Orbital decompression for thyroid eye disease. Surv Ophthalmol. 2018;63(1):86–104.

16. Moore RF. Exophthalmus and limitation of eye movements of Graves' disease. Lancet. 1920;2:701.

17. Trokel S, Kazim M, Moore S. Orbital fat removal. Decompression for Graves orbitopathy. Ophthalmology. 1993;100(5):674–82.

18. Adenis JP, Robert PY, Lasudry JG, Dalloul Z. Treatment of proptosis with fat removal orbital decompression in Graves' ophthalmopathy. Eur J Ophthalmol. 1998;8(4):246–52.

19. Liao SL, Huang SW. Correlation of retrobulbar volume change with resected orbital fat volume and proptosis reduction after fatty decompression for Graves ophthalmopathy. Am J Ophthalmol. 2011;151(3):465–9.

20. Li EY, Kwok TY, Cheng AC, Wong AC, Yuen HK. Fat-removal orbital decompression for disfiguring proptosis associated with Graves' ophthalmopathy: safety, efficacy, predictability of outcomes. Int Ophthalmol. 2015;35(3):325–9.

21. Wu CH, Chang TC, Liao SL. Results and predictability of fat removal orbital decompression for disfiguring graves exophthalmos in an Asian patient population. Am J Ophthalmol. 2008;145(4):755–9.

22. Cheng CL, Seah LL, Khoo DH. Ethnic differences in the clinical presentation of Graves' ophthalmopathy. Best Pract Res Clin Endocrinol Metab. 2012;26(3):249–58.

23. Boboridis KG, Uddin J, Mikropoulos DG, Bunce C, Mangouritsas G, Voudouragkaki IC, Konstas AG. Critical appraisal on orbital decompression for thyroid eye disease: a systematic review and literature search. Adv Ther. 2015;32(7):595–611.

24. Goldberg RA, Soroudi AE, McCann JD. Treatment of prominent eyes with orbital rim onlay implants: 4-year experience. Ophthalmic Plast Reconstr Surg. 2003;19:38–45.

25. Wakelkamp IM, Baldeschi L, Saeed P, Mourits MP, Prummel MF, Wiersinga WM. Surgical or medical decompression as a first-line treatment of optic neuropathy in Graves' ophthalmopathy? A randomised controlled trial. Clin Endocrinol (Oxf). 2005;63(3):323–8.

26. Mourits MP, Koornneef L, Wiersinga WM, Prummel MF, Berghout A, van der Gaag R. Orbital decompression for Graves' ophthalmopathy by inferomedial, by inferomedial plus lateral, and by coronal approach. Ophthalmology. 1990;97(5):636–41.

27. Kalmann R, Mourits MP, van der Pol JP, Koornneef L. Coronal approach for rehabilitative orbital decompression in Graves' ophthalmopathy. Br J Ophthalmol. 1997;81(1):41–5.

28. Shepard KG, Levin PS, Terris DJ. Balanced orbital decompression for Graves' ophthalmopathy. Laryngoscope. 1998;108(11 Pt 1):1648–53.

29. Goldberg RA, Perry JD, Hortaleza V, Tong JT. Strabismus after balanced medial plus lateral wall orbital decompression for dysthyroid orbitopathy. Ophthalmic Plast Reconstr Surg. 2000;16:271–7.

30. Goldberg RA, Kim AJ, Kerivan KM. The lacrimal keyhole, orbital door jamb, and basin of the inferior orbital fissure. Three areas of deep bone in the lateral orbit. Arch Ophthalmol. 1998;116(12):1618–24.

31. She YY, Chi CC, Chu ST. Transnasal endoscopic orbital decompression: 15-year clinical experience in southern Taiwan. J Formos Med Assoc. 2014;113(9):648–55.

32. Otto AJ, Koornneef L, Mourits MP, Deen-van L. Retrobulbar pressures measured during surgical decompression of the orbit. Br J Ophthalmol. 1996;80(12):1042–5.

33. Harvey JT. Orbital decompression for Grave's disease leaving the periosteum intact. Ophthalmic Plast Reconstr Surg. 1989;5(3):199–206.

34. Jellema HM, Braaksma-Besselink Y, Limpens J, von Arx G, Wiersinga WM, Mourits MP. Proposal of success criteria for strabismus surgery in patients with Graves' orbitopathy based on a systematic literature review. Acta Ophthalmol. 2015;93(7):601–9.

35. Terwee CB, Dekker FW, Mourits MP, Gerding MN, Baldeschi L, Kalmann R, Prummel MF, Wiersinga WM. Interpretation and validity of changes in scores on the Graves' ophthalmopathy quality of life questionnaire (GO-QOL) after different treatments. Clin Endocrinol. 2001;54(3):391.

36. McCord CG Jr. Orbital decompression for Graves' disease. Exposure through lateral canthal and inferior fornix incision. Ophthalmology. 1981;88(6):533–41.

37. Sasim IV, de Graaf ME, Berendschot TT, Kalmann R, van Isterdael C, Mourits MP. Coronal or swinging eyelid decompression for patients with disfiguring proptosis in Graves' orbitopathy? Comparison of results in one center. Ophthalmology. 2005;112(7):1310–5.

38. Paridaens D, Hans K, van Buitenen S, Mourits MP. The incidence of diplopia following coronal and translid orbital decompression in Graves' orbitopathy. Eye. 1998;12:800–5.

39. Jacobs SM, McInnis CP, Kapeles M, Chang SH. Incidence, risk factors, and management of blindness after orbital surgery. Ophthalmology. 2018;125:1100–8.

Open Access This chapter is licensed under the terms of the Creative Commons Attribution 4.0 International License (http://creativecommons.org/licenses/by/4.0/), which permits use, sharing, adaptation, distribution and reproduction in any medium or format, as long as you give appropriate credit to the original author(s) and the source, provide a link to the Creative Commons license and indicate if changes were made.

The images or other third party material in this chapter are included in the chapter's Creative Commons license, unless indicated otherwise in a credit line to the material. If material is not included in the chapter's Creative Commons license and your intended use is not permitted by statutory regulation or exceeds the permitted use, you will need to obtain permission directly from the copyright holder.

Orbital Cellulitis

Maarten P. Mourits

Learning Objectives

- The term subperiosteal abscess is a misnomer. It should be called more appropriately subperiosteal empyema.
- For Chandler's classification, there seems to be little support.
- Orbital abscess formation as a complication of orbital cellulitis is rare.
- If during inspection, the eye itself cannot be visualized because of severe swelling and tightness of the eyelids, the condition must be regarded as a retroseptal orbital cellulitis and managed accordingly.

Introduction: Classifications and Fallacies

Rootman [1] and others distinguish several types of orbital inflammation:

1. **Specific inflammations, characterized by a specific pathogen or a specific clinical constellation and/or a specific histopathology:**
   (a) Caused by bacteria, viruses, fungi or parasites
   (b) With predominant vasculitis (granulomatosis with polyangiitis or Wegener's disease)
   (c) With predominant granulomatosis (sarcoidosis, xanthogranulomatosis)
   (d) Transitional (Kimura's disease, Sjögren's disease)
   (e) Autoimmune (Graves' disease, IgG4-related disease)
2. **Non-specific inflammation:**
   (a) Idiopathic orbital inflammation

In this chapter, we will mainly focus on infectious orbital inflammations caused by bacteria, which are called orbital cellulitis (OC). Two types of OC can be distinguished: (1) the rather innocent preseptal orbital cellulitis (pOC) and (2) the potentially life-threatening retroseptal orbital cellulitis (rOC). The orbital septum is a thin, but firm, fibrous structure stretched out between the orbital rim and the transition of the eyelid retractors to the tarsal plate (Fig. 19.1). The orbital septum separates the orbital contents from the skin. A pOC is hence no more than an inflammation of the eyelid, whereas a rOC is an inflammation of the orbit itself.

Patients with OC are seen by many different medical doctors: general practitioners, pediatricians, ear, nose and throat specialists, ophthal-

The original version of the chapter has been revised. A correction to this chapter can be found at https://doi.org/10.1007/978-3-031-40697-3_23

M. P. Mourits (✉)
Amsterdam University Medical Centers, Location AMC, Amsterdam, The Netherlands

© The Author(s) 2023, corrected publication 2024
P. J. J. Gooris et al. (eds.), *Surgery in and around the Orbit*,
https://doi.org/10.1007/978-3-031-40697-3_19

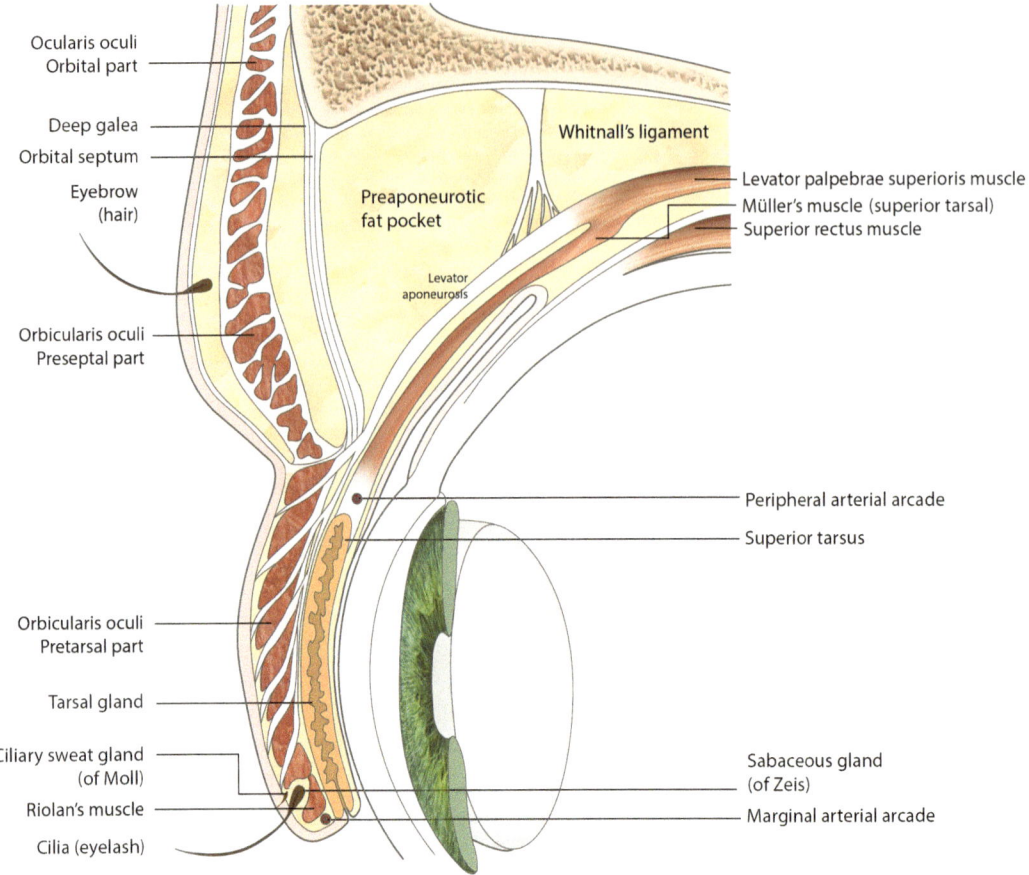

Fig. 19.1 Orbital septum, the anterior border of the orbital cavity

mologists, oromaxillary surgeons, neurosurgeons, and radiologists. The literature on OC appears in all of their specialized journals, and this may contribute to the erroneous ideas that exist in this field of medicine.

The most important fallacy is that no distinction is made between an orbital abscess and a subperiosteal empyema, for which then the misnomer subperiosteal abscess is used. As will be explained, subperiosteal empyema and an orbital abscess are extensions and complications of OC, but differ significantly in their nature and require different therapeutical approaches.

An abscess is a collection of pus that has been accumulating within a tissue [2]. It is a defensive reaction to prevent the spread of infectious material to other parts of the body. An abscess is composed of an abscess wall (or: capsule), which is formed by adjacent healthy cells in an attempt to keep the pus from infecting neighboring structures. However, such an encapsulation may prevent penetration of antibiotics and immune cells from attacking bacteria present within the pus or from reaching the causative organism or foreign object. Therefore, treatment of an abscess consists of opening the abscess wall and draining its contents. In contrast, an empyema is a collection of pus in an extant space, such as the space between the bone and the periosteum. There is no capsule that hampers antibiotics to reach their target.

Another misconception, in our understanding, is the classification of the progression of the inflammation as proposed by Chandler et al. [3]: I. preseptal cellulitis > II. diffuse infiltration of the orbit (OC) > III. subperiosteal abscess > IV. orbital abscess > V. cavernous sinus thrombosis. In our series, only 2 out of 68 patients with pOC developed a rOC [4], i.e., 2.9%. It is more plausible to assume that bacteria spread to the most adjacent structures. So, infections of the ethmoid sinuses spread primarily to the orbit, whereas infections of the sphenoid sinus spread to the cavernous sinus and infections of the frontal sinus spread preferentially to the orbit and to the brain.

We have defined the several presentations of OC as follows [4]

1. Diffuse OC: Diffuse or localized infiltration of the orbital fat without fluid collection or rim enhancement (Fig. 19.2).
2. Orbital subperiosteal empyema: Fluid collection between the orbital wall and the periorbita (Fig. 19.3).
3. Orbital abscess: Fluid collection with peripheral enhancement and intraorbital extension through a defect of the periorbita (Fig. 19.4).

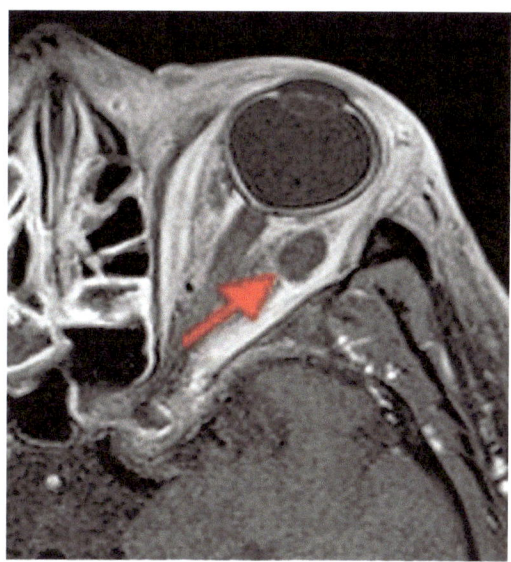

Fig. 19.4 Orbital abscess: fluid collection with peripheral enhancement and intraorbital extension through a defect of the periorbita

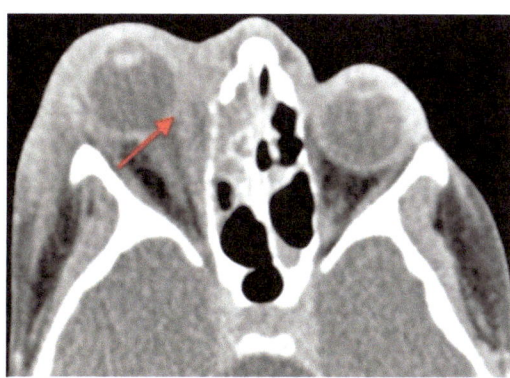

Fig. 19.2 Diffuse infiltration of the medial part of the orbit secondary to ethmoidal sinus disease

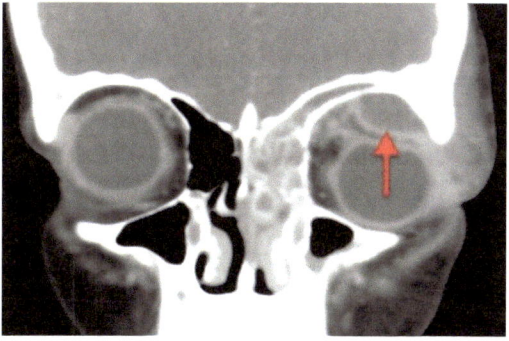

Fig. 19.3 Orbital subperiosteal empyema: Fluid collection between the orbital wall and the periorbita

Epidemiology and Presentation

OC is a disease of all ages. Over a period of 26 years, 53 adults with OC were seen at a university clinic in Split, Croatia [5]. For 6 years, 28 children with OC were seen at a university clinic in Marrakech, Morocco [6]. In the University Medical Center of Amsterdam, we have seen 116 patients, both children and adults, with OC during a period of 7 years [4]. Sixty-eight of them had pOC, 48 had rOC. Seventeen patients (25%) with pOC were younger than 9 years of age; another 17 (35%) with rOC fell in the same age-group. In conclusion, OC—albeit pOC or rOC—is rather rare. The reported incidence is 0.1–3.5 per 100,000 individuals [7]. It is assumed that children suffer from OC more often than adults. This is not in accordance with our experience, but this may be due to referral bias. In our series, 47% of the patients with pOC and 77% of the patients with rOC were male.

Clinical symptoms of pOC are diffuse eyelid swelling, eyelid redness, chemosis, and conjunctival redness. Symptoms of rOC are in addition: pain, exophthalmos, eye motility impairment, a relative afferent pupil defect, retinal vascular

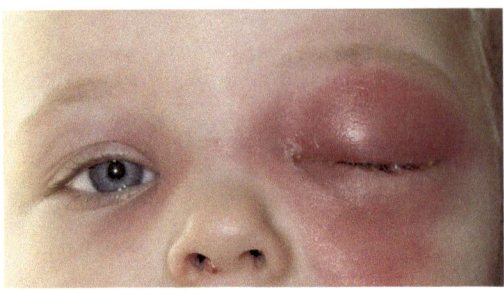

Fig. 19.5 Young child with swollen ptotic and red eyelid suggestive for orbital cellulitis

involvement, loss of visual functions, and a swollen optic nerve head as the result of compression of the optic nerve. If the eyelids are swollen and tight to such an extent that the eye itself cannot be inspected and the abovementioned symptoms cannot be assessed, the condition must be regarded as a pOC (Fig. 19.5).

Etiology, Differential Diagnosis and Complications

The most frequent cause of OC is an extension of paranasal sinus disease. Inflammation of the ethmoid sinus can co-exist with signs of pOC or with a rOC. The medial wall of the orbit is paper-thin (hence, its name lamina papyracea). Extension of infectious material along vessels through the openings in this wall or retrograde spread by the interconnecting valveless venous system of the orbit and the sinuses easily results in OC. pOC is also seen as a complication of dermatitis. The diagnosis is based upon the presence of the clinical symptoms in combination with (a history of) upper respiratory disease or dermal disease. MRI scans or CT scans of the orbit and paranasal sinuses are indicated in case

of doubt and to differentiate between the possible sources of origin and the different orbital presentations. A peculiar form is cellulitis of odontogenic origin, which is feared for its bad outcome [8].

In general, a distinction is made between OC in adults and in young children. According to Harris et al. [9] the populations of causative bacteria in children under the age of nine are usually monomicrobial in nature and less aggressive than those found in older patients. Therefore, they would better respond to antibiotic therapy. In odontogenic cellulitis, in contrast, a polymicrobial infection consisting of both aerobic and anaerobic bacteria is responsible for the bad outcome [10].

The eyelid swelling and redness as seen in pOC must be differentiated from severe conjunctivitis, secondary to severe uveitis or endophthalmitis, and secondary to eyelid or orbital surgery. A mosquito bite, allergic reactions, or malignancies such as a rhabdomyosarcoma or non-Hodgkin lymphoma can also be mistaken for OC.

Apart from blindness, extension to the cavernous sinus resulting in thrombosis and extension to other intracranial loci are feared complications of OC. Cavernous sinus thrombosis will be discussed at the end of this chapter. In our series, we have seen three children and one adult with intracranial spread of OC. Interestingly, they had a short history of disease (less than 4 days) and presented with manifestations of cerebral spread, already at their admission to the hospital (Figs. 19.6 and 19.7). They all survived and none had long-term complications.

Other long-term complications are diplopia due to motility impairment and eyelid abnormalities (retraction, entropion, ptosis, and lasting swelling.

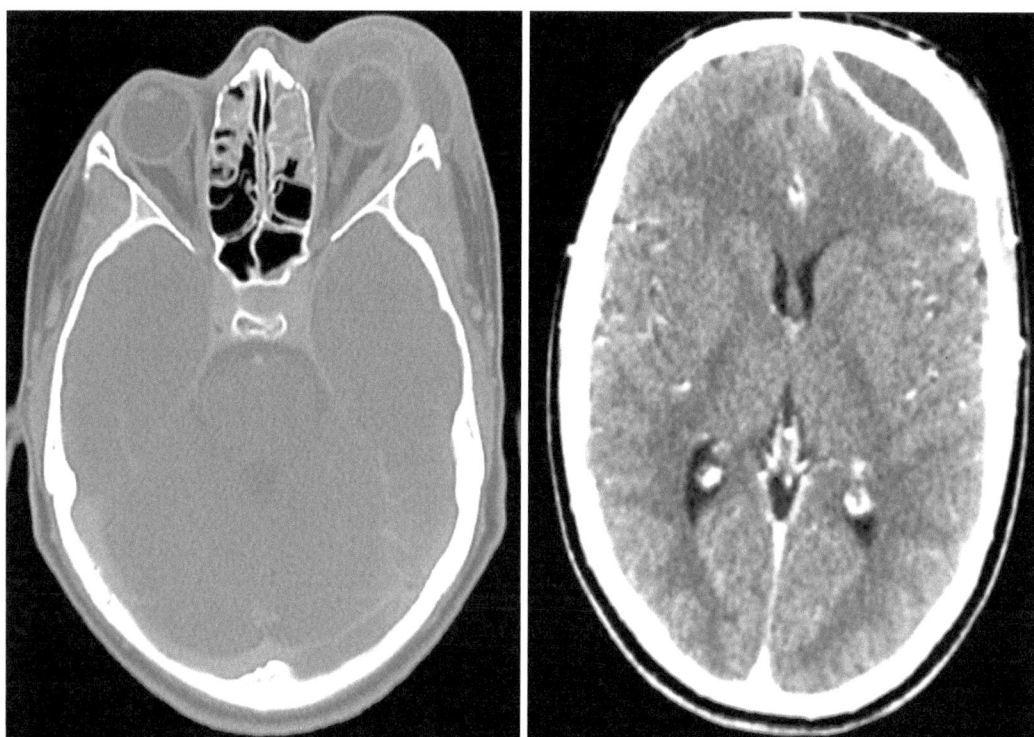

Fig. 19.6 and 19.7 CT scans showing orbital cellulitis and cerebral involvement

Treatment and Outcome

Since classical times, the adage "ubi pus, ibi evacua" (pus must be surgically drained wherever it shows itself) has been cherished in western medicine and through empowerment of time modern-age, medical doctors act almost automatically to the term abscess with an intention to operate. In our series, we found many cases in which surgery had been performed without clear indications. In the era of antibiotics, the need for a surgical approach of OC has dramatically decreased. Souliere and Harris demonstrated that medical treatment alone is safe in most children younger than 9 years of age [11, 12]. In our retrospective series, 68 patients with pOC were treated with antibiotics (74% orally, and 26% intravenously administered). Five patients (7%) underwent orbital surgery (incision and drainage, the indication could not be retrieved). All patients with pOC recovered completely without any com-plication. We therefore suggest ambulatory oral antibiotics as an initial treatment for patients with pOC plus careful watching. We found that only two patients (3%) with pOC developed a rOC.

Of 48 patients with rOC, 83% had co-existing paranasal sinus disease. All patients were admitted to the hospital. Nineteen patients (40%) were treated with intravenously administered antibiotics alone. The remaining 29 patients (60%) underwent surgery as well (21 underwent external orbitotomy with or without sinus surgery; 8 underwent sinus surgery alone). Retrospective evaluation between the two groups (antibiotics alone versus antibiotics plus surgery) revealed no significant changes in age, gender, duration of disease prior to admission, presumed origin of the disease, intraorbital localization of the inflammation, and cultured bacteria. In other words, the criteria for surgery could not be clarified. Was it the term subperiosteal abscess that prompted surgery?

In our initial radiological reports, orbital abscess was mentioned in 16 patients. Re-examination of the MRI scans and CT scans, using the abovementioned criteria, resulted in 12 subperiosteal empyema's and only four true orbital abscesses in a series of 48 patients with rOC (8%).

Three out of 48 patients (6.5%) experienced permanent blindness due to the inflammation. In contrast, one 58-year-old woman without any light perception for 24 subsequent hours left the hospital with normal visual functions. Other long-term complications were reduced motility in five patients, ptosis in two patients, and lag-ophthalmos in another two patients. These complications seemed to be caused by excessive fibrosis secondary to the inflammation. A prudent use of adjuvant systemic steroids may therefore be considered [13].

In conclusion, in patients with rOC we advise to start with intravenously administered antibiotics. If the response is not sufficient within the next 48–72 h, surgery can be considered. Immediate surgery is advised in cases with: (1) a tight orbit with signs of compressive opticopathy, due to raised intraorbital pressure [14] as a consequence of space-taking swellings in the orbit, such as a very large subperiosteal empyema (compare with retrobulbar hemorrhage), (2) true orbital abscess, and (3) odontogenic OC.

Cavernous Sinus Thrombosis and Mucormycosis

Cavernous sinus thrombosis (CST) is notorious for its bad prognosis in terms of mortality and morbidity, but frequencies of these are based on old series. Using the files of the University Medical Center at Amsterdam during the period 2005–2017, we found only 12 patients with CST. Eleven survived and nine recovered without any permanent deficits [15]. Patients presented with eyelid swelling, chemosis, proptosis, high rates of impaired ocular motility, and signs of optic neuropathy. In contrast to "simple" OC, patients with CST more often had bilateral eye symptoms, high fever, and severe headaches. CT scans—or preferably contrast-enhanced MRI scans—show direct and indirect signs of CST. Direct signs are expansion of and filling defects in the cavernous sinus. Indirect signs, caused by venous obstruction, are dilatation of the superior ophthalmic vein, exophthalmos, and increased dural enhancement of the border of the cavernous sinus. An associated sign is secondary thrombosis. In our series, 50% of the CST was caused by extension of an infection of the sphenoid sinus. Other causes were otitis media, fungal rhinosinusitis, pharyngeal infection, and meningitis. Treatment consists of intravenously administered antibiotics in combination with surgical (endoscopic) drainage of the initially infected spaces or paranasal sinuses. In addition, selective anticoagulation therapy should be considered.

Rhino-orbital mucormycosis is a rare and opportunistic fungal infection, which is seen in patients with diabetes mellitus or ketoacidosis as well as in immunocompromised individuals. At the early stage of the infection, occlusion of blood vessels and subsequent thrombosis and necrosis can lead to a blind and immobile eye. Immediate treatment of the underlying disease together with extensive surgical debridement (often including orbital exenteration) and systemic antifungal drugs can be life-saving (Fig. 19.8).

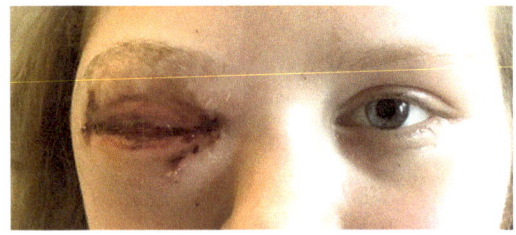

Fig. 19.8 A 20-year-old woman with status after eyelid-sparing exenteration for mucormycosis after ketoacidosis in diabetes mellitus

References

1. Rootman J. Inflammatory diseases. In: Rootman J, editor. Diseases of the orbit: a multidisciplinary approach. 2nd ed. Philadelphia: Lippincott Williams & Wilkins; 2003. p. 455–506.
2. Singer AJ, Talan DA. Management of skin abscesses in the era of methicillin-resistant *Staphylococcus aureus*. N Engl J Med. 2014;370:1039–47.
3. Chandler JR, Langenbrunner DJ, Stevens ER. The pathogenesis of orbital complications in acute sinusitis. Laryngoscope. 1970;80:1414–28.
4. Van Der Veer EG, Van Der Poel NA, De Win MM, Kloos RJ, Saeed P, Mourits MP. True abscess formation is rare in bacterial orbital cellulitis: consequences for treatment. Am J Otolaryngol. 2017;38:130–4.
5. Ivanišević M, Ivanišević P, Lešin M. Epidemiological characteristics of orbital cellulitis among adult population in the Split region, Croatia. Wien Klin Wochenschr. 2019;131:205–8.
6. Daoudi A, Ajdakar S, Rada N, Draiss G, Hajji I, Bouskraoui M. Orbital and periorbital cellulitis in children. Epidemiological, clinical, therapeutic aspects and course. J Fr Ophtalmol. 2016;39:609–14.
7. Murphy C, Livingstone I, Foot B, Murgatroyd H, MacEwen CJ. Orbital cellulitis in Scotland: current incidence, aetiology, management and outcomes. Br J Ophthalmol. 2014;98:1575–8.
8. Yan W, Chakrabarti R, Choong J, Hardy T. Orbital cellulitis from odontogenic origin. Orbit. 2015;34:183–5.
9. Harris GJ. Age as a factor in the bacteriology and response to treatment of subperiosteal abscess of the orbit. Trans Am Ophthalmol Soc. 1993;91:441–516.
10. Giunta Crescente C, Soto De Facchin M, Acevedo Rodríguez AM. Medical-dental considerations in the care of children with facial cellulitis of odontogenic origin. A disease of interest for pediatricians and pediatric dentists. Arch Argent Pediatr. 2018;116:e348–53.
11. Souliere CR, Antoine GA, Martin MP, Blumberg AI, Isaacson G. Selective non-surgical management of subperiosteal abscess of the orbit: computerized tomography and clinical course as indication for surgical drainage. Int J Pediatr Otorhinolaryngol. 1990;19:109–19.
12. Garcia GH, Harris GJ. Criteria for nonsurgical management of subperiosteal abscess of the orbit: analysis of outcomes 1988–1998. Ophthalmology. 2000;107:1454–6.
13. Mahalingam S, Luke L, Pundir J, Pundir V. The role of adjuvant systemic steroids in the management of periorbital cellulitis secondary to sinusitis: a systematic review and meta-analysis. Eur Arch Otorhinolaryngol. 2020;278:2193. https://doi.org/10.1007/s00405-020-06294-z.
14. Harris GJ. Subperiosteal abscess of the orbit. Arch Ophthalmol. 1983;101:751–7.
15. Van Der Poel NA, Mourits MP, De Win MML, Coutinho JM, Dikkers FG. Prognosis of septic cavernous sinus thrombosis remarkably improved: a case series of 12 patients and literature review. Eur Arch Otorhinolaryngol. 2018;27:2387.

Open Access This chapter is licensed under the terms of the Creative Commons Attribution 4.0 International License (http://creativecommons.org/licenses/by/4.0/), which permits use, sharing, adaptation, distribution and reproduction in any medium or format, as long as you give appropriate credit to the original author(s) and the source, provide a link to the Creative Commons license and indicate if changes were made.

The images or other third party material in this chapter are included in the chapter's Creative Commons license, unless indicated otherwise in a credit line to the material. If material is not included in the chapter's Creative Commons license and your intended use is not permitted by statutory regulation or exceeds the permitted use, you will need to obtain permission directly from the copyright holder.

Peter J. J. Gooris, Gertjan Mensink, Rob Noorlag, and J. Eelco Bergsma

Learning Objectives
- Be alert of post extractionem signs and symptoms
- Do not always consider removal of dentition as final treatment as the infection may have developed into an independent process
- Watch carefully when, after the removal of an infected tooth-molar the patient returns with complaints in or around the eye
- If there is suspicion of orbital involvement of an original dentogenic infection, do not hesitate to employ early intervention

P. J. J. Gooris
Department of Oral and Maxillofacial Surgery, Amphia Hospital Breda, Breda, The Netherlands

Amsterdam University Medical Centers, Amsterdam, The Netherlands

Department of Oral and Maxillofacial Surgery, University of Washington, Seattle, WA, USA

G. Mensink (✉)
Department of Oral and Maxillofacial Surgery, Amphia Hospital Breda, Breda, The Netherlands

University Medical Center Leiden, Leiden, The Netherlands
e-mail: gmensink1@amphia.nl

R. Noorlag
Department of Oral and Maxillofacial Surgery, University Medical Center Utrecht, Utrecht, The Netherlands

J. E. Bergsma
Department of Oral and Maxillofacial Surgery, Amphia Hospital Breda, Breda, The Netherlands

Amsterdam University Medical Centers, Amsterdam, The Netherlands

Acta dental school, Amsterdam, The Netherlands

Introduction

A 50-year-old female, otherwise healthy patient visited the dentist because she suffered from mild toothache problems in the left upper jaw since 6 days. The dentist diagnosed a deeply carious non-vital upper molar which was subsequently removed. The molar removal did not result in an apparent oro-antral communication. In the days following the removal, the patient developed swelling of the ipsilateral side of the face, reason for the dentist to start antibiotic therapy: amoxicillin 500 mg 3 times per day orally. Despite the antibiotics, the infection continued to spread progressively, including swelling around the eye. Once fever developed, the patient decided to visit the general practitioner (GP). The GP directly referred the patient to the outpatient clinic of oral and maxillofacial surgery for further evaluation and treatment.

© The Author(s) 2023
P. J. J. Gooris et al. (eds.), *Surgery in and around the Orbit*,
https://doi.org/10.1007/978-3-031-40697-3_20

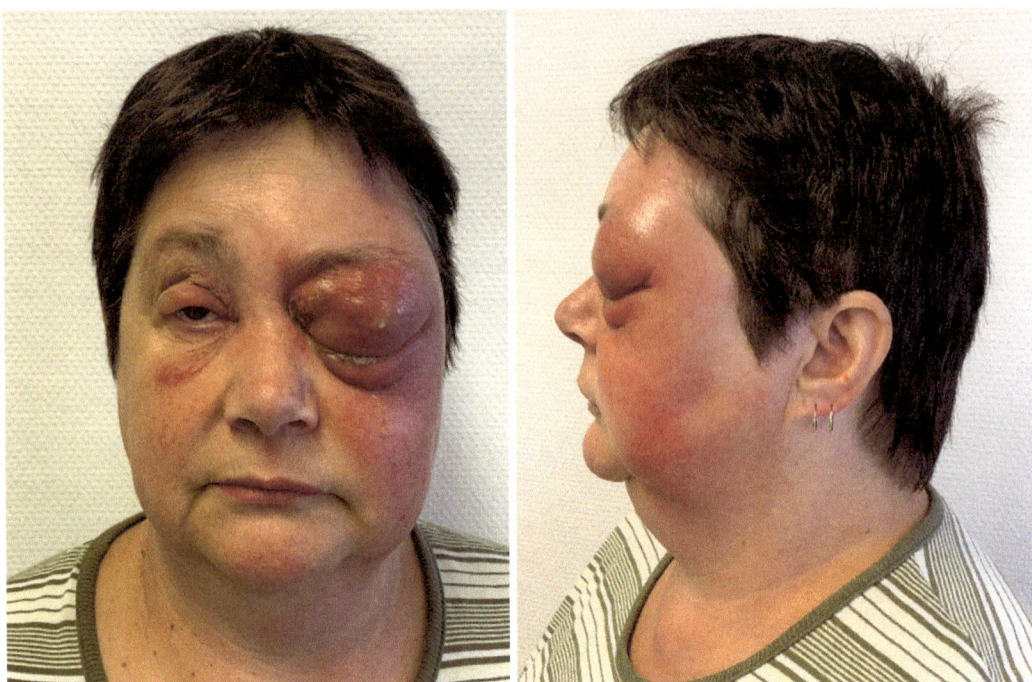

Fig. 20.1 and 20.2 Presentation of a 50-year-old patient with periorbital swelling and proptosis in a frontal (Fig. 1) and a lateral (Fig. 2) view

Physical Examination

On her visit to the clinic, we saw a moderately sick woman, in mild distress, with a body temperature of 37.8 °C, a blood pressure of 140/85, blood values showed a CRP 185, and leucocytes 16.8, including a left shift. Intraorally, there was an undisturbed alveolus healing without discharge. An oro-antral communication could not be demonstrated.

Visual inspection showed an erythematous swelling of the left half of the face with impressive ipsilateral eyelid involvement (Figs. 20.1 and 20.2). On palpation, there was a tense swelling of the left upper eyelid. There was spontaneous discharge of purulent material from the medial canthal region. The patient was not able to open the eye spontaneously and after manual opening of the eyelids, the pupil was not reactive to direct light stimulus, and in addition, the patient denied vision. The globe felt firm on pressure while the iris showed a cloudy aspect. There was extreme limitation of globe motility, and ductions were not possible.

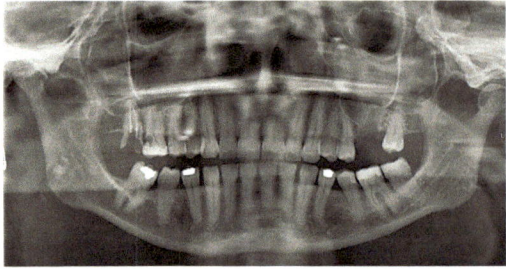

Fig. 20.3 Panoramic X-ray with obscuration of the left maxillary sinus. Note the extraction alveolus of the upper left first molar and the close relationship with the maxillary sinus

Radiological Examination

The orthopantomogram showed obscuration of the left maxillary sinus (Fig. 20.3). Additional computer tomography (CT) examination of the head and neck area showed a fluid level of the maxillary sinus on the left-hand side, which is consistent with an inflammatory process cranial of the alveolus of the removed upper left first molar. The inflammation extended, through the

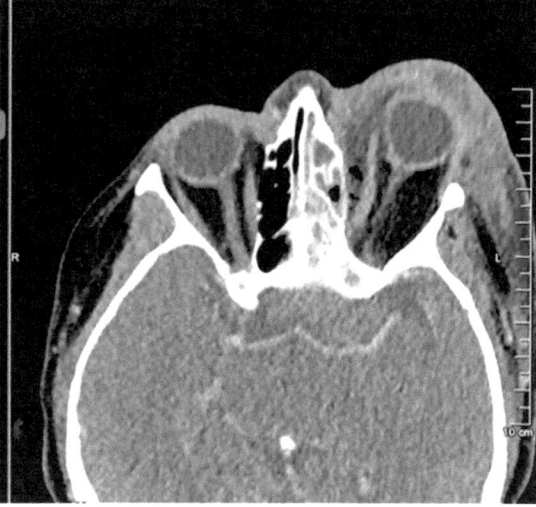

Fig. 20.4 Coronal and axial slide of the CT scan. The oral-antral communication is visible with partial obscuration of the maxillary sinus to flow out the nasal infundibu-lum to the ethmoid sinus and resulting in a spread of the abscess along the medial wall of the orbit (lamina papyracea) into the orbit

nasal infundibulum superiorly into the left ethmoid sinus and, after perforation of the extremely thin lamina papyracea, in the orbital space (Fig. 20.4). In the orbit, a pus pocket was visible under the inferior rectus muscle with gas formation. Another pus pocket was diagnosed in the upper eyelid which was connected to the supra- and retrobulbar space. There was no evidence of intracranial spread-extension of the infection (Fig. 20.4).

Therapy

The patient was subsequently taken to the operating theatre to carry out an emergency incision and evacuation of the pus pockets under general anaesthesia, leaving drains both intraorally at the location of alveolus upper left first molar where an oro-antral opening was created for adequate drainage of the maxillary sinus and at the location of the lower and upper eyelid in the orbit (Fig. 20.5).

The drainage material was sampled and sent for microbiology culture. The culture outcome showed growth of *Streptococcus anginosus* and Bacteroides, both sensitive to Augmentin (amoxicillin + clavulanic acid). Due to insuffi-cient clinical improvement of the swelling and pupil reaction, the patient was transferred to an academic hospital. Unfortunately, it turned out that bulb rupture had already taken place and vision of the left eye had been lost. It was then decided to perform evisceration of the left eye while retaining the sclera and the extraocular eye musculature (Fig. 20.6). After extensive intravenous antibiotic treatment during a period of 2 weeks hospitalization, the patient made a good clinical recovery. At a later stage, an acrylic implant was placed in the vacant eye cavity (Figs. 20.7 and 20.8).

Contemplation

In this specific case, the periapical infectious process had developed into a maxillary sinusitis. In time, the isolated sinus infection extended along the nasal infundibulum into the ethmoid cells. From here, perforation of the lamina papyracea occurred resulting in further spread of the infection into the orbital cavity and upper and lower eyelids. Especially the impressive swelling of the upper eyelid that obscured the eye accordingly was of most concern to the patient. There were no obvious complaints from the mouth; no residual

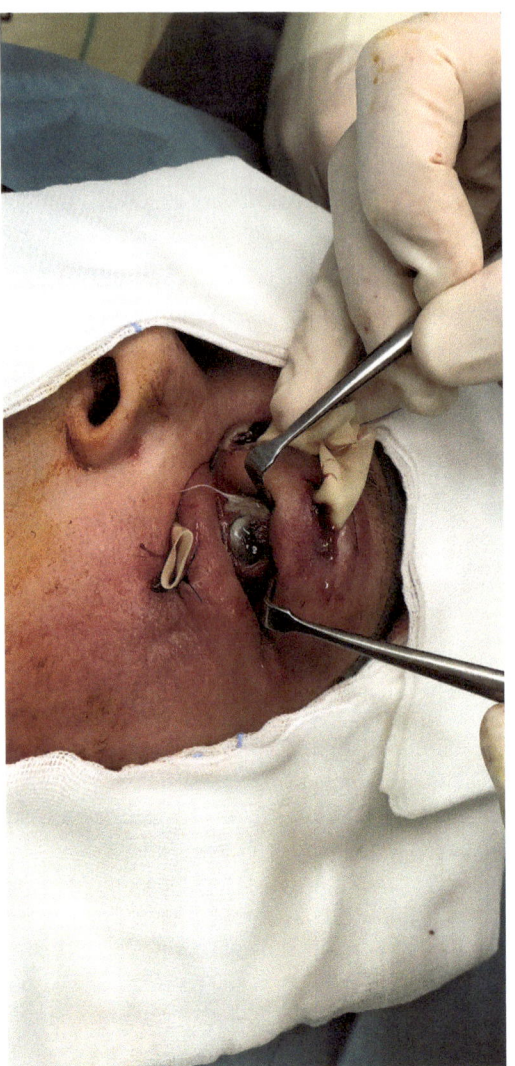

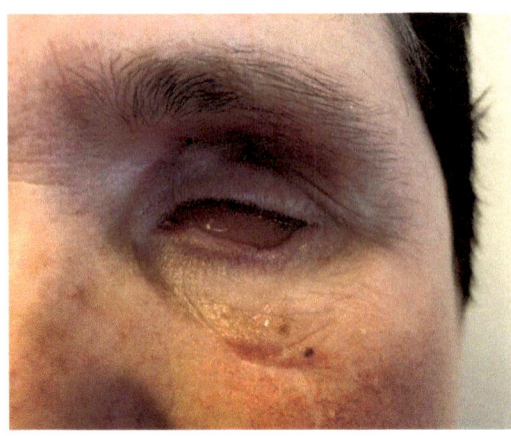

Fig. 20.6 Post operative image without the implant, orbital cavity visible

swelling or discharge was reported or seen after the molar removal. The purulent infection which was insufficiently responsive to antibiotic treatment eventually resulted in a pre- and retroseptal and retrobulbar abscess. The course of the illness was further complicated by complete loss of vision, probably due to compression of the vaso nervorum and direct compression of the optic nerve. Bulb rupture resulted as a consequence of increased pressure in and around the eye cavity.

Fig. 20.5 Intra-operative image of the drainage of the abscess of the lower and upper eyelid. Through these access points, the abscess could be released of the pus at the upper eye lid, and the lower eye lid to the orbital floor

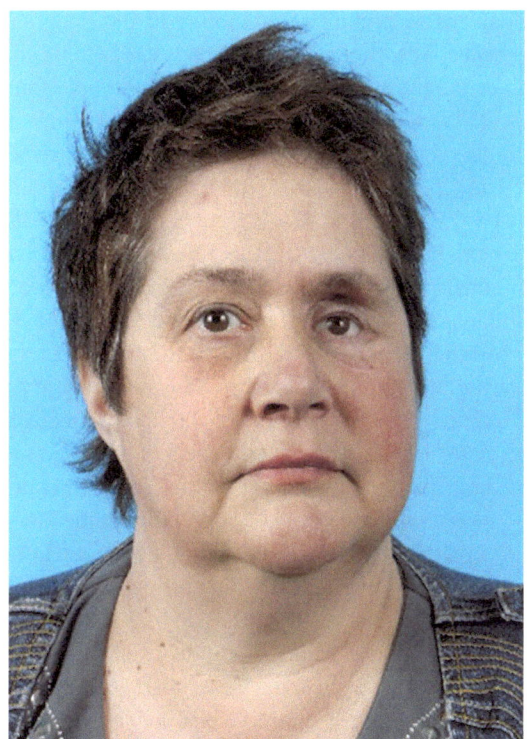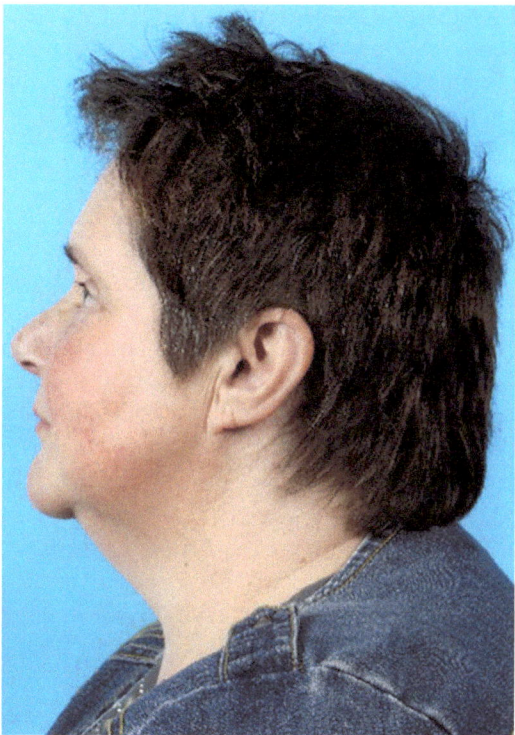

Fig. 20.7 and 20.8   Postoperative image with eye implant in situ

Discussion

Orbital infections (cellulitis and abscess) are infections of orbital tissue with various causes such as trauma, as a complication after eye surgery, tear duct infections, eyelid infections and dental infections. In 1970, Chandler developed a classification method for infections of the orbit which, however, is currently open to debate (Table 20.1) [1]. An apparent distinction here is a preseptal infection (stage I), which often can be treated with antibiotics, compared to retroseptal infection (other stages), which requires more frequent surgical intervention. An odontogenic cause of an orbit infection is very rare. However, should this complicating pathway occur, there is usually retroseptal abscess forma-

Table 20.1   Chandler's classification system for infections of the orbit

Stadium I	Preseptal cellulite	Infection and oedema anterior of the orbital septum (preseptal)
Stadium II	Orbit cellulitis	Extension of infection posterior of orbital septum (retroseptal)
Stadium III	Subperiosteal abscess	Abscess between the peri-orbita and the orbital wall
Stadium IV	Intraorbital abscess	Abscess within orbita content
Stadium V	Sinus cavernosus thrombosis	Extension of inflammation and posterior phlebitis with thrombus formation to sinus cavernosus with bilateral eye abnormalities

tion (Stage III or IV) requiring surgical intervention. Successively, the aetiology, symptoms, treatment and outcomes of odontogenic orbit infections will be discussed.

Aetiology

The immune status of the patient appeared to have a predictive role for the causative agent of an odontogenic orbital infection. In patients with a normal functioning immune system (immune competence), there is usually a *bacterial* agent causing the infection [2]. The most common pathogens are anaerobic bacteria (mouth flora), *S. aureus* (in some cases multidrug-resistant *Staphylococcus aureus*, MRSA), *S. epidermidis*, haemolytic Streptococci, *Pseudomonas aeruginosa* and *S. milleri*. In many of the infections, a combination of bacteria are responsible for the infection [2, 3]. In immunocompromised patients, such as patients with disordered diabetes mellitus, chemotherapeutic treatment or use of immunosuppressants, in addition to bacteria, *fungi* are the secondary cause of orbital infection. Ketoacidosis in a dysregulated diabetes patient leads to a structural change of neutrophils, and both immunosuppressants and chemotherapy reduce the quality and quantity of functioning neutrophils and macrophages. This reduces the immune response in these patients to withstand fungal infections. Common pathogens in immunocompromised patients are Aspergillus and Mucorales [2].

Spread of the Infection

The route along which a dental infection can spread is largely determined by the relationship between the radix, bone, adjacent muscle attachments and fascial blades, air spaces and gravity. In odontogenic orbit infections, four different routes have been identified in which an odontogenic infection can spread towards the orbit [3] (Fig. 20.9). The most common route is from the

apices of premolars or molars to the maxillary sinus. The infection then spreads easily to the air-bearing adjacent ethmoid cells (ethmoid sinus) via the infundibulum nasi. Here the infection can spread to the orbit without too much resistance via perforation of the paper-thin lamina papyracea. Also, after having spread to the adjacent maxillary sinus, the infection may travel superiorly to the orbital floor. The infection can spread to the orbit through bone erosion of the orbital floor or via the infraorbital canal [4] (Fig. 20.9a). The second route is along the fossa canina between the bone (the buccal cortical plate) and the periosteum (periosteum) towards the periorbital tissue. In the third route, infection spreads from the molar region through the pterygopalatine fossa and the communicating infratemporal fossa to the orbit through the cranial inferior orbital fissure (Fig. 20.9b). The last route is through the facial plexus veins. Because the facial veins, eyes and paranasal sinuses are connected without the presence of valves, an infection or a septic thrombo-embolus can spread to the orbit as a result of haematogenic regurgitation [3, 5] (Fig. 20.9c). With this extension pathway: watch for early signs of septic cavernous sinus thrombosis.

Symptoms

Proptosis and periorbital oedema are the most common symptoms of an odontogenic orbit infection [3, 4]. When the infection spreads post-septically, it can result in increased eye pressure, manifested in decreased eye tracking movements, loss of vision or retrobulbar pain, due to pressure on the vasa nervorum orbita [4]. A disturbed pupil reaction is much rarer; this probably occurs as a result of pressure on the ciliary ganglion (efferent pathway), here the indirect pupil response should also be tested. If disturbed, this indicates compression or infection of the optic nerve (afferent trajectory). For the diagnosis, additional imaging by contrast CT scan is indicated to visualize the infection/abscess with

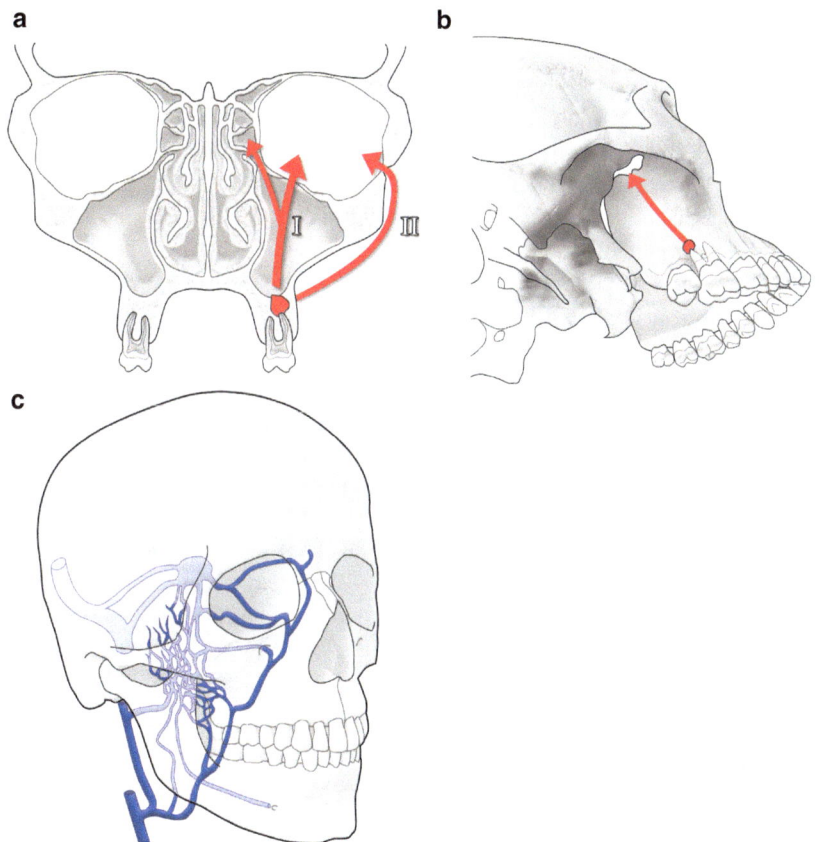

Fig. 20.9 Spread route odontogenic infection: (**a**) I via maxillary sinus—infundibulum nasi—ethmoid sinus—perforation lamina papyracea—orbit. II via maxillary sinus—erosion orbital floor—orbit. III via subperiosteal route along the canine fossa or buccal vestibule—perfo- rate eyelid septum—orbital septum—orbit. (**b**) Via pter- ygopalatine fossa—infra-temporal fossa—infra- orbital-fissure—orbit. (**c**) Via retrograde venous flow through valveless facial plexus—orbit/cavernous sinus (Courtesy of Serge Steenen)

edge staining. In addition to periapical radiolu- cency at the apex of an odontogenic focus, this often shows ipsilateral obscuration of the maxil- lary sinus. In addition, in the case of a subperios- teal abscess (stage III) or orbit abscess (stage IV), the scan shows the exact pus location [3] (see Chap. 4).

Therapy

In addition to treatment of the odontogenic focus through extraction of the responsible tooth, direct antibiotic treatment with broad spectrum antibi- otics is the cornerstone of the treatment of odon- togenic orbital infections. Treatment consists of a combination of antibiotics (augmentin, cefurox- ime and/or metronidazole) for adequate cover- age, which can be specifically adjusted after cultivation of the responsible bacteria [2, 3, 5]. In addition, corticosteroids (prednisone or dexa- methasone) can also be given in the case of orbital cellulitis. The anti-inflammatory medica- tion is said to have a rapid effect on pain, fever, periorbital swelling and reduction of proptosis, although the effect on vision in the long term has not been demonstrated. This makes both the dose and timing of corticosteroid treatment a subject of discussion [6]. In immunocompromised patients, local and systemic antifungal treatment should also be started and any dysregulated glucose level should be treated [2].

In the case of a subperiosteal abscess (stage III) or orbit abscess (stage IV), surgical intervention is necessary. In addition, with an odontogenic orbital infection without evident abscess, one should be wary of a more fulminant course than with an average orbital infection. In cases of an orbital cellulitis (stage II), in addition to antibiotic treatment, surgical treatment is also indicated in a few cases [5]. The approach to an orbital abscess depends on the location of the abscess on the CT scan, but in most cases, it will be an open approach through the conjunctiva (with medial or lateral canthotomy), through the lower eyelid to the orbital floor. In addition, the maxillary sinus often needs to be drained, which can be done via the alveolus of the causal tooth or via a Caldwell Luc approach or a transnasal approach. In the case of a medial orbital abscess, an endoscopic approach is sometimes used with removal of the anterior part of the ethmoid sinus and lamina papyracea by an ENT surgeon [3, 4]. In a rare case of rapidly progressive infection, strong pressure on the eyeball with pain, proptosis and loss of vision, an acute canthotomy is indicated. At the level of the lateral canthus, a sharp preparation is carried out along the temporal part of the orbit in a dorsal direction, so that the dorsal pressure-increasing moment can be acutely relieved or eliminated.

Outcome

If an odontogenic infection of the orbit is recognized too late or left untreated, this can lead to loss of vision. Analysis of 24 patients with an orbital infection showed that half retained moderate to good vision, but the other half retained only light perception or no vision at all. The need for (multiple) surgical intervention by means of incision and drainage is associated with a higher risk of permanent loss of vision. In addition, a more serious visual impairment at presentation such as limited light perception (afferent) or even no light perception appears to be associated with a higher risk of permanent loss of vision [7]. If surgical intervention is necessary, treatment within 24 h seems to give a higher chance of complete visual recovery than treatment after 2–7 days [3]. Loss of the eye, as in this case, is very rare.

Conclusion

Although rare, odontogenic infections may spread into the orbit and can result in loss of vision as a result of orbital abscess development and subsequent necessary globe evisceration.

Despite the removal of the involved molar tooth, the periapical infection had already developed into an independent infectious process.

Although the patient had been treated, fulminant postoperative complications still may occur. Do not underestimate post extractionem signs and or symptoms; once complaints of eye involvement are reported, immediate evaluation and appropriate targeted treatment should be carried out.

References

1. Chandler JR, Langenbrunner DJ, Stevens ER. The pathogenesis of orbital complications in acute sinusitis. Laryngoscope. 1970;80:1414–28.
2. Leferman CE, Ciubotaru AD, Ghiciuc CM, Stoica BA, Gradinaru I. A systematic review of orbital apex syndrome of odontogenic origin: proposed algorithm for treatment. Eur J Ophthalmol. 2021;31:34–41.
3. Procacci P, Zangani A, Rossetto A, Rizzini A, Zanette G, Albanese M. Odontogenic orbital abscess: a case report and review of literature. Oral Maxillofac Surg. 2017;21:271–9. https://doi.org/10.1007/s10006-017-0618-1.
4. Geusens J, Dubron K, Meeus J, Spaey Y, Politis C. Subperiosteal orbital abscess from odontogenic origin: a case report. Int J Surg Case Rep. 2020;73:263–7. https://doi.org/10.1016/j.ijscr.2020.07.014.

5. DeCroos FC, Liao JC, Ramey NA, Li I. Management of odontogenic orbital cellulitis. J Med Life. 2011;4:314–7.

6. Pushker N, Tejwani LK, Bajaj MS, Khurana S, Velpandian T, Chandra M. Role of oral corticosteroids in orbital cellulitis. Am J Ophthalmol. 2013;156:178–183.e1.

7. Youssef OH, Stefanyszyn MA, Bilyk JR. Odontogenic orbital cellulitis. Ophthalmic Plast Reconstr Surg. 2008;24:29–35.

Open Access This chapter is licensed under the terms of the Creative Commons Attribution 4.0 International License (http://creativecommons.org/licenses/by/4.0/), which permits use, sharing, adaptation, distribution and reproduction in any medium or format, as long as you give appropriate credit to the original author(s) and the source, provide a link to the Creative Commons license and indicate if changes were made.

The images or other third party material in this chapter are included in the chapter's Creative Commons license, unless indicated otherwise in a credit line to the material. If material is not included in the chapter's Creative Commons license and your intended use is not permitted by statutory regulation or exceeds the permitted use, you will need to obtain permission directly from the copyright holder.

Maarten P. Mourits

Learning Objectives
- The diagnosis of necrotizing fasciitis can be troublesome, but an early start of treatment is essential to prevent death. So, in case of doubt, go for the worst-case scenario.

Introduction

The infection, which sometimes starts at a small skin lesion (varicella for example) or after surgery, spreads along the fascia of the muscles and causes necrosis of the skin, subcutaneous fat, fascia, and sometimes the underlying muscles. In common parlance, bacteria causing NF are therefore called "flesh-eating bacteria." NF is a rare disease in which a multitude of factors play a role. Large series of patient cases are lacking and hence reports on incidence, etiology, prognosis, mortality rate etcetera are often conflicting. NF seems to be more common in men than in women. It occurs more frequently in African and Asian countries. It is extremely rare in children.

Many risk factors (diabetes mellitus, burns, intravenous drug use, varicella, immunosuppression, malnutrition, age >60 years, decreased kidney function, malignancy, and arterial vasculopathy) have been suggested, but diabetes mellitus appears to be the most important. From a systematic literature search including 1463 patients, it is evident that in almost 50% of cases diabetes mellitus is a comorbidity [1].

NF must be differentiated from erysipelas (which has a less fulminant course) and gas gangrene by means of culturing micro-organisms from infected tissues. In addition, orbital cellulitis must be distinguished from PONF. Signs that help to come to a right and early diagnosis are pain that is out of proportion, failure to respond to broad-spectrum antibiotics, and the presence of bullae and necrosis [2].

Presentation

Necrotizing fasciitis (NF) is a life-threatening bacterial or fungal infection of the subcutaneous tissues. Most often, the extremities and the perineal region are involved [1], but occasionally the eyelids and other parts of the face are the target of the infection (Figs. 21.1, 21.2, 21.3 and 21.4). In that case, we speak of periocular necrotizing fasciitis (PONF). Characteristic for the disorder is extreme pain together with swelling and redness of the skin. Within hours to days, the area turns from red to purple and then to black. Bullae may arise, and crepitations can be

M. P. Mourits (✉)
Amsterdam University Medical Centers, Location AMC, Amsterdam, The Netherlands

© The Author(s) 2023
P. J. J. Gooris et al. (eds.), *Surgery in and around the Orbit*,
https://doi.org/10.1007/978-3-031-40697-3_21

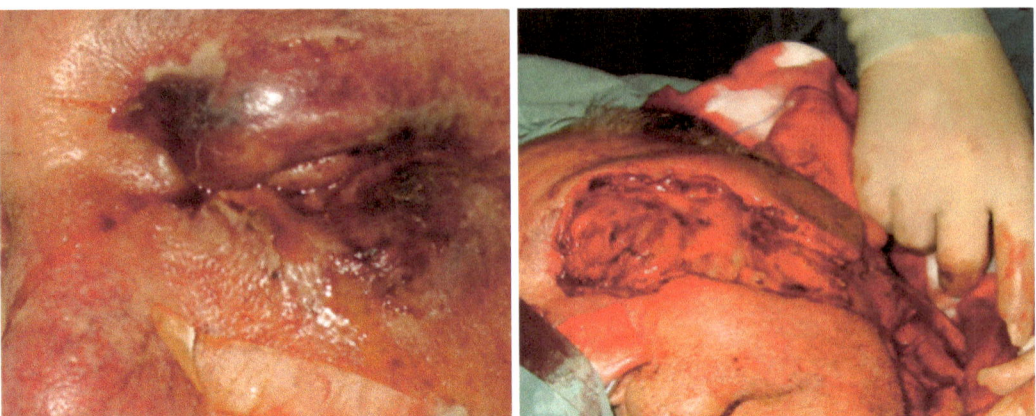

Fig. 21.1 and 21.2 A 68-year-old male with PONF of the left part of his face starting around the orbit at admission and 72 h later after debridement

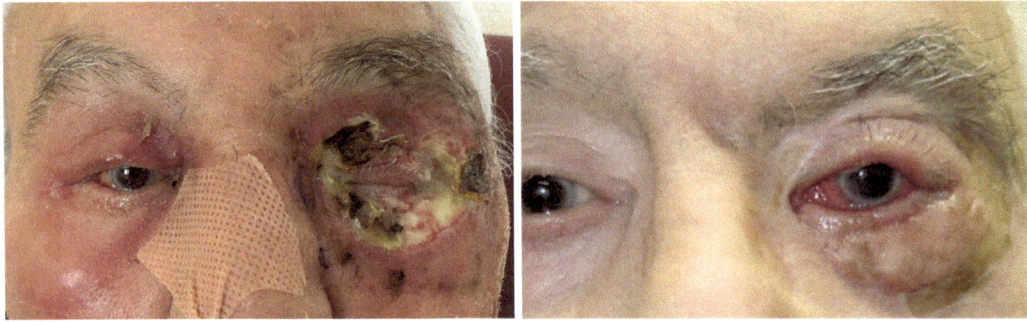

Fig. 21.3 and 21.4 A 84-year-old male with PONF of the left orbit after debridement and after rehabilitative surgery

felt. The inflamed area is relatively painless, whereas the adjacent parts are extremely painful. Next, the patient becomes seriously ill and somnolent with high fever, nausea and diarrhea, leukocytosis, and high sedimentation rates. Release of toxins can cause toxic shock syndrome associated with multi-organ failure an acute respiratory distress syndrome. Without treatment, the disease is fatal within 24–72 h.

Etiology

Four types of NF can be distinguished as follows

1. Type 1: Caused by a mixed flora of anaerobic and aerobic bacteria. This type is seen most often in patients with diabetes mellitus. Crepitations may be present. It is assumed that 70–80% of all cases belong to this type. The prognosis is relatively good.

2. Type 2: Caused by group A β-hemolytic Streptococcus (i.e., *Streptococcus pyogenes*) and by many other types of bacteria. *Streptococcus pyogenes* is found in 20–30% of cases of NF, but in up to 85% of patients with PONF [3]. This type of NF is especially seen in previously healthy individuals and is known for its fulminant course with toxic shock syndrome. Untreated, it is almost always lethal. However, PONF caused by *Streptococcus pyogenes* has a much better prognosis.

3. Type 3: Caused by Vibrio species from seawater. This type is seen especially in Asian countries and has a high mortality rate.

4. Type 4: Caused by fungi (e.g., Candida). This type also has a high mortality rate.

Treatment

Treatment initially consists of intravenous administration of broad-spectrum antibiotics and surgical removal (debridement) of the necrotic tissues. Every 2 h the boundaries of the inflamed tissues have to be demarcated with a dermo-marker in order to follow progression of the disease. Because of obliteration of blood vessels and tissue necrosis, the antibiotics not always reach their targets and this explains the initial failure of medical treatment. So, it is important to remove the necrotic tissues until healthy tissue is attained. This must be repeated until the inflammation is gone. It is not exceptional that at least five sessions are required [3]. It is evident that antibiotic treatment should be adjusted accordingly as soon as the causative pathogens have been identified. In addition to local measures, general (hemodynamic, respiration) measures must be taken [4]. Once the inflammation is gone, reconstruction of the tissue defects with free or pedicled flaps can be undertaken (Fig. 21.4).

References

1. Freeman HP, Oluwole SF, Ganepola GA, Dy E. Necrotizing fasciitis. Am J Surg. 1981;142:377–83.
2. Goh T, Goh LG, Ang CH, Wong CH. Early diagnosis of necrotizing fasciitis. Br J Surg. 2014;101:e119–25.
3. Rajak SN, Figueira EC, Haridas AS, Satchi K, Uddin JM, McNab AA, Rene C, Sullivan TJ, Rose GE, Selva D. Periocular necrotising fasciitis: a multicentre case series. Br J Ophthalmol. 2016;100:1517–20.
4. Tambe K, Tripathi A, Burns J, Sampath R. Multidisciplinary management of periocular necrotising fasciitis: a series of 11 patients. Eye. 2012;26:463–9.

Open Access This chapter is licensed under the terms of the Creative Commons Attribution 4.0 International License (http://creativecommons.org/licenses/by/4.0/), which permits use, sharing, adaptation, distribution and reproduction in any medium or format, as long as you give appropriate credit to the original author(s) and the source, provide a link to the Creative Commons license and indicate if changes were made.

The images or other third party material in this chapter are included in the chapter's Creative Commons license, unless indicated otherwise in a credit line to the material. If material is not included in the chapter's Creative Commons license and your intended use is not permitted by statutory regulation or exceeds the permitted use, you will need to obtain permission directly from the copyright holder.

Part VIII

Orbital Soft Tissue Surgery

Maarten P. Mourits

Learning Objectives

- Observation of the eyelids, globe position, and pupillary reflex may uncover life-threatening conditions.
- Upper eyelid surgery can best be approached from the skin crease.
- Lower fornix incision as done in orbital fracture surgery carries the risk of creating a cicatricial entropion.
- Intermittent eyelid edema is seen in patients with the blepharochalasis syndrome, a diagnosis that is often missed.

Introduction: Anatomy

The eyelids are composed of two lamellae, the outer (or: anterior) lamella consists of the skin and the orbicularis muscle, whereas the inner (or: posterior) lamella consists of a firm plate, called the tarsal plate, and the conjunctiva. The conjunctiva of the eyelids is continuous with the conjunctiva of the bulbus via a curved bobby-pin-like structure called the fornix.

Horizontally, the eyelids are medially connected via de medial canthal tendon to the periosteum of the medial ridge of the orbit. The medial canthal tendon consists of an anterior and a posterior crus, which enfold the lacrimal sac.

Laterally, the eyelids are connected via the lateral canthal tendons to a small elevation on the lateral wall just posterior to the orbital ridge, called Whitnall's tubercle. Vertically, the eyelids are connected to the lid retractors. The upper contains two retractors: the levator muscle, innervated by the oculomotor nerve, and Müller's muscle, which is innervated by orthosympathetic neurons. The origin of the levator muscle is the *annulus tendineus communis* (or: annulus of Zinn). Whitnall's suspensory ligament stretches from the fascia around the trochlea horizontally to the fascia of the lacrimal gland and supports the eyelid and the levator aponeurosis. It allows for a vector change of the levator muscle, enabling the upper eyelid to be elevated rather than directly retracted posteriorly. Anterior to the retractors lies yellow fat, called preaponeurotic fat, not to be mistaken with the fat underneath the eyebrow. In the lower eyelid, the retractors are less well developed and connected to the inferior rectus muscle. The orbital septum (Fig. 22.1) is a thin, but firm, fibrous structure stretched out between the orbital rim and the transition of the eyelid retractors to the tarsal plate. Extensions of the levator tendon, called aponeurosis, fuse with the medial and lateral canthal ligaments and insert

M. P. Mourits (✉)
Amsterdam University Medical Centers, Location AMC, Amsterdam, The Netherlands

© The Author(s) 2023
P. J. J. Gooris et al. (eds.), *Surgery in and around the Orbit*,
https://doi.org/10.1007/978-3-031-40697-3_22

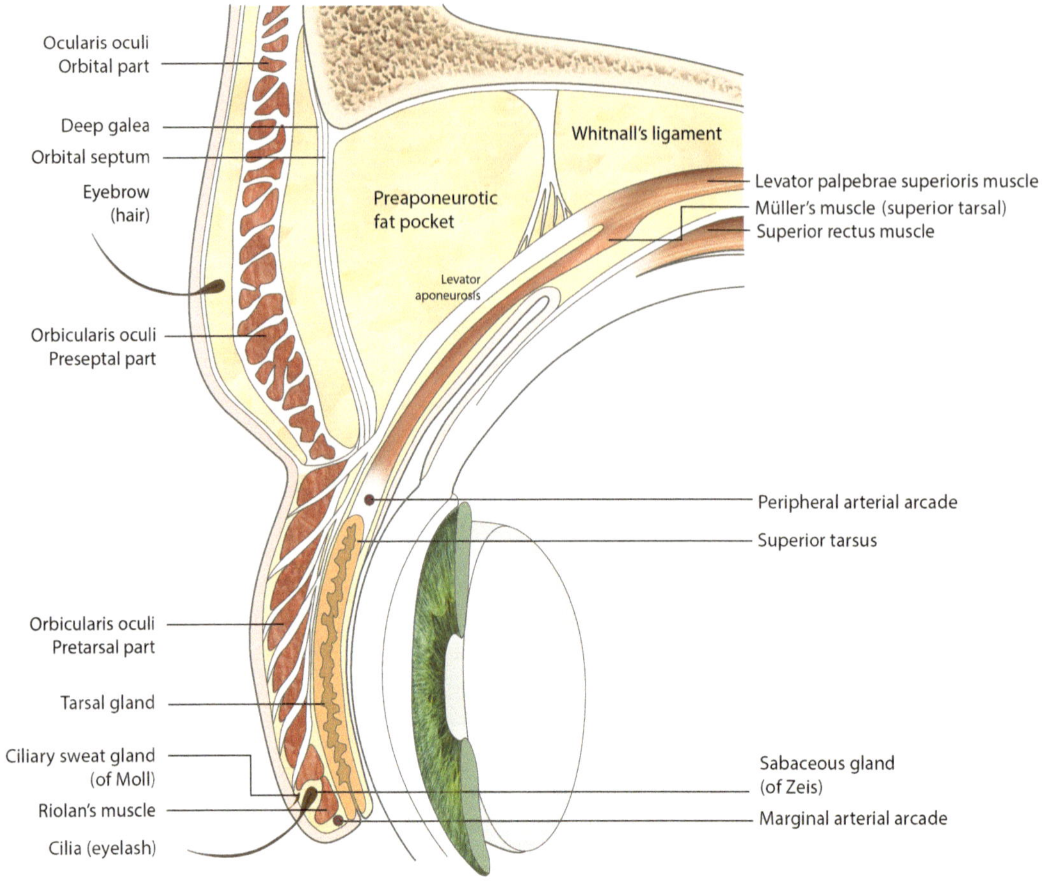

Ocularis oculi
Orbital part

Deep galea

Orbital septum

Eyebrow
(hair)

Orbicularis oculi
Preseptal part

Orbicularis oculi
Pretarsal part

Tarsal gland

Ciliary sweat gland
(of Moll)

Riolan's muscle

Cilia (eyelash)

Whitnall's ligament

Preaponeurotic
fat pocket

Levator
aponeurosis

Levator palpebrae superioris muscle

Müller's muscle (superior tarsal)

Superior rectus muscle

Peripheral arterial arcade

Superior tarsus

Sabaceous gland
(of Zeis)

Marginal arterial arcade

Fig. 22.1 Sagittal intersection of the orbit and eyelids. See the text for a description

into the skin, creating a skin or eyelid crease. This skin crease determines someone's looks. In Asian people, the skin crease is often hidden by a skin fold (and fat). Asymmetry of the skin crease, and as a consequence asymmetry of the pretarsal area, is almost always experienced as a cosmetically disturbing feature [1] (Fig. 22.2). In adults, the horizontal eyelid aperture measures approximately 35 mm and the vertical aperture 8–11 mm. The border of the inferior eyelid touches the corneal limbus at 6 o'clock, whereas the upper lid covers the cornea for about 2 mm at the 12 o'clock position [2]. The eyelid aperture is determined by the tension of the retractors on the one hand, and the tension of the orbicularis muscle at the other hand. The orbicularis muscle is innervated by the facial nerve (CN VII). The tarsal plate of the upper eyelid (8–10 mm) is about

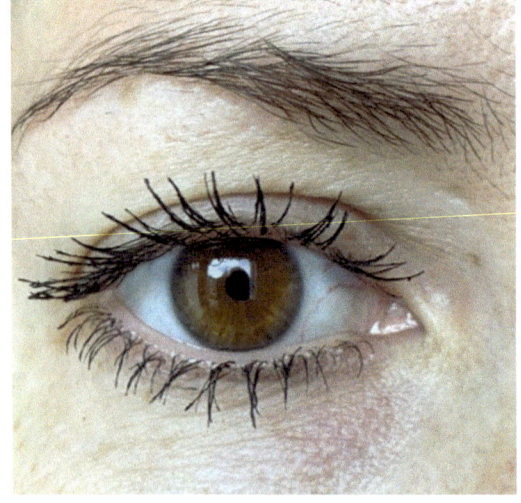

Fig. 22.2 The skin crease partly covered by a skin fold at the lateral side

twice as high as that of the lower eyelid. The eyelid borders contain rows of eyelashes. Medially to them, the lacrimal puncta arise, which are the beginning of the canaliculi transporting tear fluid to the lacrimal sac and, ultimately, to the nose. Within the eyelids, a series of small glands are found, which are essential for the permanent lubrication of the eye. The lacrimal gland is located in the upper lateral corner of the orbit and consists of an intraorbital and a palpebral part.

Some Disorders of the Eyelids and Their Management

The most common disorders of the eyelids, apart from a chalazion, are dermatochalasis, blepharoptosis, entropion, ectropion, floppy eyelids, and tumors of the eyelid. Common problems in Graves' orbitopathy are retraction of the upper and/or lower eyelids (either too high or too low), whereas in orbital fractures, entropion is regularly seen after surgery. There is an abundance of techniques to operate upon eyelids. For instance, hundreds of techniques alone exist for the correction of entropion of the lower lid. The eyelid can be approached from either the anterior (skin) side or the posterior (conjunctival) side. We prefer the anterior approach in order to avoid surgery to the posterior lamella, because lesions of the palpebral conjunctiva may be associated with temporary or lasting ocular irritation. Moreover, an anterior approach can easily be combined with skin resection (blepharoplasty). Upper eyelid surgery can best start at the eyelid crease, whereas the lower eyelids can best be approached with a subciliary incision, the scar of which becomes almost invisible after some weeks to months.

Dermatochalasis

Dermatochalasis is an excess of eyelid skin and/or fat. It is seen especially in older people and in smokers. The *bleparochalasis syndrome* [3] is a disorder of unknown etiology, in which the upper and/or lower eyelids, on one or both sides, swell due to edema. Swelling disappears after a few

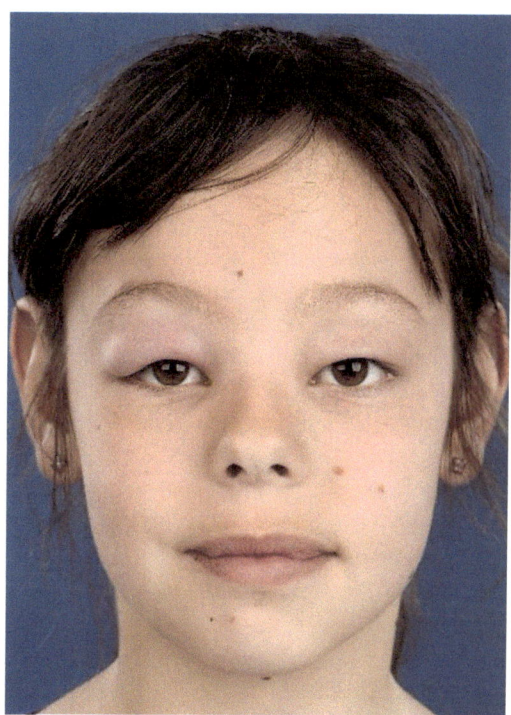

Fig. 22.3 Patient with blepharoochalasis syndrome of recent onset

days, but repeated swelling stretches the skin and finally disrupts eyelid architecture. The disorder can start at any time of life (Fig. 22.3).

Blepharoptosis

Cosmetic or functional blepharoptosis (ptosis, drooping eyelid) is one of the most common disorders in oculoplastic surgery. The ptosis can be present from birth (congenital) or appear at a later age (acquired). To examine ptosis, after excluding abnormal globe position and gaze abnormalities, the distance between the pupillary reflex and the margin of the upper eyelid is measured with a ruler, while the patient assumes the primary sitting position (body and head straight up, shoulders back, and looking forward). This distance is called margin reflex distance 1 (MRD1). (MRD2 is the reflex distance to the border of the lower eye lid.) Normally, MRD1 is approximately 3 mm. In the Netherlands, insurance companies reimburse ptosis surgery if

MRD1 is less than 1 mm. It should be noted that ptosis is not a diagnosis, but a symptom, of which the cause must be determined.

Congenital Ptosis

Ptosis becomes visible at a young age and is more often unilateral than bilateral. Sometimes, there is a long-standing family history of ptosis. Typically, an eyelid lag is present at downgaze. When the levator function is less than 5 mm, a frontalis suspension is indicated. Ptosis in young children can cause significant amblyopia. Therefore, ptosis should be corrected as soon as possible. The *blepharophimosis syndrome* is an autosomal dominant disorder including bilateral congenital ptosis, blepharophimosis, telecanthus, and epicanthus inversus [4]. It can be associated with infertility, microphthalmos, and other conditions.

Aponeurotic Ptosis

This is by far the most frequent cause of ptosis, in which innate weakness of the aponeurotic transition to the tarsal plate with our without mini-traumas is assumed to contribute to its origin. It is typically seen in patients wearing contact lenses [5], in some families and at old age. Typically, the skin crease has moved upwards, resulting in an abnormally increased distance between the eyelid border and the skin crease. A deep superior sulcus also fits the condition. Aponeurotic ptosis, because of its presentation at a somewhat older age, often presents in combination with dermatochalasis. Hence, in order to get a satisfactory surgical outcome, ptosis correction should be combined with a blepharoplasty in many cases.

Myogenic Ptosis

This is a rare form of ptosis and it is caused by external ophthalmoplegia or ocular myopathies, in which often not only the levator muscle is involved but also other extraocular and non-extraocular muscles. Hence, the risk of a postoperative corneal ulcer is much higher than in other forms of ptosis surgery. Especially the *Kearns-Sayre syndrome*—a mitochondrial muscle disorder—is of interest, because of its associated heart rhythm abnormalities, which can lead to a sudden cardiac arrest. The retina shows pigmentary alterations typical for retinitis pigmentosa. Myasthenia is another cause for ptosis. The disease itself should be treated. Surgery on the levator is not very beneficial.

Neurogenic Ptosis

A lesion of the oculomotor nerve (CN III), in particular involving the branch to the levator muscle, causes complete ptosis. Surgery is usually not possible because of the other manifestations of oculomotor nerve palsy. Damage to the ortho-sympathetic innervation causes 1–2 mm ptosis in combination with a miosis (smaller-than-normal pupil), which is called *Horner's syndrome*. An important factor causing this syndrome is a carcinoma on the apex of the lung.

Mechanical Ptosis

Tumors of the eyelid can cause ptosis by their weight, while cicatricial conjunctival conditions may draw the eyelid downwards.

Marcus-Gunn Jaw-Winking Ptosis

This is a form of ptosis occurring during movements of the mouth and is caused by aberrant nerves branches.

Brow Ptosis

This is a condition in which the eyebrow is sagging below the superior margin of the orbit. It often presents with dermatochalasis.

Lash Ptosis

This is a condition in which the lashes of the upper eyelid are directed downwards. The patient has to look through his lashes.

The choice of ptosis surgery is determined by the function of the levator muscle. The patient is asked to look down and up. The excursion of the eyelid is a proxy of levator muscle function (normally more than 15 mm). Check that the frontal muscle is not involved in elevation of the upper eyelid. When levator muscle function is more than 5 mm, levator reinsertion or plication can be done. When levator muscle function is less than 5 mm, a frontalis suspension may be effective.

Blepharoplasty

In case of an excess of skin with or without too much eyelid fat, excessive skin and fat can be excised. Although some patients report a number of subjective complaints ("heavy eyelids"), the most common indication for blepharoplasty is cosmetic. However, one must not trivialize the impact of a blepharoplasty on a patient's well-being. Dermatochalasis causes a tired look and is sometimes mistakenly associated with alcohol abuse or an exhausted physical state. Dermatochalasis is sometimes associated with a lacrimal gland prolapse. Soft lesions in the upper lateral part of the eyelids may be palpable, and these can also look like small eyelid tumors. Treatment consists of repositing and fixing the prolapsed part of the gland to the periosteum of the superior margin of the orbit [6].

Levator Reinsertion/Plication

Preferably, levator reinsertion or plication is done under local anesthesia, because during surgery the eyelid aperture has to be checked (Fig. 22.4). If general anesthesia is absolutely necessary, the result of surgery is less predictable. After having marked the skin crease with a dermomarker, a few subcutaneous injections with xylocaine plus epinephrine are given. A skin and orbicularis muscle incision are made with a surgical knife

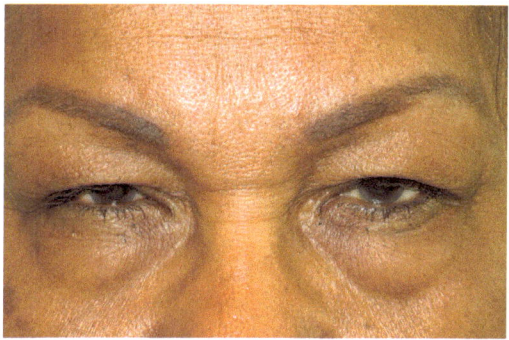

Fig. 22.4 A 54-year-old woman with dermatochalasis and bilateral blepharoptosis

or—after having wetted the skin—with a monopolar electrocautery needle along the skin crease. The septum is opened with a pair of Wescott scissors, the preaponeurotic fat becomes visible and it will be retracted with a Desmarres retractor. Next, the tarsal plate and the aponeurosis are exposed and a 6.0 Daclon suture (Daclon, because of the particular shape of its needle that easily cuts through the stiff tarsal plate) is placed through the upper part of the tarsal plate anterior to the conjunctiva and through the aponeurosis. When the suture knot is temporarily tied, the eyelid comes up. The patient is put in a supine position, and the eyelid aperture is inspected. If the eyelid is still too low, the suture is tightened; if the eyelid is too high, the suture is loosened. If the eyelid contour is still not satisfying, a second suture is placed next to the first one. The skin is closed with a 6.0 Nylon suture which can be removed after 7 days. This technique takes only 20–30 min in experienced hands and yields satisfying results in almost all patients with a levator muscle function of more than 5 mm. Only a few patients need a second procedure, during which an incision through the previous scar is made and the sutures are adapted. No absorbable sutures should be used, because of the risk or a recurrence of the ptosis.

Frontalis Suspension

After six stab incisions (three through the skin of the upper lid (2–3 mm above the lash line), one at the medial border, one at the lateral border of the eye brow, and the sixth 1–2 cm above the eye brow right above the pupil in the primary position), a silastic sling or a fascia lata strip is passed underneath the skin and sutured at the top end. At the end of the surgery, the tension of the strip should be sufficient to put the eyelid margin at the level of the limbus at 6 o'clock, with the patient lying flat. The disadvantage of this technique (that has to be taken for granted) is some degree of lagophthalmos. Hence, one has to use lubricants, sometimes forever.

(External) Browlift

A semi-ellipse of skin and subcutis just above the brow is excised between the frontal nerve medially and the bifurcation of the facial nerve laterally. The wound is sutured with 5.0 Vicryl and 5.0 Nylon (intracutaneously). This extremely successful procedure has only one disadvantage: It takes 3–6 months before the scars become less visible (Fig. 22.5).

Complications of Ptosis Surgery

Because of the enlarged eyelid aperture, there is an increased risk of dry eye and even of corneal ulcer. Therefore, lubricants must be used and slowly tapering off according to the postoperative course. In operations, in which the orbital septum is opened, there is a small risk of retrobulbar hemorrhage (see Chap. 13).

Ectropion and Entropion and Floppy Eyelids

In both ectropion and entropion, the two lamellae of the eyelid are dissociated. In ectropion, the posterior lamella moves upwards, whereas in entropion it is the anterior lamella that moves upwards. Ectropion and entropion of the upper eyelid are rare in Western countries and will not be discussed. An ectropion (outward rotation of the lid margin) of the lower lid is mostly involu-

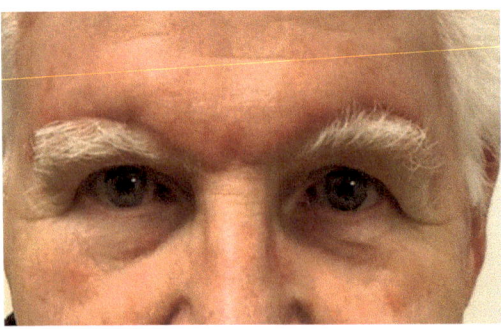

Fig. 22.5 Male, 64 years of age, 3 months after external brow lift

tional, i.e., caused by slackening of the underlying connective tissue in the lower eyelid. Other types of ectropion are congenital, cicatricial, or paralytic. The outward rotation of the eyelid can be more prominent at the medial side (so-called medial ectropion).

Involutional ectropion is seen in elderly people and often associated with horizontal laxity and retractor weakness. A commonly used therapeutic procedure is horizontal shortening of the eyelid by excising part of the lateral tarsal plate and creating a new lateral canthal tendon. This is called the lateral tarsal strip procedure. The lateral canthal tendon, however, is not always found as nicely as depicted in textbooks. Reconstruction requires that the upper eyelid, which may be also rather lax, covers the lower lid in the lateral corner. Sometimes, a small lateral tarsorrhaphy helps to reach the desired outcome. Provided that there is no significant canthal laxity, the eyelid can simply be tightened by excision of a full-thickness eyelid pentagon. This is especially effective in medial ectropion. In ectropion with predominant retractor weakness, characterized by poor eyelid movement in downgaze, reinsertion or shortening of the retractors can add to the positive outcome of surgery. *Cicatricial ectropion* is caused by a shortage of skin, as seen in a number of skin diseases (such as ichthyosis) and after trauma or surgery. Sometimes, if the traction is superficial, a Z-plasty will be sufficient, but more often a skin flap or a free skin graft is needed to correct this kind of ectropion.

Involutional entropion is typically seen in aging people. Due to the inward rotation of the eyelid, the lashes point to the cornea and can easily damage the corneal epithelium. Treatment, therefore, should not be delayed. A bandage lens to protect the cornea against the lashes can be inserted before surgery takes place. (N.B.: *Trichiasis* is a condition in which not the whole eyelid, but only one or more individual lashes point to the cornea. (Repeated) coagulation of the hair follicles of these lashes cures the problem.) Similar to involutional ectropion, horizontal laxity plays an important role. However, horizontal shortening alone will not last in the long run. The shortening has to be combined with everting sutures [7].

Cicatricial entropion is seen in diseases of the conjunctiva or after (surgical) trauma. In severe cases, the scar tissue has to be excised and replaced by mucous membrane or hard palate mucosa.

Floppy eyelids are lax eyelids that can easily be everted. The phenomenon was first described as the floppy eyelid syndrome in middle-aged obese men and related to the obstructive sleep apnea syndrome, but later it got a wider appreciation as a premature aging of the eyelid [8, 9]. Floppy eyelids easily cause chronic eye irritation. Association with different conditions has been described. Full-thickness excision of a pentagon of the eyelid can result in lasting improvement (Fig. 22.6).

Eyelid Tumors

There are numerous eyelid tumors, both benign and malignant. Basal cell skin carcinoma is most often seen in oculoplastic clinics. Mohs surgery has become immensely popular. However, basocellular carcinomas involving the eyelids and in particular the eyelid margin should not be left to a Mohs surgeon alone, who has no experience with reconstruction of (large) eyelid defects. Close cooperation from the start between a Mohs surgeon and an oculoplastic surgeon is advisable

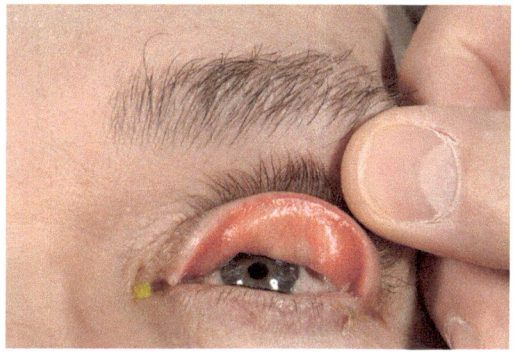

Fig. 22.6 Floppy eyelid

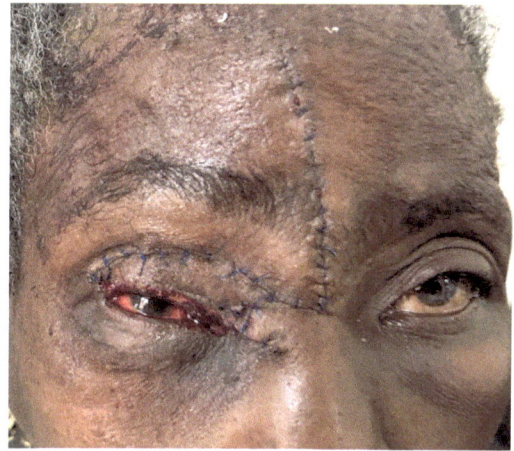

Fig. 22.7 Full upper eyelid reconstruction with oral mucosa and frontalis flap

to prevent "unforeseen" situations. Very small tumors on the margin of the eyelid can be removed by a shaving excision, larger lesions by a full-thickness excision. If primary closure is possible, it is the method of first choice. If not, pedicled flaps can be used to close the defect. A tarsoconjunctival flap (Hughes procedure) is ideal to reconstruct a lower eyelid defect, a glabella flap can be used to reconstruct medial defects, and with a frontalis flap (in combination with oral mucosa) a completely absent upper eyelid can be reconstructed (Fig. 22.7). Essential in eyelid reconstruction is the recreation of both the anterior lamella and the posterior lamella. The inner (posterior) side should always be lined with mucosa, for which buccal or labial mucosa can be taken. Stability requires cartilage, for example, auricular cartilage from the flattest part of the helix of the pinna. The advantage of a hard palate mucosal graft is that it offers both stability and mucosal lining. Replacement of the orbicularis muscle is never needed. Skin can be taken from other eyelids or from the retro-auricular area or the inner side of the arm. To prevent graft rejection, in composite reconstructions at least one layer should be vascularized, but the second layer may be a free flap. A frontalis flap is firm, so that it does not need any additional support of cartilage or another material.

Eyelid Retraction

Eyelid retraction is seen in 90% of patients with Graves' orbitopathy, but can also—although very sporadically—be seen in other disorders. If the retraction does not spontaneously disappear, the eyelid must be lengthened. This can be attained in the upper lid by a "reverse" ptosis operation. The aponeurosis and Müller's muscle are dissected from the tarsal plate until the desired position is achieved. A non-absorbable hang-back suture has to be placed between the tarsal plate and the aponeurosis to prevent overcorrection [10]. To correct lower eyelid retraction, a spacer between the tarsal plate and the retractors is often needed. This spacer can be either donor sclera or auricular cartilage.

Epilogue

This overview of eyelid disorders and their management is far from complete and highly subjective. It mirrors the experience of the author in a tertiary clinic in the Netherlands over a period of more than 30 years.

References

1. Papageorgiou KI, Ang M, Chang SH, Kohn J, Martinez S, Goldberg RA. Aesthetic considerations in upper eyelid retraction surgery. Ophthalmic Plast Reconstr Surg. 2012;28:419–23.
2. Mourits MP, Koornneef L. Lid lengthening by sclera interposition for eyelid retraction in Graves' ophthalmopathy. Br J Ophthalmol. 1991;75:344–7.
3. Custer PL, Tenzel RR, Kowalczyk AP. Blepharochalasis syndrome. Am J Ophthalmol. 1985;99:424–8.
4. Beaconsfield M, Walker JW, Collin JR. Visual development in the blepharophimosis syndrome. Br J Ophthalmol. 1991;75:746–8.
5. Hwang K, Kim JHJ. The risk of blepharoptosis in contact lens wearers. Craniofac Surg. 2015;26:373–4.
6. Eshraghi B, Ghadimi H. Lacrimal gland prolapse in upper blepharoplasty. Orbit. 2020;39:165–70.
7. Danks JJ, Rose GE. Involutional lower lid entropion: to shorten or not to shorten? Ophthalmology. 1998;105:2065–7.
8. Van Den Bosch WA, Lemij HG. The lax eyelid syndrome. Br J Ophthalmol. 1994;78:666–70.
9. Shah-Desai S, Sandy C, Collin R. Lax eyelid syndrome or 'progeria' of eyelid tissues. Orbit. 2004;23:3–12.
10. Mourits MP, Sasim IV. A single technique to correct various degrees of upper lid retraction in patients with Graves' orbitopathy. Br J Ophthalmol. 1999;83:81–4.

Open Access This chapter is licensed under the terms of the Creative Commons Attribution 4.0 International License (http://creativecommons.org/licenses/by/4.0/), which permits use, sharing, adaptation, distribution and reproduction in any medium or format, as long as you give appropriate credit to the original author(s) and the source, provide a link to the Creative Commons license and indicate if changes were made.

The images or other third party material in this chapter are included in the chapter's Creative Commons license, unless indicated otherwise in a credit line to the material. If material is not included in the chapter's Creative Commons license and your intended use is not permitted by statutory regulation or exceeds the permitted use, you will need to obtain permission directly from the copyright holder.

Peter J. J. Gooris, Maarten P. Mourits,
and J. Eelco Bergsma

Correction to:
P. J. J. Gooris et al. (eds.), Surgery in and around the Orbit,
https://doi.org/10.1007/978-3-031-40697-3

This book was inadvertently published with the following errors, and the errors have been corrected.

On Page iv, affiliation of the editor Maarten P. Mourits has been updated as follows:

Professor Emeritus

Department of Ophthalmology

Amsterdam University Medical Centers, Location AMC

Amsterdam, The Netherlands

On page 66, the legend to Figure 3.7 (b), (c) has been corrected as "(b) Lateral view of patient with unilateral Graves' orbitopathy–unaffected side, (c) Lateral view of patient with unilateral Graves' orbitopathy–affected side, exophthalmos present".

On page 144, the text "–on demand–" has been deleted from the first paragraph of the Section "Visual Pathways".

On page 147, the text "The visual field is the total area that can be seen without moving one's head and eyes" has been corrected under the Section "Visual Field".

On page 171, the text "in the sagittal plane" has been added to the first paragraph of the Section "Introduction: The (Axial) Globe Position".

On page 311, the legend to Figure 19.3 has been corrected as "Orbital subperiosteal empyema: Fluid collection between the orbital wall and the periorbital".

The updated versions of the chapters can be found at
https://doi.org/10.1007/978-3-031-40697-3
https://doi.org/10.1007/978-3-031-40697-3_3
https://doi.org/10.1007/978-3-031-40697-3_5
https://doi.org/10.1007/978-3-031-40697-3_7
https://doi.org/10.1007/978-3-031-40697-3_19

© The Author(s) 2024
P. J. J. Gooris et al. (eds.), *Surgery in and around the Orbit*,
https://doi.org/10.1007/978-3-031-40697-3_23

Open Access This chapter is licensed under the terms of the Creative Commons Attribution 4.0 International License (http://creativecommons.org/licenses/by/4.0/), which permits use, sharing, adaptation, distribution and reproduction in any medium or format, as long as you give appropriate credit to the original author(s) and the source, provide a link to the Creative Commons license and indicate if changes were made.

The images or other third party material in this chapter are included in the chapter's Creative Commons license, unless indicated otherwise in a credit line to the material. If material is not included in the chapter's Creative Commons license and your intended use is not permitted by statutory regulation or exceeds the permitted use, you will need to obtain permission directly from the copyright holder.

Index

© The Editor(s) (if applicable) and The Author(s) 2023
P. J. J. Gooris et al. (eds.), *Surgery in and around the Orbit*,
https://doi.org/10.1007/978-3-031-40697-3